Halophytic and Salt-Tolerant Feedstuffs

Impacts on Nutrition, Physiology and Reproduction of Livestock

T0203815

Halophytic and Salt-Tolerant Feedstuffs

Impacts on Nutrition, Physiology and Reproduction of Livestock

Editors

Hassan M. El Shaer
Desert Research Center
Cairo, Egypt

Victor R. Squires
International Dryland Management Consultant
Adelaide, Australia

CRC Press
Taylor & Francis Group
Boca Raton London New York

CRC Press is an imprint of the
Taylor & Francis Group, an **informa** business

A SCIENCE PUBLISHERS BOOK

CRC Press
Taylor & Francis Group
6000 Broken Sound Parkway NW, Suite 300
Boca Raton, FL 33487-2742

First issued in paperback 2019

ISBN-13: 978-1-4987-0920-0 (hbk)
ISBN-13: 978-0-367-37708-3 (pbk)

Library of Congress Cataloging-in-Publication Data

Halophytic and salt-tolerant feedstuffs : impacts on nutrition, physiology and
   reproduction of livestock / editors: Hassan M. El Shaer, Victor R. Squires.
      pages cm
   Includes bibliographical references and index.
   ISBN 978-1-4987-0920-0 (hardcover : alk. paper) 1. Halophytes as feed. 2.
   Salt-tolerant crops. 3. Animal nutrition. 4. Veterinary physiology. I. El-Shaer,
   Hassan M., editor. II. Squires, V. R. (Victor R.), 1937- editor.

   SB207.H35H35 2016
   636.08'52--dc23   2015028952

Visit the Taylor & Francis Web site at
http://www.taylorandfrancis.com

and the CRC Press Web site at
http://www.crcpress.com

Preface

Use of halophytes as animal feedstuffs attracted attention of scientists toward the latter half of the twentieth century with the appearance of scientific papers[1] on effect of brackish drinking water and salty feed on animal health and meat quality/quantity and the problems encountered in such studies were identified. Subsequently, scattered reports appeared in the literature where scientists from Australia, India, Pakistan, Middle East, Africa and North/South America working in the relevant fields have attempted replacing the regular fodder with one or another halophyte or salt tolerant feedstuff.

Later, a series of monographs and edited multi-author books* on aspects of the physiology of salt tolerant plants animals were published. Attention was also focused on the use of salt tolerant plants that included obligate halophytes, as candidate species for rehabilitation of saline lands, including abandoned irrigation areas. The role of biomass produced on such rehabilitated lands and new ways to use such biomass, especially in mixed rations involving more conventional fodders, was also investigated. As the loss of arable land to secondary salinity became more widespread and as the agronomy of salt tolerant food crops was advanced attention turned to utilizing the crop by-products, including oil seed cakes and meals. Sea water irrigation of halophytes gained traction as the search for commercially viable biosaline systems were advanced. This development generated a larger quantity of potentially usable feedstuffs both as biomass and as by-product such as seed cake.

Finding suitable forage/fodder (and even grain crops) which does not encroach upon the land under conventional crops may be useful for cattle raising and meeting the requirement of meat, poultry and dairy products is a challenge and specialized research centers were established, principally in the Gulf region, the WANA region, Pakistan, Central Asia and USA (Arizona) and Australia. Concurrently there was interest in using halophytes for remediation of damaged lands, including mine sites. The potential of salt tolerant plants, including extreme halophytes, for carbon sequestration also received attention. Some of the biomass derived from C sequestration plantations began to enter the supply chain for feedlots where guaranteed supplies of high quantities of suitable feedstuffs are required. In addition to salt tolerant and halophytic plants grown under intensive cultivation there are extensive areas of naturally-occurring shrublands, woodlands and grasslands that are used as fodder reserves or protein supplements, either grazed/bowsed by livestock or in cut-and-carry systems.

---

[1] See a list of these in the various reference lists at the end of each chapter.
* See footnote 1.

The volume of data generated and the interest in filling the 'feed gap' in many animal production systems, especially in arid and semi arid regions, has provided impetus to convene International and, regional and national symposia and spawned special issues of journals (e.g., *Small Ruminant Nutrition*) and various conference proceedings. The plethora of information generated by nutritionists, animal physiologists, veterinarians, agronomists and livestock specialists is scattered throughout reports, journal articles and some specialist monographs but lacks integration and synthesis into a coordinated body of knowledge. In this book we attempt a synthesis that considers the role and potential of salt tolerant and halophytic feedstuffs and their impacts on nutrition, physiology and reproduction of livestock, including ruminants and non ruminants such and poultry and rabbits.

The breadth and complexity of the subject matter presented here is vast. To make it easier for the reader the volume of 23 chapters is divided into 5 parts.

Part 1 of 5 chapters presents a synoptic overview of the diversity (taxonomy and life form) extent and geographic distribution of salt tolerant and halophytic feed sources and attempts to quantify the potential tonnage of such materials. Special attention is given to the vast Mediterranean and North African/Gulf region[2] where aridity and salinity are taking their toll on human and livestock populations. Intake and nutritive value of some salt-tolerant fodder grasses and shrubs for livestock are illustrated with selected examples from across the globe. Specific experience in the Gulf region with halophytic feedstuffs in mixed rations for sheep and goats receives attention here.

Part 2 of 4 chapters focuses on nutritional aspects and assesses the peculiar situation where many halophytic and salt tolerant feedstuffs have a high level of anti-nutritional compounds that inhibit digestibility and feed value. Others contain toxins that create health hazards or otherwise restrict their usefulness in livestock rations. Detailed consideration is given to what is known about such phyto-toxins and we also explore the pitfalls involved in evaluating the true value of such feedstuffs. Special attention is paid to livestock productivity (weight gain, egg production, etc.).

Part 3 of 4 chapters reviews experience on using halophytic and salt tolerant feedstuffs from around the world. Feeding trials have been conducted on a wide range of feedstuffs (trees, shrubs, grasses and forbs—even sea weed) when they were fed to small ruminants (sheep and goats), cattle, camels in various forms (fresh, dry or ensiled).

Part 4 of 6 chapters reviews and synthesizes knowledge on the physiological impacts of heavy salt and toxin loads on digestion, reproduction and health, with special attention to rumen function. Water is both an important nutrient that is vital to body functioning (excretion, temperature regulation, food and energy transport, etc.) and obviously is affected by mineral ash intake that may increase when livestock eat feedstuffs high in sodium (Na) and potassium (K) and have access to water that may contain other electrolytes and minerals such as magnesium (Mg), calcium (Ca), etc. The role of water as a regulator of feed intake, and rumen function is explored in both ruminants and non ruminants. Reproductive physiology, including impact of high salt loads on

---

[2] Variously called WANA (West and North Africa) or MENA (Mediterranean North Africa).

fertility of both, males and females, milk production and skin, hair and wool production in ruminants. Sheep, goats and camels are the main livestock considered here.

Part 5 of 4 chapters is about non-ruminants and considers the special dietary requirements of poultry and rabbits as well as the differences between ruminant and non-ruminant fermentation processes. By comparison some aspects of the special adaptation by a desert dwelling rodent that is almost wholly dependent on a halophyte for its survival is highlighted here. A special case study is presented on the processing equipment used in making salt tolerant and halophytic feedstuffs more acceptable to livestock and which also retain feed value and reduce wastage. We conclude with a brief section on uniting perspectives and some suggestions about how to cope with the remaining problems and speculate about the prospects for overcoming the challenges.

Selected bibliography on the development of interest in utilization of halophytes

1980s
Pasternak, D. and A. San Pietro. 1985. Biosalinity in Action: Bioproduction with Saline Water Springer, Dordrecht, 369p.

1990s
Ismail, S., C.V. Malcolm and R. Ahmad. 1990. A Bibliography of Forage Halophytes and Trees for Salt-affected Land: Their Use, Culture and Physiology, Department of Botany, University of Karachi, 258p.

Ungar, I.A. 1991. Ecophysiology of Vascular Halophytes. CRC Press, Boca Raton, 221p.

Lieth, H. and A. Al Masoom. 1993. Towards the Rational Use of High Salinity Tolerant Plants: Vol. 1 Deliberations about high salinity tolerant plants and ecosystems. Tasks for Vegetation Science. 27 Kluwer Academic, 536p.

Lieth, H. and A. Al Masoom. 1993. Towards the Rational Use of High Salinity Tolerant Plants: Vol. 2. Agriculture and forestry under marginal soil water conditions. Tasks for Vegetation Science 28 Kluwer Academic, 447p.

Choukr-AllAh, R., Clive V. Malcolm and Afef Hamdy. 1995. Halophytes and Biosaline Agriculture. CRC Press, Boca Raton.

Squires, V.R and A.T. Ayoub (eds.). 1995. Halophytes as a resource for livestock and for rehabilitation of degraded lands. Tasks for vegetation science Vol. 32. Kluwer Academic Dordrecht.

Wickens, G.E. 1998. Ecophysiology of Economic Plants in Arid and Semi-Arid Lands. Springer, 343p.

2000s
Lieth, H. 2000. Cash Crop Halophytes for Future Halophyte Growers. Inst. Für Umweltsystemforschung, Univ. Osnabrück, 32p.

Lieth, H. and M. Mochtchenko. 2003. Cash Crop Halophytes: Recent Studies: Ten Years After Al Ain Meeting. Springer, Dordrecht.

Zhang, Huifang. 2007. Phytoremediation of Salt-contaminated Soil by Halophytes. Ben-Gurion University of the Negev, 95p.

Zerai, D.B. 2007. Halophytes for Bioremediation of Salt Affected Lands. Ph.D. Thesis, University of Arizona, Tucson. 97p.

Öztürk, M., Y. Waisel, M.A. Khan and G. Görk. 2007. Biosaline Agriculture and Salinity Tolerance in Plants,Springer, Dordrecht. 206p.

Masters, D.G., E. Sharon S.E. Benes and H.C. Norman. 2007. Biosaline agriculture for forage and livestock production: A Review. Agriculture, Ecosystems and Environment 119: 234–248.

Chedly A., M. Münir Öztürk, M. Ashraf and C. Grignon. 2008. Biosaline Agriculture and High Salinity Tolerance, Springer, Dordrecht.

Öztürk, M., Barth, Hans-J and B. Benno Böer. 2010. Sabkha Ecosystems: Volume III: Africa and Southern Europe. Kluwer Academic.

Norman, H.C., D.G. Masters and E.G. Barrett-Lennard. 2014. Halophytes as forages in saline landscapes: interactions between plant genotype and environment change their feeding value to ruminants. Iraq Salinity Project Technical Report 17 ICARDA, 51p.

Khan, M.A., B. Böer, M. Öztürk, T.Z. Al Abdessalaam, M. Clüsener-Godt and B. Gul (eds.). 2014. Sabkha Ecosystems: Volume IV: Cash Crop Halophyte and Biodiversity Conservation Series: Tasks for Vegetation Science, Vol. 47. Kluwer Academic, 339p.

February 2015

Hassan M. El Shaer
Cairo, Egypt

Victor R. Squires
Lanzhou, China
Adelaide, Australia

Acknowledgements

We the editors are grateful for the support and assistance of the Director/President of the Desert Research Center of Mataria, Egypt and the Dean, College of Grassland Science, Gansu Agricultural University, Lanzhou China. Naturally enough in a book of this type, we rely on the input from a wide range of people (researchers, animal husbandry practitioners, livestock producers and academics). The vast breadth of the subject matter covered in this book by the authors of 23 chapters has meant that the work has benefited from the input of many individual contributors from vastly different parts of the globe. We are grateful to the contributors and reviewers for their time and effort, along with the exchange of ideas and learning experience obtained by working with such a diverse and learned group. We owe a debt of gratitude to the vast 'invisible college' of colleagues whose publications have shed light on some of most pertinent problems faced by researchers and practitioners. We are also grateful to all those people from many countries whose work has revealed important information relevant to the utilization of salt tolerant and halophytic plants that have the potential to relieve pressure on the shrinking feed base in many regions as livestock inventories rise to meet burgeoning human populations, at a time when land degradation including the accelerated salinization of land and water, limits the growth of conventional crops and forages.

The contribution of Dr. Salah Attia-Ismail of the Desert Research Center in Cairo is especially appreciated. He has contributed as author, co-author, reviewer and coordinator.

Contents

List of Contributors

Abd El-Galil, K.
Animal and Poultry Nutr. Dept., Desert Research Centre, 1 Matahaf El Mataria St., P.O.B. 11753, Mataria, Cairo, Egypt.

Abd El Ghany, W.
Desert Research Centre, 1 Matahaf El Mataria St., P.O.B. 11753, Mataria, Cairo, Egypt.

Abdalla, E.B.
Animal Production Department, Faculty of Agriculture, Ain Shams University, Cairo, Egypt.

Abou El Ezz, S.
Desert Research Centre, 1 Matahaf El Mataria St., P.O.B. 11753, Mataria, Cairo, Egypt.

Abou-Soliman, Nagwa H.I.
Desert Research Centre, 1 Matahaf El Mataria St., P.O.B. 11753, Mataria, Cairo, Egypt.

Ashour, G.
Faculty of Agriculture, Cairo University, Giza, Egypt.

Attia-Ismail, S.A.
Desert Research Center, 1 Matahaf El Mataria St., P.O. Box 11753, Mataria, Cairo, Egypt.

Badawy, M.T.
Desert Research Center, 1 Matahaf El Mataria St., P.O. Box 11753, Mataria, Cairo, Egypt.

Baioni, S.
Departamento de Agronomía-CERZOS (CONICET), Universidad Nacional del Sur, San Andrés 800, 8000 Bahía Blanca, Argentina.

Blache, Dominique
School of Animal Biology, The University of Western Australia, Nedlands WA 6009, Australia.

Brevedan, R.E.
Departamento de Agronomía-CERZOS (CONICET), Universidad Nacional del Sur, San Andrés 800, 8000 Bahía Blanca, Argentina.

Busso, C.A.
Departamento de Agronomía-CERZOS (CONICET), Universidad Nacional del Sur, San Andrés 800, 8000 Bahía Blanca, Argentina.

Degen, A.A.
Desert Animal Adaptations and Husbandry, Wyler Department for Dryland Agriculture, Institutes for Desert Research, Ben-Gurion University of the Negev, Beer Sheva, Israel, 84105.

El Shaer, H.M.
Desert Research Center, 1 Matahaf El Mataria St., P.O. Box 11753, Mataria, Cairo, Egypt.

El-Bassiony, M.F.
Desert Research Center, 1 Matahaf El Mataria St., P.O. Box 11753, Mataria, Cairo, Egypt.

El Hawy, A.S.
Desert Research Center, 1 Matahaf El Mataria St., P.O. Box 11753, Mataria, Cairo, Egypt.

Emam, K.R.S.
Animal and Poultry Nutr. Dept., Desert Research Centre, 1 Matahaf El Mataria St., P.O.B. 11753, Mataria, Cairo, Egypt.

Fernández, O.A.
Departamento de Agronomía-CERZOS (CONICET), Universidad Nacional del Sur, San Andrés 800, 8000 Bahía Blanca, Argentina.

Fioretti, M.
Departamento de Agronomía-CERZOS (CONICET), Universidad Nacional del Sur, San Andrés 800, 8000 Bahía Blanca, Argentina.

Hajer, Guesmi
Laboratoire des Ressources Animales et A, Institut National Agronomique de Tunisie, 43 Av. Ch. Nicolle, 1082, Tunisia.

Hulm, E.
CSIRO Agriculture Flagship., Private Bag 5, Wembley, Western Australia, 6913.

Kam, Michael
Desert Animal Adaptations and Husbandry, Wyler Department for Dryland Agriculture, Institutes for Desert Research, Ben-Gurion University of the Negev, Beer Sheva, Israel, 84105.

Kamel, Hessini
Laboratoire des Plantes Extrêmophiles, Centre de Biotechnologie de Borj Cedria, BP 901, 2050 Hammam Lif, Tunisia.

Khidr, R.K.
Animal and Poultry Nutr. Dept., Desert Research Centre, 1 Matahaf El Mataria St., P.O.B. 11753, Mataria, Cairo, Egypt.

Laborde, H.E.
Departamento de Agronomía-CERZOS (CONICET), Universidad Nacional del Sur, San Andrés 800, 8000 Bahía Blanca, Argentina.

Masters David G.
School of Animal Biology M085, The University of Western Australia, 35 Stirling Highway, Crawley, Western Australia, 6009, Australia.

Nizar, Moujahed
Laboratoire des Ressources Animales et A, Institut National Agronomique de Tunisie, 43 Av. Ch. Nicolle, 1082, Tunisia.

Norman, H.C.
CSIRO Agriculture Flagship., Private Bag 5, Wembley, Western Australia, 6913.

Ramadan, W.
Desert Research Center, 1 Matahaf El Mataria St., P.O. Box 11753, Mataria, Cairo Egypt

Revell, D.K.
Revell Science, Agricultural Science and Natural Resource Management, Duncraig WA 6023, Australia.

Riley, J.J.
Associate Professor (retired), Soil, Water and Environmental Science Department, College of Agriculture and Life Sciences, The University of Arizona, Tucson, Arizona, 85721 USA.

Shawket, S.M.
Desert Research Center, 1 Matahaf El Mataria St., P.O. Box 11753, Mataria, Cairo, Egypt.

Squires, Victor R.
Visiting Professor, College of Grassland Science, Gansu Agricultural University, Lanzhou and University of Adelaide, Australia (retired).

Wilmot, Matt G.
CSIRO Agriculture Flagship., Private Bag 5, Wembley, Western Australia, 6913, Australia.

Zaki, Engy F.
Desert Research Center, 1 Matahaf El Mataria St., P.O. Box 11753, Mataria, Cairo, Egypt.

List of Tables

List of Acronyms, Abbreviations and Equivalents

ADF	Acid Detergent Fibre
ADIN	Acid Detergent Insoluble Nitrogen
ADL	Acid Detergent Lignin
ADMR	Average Daily Metabolic Rate
AFRC	Agricultural and Food Research Council (UK)
AIA	Acid-insoluble ash
AOAD	Arab Organization for Agriculture Development
BW	Body Weight
CBG	Crushed Barley Grains
CDS	Crushed Date Seeds
CPD	Crude Protein Digestibility
CSIRO	Commonwealth Scientific Industrial Research Organization (Australia)
DCP	Digestible Crude Protein
DMI	Dry Matter Intake
DOM	Digestible Organic Matter
DPLS	Digestible Protein Leaving the Stomach
DRC	Desert Research Center (Egypt)
dS/m	deci Siemens per metre—a measure of salinity
FAO	Food and Agriculture Organization of the UN
FWI	Feed Water Intake
GEF	Global Environment Facility
HSTF	Halophytes and Salt Tolerant Forages
ICARDA	International Center for Agriculture in Dry Areas
ICBA	International Center for Biosaline Agriculture
ICRAF	International Center for Research in Agroforestry
IFAD	International Fund for Agricultural Development
IVOMD	*In vitro* Organic Matter Digestibility
MCP	Microbial Crude Protein
MENA	Middle East and North Africa region
MER	Maintenance Energy Requirement
Mha	Millions of hectares
NDF	Neutral Detergent Fiber
NFE	Nitrogen-free Extract

NGO	Non Government Organizations
nM	nM = nanomolar = nannomoles per litre = 10^{-9} moles per litre
NIRS	Near Infrared Reflectance Spectroscopy
OM	Organic Matter
ppm	Parts per million
RDP	Rumen degradable protein
SCA	Standing Committee on Agriculture (Australia)
TAC	Total Antioxidant Capacity
TDN	Total Digestible Nutrients
UDP	Undegraded Dietary Protein
UN	United Nations
UNCBD	United Nations Convention on Biodiversity
UNCCD	United Nations Convention on Combating Desertification
UNDP	United Nations Development Program
UNEP	United Nations Environment Program
UNFCC	UN Framework Agreement on Climate Change
USDA	United States Department of Agriculture
USAID	US agency for International Development
WANA	West Asia North Africa region
WB	World Bank
WHO	World Health Organization

Equivalents

100 hectares equals 1 km^2
1 feddan equals 0.42 hectares
Sabkha marine and continental salt flats equivalent to playa
1 dS/m = 1000 EC (or µS/cm) = approximately 640 mg/kg (or ppm)
Tibin = shredded barley straws

About the Editors

Dr. Hassan M. El Shaer is an Egyptian scientist specialising in animals and rangelands. He has earned his reputation from a life time of work at the Desert Research Center (DRC) Cairo, Egypt. He has a PhD in Animal Science from Ain Shams University, Cairo, Egypt and did his Post-doctoral research at the Animal and Biochemistry Department, University College of North Wales, Bangor, UK (1985–1986), as well as at the Department of Veterinary Physiology, Pharmacology and Toxicology School of Veterinary Medicine, Louisiana State University, USA (1989–1990).

Dr. El Shaer was the Vice President of the DRC for four years and has been active in research along with serving as a mentor to many young scientists. He is the Director of the Excellency Center for Saline Agriculture (ECSA), Cairo. He has been working as a Consultant for WFP-UN/MALR, Cairo and the Technical Consultant for the Academy of Scientific Research and Technology in Egypt. As an expert in his field, he is a member of many National and International committees and scientific studies, including the FAO and UNESCO on "Livestock and drought management in some Arabic countries" and the International Center for Biosaline Agriculture (ICBA) Dubai, UAE in the field of "Utilization and feeding salt tolerant plants".

Along with acting as a supervisor to younger scientists Dr. El Shaer is author of numerous research reports. He has contributed as both author and co-author in several books edited by international publishers, he is also credited with over 120 research papers and articles published in national and international journals and conferences. He is the Principal Investigator (PI) for 11 National R&D Projects, and Co-PI for other four projects; also the PI for 12 International projects funded by USAID, EU, IFAD, ICBA and OFED. Dr. El Shaer has travelled extensively within North Africa and the Middle East as part of his research missions and has been an invited speaker at many international workshops and conferences in the Gulf region, North Africa, China and Central Asia.

Dr. Victor R. Squires is an Australian who pursued the study of animal husbandry and rangeland ecology from a young age. He has a Ph.D. in Rangeland Science from Utah State University, USA. He was Dean of the Faculty of Natural Resource Management at the University of Adelaide, where he worked for 15 years following a 22 year career in Australia's CSIRO. He is an author/editor of 13 books including "Halophytes as resource for livestock and for rehabilitation of Degraded lands" which was published in 1994 and numerous research papers on aspects of range/livestock relations. Since retirement from the University of Adelaide, Dr Squires has been a Visiting Fellow in the East West Center, Hawaii, and is currently an Adjunct Professor in the University of Arizona, Tucson and at the Gansu Agricultural University, Lanzhou, China. He has been a consultant to World Bank, Asian Development Bank and various UN agencies in Africa, China, Central Asia and the Middle East. He was awarded the *2008 International Award and Gold Medal for International Science and Technology Cooperation* by the Government of China and in 2011 was awarded the *Friendship Award* by the government of China. The Gold Medal is the highest award for foreigners. In 2015 Dr Squires was honoured by the Society for Range Management (USA) with an *Outstanding Achievement Award.*

PART 1

Extent and Geographic Distribution of Salt Tolerant and Halophytic Feedstuffs

The 5 chapters here present a synoptic overview of the diversity (taxonomy and life form), extent and geographic distribution of salt tolerant and halophytic feed sources and attempts to quantify the potential tonnage of such materials. Special attention is given to the vast Mediterranean and North African/Gulf region[1] where aridity and salinity are taking their toll on human and livestock populations. Intake and nutritive value of some salt-tolerant fodder grasses and shrubs for livestock are illustrated with selected examples from across the globe. Specific experience in the Gulf region with halophytic feedstuffs in mixed rations for sheep and goats receives attention here. Some chapters take a regional approach and provide a review and synthesis of past, current and on-going work on production and utilization of salt tolerant and halophyte feedstuffs in several major geographic regions (North Africa, the middle East, Latin America, Australia, Central Asia) as a complement to the more detailed chapters presented in the later parts.

---

[1] Variously called WANA (West and North Africa) or MENA (Middle East North Africa).

1

Global Distribution and Abundance of Sources of Halophytic and Salt Tolerant Feedstuffs

Victor R. Squires[1],* and H.M. El Shaer[2]

SYNOPSIS

There is a rich data set on the productivity of halophytic and salt tolerant plants. Complications occur when assessing forage biomass as opposed to net primary productivity. According to the best estimates, about 4–5 billion tons of halophytic feedstuffs are produced annually. The question is whether commercial use of fodder from halophytic plants and salt tolerant plants is viable?

It is certain that free ranging livestock will continue to utilize the often sparse plants on which traditional pastoralism depends but whether there can be an intensification of the animal husbandry to incorporate feedlots using halophytic feedstuffs, including processed feedstuffs, is still an open question. To be of value to processors, large quantities of raw material that has a guaranteed delivery schedule is required to justify the investment in developing the processing facility. This is much more likely to occur where plantations are established and where irrigation water is available to ensure high and reliable yields.

This chapter describes the link between saline soils and the global distribution and abundance of halophytic and salt tolerant plants. We

[1] Gansu Agricultural University, Lanzhou, China.
[2] Desert Research Center, Egypt.
* Corresponding author: dryland1812@internode.on.net

attempt to quantify the potential supply of feedstuffs that is available for grazing/browsing and for processing into feed supplements that can be used as substitutes for more conventional roughages and feed components.

Keywords: salinity, adaptation, soil, vegetation, camels, sheep, feed processing, mixed rations, salt tolerance, water, taxonomy, flora, feed supplements, saline soils.

1. Diversity and distribution of halophytes and salt tolerant plants

There is no single taxonomic group (genus or family) which makes up the bulk of halophytes (Aronson 1989). They are represented by several species of trees, shrubs forbs and grasses. If we also include salt tolerant plants then the number increases considerably (O'Leary and Glenn 1994). For example, in the Flora of the Arab region, Batanouny (1994) listed 150 species from 55 Genera and 22 Families. Many species of salt tolerant plants have been identified and they fall into various taxonomic groups (Ozturk et al. 2010). Table 1.1 is a typical list of species with their families and life form. Similar lists can be produced for North Africa, the Mediterranean, and the Gulf region.

The life form spectrum exhibits a wide range of variation. The Family Asteraceae is represented by the highest number of species (8 genera, 11 species) many of which are short-lived ephemerals, followed by Chenopodiaceae (4 genera, 9 species), Fabaceae (2 genera, 7 species) and Poaceae (4 genera, 7 species), Zygophyllaceae (4 genera, 7 species), Amaranthaceae (4 genera, 4 species), Solanaceae, Euphorbiaceae, Brassicaceae and Boraginaceae (3 species), Labiatae (2 species) whereas, 16 families including Papaveraceae, Acanthaceae, Asclepiadaceae, Nyctaginaceae, Convolvulaceae, Cyperaceae, Ephedraceae, Urticaceae, Molluginaceae, Juncaceae, Cucurbitaceae, Resedaceae, Polygonaceae, Portulacaceae, Caryophyllaceae and Tamaricaceae, are represented by a single species each (Table 1.1).

2. Types of feedstuffs derived

Feedstuffs derived from such plants (or plant parts) fall into a number of categories and the first division is between those that are grazed/browsed by livestock (some of them naturally occurring, Fig 1.1) and others that are planted for their value as fodders.

In some situations livestock can take in the halophytic feedstuffs *in situ* but it is also common to cut and carry them and feed them to penned animals. Yet another group (the main focus of this book) are used for hay/silage or made into concentrates like pellets, meals or cakes. This chapter will consider all categories.

2.1 Feedstuffs derived from halophytic and salt tolerant plants fed in situ

The most common approach is to develop plantations of species such as *Atriplex, Salicornia, Haloxylon* or to grow salt tolerant grasses like *Sorghum bicolor, Distichlis, Chloris* or salt tolerant dicots like *Suaeda, Salsola,* etc. so that these may be grazed/ browsed by ruminant livestock (Fig. 1.2).

Table 1.1. Dryland plants show great diversity in terms of Life form, and Family.

Species	Family	Life form*
Acacia asak (Forssk.) Willd	Fabaceae	Ph.
Acacia gerrardii Benth	Fabaceae	Ph.
Acacia ehrenbergiana Hayne	Fabaceae	Ph.
Acacia mellifera (Vahl) Benth	Fabaceae	Ph.
Acacia tortilis (Forssk.) Hayne	Fabaceae	Ph.
Aerva javanica (Burm. f.) J.E. Schult.	Amaranthaceae	Ch.
Aerva lanata (L.) J.E. Schult.	Amaranthaceae	Ch.
Alternanthera pungens Kunth.	Amaranthaceae	Th.
Argemone mexicana L.	Papaveraceae	Ch.
Atriplex leucoclada Boiss.	Chenopodiacea	Th.
Atriplex suberecta Verd.	Chenopodiacea	Th.
Bidens biternata L.	Asteraceae	Ch.
Blepharis ciliaris (L.) B.L. Burtt	Acanthaceae	Th.
Caralluma subulata (Decne.) A. Berger	Asclepiadaceae	Ch.
Chenopodium album L.	Chenopodiaceae	Th.
Chenopodium ambrosioides L.	Chenopodiaceae	Th.
Chenopodium murale L.	Chenopodiaceae	Th.
Chenopodium opulifolium L.	Chenopodiaceae	Th.
Cichorium bottae L.	Asteraceae	Th.
Commicarpus grandiflorus (A. Rich.) Standl	Nyctaginaceae	Ph.
Convolvulus arvensis L.	Convolvulaceae	Th,
Conyza bonariensis (L.) Cronquist	Asteraceae	Th.
Cynodon dactylon (L.) Pers.	Poaceae	G.
Cyperus laevigatus (L.)	Cyperaceae	H.
Datura innoxia L.	Solanaceae	Th.
Ephedra foliate Boiss. ex. C.A. Mey	Ephedraceae	Ph.
Euphorbia hirta L.	Euphorbiaceae	Th.
Euphorbia lathyris L.	Euphorbiaceae	Th.
Fagonia boveana (Hadidi) Hadidi & Garf	Zygophyllaceae	Ch.
Fagonia indica Burm. f.	Zygophyllaceae	Ch.
Farsetia longisiliqua Decne.	Brassicaceae	Ch.
Forsskaolea tenacissima L.	Urticaceae	Ch.
Glinus lotoides L.	Molluginaceae	Ch.
Heliotropium arbainense Frense	Boraginaceae	Ch.
Heliotropium bacciferum Forssk.	Boraginaceae	Ch.
Heliotropium curassavicum L.	Boraginaceae	Ch.
Heliotropium longiflorum L.	Boraginaceae	Ch.
Imperata cylindrical (L.) Raeusch	Poaceae	H.
Indigofera spinosa Forssk.	Fabaceae	Ph.
Juncus rigidus Desf.	Juncaceae	H.
Kedrostis foetidissima (Jacq.) Cogn	Cucurbitaceae	H.
Lactuca serriola L.	Asteraceae	Th.
Launaea spinosa (Forssk.) Sch.Bip. ex Kuntze	Asteraceae	Th.

Table 1.1. contd....

Table 1.1. contd.

Species	Family	Life form*
Leptochloa fusca (L.) Kunth.	Poaceae	Th.
Leptadenia pyrotechnica (Forssk.) Decne.	Brassicaceae	Ph.
Lycium shawii Roem. & Schult.	Solanaceae	Ph.
Mentha longifolia (L.) Huds.	Labiatae	Ch.
Nitraria retusa Forrssk. (Ascherson)	Zygophyllaceae	Th.
Ochradenus baccatus Delile	Resedaceae	Ph.
Otostegia fruticosa (Forssk.) Penz.	Labiatae	Ch.
Panicum coloratum L.	Poaceae	Ch.
Peganum harmala L.	Zygophyllaceae	Ch.
Pennisetum divisum (J.F. Gmel.) Henrard	Poaceae	Ch.
Pennisetum setaceum (Forssk.) Chiov.	Poaceae	Ch.
Pluchea dioscoridis (L.) DC	Asteraceae	Th.
Polygonum equisetiform Sm.	Polygonaceae	H.
Polypogon monspielensis (L.) Desf.	Poaceae	H.
Portulaca oleracea L.	Portulacaceae	Ch.
Pulicaria crispa (Forssk.) Oliv	Asteraceae	Ch.
Pupalia lappaceae (L.) Juss.	Amaranthacea	Ch.
Ricinus communis L.	Euphorbiaceae	Ph.
Salsola imbricata Forssk.	Chenopodiaceae	Ch.
Salsola spinescens Moq.	Chenopodiaceae	Ch.
Senna italica Miller	Fabaceae	Ch.
Sisymbrium irio L.	Brassicaceae	Th.
Silybum marianum (L.) Gaertn.	Asteraceae	Ch.
Sonchus oleraceus L.	Asteraceae	Ch.
Spergularia marina (L.) Griseb.	Caryophyllaceae	Th.
Suaeda monoica Forssk. ex. J.F. Gmel.	Chenopodiaceae	Ph.
Tamarix nilotica (Ehrenb.) Bunge	Tamaricacea	Ph.
Tagetes minuta L.	Asteraceae	Ch.
Tribulus parvispinus Presl	Zygophyllaceae	Ch.
Tribulus terrestris L.	Zygophyllaceae	Ch.
Withania somnifera (L.) Dunal.	Solanaceae	Ch.
Xanthium strumarium L.	Asteraceae	Ch.
Zygophyllum simplex L.	Zygophyllacea	Th.

Source: Farrag (2012).
*(Th = Therophytes; H = Hemicryptophytes; G = Geophytes; Ch = Chaemophytes; Ph = Phanerophytes.

Halophytic forage/browse can play an important role as a protein supplement. Even as little as 20 g. of green leaf per day can be enough to improve the digestibility of high energy materials like straw, hay or dry grass (Squires 1980; Wilson 1994). Because the halophytic fodders are of low palatability the intake by the livestock tends to be self limiting. There is a great disparity between the proportion of halophyte on offer to free ranging sheep and the amount eaten (Squires 1980; Graetz and Wilson 1980) but there is variation between seasons of the year in the winter rainfall region

Fig. 1.1. Naturally occurring halophytic plants like *Atriplex, Sclerolaena, Salsola, Nitraria* and *Maireana* cover large areas of semi-arid inland Australia (Photo V. Squires).

Fig. 1.2. Sheep can utilize *Atriplex nummularia* plantations such as this one that was established on a highly saline soil in South Australia (Photo V. Squires).

of southern Australia. The presence of salts, terpenes, tannins and other compounds make many halophytes unpalatable (Attia-Ismail, this volume). These anti-nutritional compounds also affect the digestibility and feed value (Masters, this volume). In some instances fresh fodder may be cut and carried to livestock outside the plantation. This is more common in situations where livestock are penned for the night.

2.2 Feedstuffs from halophytic and salt tolerant plants fed to penned livestock

Feeding these feedstuffs to penned animals expands the possibilities of regulating the proportion in the ration and also the form in which it is presented. Some plant material can be used whole but most commonly it is cut or macerated (e.g., in a hammer mill) to make the pieces smaller and to improve intake. In more sophisticated arrangements the materials can be pelleted and augmented with crushed grains, mineral supplements and other compounds. Some products, such as the residues from oil seed crops like *Salicornia bigelovii* can be pressed into cakes. Pellets can vary in size and degree of firmness depending on which livestock species is being fed. Poultry and rabbits require smaller and softer pellets. Special equipment for processing halophytic feedstuffs are described by El Shaer, this volume. For penned animals there is much less wastage if processed feedstuffs are supplied. Such feedstuffs can be more easily delivered in the desired quantities to individuals', e.g., pregnant and lactating dairy cows or sheep/goats, camels, etc.

3. Quantifying feed supplies and assessing reliability and continuity

3.1 World distribution of halophytic plants

Halophytic plants grow in many arid and semi arid regions around the world (Squires 1994). Those that have greatest value for livestock are grown in the mid latitudes but halophytic plants occur on every continent (except Antarctic), and on some islands. Many halophytic species grow in natural *salinas* which are rapidly deteriorating through the process of desertification (Heshmati and Squires 2013). The world distribution of these soils is discussed by Szabolcs 1994.

Table 1.2 shows the distribution of salt-affected soils in the world. Clearly, not a single continent is free of these soils (Kovda and Szabolcs 1979).

The soil might be saline soils (solonchak) or sodic soils (solonetz) (Szabolcs 1994). Table 1.3 shows the defining characteristics of each of the recognized categories.

These are the soils that usually support halophytic plant species. Because the pattern of distribution of halophytic plants has remained more or less intact, halophytic plants offer an immensely valuable and diverse resource that can provide forage/browse and feedstuffs for the world's burgeoning livestock populations. Large-area plantations of halophytic or salt tolerant species has occurred in many places worldwide and as Glenn et al. (1997) demonstrated, the potential areas on the global drylands and coastal deserts where halophytes could be grown exceed 150 Mha (Glenn et al. 1997; Squires 1998; Squires and El Shaer, this volume). To be commercially viable, feed processors need a reliable supply and a minimum volume to justify the expenditure associated with setting up a facility to provide the feedstuff in the desired configuration. At the local householder level there tends to opportunistic reliance on halophytic feedstuffs and continuity of supply in large volumes is not so important.

Table 1.2. Areas of salt-affected soils in different regions of the world.

Region	Area (ha x 10³)
N. America	15,755
Mexico & Central America	1,965
S. America	129,163
Africa	80,538
S. Asia	87,608
North and Central Asia	211,686
South-East Asia	19,983
Australia	357,330
Europe	50,804

Adapted from Kovda and Szabolcs (1979).

Table 1.3. Classification of salty soils

Soil association	Chemical characteristics
Saline (Solonchak)	Encrustation of white efflorescence. Sodium < 15% of exchangeable cations. Chief anions are chloride and sulphate, rarely nitrate. Soluble carbonate in low proportion only. Conductivity of the saturation extract not less than 4 mmhos cm^{-1}, pH < 8,5
Saline-sodic (Solonetz-Solonchak)	White encrustation on peaks and mixture of black and white in depressions. Pre-dominant anions, chloride, sulphate and carbonate. High proportion of sodium sand low proportions of calcium cations, pH around 8.5
Non-saline sodic Solonetz or alkaline	(A) horizon saturated with sodium in absorbed state. (B) Horizon generally of Columnar or prismatic structure in heavy soils. Exchangeable sodium greater than 15%. Predominant anions are carbonates and bicarbonates; pH up to 10

4. Soil salinity and its effect on halophyte and salt tolerant plants' distribution

The most common salts present in saline soils are sodium, potassium, calcium and magnesium in the form of chlorides, sulfates or bicarbonates. Sodium and chloride are present in the highest percentage, particularly in highly saline soils. The most common sodic salts present in sodic soils are the carbonates particularly sodium. Ionic distribution may affect the distribution of the halophytic species (Egan and Ungar 2000).

The reaction of plants to salt stress is called salt resistance. Resistance may be represented either by salt tolerance or salt avoidance. Salt tolerance involves processes of adaptation, either physiological or biochemical, in order to maintain protoplasmic viability as cells accumulate electrolytes. Salt avoidance involves similar processes, but include structural along with physiological adaptation, in order to minimize salt concentrations of the cells or physiological exclusion by root membranes (Flowers et al. 1977).

Aronson (1989) estimated that about one-third of the approximately 350 families of flowering plants contain halophytes and about 50% of the genera belong to 20 of these families (Flowers et al. 1986). Therefore, it appears that halophytes are widely distributed among several families of flowering plants. Based on these studies and those by Chapman (1975) and Flowers et al. (1977), it is clear that there is no family in which more than half of the species are halophytes. The fact that the limited number of halophytic species is spread among so many different families indicates that halophytism, even though a trait controlled by several genes, is not such a complex characteristic that only arises once during evolution. This polyphyletic origin of halophytism (Flowers et al. 1986) suggests that it is a characteristic that could be introduced into species not yet possessing it, particularly species in the same families that already contain halophytic species. Both climate and soil characteristics have influence on the distribution of halophytes (Szabolcs 1994). He mentioned that salt affected soils are distributed not only in the deserts and semi-deserts, but also in fertile alluvial plains, river valleys and coastal areas, close to highly populated areas and irrigation systems. Salt affected soils are most appropriate for the growth of halophytes; although they occur in completely different soils. A wide range of plant genera inhabits salt-affected soils and are referred to as halophytes. Not all of the halophyte plants are able to germinate in their native saline habitats without early leaching of the soil (Batanouny 1993). He concluded that the correlation between salt tolerance of a plant and its tolerance at other stages is not obligatory. This has been noted in certain halophytes (Waisel 1972). The water relations of their habitats therefore, affect germination of halophytes. Water content (= water budget) is a critical factor for the growth of halophytes. Therefore, Le Houerou (1994) classified saline soils with respect to water budget into permanently wet, quasi-permanently dry and some may have a permanent or a temporary water table or not. The soil's chloride content affects the germination of halophytes. This is called the ionic budget. However, vegetative reproduction is a important means of reproduction in many halophytic species (e.g., *Limonium vulgare, L. humile*, and *Tamarix aphylla*).

The relationship between soil salinity and halophyte distribution have been investigated in several parts of the world in Egypt (Abd El-Ghani and Amer 2003; Abd El-Ghani and El-Sawaf 2005), Australia (Bui and Henderson 2003; Crowley 1994), USA (Egan and Ungar 2000; Omer 2004), China (Li 1993; Liu et al. 2003; Toth et al. 1995), Iran (Jafari et al. 2003, 2004), Italy (Silvestri et al. 2005), Pakistan (Khan et al. 2006a). However, salinity gradient, moisture and available nutrient in soil are the important factors in controlling the distribution of vegetation. Results of Li et al. (2008) showed that salinity, pH, moisture and available nitrogen were the major soil factors responsible for variations in the pattern of vegetation. Ungar (2000) indicated that species like *Puccinellia nuttalliana, Distichlis stricta, Hordeum jubatum, Suaeda depressa, Salicornia rubra, Atriplex triangularis, A. patula* and *Triglochin maritime* are characteristic species of soils dominated by anions of chloride, sulfate and carbonate.

5. Tolerance to salinity

Halophytes are plants that grow inevitably in saline environment. Tolerance is the ability of a plant species to adapt to the varying effects of presence of certain ions.

Growth, chemical composition and other botanical parameters are the net result of the growing conditions, i.e., soil, water and the plant. This relationship is not as simple as it appears; it is a very complex relationship. Most halophytes differ in their response to salt concentrations in the soil at varying degrees and at different stage of growth. Other biotic and abiotic factors may play a significant role in the degree of tolerance of halophytic plants and salt tolerant plants to varying degrees of salinity. The tolerance is then the adjustment actions taken by the plant to these conditions. Therefore, the salt tolerance may, partially, be controlled by the tolerance of the specific species ability to regulate the high salt tolerance.

Ungar (2000) has collected some results from the literature regarding the ability of germination rate of specific halophytes. He has discussed this aspect in detail and also on the different tolerance levels of each plant discussed. Flowers and Colmer 2008 has shown the tolerance of some halophytic plants to different seawater irrigation levels (Table 1.5). The striking fact about tolerance of halophytes to salinity mentioned

Table 1.4. Three categories of land that has potential for growing halophytes. 1. **Coastal deserts** usable for sea water irrigation (< 100 m elevation, presently unused for agriculture or settlement, within reach of seawater irrigation). 2. **Inland salinized regions** suitable for brackish water irrigation. 3. **Arid irrigation areas** where secondary salinity is a problem.

Potential sites for Growing halophytes	Area km²
COASTAL DESERTS	
- India/Pakistan	
- Iran/Iraq	
- Near East/Africa	
- Australia	
- N. & S. America	
SUB TOTAL	494,500
INLAND SALINIZED BASINS	
- Eurasia	
- Subcontinent	
- Africa	
- Middle East/Gulf	
- Australia	
- N & S America	
- China	
- Central Asia	
- SUB TOTAL	631,250
ARID IRRIGATION AREAS (Secondary)[1]	
Saharan/Middle Eastern	
Indus valley	
Kalahari	
Peru	
Mexico	
Central Asia	
SW USA	
Mongolia	
- SUB TOTAL	376,980

[1] About 20% usable for halophytes.

Table 1.5. Tolerance of some halophytes to different seawater irrigation levels.

Genus and species	Seawater (%)	Mean water EC, dS/m*	Synthetic comment
		Surviving	
Chamaerops humilis	40	18	Very tolerant
Citrus macrophylla	-	-	Sensitive
Elytrigia elongate	60	27	Excellent
Ficus carica	33	15	Tolerant
Nicotiana glauca	-	-	Sensitive
Olea europea	40	18	Very tolerant
Pistacia vera	33	15	Tolerant
Punica granatum	33	15	Tolerant
Sesuvium portulacastrum	100	45	Exceptional
Severinia buxifolia	33	15	Tolerant
Spartina alterniflora	100	45	Exceptional
Vetiveria zizanioides	60	27	Excellent

*dS/m = deciSiemens/meter—a measure of electrical conductivity.

by Kumar Parida and Bandhu Das (2005) is that they can tolerate salinity levels above sea water (of 45–50 dS/m). According to these authors the plants can do this by employing different mechanisms of defense. Perennial grasses are reported to be highly salt tolerant. Some grasses grew in soil salinity ranges between 300 to 500 mM NaCl (Peng et al. 2004) while others could not survive in salt concentrations above 300 mM NaCl (La Peyre and Row 2003). Some other grasses could survive in up to 1000 mM NaCl (Bodla et al. 1995). Some data on salinity tolerance are shown in Table 1.5.

6. Sources of fodder and potential yields

6.1 Irrigated

Under irrigation, halophyte crops can perform as well as conventional crops when irrigated with brackish water. Intensive breeding and selection programs have been carried out to improve salt tolerance of traditional and forage crops (Khan et al. 2014). A few improved cultivars have been released, but in general results have been disappointing. Few conventional crops can be irrigated with water even 10% as salty as sea water, and drainage water often exceeds this level of salinity. More rapid progress has been made in domesticating wild halophytes as domestic crops (Koyro et al. 2011; Ventura et al. 2015) the most promising halophytes is an annual, succulent salt marsh plant *Salicornia bigelovii*. It produces an oilseed (Weber et al. 2006) that can be pressed to provide human-edible oil and protein meal for animal diets while the straw can be used for fodder (Swingle et al. 1996; Riley, this volume).

Salicornia bigelovii is a native to coastal areas of the eastern and southern USA, as well as southern California, Belize, and coastal Mexico (both the east and west coasts). It can be grown using undiluted sea water in coastal desert areas. Using sea water for irrigation requires special irrigation techniques, since the soil must be kept constantly moist and an excess of sea water (above consumptive use) must be added to keep salts from accumulating in the root zone. Sea-water grown *Salicornia* yields as much as biomass as conventional crops but needs 35% more water per unit of biomass. In low lying coastal regions this disadvantage can be outweighed by the lower lift required to irrigate with sea water compared to deep wells. Although sea water irrigation produced large quantities of saline drainage water, in coastal regions it can be drained back to the sea by gravity, whereas inland irrigation basins often have no outlet. Sea-water irrigated halophytes and salt tolerant plants have been grown successfully in coastal deserts in North Africa, the Gulf region and in the Nile Delta (Fahmy et al. 2010). They are particularly suited to coastal areas which lack fresh water for crop production, or as an attempt to conserve fresh water for human use. Halophytes have been shown to be cost effective as crops to reuse saline water and soils within existing irrigation districts. Even seaweeds (*Ruppia maritima* and *Chaetomorpha linum*) have been used as feedstuffs for sheep (Rjiba Ktita et al. 2010).

6.2 Non-irrigated

Large areas of sub-humid to semi arid land can be used to grow fodder/forage for livestock. In Algeria, Libya, Tunisia and other north African countries as well as in the Gulf region (El Shaer 2009).

Halophytes such as *Atriplex nummularia* and *A. canescens* have been planted over large areas as fodder reserves in many countries. Salt tolerant fodder species such as *Haloxylon* spp. occur widely in Central Asia where extensive woodlands occur and have been planted in Iran, often in association with *Atriplex canescens*. *Kochia scoparia* and *K. prostrata* are other salt tolerant species that have been used extensively in Central Asia (Gintzburger et al. 2003). Planting is often laborious if transplants are used and of course nursery facilities are required to produce the thousands of plants that are required to vegetate extensive areas. Notwithstanding this, over 50,000 ha were re-vegetated using transplanted *A. canescens* and *Haloxylon ammodendron* seedlings in a major project in the South Khorasan province of Iran in the 1990s. Some *Atriplex* can be planted from seed often in conjunction with machinery that creates a seedbed and micro-catchment to trap and store moisture (Fig. 1.3).

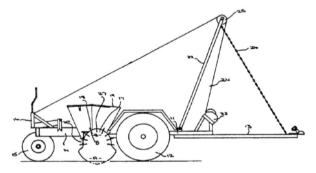

Fig. 1.3. This is an example of a pitting implement intended to provide a series of spaced pits on the surface of the ground where by the pits are intended to receive seed and provide an environment which promotes germination of the seed (After Gintzburger et al. 2000).

Halophytes occur in extensive natural stands around the shores of the Caspian sea in Iran and Turkmenistan (Heshmati 2013) and over millions of ha of inland Australia (Squires 1989), parts of western USA, northern Mexico (O'Leary and Glenn 1994) and in western China and parts of Inner Asia (Zhang et al. 1994). Natural stands provide browse for livestock and act as drought reserves. Because of their low palatability halophytes are rarely eaten when alternative fodders are available (Wilson 1994) but can provide a maintenance ration in dry times (Warren et al. 1995; Jin 1994). The intake tends to self-limiting. Not all livestock species or breeds are equally deterred by the low palatability. Camels and goats are less likely to be put off than cattle or horses (Safinaz and El Shaer, this volume).

Satisfactory performance of small ruminants fed on *A. nummularia,* one of the most widely used forage halophytes, has been reported in numerous research studies (reviewed by Ben Salem et al. 2010). There is some controversy about the nutritive value of halophytic feedstuffs with a great disparity in value as assessed under field conditions and in the laboratory (El Shaer, this volume and Masters, this volume). Livestock often avoid halophytic plants in natural rangelands because halophytic plants tend to be less palatable than other forage on offer, but when halophyte biomass is incorporated into mixed rations, it can completely replace conventional roughage (hays, etc.) in fattening diets. The animals need no special inducements to eat halophyte material if it is mixed with other diet ingredients (Swingle et al. 1998). Livestock fed on halophytic feedstuffs do however consume more feed and water per unit of weight gain due to the high salt content. Notwithstanding this, the practice of incorporating halophytic feedstuffs into mixed rations has been demonstrated to have merit (Squires and Ayoub 1995; Swingle et al. 1995; Swingle et al. 1998; El Shaer, this volume) and is more fully explored in this book.

7. Quantifying sources of forage and potential yields

There is a rich data set on the productivity of halophytic and salt tolerant plants (Lieth and Mochtchenko 2011; Squires and Ayoub 1994; Choukra-allah et al. 1995; Khan and Ansari 2008; Azhar et al. 2003). Complications occur when assessing forage biomass as opposed to net primary productivity. Palatability of small twigs varies with the animals eating it. If livestock are adapted and tolerant then even quite woody branches and large twigs may be consumed, although its nutritive value will be low. For some species, like cultivated annual halophytes such as *Salicornia bigelovii,* almost 100% of the plant can be considered to be forage/fodder biomass, but for others it can be as low as 20%.

Regional variations in type, quantity and potential of feedstuffs are dealt with in the various country/region-specific overviews in this book. Special attention is given to North Africa including Egypt, and Australia, South America, Gulf region, Central Asia, the Caspian region and the Indian subcontinent. The biomass production and quality of the natural vegetation in such areas vary considerably from season to season and from area to area depending on several factors, which are mainly environmental (Qadir et al. 2008). The average area of salt-affected soils might be calculated on the basis of FAO averages to be 450 Mha. That means we are talking about 4–5 billion tons of biomass. They are widely distributed throughout several regions of the world due to the presence of numerous saline areas along the sea shores and inlands (littoral salt marshes and inland salt marshes). Halophytes represent a major part of natural rangelands particularly the perennials and shrubby ones. There are several possibilities to utilize halophytes in economic ways. In fact halophytic plants have many uses: they can be used as animal feed ingredients, vegetables, and drugs, as well as for sand dune fixation, wind shelter, soil cover, cultivation of swampy saline lands, laundry detergents, paper production, and so on.

8. Is commercial use of fodder from halophytic plants and salt tolerant plants viable?

This is the most important question. Small scale, local plantations will continue to prosper as freshwater resources are rapidly declining and the efforts to obtain more from greater depths or from more distant places are limited by energy constraints. The dryland regions of the world are likely to face severe hardships in the future unless there are viable alternatives on offer (Qadir et al. 2008; Lieth and Moschtchenko 2014). The problem is particularly pressing for agricultural and animal husbandry systems in the dry tropical and subtropical region of the world. Burgeoning human populations put extra stress on the system. The development of agricultural systems based on saline soil and irrigation water has received attention for some years (see Riley, this volume; Lieth and Moschtchenko 2014).

8.1 Problems and prospects

To be of value to processors, large quantities of raw material that has a guaranteed delivery schedule is required to justify the investment in developing the processing facility. This is much more likely to occur where plantations are established and where irrigation water is available to ensure high and reliable yields. Depending on the plant material, a range of products may be developed. The most common are **pellets** that are a mixture of the crushed halophyte and additives including minerals, etc. that improve its palatability and nutritive value. The size and composition of the pellets can be tailor-made to suit the livestock species that can range from large ruminants to sheep, goats and non ruminants such as rabbits and poultry. For beef feedlots a mixed ration that is high in protein and energy is required. Grain and oilseed residue such as cotton seed meal is added. Du Toit et al. (2004) studied the effects of type of carbohydrate supplementation on intake and digestibility of *Atriplex nummularia* and found that were large differences. Some halophytic materials such as *Salicornia* oilseed cake can substitute for cotton meal. Roughage in the form of stalks of some halophytes and salt tolerant plants such *Distichlis* spp. and *Chloris* spp. can be added up to 30% of the total, as a substitute for more conventional roughages such as grass hays, as elaborated in this book.

    Special harvesting equipment (forage harvesters) and forage choppers are used in preparing halophytes and other fodders as silage. Both palatability and digestibility are enhanced when the plants are ensiled. It also provides an opportunity to incorporate additives and makes feeding easier while reducing wastage (see also El Shaer, this volume).

9. Summing up

The shortages of animal feeds are the main constraint to increasing indigenous animal production and improve livestock productivity. It is especially a common characteristic of the arid and semi-arid regions of the world. Animal husbandry, as the main income resource for pastoralists, is based mostly on the natural vegetation for rearing sheep, goats and other herbivores. Increasingly, as populations, rise and living standards improve there is more demand for livestock products, including poultry and non ruminant animals such as rabbits. Some of this increase will come from value adding in the food supply chain. Feedlots, ranging from small scale to large commercial enterprises are creating a demand for more fodder. Feedstuffs derived from halophytes and salt tolerant plants have potential to fill the gaps in a number of regions throughout the world (see also Chapter 22).

References and Further Readings

Abdelly, C., M. Ozturk, M. Ashraf and C. Grignon (eds.). 2008. Biosaline Agriculture and High Salinity Tolerance. BirkhuserVerlag, Switzerland.

Abd El-Ghani, M.M. and W.M. Amer. 2003. Soil-vegetation relationships in a coastal desert plain of southern Sinai, Egypt. J. Arid Environ. 55: 607–628.

Abd El-Ghani, M.M. and N.A. El-Sawaf. 2005. The coastal roadside vegetation and environmental gradients in the arid lands of Egypt. Community Ecol. 6: 143–154.

Al-Shorepy, S.A., G.A. Alhadrami and A.I. El Awad. 2010. Development of sheep and goat production system based on the use of salt-tolerant plants and marginal resources in the United Arab Emirates. Small Ruminant Research 91(1): 39–46.

Ahmed, R. and K.A. Malik (eds.). 2002. Prospects of Saline Agriculture. Kluwer Academic Press, Netherlands. Tasks for Vegetation Science 37: 460.

Abu-Zanat, M.M.W. and M.J. Tabbaa. 2006. Effect of feeding *Atriplex* browse to lactating ewes on milk yield and growth rate of their lambs. Small Ruminant Research 64(1-2): 152–161.

Aronson, J. 1989. HALOPH: Salt Tolerant Plants for the World—A Computerized Global Data Base of Halophytes with Emphasis on Their Economic Uses. University of Arizona Press, Tucson.

Ashraf, M., M. Ozturk, Athar, Habib-ur-Rehman (eds.). 2009. Salinity and Water Stress: Improving Crop Efficiency. Tasks for Vegetation Science, Vol. 44, Kluwer Academic, Dordrecht.

Attia-Ismail, S.A. (in press). Plant secondary metabolites of halophytes and salt tolerant plants, pp. 127–142, this volume.

Attia-Ismail, S.A., H.M. Elsayed, A.R. Askar and E.A. Zaki. 2009. Effect of different buffers on rumen kinetics of sheep fed halophyte plants. Egyptian J. of Sheep and Goats Sciences 1: 1–11.

Attia-Ismail, S.A., A.A.M. Fahmy and R.T. Fouad. 1994. Improving nutritional value of some roughages with mufeed liquid supplement. Egypt. J. Anim. Prod. 31: 161–174.

Attia-Ismail, S.A., M.F. Afaf and A.A. Fahmy. 2002. Mineral, nitrogen and water utilization of salt plant fed sheep as affected by monensin. Egypt. J. Feeds and Nutr. 6: 151–161.

Azhar, A., M. Huq, M.A. Khan, C. Lewis, A. Shibli, A. Siddiqui and S.H. Zaidi. 2003. Technology and Development in New Millennium. Karachi University Press, Karachi (Pakistan), pp. 320.

Batanouny, K.H. 1993. Adaptation of plants to saline conditions in arid regions. pp. 387–401. *In*: H. Lieth and A al Masoom (eds.). Towards rational use of high salinity tolerant plants Vol.1 Kluwer Academic, Dordrecht.

Batanouny, K.H. 1994. Halophytes and halophytic plant communities in the Arab region: Their potential as a rangeland resource. pp. 139–164. *In*: V.R. Squires and A.T.Ayoub (eds.). Halophytes as a Resource for Livestock and for Rehabilitation of degraded lands. Tasks in Vegetation Science 32, Kluwer Academic, Dordrecht.

Batanouny, K.H. 2001. Halophytes and halophytic plant communities in the Arab region. Kluwer-Academic Pub., Amsterdam 1: 136–139.

Ben Salem, H., H.C. Norman, A. Nefzaoui, D.E. Mayberry, K.L. Pearce and D.K. Revel. 2010. Potential use of oldman saltbush (*Atriplex nummularia* Lindl.) in sheep and goat feeding. Small Ruminant Research 91(1): 28–39.

Bodla, M.A., M.R. Chaudhry, S.R.A. Shamsi and M.S. Baig. 1995. Salt tolerance in some dominant grasses of Punjab. pp. 190–198. *In*: M.A. Khan and I.A. Ungar (eds.). Biology of Salt Tolerant Plants. Book Crafters, Michigan, USA.

Bui, E.N. and B.L. Henderson. 2003. Vegetation indicators of salinity in northern Queensland. Austral Ecology 28(5): 539–552.

Chapman, V.J. 1975. Salt marshes and salt deserts of the world. Springer-Verlag, Berlin/New York

Choukr-Allah, R., C. Malcolm and A. Hamdy (eds.). 1995. Halophytes and Biosaline Agriculture, Marcel Dekker, New York, pp. 221–236.

Crowley, G.M. 1994. Quaternary soil salinity events and Australian vegetation history. Quat. Sci. Rev. 13: 15–22.

Du Toit, C.J.L., W.A. Van Niekerk, N.F.G. Rethman and R.J. Coertze. 2004. The effect of type and level of carbohydrate supplementation on intake and digestibility of *Atriplex nummularia* cv. De Kock. South African Journal of Animal Science 34(5): 35.

Egan, T.P. and I.A. Ungar. 2000. Similarity between seed banks and above-ground vegetation along a salinity gradient. J. Vegetation Sci. 11:189–194.

El Shaer, H.M. 2009. Potential role of sabkhas in Egypt: an overview. pp. 221. *In*: M. Ashraf (ed.). Salinity and water stress. Springer, Dordrecht.

El Shaer, H.M. and S.A. Attia-Ismail. (in press). Halophytic and salt tolerant feedstuffs in the Mediterranean basin and Arab region: An overview, pp. 21–36, this volume.

Fahmy, A.A., K.M. Youssef and H.M. El Shaer. 2010. Intake and nutritive value of some salt-tolerant fodder grasses for sheep under saline conditions of South Sinai, Egypt. Small Rumin. Res. 91: 110–115.

Farrag, H.F. 2012. Floristic composition and vegetation-soil relationships in Wadi Al-Argy of Taif region, Saudi Arabia. International Research Journal of Plant Science 3(8): 147–157.

Flowers, T.J. and T.D. Colmer. 2008. Salinity tolerance in halophytes. New Phytologist 179: 945–963.

Flowers, T.J., M.A. Hajiheri and N.J.W. Clipson. 1986. Halophytes. The Quart. Rev. Biol. 61: 313.

Flowers, T.J., R.F. Troke and A.R. Yeo. 1977. The mechanism of salt tolerance in halophytes. Ann. Rev. Plant Physiol. 28: 89.

Gintzburger, G., M. Bounejmate and A. Nefzaoui (eds.). 2000. Fodder shrub development in arid and semi-arid zones. Volume 2. Proceedings of the Workshop on Native and Exotic Fodder Shrubs in Arid and Semi arid Zones, 27 October-2 November 1996, Hammamet, Tunisia, pp. 518–523.

Gintzburger, G., K.N. Toderich and M.M. Mardonov. 2003. Rangelands of the arid and semi arid zones in Uzbekistan, CIRAD, Montepellier, p. 431.

Glenn, E.P., J. Riley, N. Hicks and S. Swingle. 1995. Seawater irrigation of halophytes for animal feed. pp. 221–236. *In*: R. Choukr-Allah, C. Malcolm and A. Hamdy (eds.). Halophytes and Biosaline Agriculture, Marcel Dekker, New York.

Glenn, E.P., V.R. Squires and J.J. Brown. 1997. Saline soils in the drylands: extent of the problem and prospects for utilisation. pp. 144–147. *In*: World Atlas of Desertification (2nd ed.) Edward Arnold/UNEP.

Glenn, E.P., R.S. Swingle, J.J. Riley, C. Mota, M.C. Watson and V.R. Squires. 1994. North American Halophytes : potential use in animal husbandry. pp. 165–174. *In*: Victor Squires and A.T. Ayoub (eds.). Halophytes as a Resource for Livestock and for Rehabilitation of Degraded Land, Tasks for Vegetation Science.

Glenn, E.P., V.R. Squires and J.J. Brown. 1997. Saline soils in the drylands: extent of the problem and prospects for utilisation. pp. 144–148. *In*: World Atlas of Desertification (ibid).

Graetz, R.D. and A.D. Wilson. 1980. Comparison of the diets of sheep and cattle grazing a semi-arid chenopod shrubland, Aust. Rangel. J. 2(1): 67–75.

Gul, B., R. Ansari, H. Ali, M.Y. Adnan, D.J. Weber, B.L. Nielsen, H.W. Koyro and M.A. Khan. 2014. The sustainable utilization of saline resources for livestock feed production in arid and semi-arid regions: A model from Pakistan. Emir. J. Food Agric. 26(12): 1032–1045.

Heshmati, G.A. 2013. Indigenous Plant species from the Drylands of Iran, distribution and potential for habitat maintenance and repair. pp. 355–375. *In*: G.A. Heshmati and V.R. Squires (eds.). Combating Desertification in Asia, Africa, and the Middle East. Springer, Dordrecht.

Jafari, M., M.A.Z. Chahouki, A. Tavili and H. Azarnivand. 2003. Soil-vegetation in Hon-e-Soltan region of Qom Province, Iran. Pak. J. Nut. 2: 329–334.

Jafari, M., M.A.Z. Chahouki, A. Tavili and H. Azarnivand. 2004. Effective environmental factors in the distribution of vegetation types in Poshtkouh rangelands of Yazd Province (Iran). J. Arid Environ. 56: 627–641.

Jin, Q. 1994. *Alhargi sparsifolia* Schap.: a potentially utilizable forage in saline soils. *In*: Victor Squires and A.T. Ayoub (eds.). Halophytes as a Resource for Livestock and for Rehabilitation of Degraded Land. Tasks in Vegetation Science Series, Kluwer Academic Press, The Netherlands.

Khan, M.A. and R. Ansari. 2008. Potential use of halophytes with emphasis on fodder production in coastal areas of Pakistan. Biosaline Agriculture and High Salinity Tolerance. Edited by C. Abdelly, M. Ozturk, M. Ashraf and C. Grignon, BirkhuserVerlag, Switzerland, pp. 163–175.

Khan, M.A., R. Ansari, B. Gul and M. Qadir. 2006a. Crop diversification options for salt-prone land resources. *In*: Perspectives in Agriculture, Veterinary Science, Nutrition and Natural Resources. CABI, International, Vol. 1 number 48. USA.

Khan, M.A. and M. Qaiser. 2006b. Halophytes of Pakistan: distribution, ecology, and economic importance. pp. 129–153. *In*: M.A. Khan, H. Barth, G.C. Kust and B. Boer (eds.). Sabkha Ecosystems. Vol. II: Springer, Netherlands.

Khan, M.A., B. Boer, G.S. Kust and H.J. Barth. 2010. Sabkha Ecosystems: Volume II: West and Central Asia (Tasks for Vegetation Science), Kluwer Academic, Dordrecht.

Khan, M.A., B. Böer, M. Öztürk, T.Z. Al Abdessalaam, M. Clüsener-Godt and B. Gul (eds.). 2014. Sabkha Ecosystems: Volume IV: Cash Crop Halophyte and Biodiversity Conservation (Tasks for Vegetation Science), Kluwer Academic, Dordrecht.

Khidr, R.K., K. Abd El-Galil and K.R.S. Emam. (in press). Utilization of halophytes and salt tolerant feedstuffs for poultry and rabbits, pp. 361–372, this volume.

Kovda, V.A. and I. Szabolcs. 1979. Modeling of soil salinization and alkinization. Agrokemia Talajtan (Suppl).

Koyro, H.-W., T. Hussaina, B. Huchzermeyer and M.A. Khan. 2011. Photosynthetic and growth responses of a perennial halophytic grass *Panicum turgidum* to increasing NaCl concentrations. Environ. Exper. Bot. 91(2013): 22–29.

Kumar Parida, A. and A. Bandar Das. 2005. Salt tolerance and salinity effects on plants: a review Ecotoxicology & Environ. Safety 60(3): 324–349.

Le Houerou, H.N. 1994. Halophytes and halophytic plant communities in Inner Asia. *In*: V.R. Squires and A.T. Ayoub (eds). Halophytes as Resource for Livestock and for Rehabilitation of Degraded Lands. Tasks in Vegetation Science 32 Kluwer Academic, Dordrecht

Le Peyre, M.K. and S. Row. 2003. Effects of salinity changes on growth of *Ruppia maritima* L. Aquat. Bot. 77: 235–241.

Li, X.D. 1993. Canonical analysis and the principal components analysis of plant community with its environmental factors in the Yellow River Delta. Acta Botanica Sin. 35(Supplement): 139–143 [In Chinese with English abstract].

Li ,W.-Q., X.J. Liu, M.A. Khan and B. Gul. 2008. Relationship between soil characteristics and halophytic vegetation in coastal region of north China. Pak. J. Bot. 40(3): 1081–1090.

Lieth, H and M. Mochtc/chenko. 2011. Cash Crop Halophytes: Recent Studies 10 Years After Al Ain Meeting. Tasks for Vegetation Science 38, Kluwer Academic, Dordrecht.

Lieth, H. and A.A. Al Massoom. 1993. Toward the rational use of high salinity tolerant plants and ecosystems. Vol. 1 and 2. Kluwer Academic, Dordrecht.

Lieth, H., M. Mochtchenko, M. Lohmann, H.-W. Koyro and A. Hamdy (eds.). 1999. Halophyte uses in different climates 1: Ecological and ecophysiological studies, in Progress in biometeorology, Vol. 13. Publishers Backhuys, The Netherlands.

Lieth, H. and U. Menzel. 1999. Halophyte Data base Vers. 2 (Annex 4). pp. 159–258. *In*: H. Lieth, M. Moschenko, M. Lohmann, H.-W. Koyro and A. Hamdy (eds.). Progress in Biometeorology, Volume 13. Halophyte uses in different climates I. Ecological and ecophysiological studies. Publisher Backhuys, Leiden.

Liu, X.J., W.Q. Li and Y.M. Yang. 2003. Studies on the nutrient characteristics of soil and halophyte in coastal saline soil of Hebei Province. Eco-agr. Res. 11(2): 76–77 [In Chinese with English abstract].

Masters, D.G., S.E. Benes and H.C. Norman. 2007. Biosaline agriculture for forage and livestock production. Agri. Ecosys. Environ. 119: 234–248.

Masters, D.G. (in press). Assessing feed value of halophytes, pp. 89–105, this volume.

McArthur, E.D. and D.J. Fairbanks (compilers). 2000. Proceedings: Shrubland Ecosystem Genetics and Biodiversity; 2000 June 13–15; Provo, UT. Proceedings RMRS-P-21. Ogden, Ut: U.S. Department of Agriculture, Forest Service Rocky Mountain Research Station (USA).

O'Leary, J.W. and E.P. Glenn. 1994. Global distribution and potential of halophytes. pp. 7–17. *In*: V.R. Squires and A.T. Ayoub (eds.). Halophytes as a resource for livestock and for rehabilitation of degraded lands. Kluwer Academic, Dordrecht.

Omer, L.S. 2004. Small-scale resource heterogeneity among halophytic plant species in an upper salt marsh community. Aquat. Bot. 78: 337–348.

Osman, A.E and M.A. Shalla. 1994. Use of edible shrubs in pasture improvements under Mediterranean environment in northern Syria. pp. 255–258. *In*: Victor Squires and A.T. Ayoub (eds.). Halophytes as a Resource for Livestock and for Rehabilitation of Degraded Land. Tasks in Vegetation Science Series, Kluwer Academic Press, The Netherlands.

Öztürk, M., B. Böer, H.J. Barth, S.W. Breckle, M. Clüsener-Godt and M.A. Khan. 2010. Sabkha Ecosystems: Volume III: Africa and Southern Europe. Kluwer Academic (Tasks for Vegetation Science) Dordrecht.

Peng, Y.H., Y.F. Zhu, V. Mao, S.M. Wang and W.A. Su. 2004. Alkali grass resists salt stress through high [K^+] and endodermis barrier to Na^{++} J. Exp. Botany 55: 939–949.

Qadir, M., A. Tubeileh, J. Akhtar, A. Larbi, P.S. Minhas and M.A. Khan. 2008. Productivity enhancement of salt-prone land and water resources through crop diversification. Land Degrad. Dev. 19: 429–453.

Rabhi, M., S. Ferchichi, J. Joini, M.H. Hamrouni, H.-W. Koyro, A. Ranieri, C. Abdelly and A. Smaoui. 2010. Phytodesalinisation of a salt affected soil with halophyte *Sessuvium portulacastrum* L. to arrange in advance the requirement for the successful growth of a glycophytic crop. Biores. Technol. 101: 6822–6828.

Riley, J.J. (in press). Review of Halophyte Feeding Trials with Ruminants, pp. 117–217, this volume.

Rjiba Ktita, S., A. Chermiti and M. Mahouachi. 2010. The use of seaweeds (*Ruppia maritime* and *Chaetomorpha linum*) for lamb fattening during drought. Small Ruminant Research 91(1): 116–199.

Safinaz, M.S. and M.H. El Shaer. (in press). Halophyte and salt tolerant plants feeding potential to dromedary camels, pp. 247–260, this volume.

Silvestri, S., A. Defina and M. Marani. 2005. Tidal regime, salinity and saltmarsh plant zonation Estuar. Coast. Shelf. Sci. 62: 119–130.

Squires, V.R. 1989. Australia: distribution and characteristics of shrublands. pp. 61–92. *In*: C.M. McKell (ed.). Biology and Utilization of Shrubs. Academic Press, New York.

Squires, V.R. 1980. Livestock Management in the Arid Zone. Inkata Butterworth Heinemann, Melbourne. 271 p.

Squires. 1994. Overview of problems and prospects for utilizing halophytes as a resource for livestock and for rehabilitation of degraded lands. pp. 1–6. *In*: V.R. Squires and A.T. Ayoub (eds.). Halophytes as a Resource for Livestock and for Rehabilitation of degraded lands. Tasks in Vegetation Science 32, Kluwer Academic, Dordrecht.

Squires, V.R. 1998. Prospects for increasing carbon storage in desert soils and the likely impacts on mitigating global climate change. pp. 19–30. *In*: S. Omar, R. Misak and Al Ajmi (eds.). Sustainable Development in Arid Zones, Vol. 1. Assessment and Monitoring of Desert Ecosystems, Balkema. Rotterdam, Brookfield.

Squires, V.R. 2006. Range and animal production in the arid lands of Australia. Secheresse 17(1–22).

Squires, V.R. (1998). Dryland Soils: their potential as a sink for carbon and as an agent in mitigating climate change. Advances in GeoEcology 31: 209–215.

Squires, V.R. and A.T. Ayoub (eds.). 1995. Halophytes as a Resource for Livestock and for Rehabilitation of Degraded Land, Tasks in Vegetation Science Series, Kluwer Academic Press, The Netherlands

Squires,V.R. and H.M. El Shaer (in press). Unifying Perspectives: Halophytes and Salt-tolerant feedstuffs and their role in livestock production systems, pp. 406–416, this volume.

Squires, V.R. and E.P. Glenn. 1997. Carbon sequestration in drylands. pp. 140–143. *In*: World Atlas of Desertification (2nd ed.). Edward Arnold/UNEP.

Swingle, R.S., E.P. Glenn and J.J. Riley. 1995. Halophytes in mixed feeds for livestock. pp. 97–100. *In*: V.R. Squires and A.T. Ayoub (eds.). Halophytes as a Resource for Livestock and for Rehabilitation of Degraded Land, Tasks in Vegetation Science Series, Kluwer Academic Press, The Netherlands.

Szabolcs, I. 1994. Salt affected soils as the ecosystem for halophytes. pp. 19–24. *In*: V.R. Squires and A.T. Ayoub (eds.). Halophytes as a Resource for Livestock and for Rehabilitation of degraded lands. Tasks in Vegetation Science 32, Kluwer Academic, Dordrecht.

Swingle, R.S., E.P. Glenn and V.R. Squires. 1996. Growth performance of lambs fed mixed diets containing halophyte ingredients. Animal Feed Sci. Technol. 63: 137–148.

Toth, T., S. Matsumoto, R. Mao and Y. Yin. 1995. Precision of predicting soil salinity based on vegetation categories of abandoned lands. Soil Sci. 160: 218–231.

Ventura, Y., A. Eshel, D. Pasternak and M. Sagi. 2015. The development of halophyte-based agriculture: past and present. Annals Bot. 115: 529–540.

Waisel, Y. 1972. Biology of halophytes. Academic Press, New York, NY.

Warren, B., T. Casson and D.H. Ryall. 1994. Production from grazing sheep on revegetated saltland in Western Australia. pp. 263–266. *In*: Victor Squires and A.T. Ayoub (eds.). Halophytes as a Resource for Livestock and for Rehabilitation of Degraded Land. Tasks in Vegetation Science Series, Kluwer Academic Press, The Netherlands.

Weber, D.J., R. Ansari, B. Gul and M.A. Khan. 2006. Potential of halophytes as source of edible oil. Journal of Arid Environment 68: 315–321.

Wilson, A.D. 1994. Halophytic shrubs in semi-arid regions of Australia: Value for grazing and land stabilization. pp. 101–114. *In*: Victor Squires and A.T. Ayoub (eds.). Halophytes as a Resource for Livestock and for Rehabilitation of Degraded Land. Tasks in Vegetation Science Series, Kluwer Academic Press, The Netherlands.

Yensen, N.P. 2006. Halophyte uses for the twenty first century. pp. 367–396. *In*: M.A. Khan and D.J. Weber. (eds.). Ecophysiology of High Salt Tolerant Plants. Springer, Netherlands.

Yiang, Z. 1989. *Alhargi sparsifolia* Schap. *In*: Flora of Chinese Fodder. Agriculture Press, Beijing (in Chinese).

Zhang, L.Y., Y. Xia and Y. Zou. 1994. Halophytes and halophytic plant communities in Inner Asia. pp. 114–122. *In*: V.R. Squires and A.T. Ayoub. Halophytes as a Resource for Livestock and for Rehabilitation of Degraded Land. Tasks in Vegetation Science Series, Kluwer Academic Press, Dordrecht.

Halophytic and Salt Tolerant Feedstuffs in the Mediterranean Basin and Arab Region: An Overview

H.M. El Shaer* and S.A. Attia-Ismail

SYNOPSIS

The Mediterranean area, in its broadest sense, refers to the area from the Aral Sea to the Atlantic Ocean. This chapter discusses the nature and ecology of the various plant species and the habitats in which they grow. This chapter also provides an overview of the region and examines the utilization, problems and future prospects of salt tolerant and halophyte forages/feedstuffs.

Keywords: taxonomy, ecology, nutritive value, anthropogenic salines, sabkha, palatability, feed value, Red Sea, Mediterranean Sea, Arab Sea, Indian Sea, Gulf region, North Africa, Pakistan, Syria, Iran, Gulf Countries, Jordan, Iraq, Saudi Arabia, Yemen, Sudan, Algeria, Tunisia, Morocco, Libya Egypt, rumen function.

1. Introduction

Natural resources have been diminishing due to the pressure of increasing demands resulting from the ever-increasing population of the world. Inevitably, marginal and

Animal Nutrition, Desert Research Center, Cairo, Egypt.
* Corresponding author: hshaer49@hotmail.com

long-neglected natural resources have to be re-assessed in preparation for utilization. It is estimated that 7–10% of the world land area is salt affected (Dudal and Purnell 1986). Halophytes (salt plants) survive salt concentrations around 200 mM NaCl or more in order to reproduce in environments where they constitute about 1% of the world's flora (Flowers and Colmer 2008). They grow in many arid and semi-arid regions around the world and are distributed from coastal areas to mountains and deserts. Halophytes belong to several taxonomic groups. The word halophyte, then, does not imply any reference to being a particular taxon or belonging to any specific geographic or physiogeographic area (Squires and Ayoub 1994).

1.1 Nature and ecology of halophytes

Le Houerou (1994) concluded that the nature and ecology of halophytes are very complex. They do not necessarily need salinity to grow. Halophytes constitute around 6,000 species in the world. In a previous study, Le Houerou (1993) divided halophytes into "optional or facultative halophytes" implying those that do not necessarily need salinity to grow and "obligate or true halophytes" to refer to those that need saline conditions for their growth. Of the former he mentioned *Atriplex* spp., *Maireana* spp., *Tamarix* spp., *Salsola* spp., *Limonium* spp., and *Puccinellia* spp., and of the latter *Halocnemum* spp., *Arthrocnemum* spp., *Salicornia* spp., *Suaeda* spp., etc. A third class was the "preferential halophytes"; examples of them are some *Atriplex* spp., some *Maireana* spp. and some *Tamarix* spp. There are no clear delineations, however, between these different classes (Le Houerou 1994).

The reaction of plants to salt stress is called Salt resistance. Resistance may comprise of either salt tolerance or salt avoidance. Salt tolerance involves processes of adaptation either physiological and biochemical in order to maintain protoplasmic viability as cells accumulate electrolytes. Salt avoidance involves similar processes, but also structural along with physiological adaptation, in order to minimize salt concentrations of the cells or physiological exclusion by root membranes.

Aronson (1989) estimated that about one-third of the approximately 350 families of flowering plants contain halophytes and about 50% of the genera belong to 20 of these families (Flowers et al. 1986). Therefore, it appears that halophytes are widely distributed among several families of flowering plants. Based on these studies and those by Batanouny (1996) and Flowers et al. (1977), it is clear that there is no family in which more than half of the species are halophytes. The fact that the limited number of halophytic species is spread among so many different families indicates that halophytism, even though a trait controlled by several genes, is not such a complex characteristic that only arose once during evolution. This polyphyletic origin of halophytism (Flowers et al. 1986) suggests that it is a characteristic that could be introduced into species not yet possessing it, particularly species in the same families that already contain halophytic species. Both climate and soil characteristics have influence on the distribution of halophytes (Szabolcs 1994). He mentioned that salt affected soils are distributed not only in the deserts and semi-deserts, but also in fertile alluvial plains, river valleys and coastal areas, close to highly populated areas and irrigation systems. Salt affected soils are most appropriate for the growth of halophytes; although they occur in completely different soils. A wide range of plant genera inhabits salt-affected soils and are referred to as halophytes. Not all of the halophyte plants are able to germinate in their native saline habitats without prior leaching of the soil (Batanouny 1993). He concluded that the correlation between salt tolerance of a plant and its tolerance at other stages is not obligatory. This has been noted in certain halophytes (Waisel 1975). The water relations of their habitats therefore, affect germination of halophytes. Water content (= water budget) is a critical factor for the growth of halophytes. Therefore, Le Houerou (1994) classified saline soils with according to the water budget into permanently wet, quasi-permanently dry and permanent or a temporary table or not. The soil chloride content affects the germination of halophytes. This is called ionic budget. However, vegetative reproduction is a major means of reproduction in many halophytic species (e.g., *Limonium vulgare*, *L. humile*, and *Tamarix aphylla*).

2. Halophytes in the Mediterranean basin and Arab region

As mentioned above the Mediterranean area refers to the area from the Aral Sea to the Atlantic Ocean. The halophyte forage species in the Mediterranean Basin are very diverse in terms of plant systematics, biology and ecology. They constitute around 700 species. As is the case with halophytes in any other area, they include both perennial and annual species. They are also divided into shrubs and trees.

Table 2.1 shows different proportions of different halophytes in the Mediterranean Basin. *Chenopodiaceae* dominates the plant communities either alone or mixed with other types of plants. *Halocnemum strobilaceum, Arthrocnemum indicum, Salicornia,*

Table 2.1. Proportions of different halophytes in the Mediterranean Basin.

Family	Percentage
Chenopodiaceae	27.5
Graminaceae	15
Compositeae	6
Caryophyllaceae	5
Leguminoseae	5
Zygophyllaceae	5
Aizoaceae	3
Frankeniaceae	3
Tamaricaceae	3
Cyperaceae	3
Plantaginaceae	2
Ombellifereae	2
Crucifereae	2
Boraginaceae	0.5
Convolvulaceae	0.5
Gentianaceae	0.5
Juncaginacea	0.5
Other families	6

Adapted from (Le Houerou 1969a, 1986, 1993a; Le Houerou et al. 1975).

Salsola, Sueda, and *Atriplex* are examples of that dominance. Table 2.2 shows the number of halophytes in some Asian–Mediterranean Basin countries. Batanouny (1994) reviewed halophytic plant communities in the Arab region. The area partially includes a large proportion of the Mediterranean Basin, part of west Asia, and part of east and West Africa. Therefore, it has a wide variety of halophytes. The halophytes in the Arab area represent less than 5% of the flora of the area (Khan et al. 2014; Batanouny 1996). However, the floras comprise some 150 halophytic species in about 55 genera and about 22 families. The salinity of the soils in both the Mediterranean Basin and the Arab world varies from place to place with different types of salt distribution in the profile. A few annual species, usually succulents, are recorded, e.g., *Salsola europaea, Binertia cyclopetra, Sueda aegyptiaca, Sueda salsa,* and *Halopeplis amplexicaulis.* The main halophyte families are Chenopodiaceae, Gramineae, Aizoaceae, Avicenniaceae, Caryophyllaceae, Compositae, Convolvulaceae, Cynomoriaceae, Cyperaceae, Frankeniaceae, Juncaeae, Leguminosae, Nitrariaceae, Plantaginaceae, Plumbaginaceae, Rhizophoraceae, Salvadoraceae, Tamaricaceae, Typhaceae, and Zygophyllaceae.

Table 2.2. Number of halophyte species in some Asian–Mediterranean Basin countries.

Country	No. of halophytic species	Total flora	% of halophytes in the flora
Iran	354	6170	5.7
Iraq	135	1200	10.4
Jordan	260	2100	12.4
Kuwait	80	450	23.0
Palestine	300	2800	10.7
Qatar	70	435	16.0
Saudi Arabia	250	2200	11.4
Syria	260	3460	7.5
United Arab Emirates	70	450	15.6

Adapted from Le Houerou (1993).

2.1 Halophytes distribution in the Mediterranean basin and Arab region

Halophytic communities are mainly located in several saline areas as reported by Batanouny (1993):

1. The coastal salines:
   They are prominent and form an important ecosystem at the Red Sea, Mediterranean Sea, Arab Sea, Indian Sea and Gulf areas. Extensive Sabkhas are present at levels slightly below or above sea level. The most common halophytic species are: *Nitraria retusa, Tamarix* spp., *Zygophyllum album* and *Atriplex halimus* (Khan et al. 2014).

2. Great depressions and oases:
   Salt marsh plants are widely distributed in oases and depressions. The common species in such habitats include *Tamarix* spp., *Juncus* spp. and *Alhagi marorum.*

3. Inland saline depressions:
   The saline depressions in the desert or in the wadies receive runoff with dissolved salts. These depressions support different halophytes species. All over the deserts in the Arab countries such salines occur in wadis, e.g., in Saudi Arabia, Sahara and eastern desert of Egypt. These depressions support different halophytes such as *Nitraria retusa, Atriplex halimus, Zygophyllum album* and *Tamarix* spp.

4. Anthropogenic salines:
   They are salines formed due to human activities as a result of heavy irrigation of the cultivated lands and the lack of drainage devices. Reeds (represented by *Phragmites* and *Typha*) and rushes (*Juncus acutus* and *Juncus rigidus*) form an extremely important component of such salines.

   In general, the numbers of halophytic species in some of the Asian–Mediterranean basin countries are shown in Table 2.2. The ranking order of halophytic families with respect to the number of species in the region is: *Chenopodiaceae, Poaceae, Asteraceae, Brassicaceae, Plumbaginaceae, Cyperaceae, Tamaricaceae, Zygophyllaceae, Polygonaceae* and other families.

However, three broad groups of halophytes are classified in the region:

1. Samphires (glassworts):
   These include the genera *Salicornia, Arthrocnemum, Halocnemum, Halosarcia* and *Allenrolfea*. These plants occur on highly saline sites which are in many cases waterlogged during certain times of the year. The stems of these leafless plants are succulent and highly saline but some samphires are eaten by sheep when other feed is scarce. Some samphires develop a substantial woody frame.

2. Saltbushes:
   These include the genera *Atriplex, Chenopodium, Rhagodia* and *Halimione*. They are highly salt-tolerant but occur in less waterlogged situations than the samphires. Saltbushes are used mainly as animal feed and vary greatly in palatability. They are also used as fuel and some species produce a strong woody frame.

3. Bluebushes and saltworts:
   These include the genera *Salsola, Kochia, Maireana, Sarcobatus, Suaeda* and *Enchylaena*. These shrubs have succulent leaves which vary greatly in their palatability to animals. They range in salt and water logging tolerance from those which occur in association with samphires and saltbushes to those which are less salt and water logging-tolerant than most saltbushes.

In addition to these groups, reeds and rushes constitute an extremely important component of the wetlands in the region mainly in rivers, estuaries (such as in Damietta and Rosetta in Egypt) and coastal and inland reed swamps. The reeds are represented by *Phragmites* and *Typha* and the rushes are represented by two *Juncus* species: *J. acutus* and *J. rigidus*. It is noted that *Phragmites* is the species with highest tolerance to soil salinity followed by *Juncus rigidus, J. acutus* and *Typha domingensis; Juncus rigidus* tolerate higher soil salinity than *Juncus acutus* (Serag et al. 2002). Reeds and rushes provide habitat for a wide range of organisms from invertebrates to birds (e.g., nesting sites, substratum, feeding materials) as reported by Zahran and Willis (2002).

In general, the halophytic plant species cover huge areas of rangelands in Pakistan, Syria, Iran, Gulf Countries, Jordan, Iraq, Saudi Arabia, Yemen, Sudan, Algeria, Tunisia, Morocco, Libya and Egypt (Batanouny 1993). These rangelands are mainly used by sheep, goat and camels throughout the year, as defined by Batanouny (1993), such as: *Atriplex halimus, A. mollis, A. portulaccoides, A. glauca, A. nummularia, Suaeda fruticosa, S. brevifolia, S. mollis, Salicornia arabica, Limoniastrum monopetalum, Limoniastrum guyonianum, Traganum nudatum, Salsola vermiculata var. villosa, Salsola sieberi, S. tetrandra, Arthrocnemum indicum, Salicornia fruticosa, Inula crythmoide, Halocnemum strobilaceum, Tamarix* spp., and *Nitraria retusa*. Some non-shrubby perennial halophytes exhibit a fair to good palatability, e.g., *Nitraria retusa, Suaeda fruticosa, Plantago crassifolia, Spergularia media, Sp. marginata, Hedysarum carnosurn, Puccinellia* spp., and *Spartina patens* while others are almost unpalatable, e.g., *Aster tripolium, Heliotropium curassavicum, Suaeda maritime, Juncus* spp., *Schoenus nigricans, Cyperus* spp., *Scirpus* spp., *Phragmites* spp., *Typha* spp., *Arundo plinii, A. donax, Saccharum ravennae,* and *Ruppia* spp.; most of these are hygro-halophytes (Batanouny 1993). Most annual halophytes have a very low palatability, if any, and produce little phytomass, e.g., *Hordeum maritimum, Polypogon, Sphenopus, Lepturus, Pholiurus, Psilurus. Eremopyrum, Frankenia, Aizoon, Mesembryanthemum, Cressa, Zygophyllum, Tetradiclis. Halopeplis, Halogeton, Schanginia, Suaeda, Salsola* and *Salicornia.*

2.2 Potential uses of halophytes

A number of halophytes have been used for different purposes. Aronson (1989) enumerated 1560 halophyte species which are already in use. There are, however, several broad situations in which halophytes are used by livestock. Halophytes are on trial under irrigation, mainly in developing countries (Squires 1994). Other uses of halophytes are: land rehabilitation, utilization of desert and saline water in irrigation, feedstuff component, medicinal materials, fuel wood, shade and shelter, sequestration of carbon dioxide, etc. Some halophytes are used as building materials (e.g., *Avicennia marina, Prosopis tamarugo*), others as wood for furniture, timber, charcoal, fire wood (e.g., *Tamarix* spp.). Many halophytes are used in medical purposes as drugs; these include *Annona glabra, Gomphrena globosa, Juncus acutus,* and *Salsola kali.* Moreover, halophytes have been reported to be used as fertilizers (e.g., *Sesbania speciosa* and *Zostera marina*). Many of them are edible and utilized as vegetables such as *Aster tripolium, Suaeda glauca,* and also *Salicornia fruticosa,* which is used for oil production. Other uses of halophytes include the utilization in reclamation of saline lands, laundry detergent, paper production, herbal tea, sea floor fixation, as a green cover, as ornamental plants and as hedges.

3. Prospects of halophytes as animal fodder

Halophytes can play a significant role in the well being of different peoples. The way in which halophytes are assessed will very much depend on which system dominates. Any evaluation must depend on viewing performance in the context (biological/economic) in which it occurs. The shortage of animal feeds is the main constraint to increase indigenous animal production. It is a common characteristic in arid and semi-arid regions which is considered the main constraints to improve livestock productivity. Animal husbandry, as the main income resource for nomads, is based mostly on the natural vegetation for rearing sheep, goats and other herbivores. However, unpalatable halophytes are widely distributed throughout the world. Halophytic plants such as *Atriplex* spp., *Nitraria retusa, Salsola* spp., are generally considered extremely valuable as a fodder reserve during drought. Most of countries in these regions import large amounts of feed materials to cover the nutritional gap of animal and poultry which puts a heavy burden to on the local governments and the farmers as well. It consequently decreases the net profits of animal investments operations due to the high costs of such feed materials. Therefore, intensive efforts have been directed to find alternative feed resources from halophytes.

3.1 Palatability and feeding values of halophytes

Palatability is defined as the ratio between the amount of feed ingested by herbivores and the amount on offer for a given period of time. Preference is the order in which forage species are selected by herbivores within a given plant community or population, or at a grazing site, at a given time (Le Houérou et al. 1982; Glenn et al. 1999). The palatability and feeding values of individual halophytes or any other types of rangelands vary widely from virtually zero to very high. In almost any forage population of a given species of browse, either natural or planted, there are various degrees of palatability from one plant individual to the next, ranging from high palatability to low palatability. Palatability also depends on the relative abundance of the species under consideration on the rangeland; all other conditions being equal the palatability of a given taxon is inversely related to its abundance on the range, except for a few species which are specially relished in all circumstances. In addition to the internal plant factors, there are also internal animal factors such as: species and race of livestock, age, physiological and health status, feeding habits, nutritional status.

The feeding value of any halophytic range could also affect its palatability for the same animal. For instances: the content of crude fiber in forage plays an important role in its selection by livestock. Forages with high fiber content are usually better accepted by cattle than by sheep and goats; but this, in turn, depends on the proportions of the various components of fiber: cellulose, hemicelluloses, acid detergent fiber (ligno-cellulose), neutral detergent fiber (cell walls), etc. But there are many other criteria for selection such as organoleptic qualities of the forage; the latter have hardly been explored in research. Mineral content may, also, be an important factor either limiting or favoring (usually limiting in low rainfall areas and favorable in high rainfall areas, when silica-free minerals are concerned). Finally, the overall balance of the diet plays a major role in forage selection. It has been shown (Skouri 1975) that the amount of highly fibrous material ingested depends to a large extent on the amount of protein in the overall diet. Besides these intrinsic or internal plant and animal factors, there are also extrinsic or environmental criteria influencing diet selection. Some of those have been referred to above, such as the relative abundance of a given taxon on the range and the botanical composition of the forage available. The palatability of a given taxon of browse for a given type of animal depends to a large extent on the plant community or association in which this taxon is being browsed, since most of them occur in many plant associations under various ecological and bioclimatic conditions.

Nevertheless, a number of halophytic species, including common and dominant ones are ignored by livestock such as *Halocnemum strobilaceum, Halopeplis* spp., *Salsola baryosma, Limonium* spp., *Limoniastrum* spp., *Mesembryanthemum* spp., *Aizoon* spp., *Cressa cretica, Frankenia* spp., *Blackia inflata, Ammi visnaga, Centaurium spicatum,* etc. Others are more or less grazed depending on the local situation of offer and demand, on phenological stage, season, animal species, etc. and are such as: *Arthrocnemum indicum, Salicornia arabica, Salsola tetrandra, S. tetragona, Salsola vermiculata var. villosa, Suaeda fruticosa,* etc. A third category should be distinguished, of species which are palatable, or highly palatable to livestock and game. Among those are *Atriplex halimus, A. portulaccoides, A. glauca, A. mollis, Salsola tetrandra,* some ecotypes of *Salsola vermiculata* (Syria), a biannual legume halophyte *Hedysarum carnosum,* and a number of exotic species introduced for fodder shrub plantations such as: *Atriplex amnicola, Atriplex nummularia* and *Kochia* (syn. *Bassia) indica* introduced from Australia and *Atriplex lentiformis* and *Atriplex canescens* from U.S.A.

Biomass production, palatability and nutritive value of halophytes vary from area to area within the same region and from season to season depending upon several environmental factors (Le Houerou 1994; El Shaer 1997). The data for common natural halophytic ranges, their nutritive values and palatability for sheep, goat and camels are presented in Table 2.1. It appears that, palatability and nutritive values vary significantly among halophyte species. In general, most halophytes contain sufficient levels of crude protein and other essential nutrients which seems to cover the nutritional requirements of animals particularly during the wet season (El Shaer 1981; Arieli et al. 1989). It was concluded, from several studies, that many of these halophytes could have a great potential for grazing since they provide animals with necessary nutrients all the year round (El Shaer 1981; Kandil and El Shaer 1988; Le Houerou 1994; Koocheki and Mohalati 1994; Glenn et al. 1999).

3.2 Value of halophytes as a feed component for livestock

The nutritive value of halophytic plants differs from species to species and several factors are involved in this, the highest forage values are found during the wet season of the year. The nutritive value decreases with growth and maturation. The digestibility is generally higher during grazing season than in drought season (El Shaer 2006). Although many halophyte taxa are high in protein content they should not be offered as the sole diet to animals for a long term. The performance of animals on the halophytic ranges depends on factors like animal species, season of the year, forage abundance, and nutritional values of forage species. In general, feeding small ruminants with good quality halophytes, either alone or in mixed rations may prove economical for the farmers.

The chemical analysis of some halophytic plants (Table 2.2) reveals that halophytic species have the potentiality as an animal fodder (Gihad and El-Shaer 1994). However, the highest forage values are found during the wet season of the year when plants are green and actively growing (Chatteron et al. 1971; Kandil and El-Shaer 1990). In comparison with other shrub species, all chenopod shrubs contain high concentrations of digestible crude protein, and Na, K, and Cl- ions. The acceptability of most chenopod shrubs to domestic animals is moderate to low when grasses and herbs are available but it increases, as these components become scarce (Wilson 1984). *Suaeda* species is considered one of the most palatable chenopod shrubs (El Shaer 1986). Factors influencing grazing and nutritive values of halophytes are the plant species, ecotypes (Rizk 1986), stage of growth (El Shaer 1981), season of use (wet season versus dry season), environmental factors (Gihad 1993), and location (Gihad and El-Shaer 1994). Halophytic plants differ in their nutritive value from one species to another. Chemical composition of saltbush species differ widely, and the differences exist even more within the same species. More exciting is that the chemical composition of the same salt plant species differs seasonally and even within the same season according to the stage of growth. Table (3) shows some of the differences that exist between some halophytic species in different seasons. There are many examples that show the differences in chemical composition and, hence, nutritive values within the same species. Ramirez and Torres (1997) found that *Acacia rigidula* and *Acacia farnesiana* were different in their nutritive values. Degen et al. (1997) reported the same outcome when they fed *Acacia saligna* and *Acacia salicina* to goats and sheep.

4. Utilization of fresh and dried halophytes as animal fodders

4.1 Intake and nutrient utilization

Voluntary feed intake (VFI) (Table 2.3) and digestibility are considered the two major components reflecting upon forage quality of grazing ruminants. The process of aging and maturation of the rangeland plants was found to be associated with a decline in digestibility and CP content, and consequently the nutritive value (El Shaer 1981; El-Bassosy 1983). The DMI and DM digestibility of halophytic forages were higher in grazing season than in drought season with both sheep and goats. Rams consumed less DMI of forages than bucks in drought season (34.4 vs. 44.1 $g/kg^{0.75}$) as reported by El Shaer (1981). Warren et al. (1990) recorded dry matter intakes of 400–800 g/ day for 4 species (*A. undulata, A. lentiformis, A. amnicola* and *A. cinerea*) fed to sheep that had dry matter digestibility of 53–62%. Le Houérou (1993) reported that sheep became adapted to saltbush and increased their intake of forage over a 3–5 month period. Therefore, it seems that animals fed on halophytes need more time to adapt to such feeds, particularly the less palatable forages (Degen and Squires, this volume).

Organic constituents of halophytes are highly digestible by goats and sheep and mixed diets containing halophyte forages are acceptable to them. Most of the halophyte shrubs are high in protein content and of moderate digestibility. On the other hand, Abou El-Nasr et al. (1996) added that sheep fed saltbush hay, fresh *Acacia* and *Acacia* hay were not able to cover their maintenance requirements of DCP and showed negative nitrogen balance which was, mainly, attributed to lower nitrogen

Table 2.3. Palatability of halophytic plants for different animal species.

Plant species	Palatability for animal species
Alhagi maurorum	Sheep, Goats
Arhthrocenemon glaucum	Camels
Atriplex halimus	All species
Atriplex leucoclada	Sheep,Goats
Atriplex nummularia	All species
Halocenemom strobilaceum	Camels
Haloxylon salicornicum	Nil
Juncus acutus	All species
Nitraria retusa	All species
Salicornia fruticosa	Camels
Salsola tetraundra	All species
Suaeda fruticosa	All species
Limoniastrum monopetahum	All species
Tamarix aphylla	Goats, Camels
Tamarix mannifera	All species
Zygophyllum album	Nil
Zygophyllum simplex	Camels
Zygophyllum decumbens	Camels, Goats

Table 2.4. Average values of chemical composition of common range plants in southern Sinai.

Plant species		CP	CF	EE	Ash	NFE
	DM	DM basis, %				
Halocnemum strobilaceum						
Wet season	28.6	6.78	14.6	2.46	35.7	40.46
Dry season	37.8	4.22	19.6	2.16	42.5	31.52
Zygophyllum album						
Wet season	30.4	7.12	14.6	2.25	26.5	49.42
Dry season	39.3	6.30	16.2	1.63	28.5	47.31
Tamarix mannifera						
Wet season	57.4	7.64	16.1	2.21	26.0	47.97
Dry season	63.1	6.26	17.5	1.78	30.9	43.47
Salsola tetrandra						
Wet season	45.0	9.73	12.4	2.61	30.1	45.1
Dry season	49.9	8.38	14.4	1.67	35.8	39.68
Nitraria retusa						
Wet season	14.6	9.10	12.8	3.01	16.2	58.87
Dry season	20.1	7.20	18.2	2.28	19.7	52.62
Atriplex halimus						
Wet season	42.1	9.21	18.7	3.2	31.4	37.49
Dry season	58.4	6.32	22.6	3.10	36.7	31.28
Suaeda fruticosa						
Wet season	30.9	11.1	10.9	3.9	25.4	48.66
Dry season	48.3	8.4	13.7	3.0	30.3	44.6

intake and live body weight loss. El Shaer et al. (1990) found that DMI from *Z. album,* *T. mannifera, A. halimus* and *H. strobilaceum* were 2.12, 10.9, 19.4 and 18.2 g kg $W^{0.75}$ for sheep and 2.94, 10.3, 15.8 and 14.2 gm kg $W^{0.75}$ for goats, respectively. In its air-dry state, *Z. album* was not eaten and the DMI from *A. halimus* and *H. strobilaceum* were 13.2 and 9.45 g DMKg$W^{0.75}$ for sheep and 21.0 and 8.23 g/Kg $W^{0.75}$ for bucks, respectively, as shown in Table 2.5. On the other hand, DMI by sheep and goats from *T. mannifera* and *A. halimus* were markedly higher ($P < 0.05$) than that from *Z. album* and *H. strobilaceum,* since the fresh *Z. album* and *H. strobilaceum* are known to be unpalatable feeds.

Table 2.5. Dry matter intake (g DM/ Kg $W^{0.75}$) of halophytic species by sheep and goats.

	Sheep (S)	Goat (G)	S/G Ratio
Fresh state:			
A. halimus	19.4	15.8	1.23
H. strobilaceum	18.2	14.2	1.28
T. mannifera	10.9	10.3	1.06
Z. album	2.12	2.94	0.72
Air-dried state:			
A. halimus	13.2	21.0	0.63
H. strobilaceum	9.45	8.23	1.15
T. mannifera	5.12	4.75	1.08
Z. album	0.00	0.00	0.00

4.1.1 Growth and feed efficiency

In Southern Sinai, Hassan et al. (1980) indicated that sheep fed a mixture of indigenous halophytic ranges in Sinai lost weight during all months studied (12 months), but losses were minimal in the spring season (26 g/day) as compared to summer season (134 g/day). Also, sheep and goats when fed halophytic plants only without feed supplement were not able to maintain their live weights even when the pasture reached its best condition (Hassan et al. 1980; Warren et al. 1990; El Shaer 1997). However, the performance of animals on the range, particularly the halophytic ranges, depends on several factors mainly the season of the year, abundance, and nutritious forage species. Similar results were obtained by El Shaer (1981) through two successive years. Rams lost weight in the drought season (–47.8 g/day) but gained weight (+24 g/day) in the grazing season, whereas, bucks gained weight in both drought and grazing seasons (22.8 and 98.1 g/day, respectively). Kids raised on the salty pasture without supplementation gained 0.36 g kg $W^{0.75}$ on the average. Meanwhile, lambs lost 2.82 g kg$W^{0.75}$. On relative gain basis, kids gained 0.77% while lambs lost 5.23% of their initial weights on the pasture.

5. Effect of halophytes on rumen function

It is proposed that the increased salt intake and, hence, the osmotic pressure of the rumen may affect the microorganism population and metabolism (Wilson 1966; Attia-Ismail, this volume; Degen and Squires, this volume) for a more detailed examination of this important topic)). Increasing osmotic pressure in the rumen is believed to be unfavorable to the growth of protozoa. Also, the increased rates of outflow are believed to contribute to decreased protozoal population (Warner and Stacy 1977). The artificial rise of rumen osmotic pressure (Bergen 1972) up to 400 mOsmol/kg caused cellulose digestion to be inhibited. Shawket et al. (2001) reported that increased salt intake increased the dilution rate of the rumen fluid phase and negatively affected the protozoal population. Kewan (2003) reported that the total protozoal count ($x10^3$) per milliliter was significantly higher ($P \leq 0.05$) for camels fed on rations that contained *Atriplex nummularia* than for those fed on *Acacia saligna* and treated rice straw rations. Total rumen protozoal count in camels fed on berseem hay ration were significantly ($P \leq 0.05$) higher than those fed on the other experimental rations. The feeding of salt plants that are present in the desert regions usually places another burden of ions in the rumen and needs more buffering capacity. Ruminants that feed on salt plants are expected to secrete more saliva and have a higher ruminal pH than those fed grains (Attia-Ismail et al. 2009).

5.1 Constraints to the utilization of halophytes as animal feeds

The poor intake of the fresh as well as the air-dried halophytic species could be attributed to three main factors: (1) high Na, Ca and silica contents, (2) higher levels of ADL and NDF and (3) many shrubs contain higher levels of plant secondary metabolites. Nutrient detergent fiber (NDF) is a good indicator for forage intake. The low NDF value (50.2%) for the fresh *P. crispus* would explain the higher intake by sheep compared to fresh *Tamarix mannifera* and *Glinus iotoides* (Kandil et al. 1990). Therefore, The low CP content (approximately 6%) and higher levels of NDF, ADL and ADF, as were shown by *T. mannifera* and *G. iotoides,* have been considered to limit the forage consumption and digestion, while *Prosopis crispus* showed a contrary trend. TDN and DCP were higher for sheep fed *P. crispus* than the other two groups. Also, only the sheep fed *P. crispus* were in a positive nitrogen balance as compared to the other groups. The feeding value of any rangelands is generally expressed in terms of voluntary intake and digestible nutrients per unit of feed. There are several factors, which could considerably limit the feeding values of halophytes such as physical and chemical defenses of the plants, lignification and salinity. The physical defense of halophytes include the so called "barbed-wire syndrome" (Gihad and El Shaer 1994), the presence of silica, fiber, spines and thorns. The chemical defense of halophytes is the presence of a chemical make-up particularly the secondary metabolites of the plants. The degree of lignification (Attia-Ismail et al. 1994) is also a matter that affects the nutritive value of not only the halophytes but also all other fodders. The salt concentration in any halophytic plant also affects its acceptability and intake by animals (Attia-Ismail et al. 2002; Fahmy et al. 2001). The animal selects the halophytic plant based on its palatability. Masters (this volume) give more detailed attention on how to assess the nutritive value of halophytes.

Some halophytes are toxic (Attia-Ismail, this volume). This toxicity results from several secondary metabolites in the plants. However, the rate of toxicity is affected by several factors such as the rate of ingestion, type and rate of microbial transformation of such metabolites in the rumen, rate of gastro-intestinal absorption, liver and kidney enzymatic activity. Alkaloids, saponins, tannins and nitrates are present in most halophytes. High concentrations of alkaloids decrease animal productivity and increases diarrhea. Tannins reduce feed intake because of their poor palatability, which is a result of the precipitation that occurs upon reaction of tannins with salivary proteins. Tannins also inhibit digestive enzymes (Streeter et al. 1993).

Some coastal halophytic plants were found to survive under seawater irrigation (Riley, this volume). These halophytes can be used for bioremediation of salt-contaminated soils and the pharmaceutical values of their products are described by Koyro et al. (2011). Several halophyte species including grasses, shrubs, and trees can remove the salt from different kinds of salt-affected and problematic soils through salt excluding, excreting, or accumulating by their morphological, anatomical, physiological adaptation in their organelle level and cellular level (Hasanuzzaman et al. 2014).

6. Conclusions and summing up

In this comprehensive overview we have outlined aspects of halophyte utilization that are dealt with in greater detail in the remainder of this volume. The key points are that as natural resources in drylands that cover 40% of the world's land area and on which over 800 million depend for livelihoods, become scarcer and more highly valued, attention has become focused on halophytes and other salt tolerant plants. As indicated here, and elsewhere in this volume, this large and diverse group of plants grow in areas as different as tidal flats that are periodically inundated with sea water, to naturally occurring saline depressions in arid inlands of many continents. Salt tolerant crops grow in irrigated lands where saline soils and water prohibit the growing of conventional crops. Vast areas of semi-arid to arid rangelands exist where halophytic and salt tolerant plants provide fodder forage to millions of livestock.

References and Further Readings

Abideen, Z., Z. Ansari and M. Ajmal Khan. 2011. Halophytes: Potential source of ligno-cellulosic biomass for ethanol production. Biomass and Bioenergy 35(2011): 1818–1822.
Abou El Nasr, H.M., H.M. Kandil, D.A. El-Kerdawy, H.S. Khamis and H.M. El Shaer. 1996. Value of processed saltbush and acacia shrubs as sheep fodder under arid conditions of Egypt. Small Ruminants Res. 24: 15–20.
Anon, A.H. 1992. Revegetation of salt affected lands of deltaic Mediterranean coast. Egypt Proj. 90009 Phase II Final report, Jan–June 1992, Supreme Council of Universities. Cairo, Egypt.
Arieli, A., E. Naim, R.W. Benjamin and D. Pasternak. 1989. The effect of feeding saltbush and sodium chloride on energy metabolism in sheep. Anim. Prod. 49: 451–457.
Aronson, J. 1989. HALOPH: a database of salt tolerant plants of the world. Office of Arid Land Studies, Univ. Arizona, Tucson, Az., pp. 77.
Attia-Ismail, S.A. (in press). Plant secondary metabolites of halophytes and salt tolerant plants. pp. 127–142, this volume.
Attia-Ismail, S.A. (in press). Rumen physiology under high salt stress. pp. 348–360, this volume.
Attia-Ismail, S.A., H.M. Elsayed, A.R. Askar and E.A. Zaki. 2009. Effect of different buffers on rumen kinetics of sheep fed halophyte plants. Journal of Environmental Science 19(1): 89–106.
Attia-Ismail, S.A., A.A.M. Fahmy and R.T. Fouad. 1994. Improving nutritional value of some roughages with mufeed liquid supplement. Egy. J. Anim. Prod. 31: 161–174.

Attia-Ismail, S.A., M.F. Afaf and A.A. Fahmy. 2002. Mineral, nitrogen and water utilization of salt plant fed sheep as affected by monensin. Egy. J. Feeds and Nutr. 6: 151–161.

Batanouny, K.H. 1993. Ecophysiology of halophytes and their traditional use in the Arab world. pp. 37. *In*: Advanced Course On "Halophyte Utilization In Agriculture" 12–26 Sept., 1993, Agadir, Morocco.

Batanouny, K.H. 1994. Halophytes and halophytic plant communities in the Arab region: Their p[potential as a rangeland resource. pp. 139. *In*: V.R. Squires and A.T. Ayoub (eds.). Halophytes as a Resource for Livestock and for Rehabilitation of Degraded Lands. Kluwer Academic Publishers.

Batanouny, K.H. 1996. Biodiversity in the rangelands of the Arab countries. In: Proc. Rangelands in a sustainable biosphere. N.E. West (ed.). 5th International Congress 1995, Salt Lake City Utah, pp. 39–40.

Bergen, W.G. 1972. Rumen osmolality as a factor in feed intake control of sheep. J. Anim. Sci. 34: 1054–1060.

Broadley, M.R., N.J. Willey and A. Mead. 1999. A method to assess taxonomic variation in shoot cesium concentration among flowering plants. Environmental Pollution 106.3: 341–49. Web of Science. Web. 24 Nov. 2012. <http://ezproxy2.library.colostate.edu:2089/science/article/pii/S0269749199001050>.

Chatteron, N.J., J.R. Goodin, C.M. McKell, R.V. Parker and J.M. Ribble. 1971. Monthly variations in the chemical composition of desert saltbush. J. Range Mngt. 12: 289.

Clüsener-Godt, M. 2014. The importance of mangroves ecosystems for nature protection and food productivity: Action of UNESCO's Man and the Biosphere Program. Int'l Conf. on Halophytes for Food Security in Dry Lands.

Degen, A.A., A. Blanke, K. Becker, M. Kam, R.W. Bemjamin and H.P.S. Makkar. 1997. The nutritive value of Acacia saligna and Acacia salicina for goats and sheep. Animal Sci. 64: 253–259. Br. Soc. Anim. Sci.

Degen, A.A. and V.R. Squires (in press). The rumen and its adaptation to salt. pp. 336–347, this volume.

Dudal, R. and M.F. Purnell. 1986. Land resources: salt affected soils. Reclamation and Revegetation Res. 5: 19.

El-Bassousy, A.A. 1983. Study of nutritive value of some range plants from Saloom to Mersa Matrouh. Ph.D. Thesis. Ain-Shams Univ. Cairo, Egypt.

El Shaer, H.M. 1981. A comparative nutrition study on sheep and goats grazing southern Sinai desert range with supplements. Ph.D. Thesis, Fac. Agric., Ain-Shams Univ., Cairo, Egypt.

El Shaer, H.M. 1986. Seasonal variations of the mineral composition of the natural pasture of southern Sinai grazed by sheep. Proc. 2nd Egyptian British Conf. Anim. and Poultry Prod. Aug, 26–28, Bangor.

El Shaer, H.M. 1997. Practical approaches for improving utilization of feed resources under extensive production system in Sinai. Proc. Int'l Symp. On Systems of Sheep and Goat Production, Bela, Italy, 25–27 Oct., 1997.

El Shaer, H.M. 2006. Halophytes as cash crops for animal feeds in arid and semi-arid regions. *In*: M. Ozturk, Y. Waisel, M.A. Khan and G. Gork (eds.). Biosaline Agriculture & High Salinity Tolerance in Plant. Birkhauser, Basel, pp. 117–128.

Fahmy, A.A., S.A. Attia-Ismail and M.F. Afaf. 2001. Effect of monensin on salt plant utilization and sheep performance. Egy. J. Feeds and Nutr. 4: 581.

Felger, R.S. 1979. Ancient crops for the twenty-first century. pp. 5–20. *In*: G.A. Ritchie (ed.). New Agricultural Crops. Westview Press, Boulder, CO.

Flowers, T.J. and T.D. Colmer. 2008. Salinity tolerance in halophytes. New Phytologist 179: 945–963.

Flowers, T.J., M.A. Hajiheri and N.J.W. Clipson. 1986. Halophytes. The Quart. Rev. Biol. 61: 313.

Flowers, T.J., R.F. Troke and A.R. Yeo. 1977. The mechanism of salt tolerance in halophytes. Ann. Rev. Plant Physiol. 28: 89.

Gihad, E.A. 1993. Utilization of high salinity tolerant plants and saline water by desert animals. *In*: H. Leith and A. al Massoom (eds.). Towards Rational use of High Salinity Tolerant Plants. Kluwer Academic Press Publishers, The Netherland.

Gihad, E.A. and H.M. El-Shaer. 1994. Utilization of halophytes by livestock on rangelands: problems and prospects. pp. 77. *In*: V.R. Squires and A.T. Ayoub (eds.). Halophytes as a Resource for Livestock and for Rehabilitation of Degraded Lands. Kluwer Academic Publishers.

Hasanuzzaman, M., K. Nahar, M.D. Alam, P.C. Bhowmik, M.D. Hossain, M.M. Rahman, M.N.V. Prasad, M. Ozturk and M. Fujita. 2014. Potential use of Halophytes to remediate saline soils. Hindawi Publishing Corporation. BioMed. Research International. Volume 2014, Article ID 589341, 12 pages.

Hassan, N.I., A.M. El-Serafy and M.F.A. Farid. 1980. Studies on pastures indigenous to Southern Sinai. I. Chemical composition and *In vitro* digestibility. J. Anim. Sci. 51(Suppl): 239.

Kandil, H.M. and H.M. El Shaer. 1988. The utilization of *Atriplex nummularia* by goats and sheep in Sinai. Proc. of the Inter .Symp. on the Constraints and Possibilities of Ruminant Production in Dry Subtropics, November 5–7, 1988. Cairo, Egypt.

Kandil, H.M. and H.M. El-Shaer. 1990. Comparison between goats and sheep in utilization of high fibrous shrubs with energy feeds. Proc. Int'l Goat Prod. Symp. Tallahassee.

Kewan, K.Z. 2003. Studies on camel nutrition. Ph.D. Thesis, Faculty of Agriculture, Alexandria University, Egypt.

Khan, M.A., B. Böer, M. Öztürk, T.Z. Al Abdessalaam, M. Clüsener-Godt and B. Gul (eds.). 2014. Sabkha Ecosystems. Vol. IV. Cash Crop Halophytes and Biodiversity Conservation (Task for Vegetation Science), Kluwer Academic, Dordrecht.

Koocheki, A. and M.N. Mohalati. 1994. Feed value of some halophytic range plants of arid regions of Iran. *In*: V.R. Squires and A.T. Ayoub (eds.). Halophytes as a resource for livestock and for rehabilitation of degraded lands. Kluwer Academic Publishers, pp. 249–253.

Koyro, H.W., M. Ajmal Khan and H. Lieth. 2011. Halophytic crops: A resource for the future to reduce the water crisis? Emir. J. Food Agric. 23(1): 001–016.

Kovda, V.A. and I. Szabolics. 1979. Modeling of soil salinization and alkalization. Agrokemia es Talajtan 28(suppl.).

Ksouri, R., W.M. Ksouri, I. Jallali, A. Debez, C. Magné, I. Hiroko and C. Abdelly. 2012. Medicinal halophytes: potent source of health promoting biomolecules with medical, nutraceutical and food applications. Crit. Rev. Biotechnol. 32(4): 289–326.

Le Houerou, H.N. 1969. Salt tolerant plants of economic value in the Mediterranean Basin. Reclamation & Revegetation Research 5: 319–341.

Le Houerou, H.N., D. Dumancicm, M. Eskileh, D. Schweisguth and T. Telahique. 1982. Anatomy and physiology of a browsing trial: a methodological approach to fodder shrub evaluation. Techn. Paper n° 28, UNTF/Lib 18, FAO, & Agric. Res. Cent.,Tripoli, Libya, 65 pp.

Le Houerou, H.N. 1993. Salt tolerant plants for the arid regions of the Mediterranean isoclimatic zone. pp. 403. *In*: H. Leith and A. El-Masoom (eds.). Towards the Rational use of High Salinity-tolerant Plants. Vol. 1., Kluwer Academic Publications, Dordrecht, The Netherlands.

Le Houerou, H.N. 1994. Forage halophytes and salt-tolerant fodder crops in the Mediterranean Basin. pp. 123. *In*: V.R. Squires and A.T. Ayoub (eds.). Halophytes as a Resource for Livestock and for Rehabilitation of Degraded Lands. Kluwer Academic Publishers.

Local Government Engineering Department (LGED) and FAO. 2006. Reducing greenhouse gas emissions by promoting bioenergy technologies for heat applications. Report No. EP/RAS/106/GEF. Country report: Bioenergy Study-Bangladesh. Available at http://www.lged-rein.org/archive_file/Bioenergy%20 Study-Bangladesh.pdf. Accessed.

Madrigal, A. 2008. Food vs. Fuel: Saltwater Crops may be Key to Solving Earth's Land Crunch. http://www.wired.com/2008/12/saltwatercrops/.

Manousaki, E., J. Kudukova, N. Papadantonakis and N. Kalogerakis. 2008. Phytoextraction and Phytoexcretion of Cd by the leaves of Tamarix Smyrnensis growing on contaminated non-saline and saline soils. Environmental Research 106.3: 326–32. Sciverse. Web. 25 Nov. 2012. <http://ezproxy2. library.colostate.edu:2089/science/article/pii/S0013935107000795.

Mondal, M.A.H and M. Denich. 2010. Assessment of renewable energy resources potential for electricity generation in Bangladesh. Renewable and Sustainable Energy Reviews 14: 2401–2413.

Qasim, M., S. Gulzar and M. Ajmal Khan. 2011. Halophytes as medicinal plants. *In* M. Ozturk, A.R. Mermut and A. Celik (eds.). Urbanisation, Land Use, Land Degradation and Environment. Daya Publishing House.

Ramirez, R.G. and R.A. Ledezma-Torres. 1997. Forage utilization from native shrubs Acacia rigidula and Acacia farnesiana by goats and sheep. Small Ruminant Res. 25: 43–50.

Riley, J.J. (in press). Review of Halophyte Feeding Trials with Ruminants, pp. 177–217, this volume.

Rizk, A.M. 1986. The phytochemistry of the flora of Qatar. Sci. and Appl. Res. Center, Univ. Qatar. pp. 33.

Serag, M.S., M.A. Zahran and A.A. Khedr. 2002. Ecology and Economic Potentialities of the Dominant Salt-Tolerant Reeds and Rushes in the Nile Delta. Proc. Inter. Sym. on "the Optimum Resources Utilization In Salt-Affected Ecosystems In Arid And Semi-Arid Regions". pp. 245–256, 8–10 April, 2002, Cairo, Egypt.

Shawkat, S.M., I.M. Khatab, B.E. Borhami and K.A. El-Shazly. 2001. Performance of growing goats fed halophytic pasture with different energy sources. Egyptian J. Nutr. Feeds 4: 215–264.

Skouri, M. 1975. Elevage et les possibilities d'accroissement des resources en proteins dans la zone mediterraneenne. Options Mediterraneennes 7: 85–63.

Squires, V.R. 1994. Overview of problems and prospects for utilizing halophytes as a resource for livestock and for rehabilitation of degraded lands. pp. 1. *In*: V.R. Squires and A.T. Ayoub (eds.). Halophytes as a Resource for Livestock and for Rehabilitation of Degraded Lands. Kluwer Academic Publishers.

Streeter, M.N., G.M. Hill, D.G. Wagner, F.N. Owens and C.A. Hibberd. 1993. Effect of bird resistant and non bird resistant sorghum grains on amino acid digestion by beef heifers. J. Anim. Sci. 71: 1648–1656.

Stuart, J. Roy, M. Tester, R.A. Gaxiola and T.J. Flowers. 2012. Plants of saline environments, in AccessScience, ©McGraw-Hill Companies, 2012, http://www.accessscience.com.

Szabolcs, I. 1994. Salt affected soils as ecosystem for halophytes. pp. 19. *In*: V.R. Squires and A.T. Ayoub (eds.). Halophytes as a Resource for Livestock and for Rehabilitation of Degraded Lands. Kluwer Academic Publishers.

Waisel, Y. 1975. Screening of salt resistance. Proc. 21st Colloquium Int. Potash Inst. Bern, pp. 143.

Warner, A.C.I. and B.D. Stacy. 1977. Influence of ruminal and plasma osmotic pressure on salivary secretion in sheep. Q.J. Expe. Physiol. 62: 133–142.

Warren, B.E., C.I. Bunny and E.R. Bryant. 1990. A preliminary examination of the nutritive value of four saltbush (*Atriplex*) species. Proc. Austr. Soc. Anim. Prod. http://www.elmogaz.com/node/175309n.

Wilson, A.D. 1966. The tolerance of sheep to sodium chloride in food or drinking water. Aust. J. Agric. Res. 17: 503–514.

Wilson, R.T. 1984. The Camel. Longmans, London, pp. 15.

World Energy Council (WEC). 2000. Energy for Tomorrow's World: Acting Now! London: Atalink Projects.

Yensen, N.P. and J.L. Bedell. 1993. Considerations for the selection, adaptation and application of halophyte crops in highly saline desert environments as exemplified by the long term development of cereal and forage cultivars of *Distichlis* (Poaceae). pp. 295–304. *In*: H. Lieth and A.A. Masoom (eds.). Towards the rational use of high salinity tolerant plants Vol. 2 Agriculture and Forestry under marginal soil and water conditions. Kluwer Academic Dordrecht.

Zahran, M.A. and A.J. Willis. 2002. Plant Life in the River Nile in Egypt, Mars Publishers, Riyadh, Saudi Arabia, 531 pp.

Potential Use of Halophytes and Salt Tolerant Plants in Ruminant Feeding: A Tunisian Case Study

Moujahed Nizar,[1,]* Guesmi Hajer[1] and Hessini Kamel[2]

SYNOPSIS

This review summarizes the main results of studies related to the fodder potential of some halophytes and salt-tolerant species in Tunisia and highlights the main constraints and opportunities for better valorizing them in ruminant feeding. In Tunisian livestock extensive systems, small ruminant herds, mobility and productivity are severely affected by several devastating factors, such as overgrazing and mismanagement of pasture, in addition to climate changes inducing frequent drought periods and soil salinization. Considering this alarming situation, new strategies and practices are being developed and applied with the aim to promote sustainable management of natural resources and to enhance livestock production. In connection with this, several native or cultivated fodder species such as xerophytes and halophytes could play an important role in the rehabilitation of pasture and the feeding of ruminants in saline lands and/or in arid and semi-arid regions, thereby increasing rural incomes and improving the livelihood of smallholder farmers. The main importance of these plants is their capacity to produce considerable amounts of biomass on saline soils and during drought conditions. The major limitations to be considered when planning halophyte plantation and/or exploitation are

[1] Institut National Agronomique de Tunisie, LRAA, 43 Av. Ch. Nicolle, Tunis, Tunisia.
[2] Centre de Biotechnologie de Borj Cedria, Hammam Lif, Tunisia.
* Corresponding author: nizar.moujahed@yahoo.fr

high salt (limiting intake and digestibility of the diet) and plant secondary metabolites, moderate energy value, low levels of biomass production and the possibility of inducing mineral imbalances. Most of these feedstuffs are best considered as a supplement rather than forage to be fed as a sole diet. Several strategies to valorize halophytes in ruminant feeding, such as (i) free grazing or cut and carry with supplementation (ii) mixing them with other fodders in shrub mixed diet (iii) using them as alternative feed supplements (iii) and (iv) in alley-cropping to supplement cereal stubble have been reported.

Keywords: cactus, barley, cut and carry, salinity, plant secondary metabolites, cereal stubble, sheep, goats, camel, *Atriplex,* concentrates, nutritive value, sabkha, eco-physiological mechanisms, chotts, salt-adapted flora, *Acacia cyanophylla, Atriplex nummularia,* rumen fermentation, chemical composition, Anti-nutritional factors.

1. Introduction

In Tunisia, livestock extensive systems are based on sheep, goats and camels and are mainly located in the central and southern areas of the country. Lands in these regions are inappropriate for forage cultivation but provide native pasture, characterized by low biomass and deficiency of essential nutrients to grazing animals. Also, mobility and productivity flocks of small ruminants are severely affected by several devastating factors, such as overgrazing and mismanagement of pasture, in addition to climate changes and human activities that induce frequent drought periods and a rapid expansion of soil salinization of lands. Soil and water salinity is the major constraint affecting dryland agriculture and animal production worldwide (Ben Salem et al. 2010). As a consequence of this degradation, the contribution of rangelands to ruminant feeding decreased drastically and currently it does not exceed 20%. This reality leads to an over-reliance on barley grain and concentrate feed, and thus to the fluctuating feed availability and prices (Nefzaoui 2002). Consequently, during the last decades, farmers use more and more supplements, mainly barley grain and commercial concentrates, which are mostly imported and costly. For example, local production of compound feed for livestock was of the order of 1.6 million tons during the year 2011, covering 23% of all livestock need. Regarding this alarming situation, and despite the policies of pasture improvement based on the establishment of fodder trees and shrubs initiated since the sixties (Nefzaoui and Chermiti 1991), new strategies and practices are being developed and applied, with the aim to promote sustainable management of natural resources and to enhance livestock production. In connection with this, several native or cultivated fodder species such as halophytes or other salt tolerant species could play an important role in the rehabilitation of pasture and constitute a major part of the feeding program of sheep, goats, and camels in the saline lands and the arid and semi- arid regions (Squires and Ayoub 1994; El Shaer 1997; Michri et al. 2014). This alternative may result in increasing rural population incomes and improving the livelihood of smallholder farmers. These plants grow in a wide variety of saline habitats, from coastal regions, salt marshes and mudflats to inland deserts, salt flats and steppes (El Shaer 2003) and have particular characteristics which enable them to evade and/ or tolerate salinity by various eco-physiological mechanisms (Michri et al. 2014).

In Tunisia, in addition to the existing native halophytes or salt tolerant plants, exotic species such as some *Atriplex* species were introduced between 1920 and 1930. Further introductions took place during the period from 1960s to 1980s under the framework of international research and development rangeland projects (Ben Salem et al. 2010). During the last decades, a growing interest has been accorded to investigate their potential-nutritive value and limits (Ben Salem et al. 1994, 2000, 2005; Moujahed et al. 2000, 2005, 2011, 2013; Moujahed and Kayouli 2005; Nefzaoui and Chermiti 1991; Nefzaoui et al. 1995). The aim of this review paper is to report the main results from studies carried out in Tunisia, that deal with the nutritional evaluation of halophytes and salt tolerant species and the opportunity that could arise by their integration in appropriate animal feeding systems.

2. Importance and repartition of halophytes in Tunisia

Halophytic vegetation is known to grow in different saline biotopes and thus could be found on sand dunes or rocky coasts, in saline depressions (sabkhas), in saline inland deserts, and in salt marshes. In Tunisia, more than 2 million hectares are occupied by sabkhas and chotts.[1] The major part of these salt zones is located in arid and semiarid regions which represent about 2/3 of the terrestrial surface of the country (Hachicha et al. 2007). Vegetation in the sabkhas and their borders is much diversified (Abdelly et al. 2011; Kawada et al. 2012). A mixture of annual salt tolerant plants which develop greatly during the years that receive rain is localized under halophyte bushes, and the land nearby (Abdelly et al. 1995). Chaib and Boukhris (1998) counted about 1630 species endemic to arid regions. These species belong to several families (Aizoaceae, Apiaceae, Asteraceae, Brassicaceae, Boraginacae, Cactaceae, Chenopodiaceae, Frankeniaceae, Juncaceae, Poaceae, Tamaricaceae, and Zygophyllaceae); but the most represented are the Chenopodiaceae and Poaceae species. Some of these plants are grazed by livestock and can provide options for livestock feeding in both arid and saline landscapes. A literature survey carried out by Atia et al. (2014) on 105 native plant species from saline and arid zones showed that the most studied species are halophytes, xerophytes and salt tolerant plants. In addition, a field survey was carried out in November 2008 and November 2009 by Kawada et al. (2012) on semi-arid zones (from Kasserine to Gafsa) and showed that these zones were dominated by shrub or perennial species. Some of the dominant species such as *Arthrophytum schmittianum, Salsola villosa, Salsola vermiculate, Salsola tetrandra, Arthrocnemum indicum* and *Halocnemum strobilaceum* are halophytes or salt-tolerant species. The richest salt-adapted flora was found in the central and southern parts of the country. However, this does not exclude the presence of some halophytes and salt tolerant plants in the north.

---

[1] A **chott** is a dry (salt) lake in the Saharan area of Africa (mainly in Tunisia, Algeria and Morocco) that stays dry in the summer but may receive and retain run-off in the cool season.

The introduction of exotic shrub species in Tunisia began during the last years of the 19th century (Le Houérou and Pontanier 1987). Much later, in addition to native halophyte plants, several species were introduced into Tunisia after the 1960s, mainly for rangeland rehabilitation, to provide further fodder resources for small ruminants and as water and soil conservation actions. This represented thousands of hectares in which *Acacia cyanophylla*, spineless cactus and various *Atriplex* spp. were planted (Nefzaoui and Chermiti 1991). As an example of fodder halophytes in Tunisia, Ben Salem et al. (2010) exhaustively reviewed the potential use of Oldman saltbush (*Atriplex nummularia*) which is one of the most important exotic species utilized on large scale (Le Houérou 1991). The authors reported that *A. nummularia* can develop under cultivation in the arid zones between the 200 and 400 mm of mean annual rainfall and like other *Atriplex* species, Oldman saltbush is able to reach water tables as deep as 10 m below ground surface. A single shrub can produce 1 kg of dry matter (DM) with only 250 kg of water.

3. Chemical and rumen fermentation characteristics of halophytes and salt tolerant species

3.1 Chemical composition

Since a long time, halophytes and other salt-tolerant plants have constituted a considerable fodder resource that has been used in extensive production systems for ruminants, mainly in arid and semi-arid regions and also in marginal lands. The growing interest has led to several research programs dealing with evaluating and including them as components in local feeding strategies. Results from several Tunisian studies on chemical composition of these plants pooled in Table 3.1 showed a wide variability in DM, ash, CP and cell wall contents, as percentage of DM. These species exhibited a large range of ash content, varying from 4 to 36.8% DM, respectively for *Stipa retorta* (Hessini et al. unpublished data) and *Limoniastrum guyonianum* (Laudadio et al. 2008). It is worthy to note the particular high content of ash in some other species such as *Frankenia thymifolia* (31.1% DM) and *Atriplex nummularia* (24 and 25.5% DM). Values higher than 30% DM are specific to territories constituting the natural habitats of camels (in the southern regions). The major part of these plants are specifically adapted halophytes species that live under saline soil conditions which are physiologically arid with sodium chloride (NaCl) concentrations higher than 1% (Laudadio et al. 2008). Also the high level of ash noted in *Atriplex* species, frequently included in feeding programs in Tunisia, is a characteristic of salt marsh plants (El Shaer 2010) and was also underlined for these species by Ben Salem et al. (2010), after reviewing several international studies. Crude protein contents ranged between 4.1% DM in *Imperata cylindrica* from the south of Tunisia (Laudadio et al. 2008) and 34.4% DM in *Arthrocnemum indicum* from the region of Sahel (Hessini et al. unpublished data). Within the most investigated halophytes in Tunisia, *Atriplex* spp. are high in CP (12.5 and 19.7% DM respectively for *Atriplex halimus* and *Atriplex nummularia*). Generally, in most of the presented plants (Table 3.1) CP reached acceptable levels allowing covering microbial requirements in the rumen (Van Soest and Jones 1968),

Table 3.1. Chemical composition of some halophyte plants in Tunisia (% DM).

Species	Families	Areas	Plant parts	DM (%)	OM	Ash	CP	CF	NDF	ADF	ADL	References
Acacia cyanophylla	Fabaceae	-	LT	38.5	-	-	13.7	21.9	52.3	30.8	13.5	Ben Salem et al. (2000)
Aeluropus lagopoïdes	Poaceae	Enfidha sabkha	-	-	88.0	12.0	26.6	-	72.3	51.4	-	Hessini et al. (unpublished data)
Aeluropus littoralis	Poaceae	Enfidha sabkha	-	-	93.4	6.6	6.9	-	83.3	54.5	-	Hessini et al. (unpublished data)
Aeluropus littoralis	Poaceae	Southern Tunisia	LS	32.5	-	14.2	8.5	28.8	56.5	38.9	17.6	Laudadio et al. (2008)
Arrhenatherum elatius	Poaceae	Enfidha sabkha	-	-	94.8	5.2	7.4	-	75.6	56.2	-	Hessini et al. (unpublished data)
Artemisia campestris	Asteraceae	-	LT	56.1	-	-	7.9	36.9	60.8	43.2	16.2	Ben Salem et al. (2000)
Artemisia campestris	Asteraceae	Kairouan	LS	55	94.2	5.8	9.8	-	56.1	45.2	12.6	Moujahed et al. (2013)
Artemisia campestris	Asteraceae	Southern Tunisia	LS	35.6	-	9.6	9.4	16.9	55.2	34.0	11.2	Laudadio et al. (2008)
Artemisia herba alba	Asteraceae	Kairouan	LS	-	93.3	6.7	9.9	-	63.7	50.2	21.6	Moujahed et al. (2013)
Artemisia herba alba	Asteraceae	-	LT	62.6	-	-	7.7	38.9	59.4	44.8	17.7	Ben Salem et al. (2000)
Arthrocnemum indicum	Chenopodiaceae	Enfidha sabkha		-	86.4	13.6	34.5	-	45.5	23.1	-	Hessini et al. (unpublished data)
Atriplex halimus	Chenopodiaceae	Southern Tunisia	LS	30.2	-	19.9	12.5	20.3	46.7	34.8	11.9	Laudadio et al. (2008)
Atriplex nummularia	Chenopodiaceae	-	LT	28.5	-	24	19.7	21.6	33.6	17.7	14.5	Saadani et al. (1989)
Atriplex nummularia	Chenopodiaceae	-	LT	28.6	-	25.5	17.8	-	44.5	22.5	10.2	Ben Salem et al. (2004)
Atriplex nummularia	Chenopodiaceae	-	LT	25	-	-	13.8	20.2	46.9	26	14.4	Ben Salem et al. (2000)
Beta macrocarpa	Chenopodiaceae	Enfidha sabkha	-	-	85.2	14.8	18.0	-	64.9	44.9	-	Hessini et al. (unpublished data)
Brachypodium distatachyum	Poaceae	Enfidha sabkha	-	-	92.2	7.8	6.4	-	80.4	40.4	-	Hessini et al. (unpublished data)
Calendula arvensis	Asteraceae	Kairouan	LS	17.7	84.4	15.6	12.5	-	30.6	22.2	4.9	Moujahed et al. (2013)
Catapodium rigidium	Poaceae	Enfidha sabkha	-	-	92.8	7.2	6.5	-	77.0	53.1	-	Hessini et al. (unpublished data)
Ceratonia siliqua	Fabaceae	-	LT	51.2	-	-	8.1	15.9	49.1	24.5	11.0	Ben Salem et al. (2002)
Frankenia thymifolia	Frankeniaceae	Southern Tunisia	LS	40.1	-	31.1	7.3	16.5	37.7	19.5	12.8	Laudadio et al. (2008)
Halocnemum strobilaceum	Chenopodiaceae	Enfidha sabkha	-	-	89.9	10.1	31.2	-	51.9	23.8	-	Hessini et al. (unpublished data)

Table 3.1. contd....

Table 3.1. contd.

Species	Families	areas	Plant parts	DM (%)	OM	Ash	CP	CF	NDF	ADF	ADL	References
Hordeum maritimum	Poaceae	Enfidha sabkha	-	-	90.1	9.9	8.1	-	78.5	57.4	-	Hessini et al. (unpublished data)
Imperata cylindrica	Poaceae	Southern Tunisia	LS	31.2	-	19.6	4.1	38.5	71.6	47.1	24.5	Laudadio et al. (2008)
Koeleria hispida	Poaceae	Enfidha sabkha	-	-	89.7	10.3	8.8	-	77.9	43.5	-	Hessini et al. (unpublished data)
Koeleria phleoides	Poaceae	Enfidha sabkha	-	-	94.5	5.5	11.2	-	78.7	59.7	-	Hessini et al. (unpublished data)
Limonia strunguyonianum	Plumbaginaceae	Southern Tunisia	LS	38.0	-	36.8	10.5	9.5	37.2	19.8	15.3	Laudadio et al. (2008)
Lolium temulentum	Poaceae	Enfidha sabkha	-	-	94.3	5.7	7	-	84.1	60.7	-	Hessini et al. (unpublished data)
Medicago minima	Fabaceae	Kairouan											
Nitraria retusa	Zygophyllaceae	Southern Tunisia	LS	23.2	-	16.9	9.6	6.7	44.1	38.4	5.7	Laudadio et al. (2008)
Phalaris bulbosa	Poaceae	Enfidha sabkha	-	-	89.5	10.5	8.1	-	81.0	48.3	-	Hessini et al. (unpublished data)
Phalaris paradoxa	Poaceae	Enfidha sabkha	-	-	90.2	9.8	7.4	-	81.4	54.3	-	Hessini et al. (unpublished data)
Pholiurus incurvatus	Poaceae	Enfidha sabkha	-	-	92.4	7.6	7.1	-	75.1	48.0	-	Hessini et al. (unpublished data)
Phragmites communis	Poaceae	Enfidha sabkha	-	-	91.3	8.7	15.3	-	81.1	47.4	-	Hessini et al. (unpublished data)
Pistacia lentiscus	Anacardiaceae	Kairouan	LS	-	95	5	9.3	-	37.8	31.5	23.9	Moujahed et al. (2013)
Pistacia lentiscus	Anacardiaceae	-	LT	49.1	-	-	9.3	15.7	44.7	27.2	13.8	Ben Salem et al. (2000)
Reseda alba	Resedaceae	Kairouan	LS	27	88.5	11.4	18.4	-	27.3	17.8	5.6	Moujahed et al. (2013)
Stipa retorta	Poaceae	Enfidha sabkha	-	-	96	4	5	-	80.0	54.5	-	Hessini et al. (unpublished data)
Suaeda fruticosa	Chenopodiaceae	Enfidha sabkha	-	-	84.1	15.9	29.1	-	48.9	30.2	-	Hessini et al. (unpublished data)

DM: Dry matter; OM: Organic Matter; CP: Crude Protein; CF: Crude Fiber; ADF: Acid Detergent Fiber; ADL: Acid Detergent Lignin; NDF: Neutral Detergent Fiber; LS: leaves and Steams; LT: leaves and Twigs.

Table 3.2. *In vitro* fermentation characteristics of some Tunisian halophytes and salt tolerant plants.

Species	Families	Areas	Plant parts	b (ml gas)	k or c* (h^{-1})	IVOMD$_{24h}$ (%)	ME (kcal kg^{-1} DM)	References
Acacia cyanophylla	*Fabaceae*	-	LT	27.2	0.045*	-	-	Ben Salem et al. (2000)
Artemisia campestris	*Asteraceae*	-	LT	37.5	0.080*	-	-	Ben Salem et al. (2000)
Artemisia campestris	*Asteraceae*	Kairouan	LS	37.1	0.085	42.6	1801	Moujahed et al. (2013)
Artemisia herba alba	*Asteraceae*	Kairouan	LS	26.7	0.055	36.3	1179	Moujahed et al. (2013)
Artemisia herba alta	*Asteraceae*	-	LT	38.6	0.054*	-	-	Ben Salem et al. (2000)
Atriplex nummularia	*Chenopodiaceae*	-	LT	28.5	0.048*	-	-	Ben Salem et al. (2000)
Calendula arvensis	*Asteraceae*	Kairouan	LS	62.5	0.085	63.6	2886	Moujahed et al. (2013)
Ceratonia siliqua	*Fabaceae*	-	LT	37.7	0.038*	-	-	Ben Salem et al. (2000)
Medicago minima	*Fabaceae*	Kairouan	LS	61.8	0.091	65.5	2935	Moujahed et al. (2013)
Pistacia lentiscus	*Anacardiaceae*	Kairouan	LS	34.1	0.040	36.2	1227	Moujahed et al. (2013)
Pistacia lentiscus	*Anacardiaceae*	-	LT	24.6	0.041*	-	-	Ben Salem et al. (2000)
Reseda alba	*Resedaceae*	Kairouan	LS	73.6	0.136	72.1	3071	Moujahed et al. (2013)

LS: leaves and stems; LT: leaves and Twigs; G: ls the gas production at time t; b: asymptotic gas production; k: The fractional rate of gas production according to the model of France et al. (1993), c: the fractional rate of gas production according to the model of Ørskov and Mc Donald (1979). ME and IVOMD24h: metabolisable energy and 24 h *in vitro* organic matter degradability calculated according to Menke and Steingass (1988).

and for some of them CP levels are higher comparatively with the usually cultivated forages in the country. Structural carbohydrate content as expressed by NDF, attained a very wide range, varying between 27.3 for *Reseda alba* (Moujahed et al. 2013) and 84.1% DM for *Lolium temulentum* (Hessini et al. unpublished data). Also, ADL content varied between 4.9 and 23.9% DM, respectively in *Calendula arvensis* and *Pistacia lentiscus* (Moujahed et al. 2013). The majority of the reported halophytes are high in total cell wall contents and in lignin, thereby limiting their use as sole forage in ruminant diets. The observed wide variation in cell wall components may indicate differences in the sampled and analyzed parts of the plants and whether the studied species are shrubby of herbaceous (Moujahed et al. 2013).

3.2 Salt and minerals compositions

Most studies on halophytes and salt tolerant plants for animal use have primarily focused on sodium, chloride, and occasionally potassium, calcium, magnesium, phosphorus, Sulfur, Zinc and Manganese (Masters et al. 2007). Several results on halophyte salt and mineral contents are pooled in Table 3.3. The study of Hessini et al. (Unpublished data) showed that among eight evaluated families, Chenopodiaceae accumulated the highest NaCl contents and Fabaceae and Asteraceae were the lowest ones. According to Masters et al. (2007), salt ions (Na^+ and Cl^-) in the high salt tolerant chenopods species (*Suaeda fruticosa, Arthrocnemum indicum*, and *Halocnemum strobilaceum*) can be accumulated to a level that exceeds the tolerance threshold for livestock (up to 10%). The high concentration of NaCl reduces feed intake and compromises animal health (Norman et al. 2013). However, these three species accumulated considerably less NaCl than most chenopod shrubs (up to 30% DM, Masters 2007). In contrast, in species with low (from Fabaceae, Anacardiaceae, Asteraceae and Poaceae families) and moderate (*Beta macrocarpa, Atriplex nummularia* and *Aeluropus lagopoïdes*) salt tolerance, sodium and chloride are accumulated under the level that could be tolerated by animals (from 0.7 to 5.4% DM; National Research Council 2005). Based on several studies on animals feeds, the NRC (2005) defined the "maximum tolerable levels" of sodium chloride in the diet (i.e., the levels that will not impair health and performance), as 3–6% for ruminants and 2–3% for swine and poultry.

Chenopods are generally more salt tolerant than halophytic grasses and legumes, so it is not surprising to find that the evaluated chenopods accumulated more NaCl than Poaceae species (Norman et al. 2012). For example, among the 32 chenopod species evaluated by Albert et al. (2000), Na^+ and Cl^- represented 67% of the solute concentration (molar in shoot water), whilst in the Poaceae the same ions averaged only 32% of the solutes in 17 species. Although excessive salt intakes depress feed intakes and causes health problems, animals could benefit from moderate levels of salt in their diets (Hemsley et al. 1975). In fact, moderate levels of NaCl in the diet could improve carcass quality of sheep (Kraidees et al. 1998). Plants can accumulate other minerals like NaCl (essentially potassium, calcium, and magnesium) to adjust osmotically (Hessini et al. 2009; Slama et al. 2015). Results found by Hessini et al. (unpublished data) are in line with those of Norman et al. (2002) who showed that potassium, calcium, and magnesium can be accumulated above or close to the maximum tolerable levels for ruminants which are 2, 1.5, and 0.6% respectively (NRC 2005). It is recommended that sulfur in the diet for sheep and cattle must be between 0.2 and 0.15% DM, respectively (SCA 2007), and plants generally have concentrations of S between 0.05 and 0.5% DM (Underwood and Suttle 1999). The percentage of Sulfur in all species listed in Table 3.3 typically exceeds recommended levels with concentrations ranging from 0.3 for *Imperata cylindrica* to 4.8% DM for *Atriplex halimus*. Sulfur is essential for synthesis of structural proteins and is a component of three amino acids (cystine, cysteine and methionine), several vitamins, insulin and coenzyme A (McDonald et al. 2002). High S concentration reduces rumen motility, decreases voluntary feed intake, causes damage to the central nervous system and induces Cu deficiency through reducing its absorption (Underwood and Suttle 1999). Results assigned in Table 3.3 relative to Hessini et al. (unpublished data) are in line with those of Norman et al. (2004, 2008) who showed that halophytic chenopods typically exceed recommended levels with concentrations of S ranging from 3.8 to 4.9 g kg^{-1} DM. Phosphorus is an essential element for plant growth and development (Agren et al. 2012), however accumulated at high concentration, it can depress food and feed quality of plants. According to the National Research Council (2005), the maximum tolerable concentration of P for cattle and sheep is 0.7%. The concentration of P for the species studied by Hessini et al. (unpublished data) was independent of Family and for almost all cases exceeded the maximum tolerable level for ruminants. Also, Zn and Mn occur in halophytes and salt tolerant plants in concentrations exceeding the recommended requirements (Table 3.3). These are also essential to ruminants. Zinc is associated with enzymes and deficiency leads to poor growth and skeletal abnormalities (McDonald et al. 2002), while Zn toxicity causes anemia and digestive disorders (Hosnedlová et al. 2007). On the other hand, Mn is associated with enzyme function, and deficiency leads to poor growth.

3.3 Anti-nutritional factors

High salt and mineral levels are not the only factors influencing voluntary feed intake and feeding value of halophytes. In responses to salt stress, these plants accumulated other solutes, rather than inorganic, for osmotic adjustment (Guo et al. 2005; Masters, this volume). Accumulated at low level, these organic compounds are beneficial for plant growth and animal health (Norman et al. 2012). However, at high rates, these compounds depress feed quality and intake. Anti-nutritive factors include alkaloids, steroids, saponins, flavonoides (Gihad and El Shaer 1994; Attia-Ismail, this volume), coumarins in *Melilotus* spp. (Macias et al. 1999), excessive oxalates in *Atriplex nummularia* (Table 3.4; Ben Salem et al. 2002), and high tannins in *Acacia* (Ben Salem et al. 2000; Moujahed et al. 2000) and *Pistacia* spp. (Ben Salem et al. 2000). The Tunisian data pooled in Table 3.4 showed a relatively wide range of variation in total phenols. Polyphenol contents ranged between 38 and 190 mg GAE. g^{-1} DM (Ksouri et al. 2007, 2012), respectively for *Suaeda fruticosa* and *Tamarix gallica* and between 1.4 and 4.8 g TAE. 100 g^{-1} DM, respectively in *Phalaris bulbosa* and *Suaeda fruticosa* (Hessini et al. unpublished data). In other shrubby species, the total polyphenol contents varied between 0.8 and 6.9 g tannic acid equivalents (TAE). 100 g^{-1} DM

Table 3.3. Mineral contents of halophyte and salt tolerant plants in Tunisia.

Species	Families	areas	Plant parts	Ca (%DM)	P (%DM)	Mg (%DM)	K (%DM)	NaCl (%DM)	S (%DM)	Zn (mg/kg DM)	Mn (mg/kg DM)	References
Aeluropus lagopoides	Poaceae	Enfidha sabkha	aerial part	0.9	-	0.4	1.3	4.1	-	-	-	Hessini et al. (unpublished data)
Aeluropus littoralis	Poaceae	Enfidha sabkha	aerial part	0.6	-	0.4	0.9	2.2	-	-	-	Hessini et al. (unpublished data)
Aeluropus littoralis	Poaceae	Southern Tunisia	LS	1200a	1752a	1053a	684a	-	1395a	9.1a	1.8a	Laudadio et al. (2008)
Arrhenatherum elatius	Poaceae	Enfidha sabkha	aerial part	0.6	-	0.3	0.9	1.3	-	-	-	Hessini et al. (unpublished data)
Artemisia campestris	Asteraceae	-	LT	0.7	0.1	-	-	0.2 >	-	-	-	Ben Salem et al. (2000)
Artemisia campestris	Asteraceae	Southern Tunisia.	LS	1282a	1105a	129a	848a	-	307a	1.5a	3.1a	Hammadi et al. (2008)
Artemisia herba alta	Asteraceae	-	LT	0.6	0.2	-	-	0.3 >	-	-	-	Ben Salem et al. (2000)
Arthrocnemum indicum	Chenopodiaceae	Enfidha sabkha	-	2.5	-	0.7	1.0	12.9	-	-	-	Hessini et al. (unpublished data)
Atriplex halimus	Chenopodiaceae	Southern Tunisia	LS	707a	1129a	906a	1134a	-	4888a	1.7a	< 0.05a	Hammadi et al. (2008)
Atriplex nummularia	Chenopodiaceae	-	LT	1.8	0.5	-	-	4.9 >	-	-	-	Ben Salem et al. (2000)
Beta macrocarpa	Poaceae	Enfidha sabkha	aerial part	3.1	-	1.5	1.5	5.4	-	-	-	Hessini et al. (unpublished data)
Brachypodium distachyum	Poaceae	Enfidha sabkha	aerial part	0.5	-	0.4	0.6	1.0	-	-	-	Hessini et al. (unpublished data)
Catapodium rigidium	Poaceae	Enfidha sabkha	aerial part	0.7	-	0.4	0.8	2.2	-	-	-	Hessini et al. (unpublished data)
Ceratonia siliqua	Fabaceae	-	LT	1.3	0.1	-	-	0.1 >	-	-	-	Ben Salem et al. (2000)

Species	Family	Location	Part									Reference
Frankenia thymifolia	*Frankeniaceae*	Southern Tunisia	LS	5047a	1834a	804a	262a	-	1376a	1.7a	5.9a	Hammadi et al. (2008)
Halocnemum strobilaceum	*Chenopodiaceae*	Enfidha sabkha	-	1.6	-	1.2	1.2	12.4	-	-	-	Hessini et al. (unpublished data)
Hordeum maritimum	*Poaceae*	Enfidha sabkha	aerial part	0.6	-	0.3	1.1	1.8	-	-	-	Hessini et al. (unpublished data)
Imperata cylindrica	*Poaceae*	Southern Tunisia	LS	222a	1902a	86a	725a	-	261a	4.2a	< 0.05a	Hammadi et al. (2008)
Koeleria hispida	*Poaceae*	Enfidha sabkha	aerial part	0.9	-	0.4	1.3	2.6	-	-	-	Hessini et al. (unpublished data)
Koeleria phleoïdes	*Poaceae*	Enfidha sabkha	aerial part	0.7	-	0.2	1.0	1.4	-	-	-	Hessini et al. (unpublished data)
Limoniastrum guyonianum	*Plumbaginaceae*	Southern Tunisia	LS	7185a	1178a	1145a	215a	-	2560a	1.6a	8.4a	Hammadi et al. (2008)
Lolium temulentum	*Poaceae*	Enfidha sabkha	aerial part	0.7	-	0.3	1.2	2.2	-	-	-	Hessini et al. (unpublished data)
Nitraria retusa	*Zygophyllaceae*	Southern Tunisia	LS	1985a	2064a	1045a	492a	-	2434a	1.5a	2.9a	Hammadi et al. (2008)
Phalaris bulbosa	*Poaceae*	Enfidha sabkha	aerial part	0.6	-	0.4	1.3	1.7	-	-	-	Hessini et al. (unpublished data)
Phalaris paradoxa	*Poaceae*	Enfidha sabkha	aerial part	0.5	-	0.3	1.8	2.6	-	-	-	Hessini et al. (unpublished data)
Pholiurus incurvatus	*Poaceae*	Enfidha sabkha	aerial part	0.9	-	0.5	1.0	2.1	-	-	-	Hessini et al. (unpublished data)
Phragmites communis	*Poaceae*	Enfidha sabkha	aerial part	0.8	-	0.4	1.7	1.6	-	-	-	Hessini et al. (unpublished data)
Pistacia lentiscus	*Anacardiaceae*	-	LT	0.8	0.1	-	-	0.4 >	-	-	-	Ben Salem et al. (2000)
Stipa retorta	*Poaceae*	Enfidha sabkha	aerial part	0.4	-	0.3	1.0	0.7	-	-	-	Hessini et al. (unpublished data)
Suaeda fruticosa	*Chenopodiaceae*	Enfidha sabkha	aerial part	1.9	-	1.3	0.7	10.4	-	-	-	Hessini et al. (unpublished data)

LT: Leaves and Twigs; LS: Leaves and Steams; a: Expressed as a *mg/100 MG DM*.

Table 3.4. Secondary compounds of halophyte and salt tolerant plants in Tunisia.

Species	Families	Areas	Plant parts	TP (%DM)	CT (%DM)	Oxalate (%DM)	References
Acacia cyanophylla	Fabaceae	-	LT	06.9[a]	4.2b	-	Ben Salem et al. (2000)
Acacia cyanophylla	Fabaceae	Zaghouan	L	-	4.3b	-	Moujahed et al. (2004)
Artemisia campestris	Asteraceae	-	LT	6.3[a]	0.4 b	-	Ben Salem et al. (2000)
Artemisia herba alta	Asteraceae	-	LT	2.6[a]	0.5 b	-	Ben Salem et al. (2000)
Atriplex nummularia	Chenopodiaceae	-	LT	0.8[a]	0.4 b	-	Ben Salem et al. (2000)
Atriplex nummularia L.	Chenopodiaceae	Zaghouan	-	0.3	< 1	3.7	Ben Salem et al. (2002)
Cakile maritima	Brassicaceae	Jerba	L	43.02[d]	-	-	Ksouri et al. (2007)
Cakile maritima	Brassicaceae	Tabarka	L	42.84[d]	-	-	Ksouri et al. (2007)
Ceratonia siliqua	Fabaceae	-	LT	11.8[a]	3.1 b	-	Ben Salem et al. (2000)
Limoniastrum guyonianum	Plumbaginaceae	Tunisian salt localities	Sh	99[c]	-	-	Ksouri et al. (2012)
Limoniastrum monopetalum	Plumbaginaceae	Tunisian salt localities	Sh	40[c]	-	-	Ksouri et al. (2012)
Mesembryanthemum edule	Aizoaceae	Tunisian salt localities	Sh	71[c]	-	-	Ksouri et al. (2012)
Pistacia lentiscus	Anacardiaceae	-	LT	16.6[a]	7.7[b]	-	Ben Salem et al. (2000)
Suaeda fruticosa	Amaranthaceae	Tunisian salt localities	Sh	38[c]	-	-	Ksouri et al. (2012)
Tamarix gallica	Tamaricaceae	Tunisian salt localities	Sh	190[c]	-	-	Ksouri et al. (2012)
Aeluropus lagopoïdes	Poaceae	Enfidha sabkha	aerial part	1.7[a]	-	0.1	Hessini et al. (unpublished data)
Aeluropus littoralis	Poaceae	Enfidha sabkha	aerial part	2.3[a]	-	0.2	Hessini et al. (unpublished data)
Arrhenatherum elatius	Poaceae	Enfidha sabkha	aerial part	2.0[a]	-	0.2	Hessini et al. (unpublished data)
Arthrocnemum indicum	Chenopodiaceae	Enfidha sabkha	aerial part	2.6[a]	-	6.8	Hessini et al. (unpublished data)
Beta macrocarpa	Chenopodiaceae	Enfidha sabkha	aerial part	3.1[a]	-	6.5	Hessini et al. (unpublished data)
Brachypodium distachyum	Poaceae	Enfidha sabkha	aerial part	1.7[a]	-	0.2	Hessini et al. (unpublished data)
Catapodium rigidium	Poaceae	Enfidha sabkha	aerial part	2.4[a]	-	0.2	Hessini et al. (unpublished data)
Halocnemum strobilaceum	Chenopodiaceae	Enfidha sabkha	aerial part	1.8[a]	-	5.1	Hessini et al. (unpublished data)

Hordeum maritimum	Poaceae	Enfidha sabkha	aerial part	1.7[a]	-	0.2	Hessini et al. (unpublished data)
Koeleria hispida	Poaceae	Enfidha sabkha	aerial part	2.9[a]	-	0.2	Hessini et al. (unpublished data)
Koeleria phleoïdes	Poaceae	Enfidha sabkha	aerial part	3.1[a]	-	0.2	Hessini et al. (unpublished data)
Lolium temulentum	Poaceae	Enfidha sabkha	aerial part	2.3[a]	-	0.2	Hessini et al. (unpublished data)
Phalaris bulbosa	Poaceae	Enfidha sabkha	aerial part	1.4[a]	-	0.1	Hessini et al. (unpublished data)
Phalaris paradoxa	Poaceae	Enfidha sabkha	aerial part	1.5[a]	-	0.2	Hessini et al. (unpublished data)
Pholiurus incurvatus	Poaceae	Enfidha sabkha	aerial part	1.7[a]	-	0.3	Hessini et al. (unpublished data)
Phragmites communis	Poaceae	Enfidha sabkha	aerial part	3[a]	-	0.2	Hessini et al. (unpublished data)
Stipa retorta	Poaceae	Enfidha sabkha	aerial part	1.6[a]	-	0.2	Hessini et al. (unpublished data)
Suaeda fruticosa	Chenopodiaceae	Enfidha sabkha	aerial part	4.8[a]	-	13.2	Hessini et al. (unpublished data)

a: Total phenols expressed as g equivalent tannic acid/100 g DM; b: Condensed tannins expressed as g equivalent catechin/100 g DM; c: Total phenols expressed as mg GAE.g⁻¹ DW; d: Total phenols mg of GAE g⁻¹ DW; LT: Leaves and Twigs+; L: Leaves; Sh: Shoots.

respectively in *Atriplex nummularia* and *Acacia cyanophylla* (Table 3.4; Ben Salem et al. 2000). Although the content of total phenols was low and varied with species, there is a positive correlation between the halophytic character of plants (high NaCl concentration) and total phenols contents. The later compounds are also accumulated by plants in response to salt stress (Ksouri et al. 2007). Condensed tannins expressed as g of catechin equivalent in each 100 g^{-1} DM ranged between 0.4 in *Artemisia campestris* and *Atriplex nummularia* and 7.7 in *Pistacia lentiscus* (Table 3.4; Ben Salem et al. 2000). When ingested at high levels, condensed tannin are known to cause astringency in the mouth, to have a protective effect against protein degradation and to negatively affect microbial activity and growth in the rumen (Leinmüller et al. 1991). Oxalates were also investigated by Hessini et al. (unpublished data). The authors found a range of variation between 0.1 to 6.8% DM, respectively in *Aeluropus lagopoïdes* and *Suaeda fruticosa*. According to Guo et al. (2005) oxalates are accumulated more by halophytic species. This secondary compound forms complexes with Ca and Mg and likely with other minerals, making them unavailable for the rumen microbes and the host animal (Nakata and Mc Conn 2007).

3.4 Rumen fermentation characteristics

In vitro rumen fermentation was investigated for several halophytes and salt tolerant species in a natural pasture from a Tunisian central region (Moujahed et al. 2013) and from a Tunisian semi-arid region (Ben Salem et al. 2000). The results presented in Table 3.2 showed that asymptotic gas production (b) ranged from 24.6 and 73.6 ml respectively in *Pistacia lentiscus* and *Reseda alba* and that the fractional rate of *in vitro* rumen fermentation (k or c) varied between 0.038 and 0.136 h^{-1} respectively for *Ceratonia siliqua* and *Reseda alba*. Calculated metabolizable energy (ME) was the highest for *Reseda alba* (3071 kcal kg^{-1} DM) and the lowest for *Artemisia herba alba* (1179 kcal kg^{-1} DM). Moujahed et al. (2013), noted positive correlations between CP content and IVOMD$_{24h}$ (r = 0.74, P < 0.0001) and also ME (r = 0.76, P < 0.0001). While, negative correlations were observed between IVOMD$_{24h}$ and NDF (r = –0.58, P < 0.0001), ADF (r = –0.65, P < 0.0001) and ADL (r = –0.72, P < 0.0001) contents. The same trend was noted with ME (r = –0.58, –0.63, –0.73, P < 0.0001, respectively for NDF, ADF and ADL).

This large diversity ranging between rich chemical compositions and high fermentative characteristics in some plants, to very poor ones is related to botanical differences, seasonal changes, environmental conditions and management practices (Le Houérou 1994; El Shaer 2004). Indeed, climate factors, such as temperature, humidity, precipitation and light intensity play an important role in controlling nutrient contents and nutritive value of plants as they affect assimilation, photosynthesis and metabolism (El Shaer 2010). This wide variability may lead to claim an eventual nutritional complementarity between plants, especially if available in natural pasture, when these species are grazed by small ruminants (Moujahed et al. 2013).

4. Palatability and intake of halophytes and salt tolerant plants

Studies on voluntary feed intake and palatability in Tunisia are rare compared with those on roughages, and halophytes were often nutritionally studied within mixed shrubs groups. Several Tunisian results on palatability, voluntary feed intake and feeding values, relative to halophytes are presented in Table 3.5. Ben Salem et al. (1994) investigated the palatability of the most important shrubs and fodder trees which are representative of the Tunisian arid zones using sheep and camels. When offered with hay, the author found that, regardless to time effect, shrubs consumption by sheep was less than hay (Table 3.5). The more palatable species were *Artemisia campestris, Acacia cyanophylla, Ceratonia siliqua, Artemisia herba alba* and *Juniperus phoenicea*. When hay was removed from the combination (day 5 to day 15), sheep increased their level of shrub intake, especially for *Atriplex halimus, Globularia alypum, Pistacia lentiscus, Rosmarinus officinalis* and *Stipa tenacissima*. The same trend was found with camel (Table 3.5). However, the authors noted that sheep and camels did not have the same preferences for shrubs and fodder trees. Camels had much greater preferences for *Ceratonia siliqua, Atriplex halimus* and *Pistacia lentiscus* than sheep had. Moreover, camels contrary to sheep, had greater intake of some shrubs than hay. Additionally, Ben Salem et al. (2000) measured palatability and intake in relationship with the nutritive characteristics of six Mediterranean shrubs on sheep and goats. The ration I7/D7 (average amount of shrub consumed over the 7-day period to average amount of shrub distributed over the 7-day period) was used as a palatability index for shrub ranking. Results from this study indicated that goats consumed more shrubs than sheep and that sheep and goat preferences for shrubs were different. Shrub intake (I7) by sheep was not related to their chemical composition. In contrast, palatability was significantly correlated to dry matter ($r = -0.70$, $P < 0.01$), ash ($r = 0.75$, $P < 0.001$), phosphorus ($r = 0.54$, $P < 0.05$) and sodium ($r = 0.67$, $P < 0.01$) contents. Goat preferred low fiber ($r = -0.59$, $P < 0.05$) and sodium rich shrubs ($r = 0.62$, $P < 0.01$). Also, sheep and goat preferences were not dependent upon condensed tannin content in shrubs. In addition, the authors found that voluntary feed intake by sheep and goats (I7) was not related to biological parameters (DM *in vitro* digestibility and *in sacco* degradability, and *in vitro* gas production). These parameters seem to have no effect on goat preferences. In contrast, sheep preferences for shrubs was correlated to DM *in vitro* digestibility (11.6, $r = 0.52$, $P < 0.05$).

These two Tunisian studies on palatability and intake of shrubs including several halophytes, highlighted the differences in animal preferences, which could have relationship with the adaptation phenomenon and history of plant consumption and also the contrasting opinions concerning the effect of the nutritive characteristics of shrub species on their palatability. Indeed, if voluntary feed intake is classically assimilated as a resultant of chemical characteristics and digestion, the interaction between taste and post ingestive feedback of animals could result in decreased consumption (Provenza 1996) and consequently palatability could be considered as the interrelation between flavor and post ingestive feedback relative to nutrient and toxins. According to Personius et al. (1987), herbivores are able to detect some toxic compounds by smell

Table 3.5. Palatability, intake and feeding value of halophytes and/or halophyte based diets.

Animal species	Plants or diets	Intake (g DMday⁻¹)	Palatability Index (%)	DOMi (g kg⁻¹BW⁰·⁷⁵)	DCPi (g kg⁻¹ BW⁰·⁷⁵)	References
Plants						
Sheep	*Acacia cyanophylla*	310	39c	-	-	Ben Salem et al. (2000)
Goats	*Acacia cyanophylla*	328	44c	-	-	Ben Salem et al. (2000)
Sheep	*Artemisia campestris*	411	28c	-	-	Ben Salem et al. (2000)
Goats	*Artemisia campestris*	426	28c	-	-	Ben Salem et al. (2000)
Lambs	*Atriplex nummularia*	1073		-	-	Ben Salem et al. (1998)
Sheep	*Atriplex nummularia*	273	55c	-	-	Ben Salem et al. (2000)
Goats	*Atriplex nummularia*	229	50c	-	-	Ben Salem et al. (2000)
Sheep	*Ceratonia silliqua*	228	31c	-	-	Ben Salem et al. (2000)
Goats	*Ceratonia silliqua*	390	52c	-	-	Ben Salem et al. (2000)
Sheep	*Pistacia lentiscus*	66	20c	-	-	Ben Salem et al. (2000)
Goats	*Pistacia lentiscus*	84	24c	-	-	Ben Salem et al. (2000)
Sheep + camels	*Acacia cyanophylla*	250	46d	-	-	Ben Salem et al. (1994)
Sheep + camels	*Artimisia campestris*	370	40d	-	-	Ben Salem et al. (1994)
Sheep + camels	*Artimisia herba alba*	240	25d	-	-	Ben Salem et al. (1994)
Sheep + camels	*Atriplex halimus*	180	29d	-	-	Ben Salem et al. (1994)
Sheep + camels	*Ceratonia silliqua*	370	38d	-	-	Ben Salem et al. (1994)
Sheep + camels	*Pistacia lentiscus*	150	25d	-	-	Ben Salem et al. (1994)
Diets						
Lambs	Atriplex + Barley	562	-	-	-	Ben Salem et al. (2005a)
Lambs	Atriplex + Cactus	438	-	-	-	Ben Salem et al. (2005a)

Lambs	Atriplex + Barley	887	-	-	-	-	Ben Salem et al. (1998)
Lambs	Atriplex + Barley	864	-	-	-	-	Ben Salem et al. (1998)
Lambs	Atriplex + Cactus + Straw	945	-	-	-	-	Ben Salem et al. (2004)
Lambs	Atriplex + Straw + Barley	915	-	-	-	-	Ben Salem et al. (2004)
Kids (Boer)	Atriplex + Rangeland + Cactus	200	-	-	-	-	Ben Salem et al. (2000)
Sheep + Goats	Acacia + hay + barley	1325	-	43.7	5.3		Moujahed et al. (2005)
Sheep + Goats	Acacia + hay + barley + PEG	1367	-	51.8	7.2		Moujahed et al. (2005)
Sheep	*Acacia cyanophylla* + *hay*	929.5	-	25.7	29.7		Moujahed et al. (2000)
Sheep	*Acacia cyanophylla* + hay + B1	1392.1	-	34.2	114.8		Moujahed et al. (2000)
Sheep	*Acacia cyanophylla* + hay + B2 (PEG)	1394.1	-	35.7	135.5		Moujahed et al. (2000)

a: Water intake expressed as a l/day; b: Water intake expressed as a g/day; c: The ratio between the average of DM intake in day 7 (I7) and the average of the amount of shrub distributed over the 7-day experimental period; d: The ratio between the average of DM intake in day 5 (I5) and the average of shrub distributed over the 5-day experimental period; B1: Urea molasses block; B2:B1 + PEG.

before eating or immediately after the first bite. Also, Ben Salem et al. (2000) found that the total extractable phenols and condensed tannins had no effect on sheep and goat preferences. This result traduces the conflicting findings relative to secondary compound effect on palatability and intake, mainly in native species. Indeed, Moujahed et al. (2000) attributed the low intake of *Acacia cyanophylla* based diets mostly to the presence of condensed tannin (4.3% DM) in relationship with astringency and the well demonstrated negative effect of tannin on rumen fermentation (Table 3.5). This effect was clearly demonstrated *in vivo* (Moujahed et al. 2000, 2005) and *in vitro* (Moujahed and Kayouli 2005) by the adding of PEG4000 in the *Acacia* based diet.

In addition to these studied factors, salt content in halophytes seems to influence voluntary intake. Results on Oldman saltbush diets from several studies reviewed and discussed by Ben Salem et al. (2010) indicated that the voluntary feed intake was variable and ranged from 37 to 115 g DM/kg $BW^{0.75}$, suggesting an effect of the high salt content and the need for a large quantity of drinking water. According to Masters et al. (2005a,b), salt is a physiological limitation to intake and sheep stop eating salty forage after they have ingested approximately 200 g of salt in a day; animals can only consume salty forage as fast as they can excrete salt, so long term adaptation to salt is unlikely (Attia-Ismail, this volume; Degen and Squires, this volume).

5. Animal performances

Studies on animal performances when integrating halophytes in the feeding systems are rare (summarized in Table 3.6). In several trials on lambs fed *Atriplex* (Ben Salem et al. 2000, 2004, 2005a) it was shown that sheep fed Oldman saltbush foliage alone decreased or, at best, maintained liveweight. However, the provision of barley or cactus as carbohydrate-energy supplement would improve considerably the growth of sheep at a rate ranging from 20 to 108 g day^{-1}. This trend was earlier noted by Le Houerou (1992) for cactus supplementation in sheep fed Oldman saltbush and receiving cactus cladodes. Moujahed and Kayouli (2005) demonstrated that lambs on *Acacia cyanophylla* supplemented with Urea-molasses blocs maintained their liveweight (about 30 g day-1 of DLW) and that DLW was increased when PEG4000 was added into the blocks. On the other hand, the effect of energy supplements on pregnant and lactating sheep and goats grazing natural halophytes was evaluated By El Shaer (1981, cited by El Shaer 2010) in Sinai (Egypt). The authors found that both animal species could not sustain themselves on the natural range species solely without barley supplementation. Similar trends were reported by El Shaer and Kandil (1990) who recommended the supplementation of soluble carbohydrate sources (i.e., barley, corn, ground date stones and molasses) to improve the performance of sheep and goats fed salt-tolerant forages and recommended to provide daily amounts of 150 and 250 g barley per head for sheep fed saltbush during wet and dry seasons, respectively. However, in some particular situations, small ruminants could have moderate performances when only grazing native pasture including halophytes. In connection with this, Moujahed et al. (2011) demonstrated that lambs grazing in spring in the native rangeland from a semi-arid region of Tunisia, containing the halophyte species cited by Moujahed et al. (2013) (Table 3.1), exhibited a substantial

Table 3.6. Performances of animals receiving halophytes and salt tolerant species diets.

Animal (breed)	Diets	Initial LW (kg)	Growth (g day^{-1})	References
Lambs (Barbarine)	Atriplex + Barley	25	108	Ben Salem et al. (1998)
Kids (Boer)	Atriplex + Rangeland + Cactus	-	60	Ben Salem et al. (2000)
Lamb (Noire de Thibar)	*Acacia cyanophylla* L. + hay + B1	19.3	33.7	Moujahed and Kayouli (2005)
Lamb (Noire de Thibar)	Acacia + hay + B2	19.3	54.6	Moujahed and Kayouli (2005)
Lambs (Barbarine)	Atriplex + Cactus + Straw	19.5	81	Ben Salem et al. (2004)
Lambs (Barbarine)	Atriplex + Straw + Barley	19.5	59	Ben Salem et al. (2004)
Lambs (Barbarine)	Atriplex + Barley	23.1	66.7	Ben Salem et al. (2005a)
Lambs (Barbarine)	Atriplex + Cactus	20.1	20.5	Ben Salem et al. (2005a)
Lambs (Barbarine)	Grazed native pasture including halophytes	29.9	128	Moujahed et al. (2011)

B1: Urea molasses block; B2: B1 + PEG4000.

LW gain (128 g day-1) without supplementation (Table 3.6). This result was mainly related to the nutritional diversity in the pasture, which included several herbaceous species with high nutritive value (Moujahed et al. 2013) and also to the nutritional complementarities between plants resulting in optimized rumen fermentation and nutrient supplies to lambs.

6. Conclusions

Based on several studies on halophytes and salt tolerant species in Tunisia, this review has highlighted the potential as well as the limits of these plants as fodder resources for small ruminants in extensive production systems in saline lands and/or in arid and semi-arid environments. The main limitations are high salt and minerals contents and in several cases it includes lignin contents, the presence of some anti-nutritive compounds, such as oxalate and tannins and their moderate or low energy value. In the most of situations, these characteristics lead to possible nutritive imbalances and low performances. However, some of them are high in nitrogen and also could be used as a source of minerals, providing, that mixed complementary shrubs are given, and supplements are provided.

Some strategies to valorize the richest salt-adapted flora (halophytes and salts tolerant species) in ruminant feeding were reviewed in this chapter, such as free grazing or cut and carry with energy (barley) or catalytic (feed blocs) supplementation, mixing them with other fodders in shrub mixed diet (cactus) and using them as alternative feed supplements (Oldman saltbush). In this connection, other promising alternatives could be proposed such as planting some halophytes in alley-cropping to supplement cereal stubble grazed in summer as is done in the context of conventional or conservation agriculture.

The cited works on small ruminant performance as related to halophytes, underline the importance of such plants to the survival of animals and even the possibility to obtain moderate LW gain, mainly when supplemented. These alternative feeding systems are cost effective and halophytes, in addition to their other benefits, could contribute to sustainability of livestock production in arid and semi-arid regions.

References and Further Readings

Abdelly, C., E. Zid, M. Hajji and C. Grignon. 1995. Biomass production and nutrition of *Medicago* species associated with halophytes on the edge of a sabkha in Tunisia. *In*: R. Choukr-allah, C.V. Malcolm and A. Hamdy (eds.). Halophytes and Biosaline Agriculture. Marcel Dekker, Inc, New York.
Abdelly, C., A. Debez, A. Smaoui and C. Grignon. 2011. Halophyte-fodder species association may improve nutrient availability and biomass production of the Sabkha ecosystem. *In*: M. Öztürk et al. (eds.). Sabkha Ecosystems, Tasks for Vegetation Science, Vol. 46. Springer, pp. 95–104.
Albert, R., G. Pfundner, G. Hertenberger, T. Kastenbauer and M. Watzka. 2000. The physiotype approach to understanding halophytes and xerophytes. pp. 69–87. *In*: S.-W. Breckle, B. Schweizer and U. Arndt (eds.). Ergebnisse Weltweiter Ökologischer Forschung. Stuttgart, Germany: Verlag Günter Heimbach.
Agren, G.I., J.A.M. Wetterstedt, F.K. Magnus and M.F.K. Billberger. 2012. Nutrient limitation on terrestrial plant growth—modeling the interaction between nitrogen and phosphorus. New Phytologist 194: 953–960.

Atia, A., M. Rabhi, A. Debez, C. Abdelly, H. Gouia, C.C. Haouri and A. Smaoui. 2014. Ecophysiological aspects in 105 plants species of saline and arid environments in Tunisia. J. Arid Land: doi: 10.1007/s40333-014-0028-2.

Attia-Ismail, S.A. (in press). Plant secondary metabolites of halophytes and salt tolerant plants. pp. 127–142, this volume.

Attia-Ismail, S.A. (in press). Rumen physiology under high salt stress. pp. 348–360, this volume.

Ben Salem, H., A. Nefzaoui and H. Abdouli. 1994. Palatability of shrubs and fodder trees measured on sheep and camels. Methodological approach and preliminary results. *In*: Cahier Options Méditerranéennes 4: 35–48.

Ben Salem, H. 1998. Effet de *l'Acacia cyanophylla* Lindl. sur l'ingestion et la digestion des régimes destinés aux ovins. Rôle des tanins et perspectives d'amélioration de sa valeur alimentaire. Thèse Doctorat, Université de Dijon, France, 252 p.

Ben Salem, H., A. Nefzaoui and L. Ben Salem. 2000. Sheep and goat preferences for Mediterranean fodder shrubs. Relationship with the nutritive characteristics. pp. 155–159. *In*: Cahier Options Méditerranéennes, No. 52.

Ben Salem, H., A. Nefzaoui and L. Ben Salem. 2002. Supplementation of *Acacia cyanophylla* Lindl. Foliage-based diets with barley or shrubs from arid areas (*Opuntia ficus-indica f. inermis* and *Atriplex nummularia* L.) on growth and digestibility of lambs. Animal Feed Science and Technology 96: 15–30.

Ben Salem, H., A. Nefzaoui and L. Ben Salem. 2004. Spineless cactus (*Opuntia ficus-indica f.* inermis) and Oldman saltbush (*Atriplex nummularia* L.) as alternative supplements for growing Barbarine lambs given straw based diets. Small Rumin. Res. 51: 65–73.

Ben Salem, H., H. Abdouli, A. Nefzaoui, A. El-Mastouri and L. Ben Salem. 2005. Nutritive value, behavior and growth of Barbarine lambs fed on old and saltbush (*Atriplex nummularia* L.) and supplemented or not with barley grains or spineless cactus (*Opuntia ficus indica f. inermis*) pads. Small Rumin. Res. 59: 229–238.

Ben Salem, H., H.C. Noman, A. Nefzaoui, D.E. Mayberry, K.L. Pearce and D.K. Revell. 2010. Potential use of Oldman saltbush (*Atriplex nummularia* Lindl.) in sheep and goat feeding. Small Rumin Res. 91: 13–28.

Chaib, M. and M. Boukhris. 1998. Flora Succinct and Illustrated of Arid Zones of Tunisia. The Association of Nature and Environment of Sfax. Tunisia: Gold Time.

Degen, A.A. and V.R. Squires (in press). The rumen and its adaptation to salt. pp. 336–347, this volume.

Kraidees, M.S., M.A. Abouheif, M.Y. Al-Saiady, A. Tag-Eldin and H. Metwally. 1998. The effect of dietary inclusion of halophyte *Salicornia bigelovii* Torr. on growth performance and carcass characteristics of lambs. Animal Feed Science & Technology 76: 149–159.

El Shaer, H.M. 1997. Sustainable utilization of halophytic plant species as livestock fodder in Egypt. pp. 171–184. *In*: Proceedings of the International Conference on "Water Management, Salinity and Pollution Control towards Sustainable Irrigation in the Mediterranean Region", September 22–26, 1997, Bari, Italy.

El Shaer, H.M. 2003. Potential of halophytes as animal fodder in Egypt. Cash Crop Halophytes: Recent Studies 38: 111–119.

El Shaer, H.M. 2004. Potentiality of halophytes as animal fodder under arid conditions of Egypt. Rangeland and Pasture Rehabilitation in Mediterranean Areas. *In*: Cahier Options Méditerranéennes 62: 369–374.

El Shaer, H.M. 2010. Halophytes and salt-tolerant plants as potential forage for ruminants in the Near East region. Small Ruminant Research 91: 3–12.

El Shaer, H.M. and H.M. Kandil. 1990. Comparative study on the nutritional value of wild and cultivated *Atriplex halimus* by sheep and goat in Sinai. Comput. Sci. Dev. Res. 29: 81–90.

France, J., M.S. Dhanoa, M.K. Theodorou, S.J. Lister, S.J. Davies and D. Isac. 1993. A model to interpret gas accumulation profiles with *in vitro* degradation of ruminant feeds. J. Theor. Biol. 163: 99–111.

Gihad, E.A. and H.M. El Shaer. 1994. Nutritive value of halophytes. *In*: V.R. Squires and A.T. Ayoub (eds.). Halophytes as a Resource for Livestock and for Rehabilitation of Degraded Lands. Kluwer Academic Publishers, pp. 281–284.

Guo, C.Y., X.Z. Wang, L. Chen, L.N. Ma and R.Z. Wang. 2015. Physiological and biochemical responses to saline-alkaline stress in two halophytic grass species with different photosynthetic pathways. Photosynthetica. 53, 1, 128–135.

Hachicha, M. 2007. Les sols salés et leur mise en valeur en Tunisie. Sécheresse 18(1): 45–50.

Hamsley, J.A., J.P. Hogan and R.H. Weston. 1975. Effect of high intake of sodium chloride on the utilization of a protein concentrate by sheep: 11. Digestion and absorption of organic matter and electrolytes. Aust. J. Res. 26: 7150–727.

Hessini, K., J.P. Martinez, M. Gandour, A. Albouchi, A. Soltani and C. Abdelly. 2009. Effect of water stress on growth, osmotic adjustment, cell wall elasticity and water use efficiency in *Spartina alterniflora*. Environmental and Experimental Botany 67: 312–319.

Hosnedlová, B., J. Trávníček and M. Soch. 2007. Current view of the significance of zinc for ruminants. A review: Agricultural tropica et subtropica 40: 57–64.

Kawada, K., K. Suzuki, H. Suganuma, A. Smaoui and H. Isoda. 2012. Plant Biodiversity in the Semi-arid Zone of Tunisia. Journal of Arid Land Studies 22-1: 83–86.

Ksouri, R., W. Megdiche, A. Debez, H. Falleh, C. Grignon and C. Abdelly. 2007. Salinity effects on polyphenol content and antioxidant activities in leaves of the halophyte *Cakile maritime*. Plant Physiology and Biochemistry 45: 244–249.

Ksouri, R., A. Smaoui, H. Isoda and C. Abdelly. 2012. Utilization of Halophyte Species as New Sources of Bioactive Substances. Journal of Arid Land Studies 22: 41–44.

Laudadio, V., V. Tufarelli, M. Dario, M. Hammadi, M.M. Seddik, G.M. Lacalandra and C. Dario. 2008. A survey of chemical and nutritional characteristics of halophytes plants used by camels in Southern Tunisia. Trop. Amin. Health Prod. 41: 209–215.

Le Houérou, H.N. 1991. Feeding shrubs to sheep in the Mediterranean arid zone: Intake, performance and feed value. In Congress International des Terre's de Parkour's, Montpellier, France, pp. 623–628.

Le Houérou, H.N. 1992. The role of saltbushes (*Atriplex* spp.) in arid land rehabilitation. In the Mediterranean basin: A review. Agrofor. Syst. 18: 107–148.

Le Houérou, H.N. 1994. Forage halophytes and salt-tolerant fodder crops in the Mediterranean Basin. pp. 123–137. *In*: V.R. Squires and A.T. Ayoub (eds.). Halophytes as a Resource for Livestock and for Rehabilitation of Degraded Lands. Kluwer Academic Publishers.

Le Houérou, H.N. and R. Pontanier. 1987. Les plantations sylvo-pastorales dans la zone aride de la Tunisie. Notes Techniques du MAB 18: 11–81.

Leinmüller, E., H. Steingass and K.H. Menke. 1991. Tannins in ruminant feedstuffs. Anim. Res. Dev. 33: 9–62.

Macias, M.L., I.S. Rojas, R. Mata and B. Lotina-Hennsen. 1999. Effect of selected coumarins on spinach chloroplast photosynthesis. J. Agr. Food Chem. 47: 2137–2140.

Masters, D.G. (in press). Assessing the feeding value of halophytes. pp. 89–105, this volume.

Masters, D.G., S.E. Benes and H.C. Norman. 2007. Biosaline agriculture for forage and livestock production. Agriculture, Ecosystems & Environment 119: 234–248.

Masters, D.G., H.C. Norman and E.G. Barrett-Lennard. 2005a. Agricultural systems for saline soil: the potential role of livestock. Asian Aust. J. Anim. Sci. 18: 296–300.

Masters, D.G., A.J. Rintoul, R.A. Dynes, K.L. Pearce and H.C. Norman. 2005b. Feed intake and production in sheep fed diets high in sodium and potassium. Aust. J. Agric. Res. 56: 427–434.

McDonald, P., R.A. Edwards, J.F.D. Greenhalgh and C.A. Morgan. 2002. Animal Nutrition, 6th ed. Pearson Education Limited, Essex UK.

Menke, K.H. and H. Steingass. 1988. Estimation of the energy feed value obtained from chemical analysis and *in vitro* gas production using rumen fluid. A. Res. and Develop 28: 7–55.

Mirchi, A.R.1., M. Yousef Elahi, M. Palizdar and M.R. Pourelmi. 2014. Determination of nutritive value of six species of halophyte plants used by camel in East South Iran. Journal of Novel Applied Sciences 3–7: 729–733.

Moujahed, N., C. Kayouli, A. Thewis, Y. Beckers and S. Rezgui. 2000. Effects of multinutrient blocks and polyethylene glycol 4000 supplies on intake and digestion by sheep fed *Acacia cyanophylla* Lindl. foliage-based diets. Animal Feed Science & Technology 88: 219–238.

Moujahed, N., H. Ben Salem and C. Kayouli. 2005. Effects of frequency of polyethylene glycol and protein supplementation on intake and digestion of *Acacia cyanophylla* Lindl. foliage fed to sheep and goats. Small Rumin. Res. 56: 65–73.

Moujahed, N. and C. Kayouli. 2005. Supplementation of *Acacia cyanophylla* Lindl. foliage-based diets with feed blocks and PEG 4000 and its effects on *in vitro* fermentation and performance in sheep. pp. 333–338. *In*: E. Molina Alcaide, H. Ben Salem, K. Biala and P. Morand-Fehr (eds.). Sustainable Grazing, Nutritional Utilization and Quality of Sheep and Goat Products. Zaragoza: CIHEAM, 2005. (Options Méditerranéennes : Série A. Séminaires Méditerranéens; n. 67).

Moujahed, N., Y. Bouaziz, Ch. Ben Mostfa, M. Ben Romdhane and C. Darej. 2011. Evaluation of the main lamb feeding system in the Central regions of Tunisia. Eighth International Symposium on the Nutrition of Herbivores (ISNH8). Aberystwyth, Wales UK, 6–9.

Moujahed, N., C. Darej, M. Taghouti, Y. Bouaziz, C. Ben Mustapha and C. Kayouli. 2013. Chemical composition and *in vitro* fermentation characteristics of range species growing in Central Tunisia. Options Méditerranéennes : Série A. Séminaires Méditerranéens; n. 107.

Nakata, P.A. and M.M. McConn. 2007. Calcium oxalate content affects the nutritional availability of calcium from Medicago truncatula leaves. Plant Science. 172: 958–961.

Nefzaoui, A. 2002. Rangeland management options and individual and community strategies of agro-pastoralists in central and Southern Tunisia. *In*: T. Ngaido, N. McCarthy and M. Di Gregorio (eds.). International Conference on Policy and Institutional Options for the Management of Rangelands in Dry Areas. Workshop summary paper. CAPRi Working Paper No. 23, Tunisia.

Nefzaoui, A. and A. Chermiti. 1991. Place et rôle des arbustes fourragers dans les parcours des zones arides et semi-arides de la Tunisie. pp. 119–125. *In*: J.-L. Tisserand and X. Alibés (eds.). Fourrages et sous-produits Méditerranéens. Zaragoza : CIHEAM, 1991 (Options Méditerranéennes: Série A. Séminaires Méditerranéens; n. 16). Fourrages et Sous-Produits Méditerranéens, 5–6 Jul 1990, Montpellier (France).

Nefzaoui, A., H. Ben Salem, H. Abdouli and H. Ferchichi. 1995. Palatability for goat of some Mediterranean shrubs. Comparison between animal browsing time and cafeteria technique. Ann. Zootech. 44(Suppl.): 117.

Norman, H.C., R.A. Dynes and D.G. Masters. 2002. Nutritive value of plants growing on saline land. 8th 24 National 25 Conference and Workshop on the Productive Use and Rehabilitation of Saline Lands (PUR$L), Perth, Australia, pp. 59–69.

Norman, H.C., C. Friend, D.G. Masters, A.J. Rintoul, R.A. Dynes and I.H. Williams. 2004. Variation within and between two saltbush species in plant composition and subsequent selection by sheep. Australian Journal of Agricultural Research 55: 999–1007.

Norman, H.C., D.G. Masters, M.G. Wilmot and A.J. Rintoul. 2008. Effect of supplementation with grain, hay or straw on the performance of weaner Merino sheep grazing old man (*Atriplex nummularia*) or river (*Atriplex amnicola*) saltbush. Grass Forage Science 63: 179–192.

Norman, H.C., D.G. Masters and E.G. Barrett-Lennard. 2013. Halophytes as forages in saline landscapes: Interactions between plant genotype and environment change their feeding value to ruminants. Environmental and Experimental Botany 92: 96–109.

NRC. 2005. National Research Council, Mineral Tolerance of Animals. The National Academies Press, Washington.

Ørskov, E.R. and Y. McDonald. 1979. The estimation of protein degradability in the rumen from determining the digestibility of feeds in the rumen. Journal Agricultural Science, Cambridge 92: 499–503.

Personius, T.L., C.L. Wambolt, J.R. Stephens and R.G. Kelsey. 1987. Crude terpenoid influence on mule deer preference for sagebrush. J. Range Manage. 40(1): 84–88.

Provenza, F.D. 1996. Acquired aversions as the basis for varied diets of ruminants foraging on rangelands. J. Anim. Sci. 74: 2010–2020.

Saadani, Y., C. Kayouli and H. Narjisse. 1989. Valeur nutritive d'un parcours mixte à *Acacia cyanophylla*, *Atriplex nummularia* et *Medicago arborea*. In Congrès International des Herbages, Nice, France, pp. 943–944.

SCA. 2007. Standing Committee on Agriculture's Nutrient Requirements of Domes-ticated Ruminants. CSIRO Publications, Melbourne, Australia.

Slama, I., C. Abdelly, A. Bouchereau, T. Flowers and A. Savoure. 2015. Diversity, distribution and roles of osmoprotective compounds accumulated in halophytes under abiotic stress. Annals of Botany: 1–1.

Squires, V.R. and A.T. Ayoub. 1994. Halophytes as a Resource for Livestock and for Rehabilitation of Degraded Lands. Kluwer Academic Publisher, Dordrecht Boston, London, 315 p.

Underwood, E.J. and N.F. Suttle. 1999. The Mineral Nutrition of Livestock, 3rd ed. CAB International, Wallingford.

Van Soest, P.J. and L.H.P. Jones. 1968. Effect of silica in forages upon digestibility. J. Dairy Sci. 51: 1644–1648.

Halophytes and Salt Tolerant Crops as a Forage Source for Livestock in South America

R.E. Brevedan,* O.A. Fernández, M. Fioretti, S. Baioni, C.A. Busso and H. Laborde

SYNOPSIS

This chapter is a broad overview of halophytic and salt tolerant feedstuffs (mainly forages from the extensive rangelands including alpine and semi desert areas) that support the pastoral livestock enterprise. Attention is focused on the productivity (as reflected in edible biomass and seasonal availability) and nutritive value of the various forages and browse. The value of the plant base to livestock is clear but the rich and varied fauna ranging from rodents to camelids and antelope-like animals must be acknowledged. The region varies from tropical north (near the Equator in Brazil) to the arid regions of southern Argentina and Chile and the high Andes.

Keywords: Andes, wildlife, rodents, cattle, rangelands, Argentina, Bolivia, Chile, Brazil, Peru, salinized, land degradation, Patagonia, bofedales, Peruvian altiplano, sylvopastoral systems.

1. Introduction

Accelerated population growth during the second half of the 20th century drove the demand for a greater food increase. At the same time, it was accompanied by

Universidad Nacional del Sur, Bahía Blanca, Argentina.
Email: hlaborde@criba.edu.ar
* Corresponding author: ebreveda@criba.edu.ar

agricultural practices that determined soil impoverishment and growth of agricultural crops in more marginal zones. At these areas, soils have low fertility or are either contaminated or salinized. In addition, advancement of agriculture towards areas with lower precipitation has determined employment of technological tools such as irrigation and fertilization. In some cases, because of their mismanagement, they are causing land degradation and even total loss of the edaphic resource.

The quality of 75% of agricultural lands has been stable since the mid 20th century (Scherr 1979). However, the remaining 25% has suffered an accelerated deterioration that ranges between 5 to 12 million ha per year. Climatic conditions since the 1990s has aggravated the problem; as a result, it was recently stated that 3 ha of soil arable are lost per minute (Anon 2006). Erosion, loss of soil organic matter and fertility, hard pans, contamination and excess salt have been the causes of the edaphic degradation. This is because productive soil surface areas have not only diminished but also they have reduced in quality.

Halophytes and salt-tolerant-plants are able to successfully develop in extremely-saline and semi-desert environments. Species that can growth in these habitats would be an essential resource in the future, and knowledge of their strategies to successively grow and develop in marginal lands will be of great value for the nutrition of animals which live in those environments (El Shaer 2010). Nutritive value of halophytes and salt-tolerant species is of relevance for ruminants or other animals which live in those environments. It is of relevance because they often constitute the only forage resource, and their importance will grow as the area of salt-affected soils increases.

2. Argentina, Chile, Perú, Bolivia, and Brazil

Argentina ranks third in the world (after Russia and Australia) in halomorphism processes (salinization, alkalinization and sodicity) according to FAO-UNESCO. These processes affect arid, semiarid and even humid areas, with and without irrigation, which, in many cases, overlap with Argentina's productive areas. Natural-saline areas, and those affected due to anthropogenic activities, currently affect an average of 30% of irrigated lands in Argentina (Lavado 2008).

2.1 Argentina

Forage quality of halophytes in Argentina has been considered by Brevedan et al. (1994).

Salt-affected areas in the humid region have soils that in general are neither saline nor totally alkaline because some of them have shallow horizons which are either neutral or even acidic. These soils have serious problems that arise when attempts are made to use them for agricultural purposes; as a result, they are utilized mostly for cattle raising (Lavado 2008). Such lands constitute the (1) Pampa Deprimida (Province of Buenos Aires) and (2) Bajos Submeridionales (Provinces of Santa Fe and Chaco).

There are 32 million hectares affected by excessive water and salts in the humid and sub-humid environments of the Pampean region (Maddaloni 1986; Cisneros et al. 2008). An important proportion of soils have not been formed under hydro-halomorphic conditions; however, they have been under alternative processes of excessive water content and salts. This has constrained their productive capacity and utilization possibilities (Cisneros et al. 2008). These soils are difficult for using in agriculture; even more, they are exposed to waterlogging. As a result, they are mostly used for livestock grazing. Even though these soils have low salinity, their salt content increases at times during the year because of capillary action. In addition, these soils have physical constraints because of their sodicity levels and eventual anaerobic problems caused by excess, undrained water (Lavado 2008).

The agricultural expansion in the early decades of the 21st century in Argentina, mostly because of the soybean crop, determined the location of livestock production areas in less favored zones. Because of this, livestock production (mostly cattle raising on rangelands) has not often reached previous production levels although there was a 19% increase in the cattle stock between 1998 and 2002 (Pensiero 2009).

The Pampa Deprimida is the most important zone for cattle raising in Argentina. Its cattle stock is 62% to a national scale. Between 25 to 30% of the area used for cattle raising are halophytic rangelands with low forage production; their soils have drainage deficiencies and important alkalinity levels. They are called "bajos salinos" where *Distichlis spicata* and *Distichlis scoparia* are dominant with a low productivity: from 500 to 2000 kg DM ha^{-1} yr^{-1} (Imaz et al. 2009; Garciandía 2010). Soil characteristics do not favor planting of temperate forage grasses from a productive standpoint—planting of megathermic pastures, the more tolerant halomorphic soils is favored instead. The introduction of *Chloris gayana* and *Panicum coloratum* were evaluated as an alternative. In the halophytic steppe the above ground net primary productivity increased by 90 and 117%, respectively, after sowing of these species. However, and even though its nutritional quality had not increased overall, rangeland plant cover doubled after the introduction of *Chloris gayana* and *Panicum coloratum*. The soil organic matter content and infiltration rate increased and the soil under these species also showed a slight pH reduction on the shallower layers which would not affect the soil sodic characteristics. As a result, *C. gayana* and *P. coloratum* could be considered as promising alternatives to improve halomorphic environments (Otondo 2011).

The species with most potential for use in the Pampa Deprimida are: *Thinopyrum ponticum, A. scabrifolium, Lotus tennuis, Melilotus albus, M. officinalis,* and *Chloris gayana.* Collection and characterization programs for native and naturalized forages of Argentina were developed by research groups of the Universidad Nacional del Litoral and the Facultad de Agronomía de la Universidad de Buenos Aires. The species selected for saline environments were: *Elymus scabrifolius, Trichloris crinita, Trichloris pluriflora, Sporobolus indicus, Sporobolus phleoides, Pappophorum caespitosum, Pappophorum philippianum, Thinopyrum ponticum* and *Melilotus albus* (Zabala et al. 2009). Research demonstrated that salinity tolerance exists in *Elymus scabrifolius,* and that the physiological basis for this tolerance is associated with sodium exclusion from the leaf (Zabala et al. 2009).

The Province of Santa Fe has approximately 44% of its surface area with halo-hydromorphic soils (Mosconi et al. 1981; Espino et al. 1983). The best adapted species in the Bajos Submeridionales (28–30° S and 60–61° 30'W) are: *Agropyron scabrifolium, Melilotus albus, Chloris gayana, Agropyron elongatum,* and *Panicum coloratum* var. makarikariense (Maddaloni 1986).

Spartina argentinensis is an unpalatable species to domestic livestock, which occupies extensive areas in the Bajos Submeridionales. These prolongate into gullies towards the center-south of the Province of Santa Fe, in the periphery of the Laguna Mar Chiquita, and into large drainage channels in the center-south of the Province of the Chaco. These lowland areas, where cattle production on rangeland vegetation is the main economic activity, have soils with a high salt and sodium concentration, pH greater than 8 and drainage problems. Burning is a common practice since regrowth has lower lignin concentrations, which increases palatability; digestibility and crude protein values (Feldman 2003).

Even though *Spartina argentinensis* is a strongly dominant species, it coexists with various companion species (Lewis et al. 1990) that can be considered promising forages: *Bothricloa lagonoides, B. saccharoides, Chloris canterae, C. ciliata, C. halophila, Diplacne uninerva, Leptochloa cloridiformis* and *Paspalum* species (Echeverría et al. 2001).

The lack of forage availability during the dry season in most parts of the semiarid west of Argentina is a severe constraint to cattle raising, the unique viable productive activity in the region (Morello and Saravia Toledo 1959; Abril et al. 2000). Productivity of cattle raising and rangelands is low. Rangelands are dominated by unpalatable shrubs most of the times (e.g., *Larrea divaricata, Junellia seriphioides*, etc.); palatable perennial grasses are rare, and they only grow under the protection of spiny shrubs such as *Prosopis* spp., *Condalia microphylla* and *Bougainvillea spinosa* (Guevara et al. 1997).

This is the result of poor management: overgrazing, permanent overstocking, uncontrolled fires, and in most cases the absence of fencing and scarcity of watering points on these extensive rangelands (Guevara et al. 1981). Any plant species that can increase forage availability during the dry season would allow a significant increase of productivity in areas chronically affected by rural poverty.

Cattle diet in the rainy season can include about 10% browse, 84% grasses and 6% forbs while in the dry season these proportions average 49, 45 and 6%, respectively (Guevara et al. 1996). The crude protein content of the grasses is 10–12% in the green stage, but in the mature stage it seldom drops below 6% (Wainstein and González 1963, 1969, 1971a,b; Wainstein et al. 1979). As it is usually the case, there isn't a direct relationship between the feed value determined by chemical analysis and the forage selection conducted by grazing animals (Guevara et al. 1997).

Diets of domestic (goats, horses, cattle) and wild (European hare, *Lepus europaeus*) herbivores were compared in Mendoza, under poor and good range conditions in an area of 10,000 ha during summer and winter (Guevara et al. 2010). Grasses were the main component of the diet of cattle in summer; and of horses during both seasons under good rangeland conditions. Under these conditions, all herbivores consumed a higher proportion of grasses in summer than in winter. Shrubs were the major component of the diet in both seasons and range conditions in goats and the European hare, and during winter in cattle. During winter, the greatest animal competition for foods was between horses and cattle under good rangeland conditions, and between goats and the European hare under poor conditions. Interspecific competition for food was low (Guevara et al. 2010).

Halophytic shrubs of the Chenopodiaceae family are very well adapted to saline areas in arid and semiarid environments. *Atriplex* spp., *Salicornia* spp., *Suaeda* spp. and *Kochia* spp. are both drought and salt tolerant (Masters et al. 2007). Standing edible biomass will depend upon range management.

Species of the Chenopodiaceae family are more tolerant to salinity than halophytic grasses and legumes most of the times. Many species of that family have an optimum growth with NaCl concentrations between 25 and 200 mM. Halophytic grasses, on the other hand, show an optimum growth with less than 100 mM NaCl, and diminished growth under any increase of salinity (Norman et al. 2013).

Rangeland grasses are the main food source for cattle grazing in the Monte. Precipitation regimes determine two periods for warm-season grass growth which are clearly different: one of these periods is of active growth (November to March: mid-spring to early fall) and the other period is of dormancy (April to October: mid-fall to early spring). The situation is *vice versa* for cool-season grasses: active growth from late February to mid-December (late summer to late spring), and dormancy during mid-December to late February (late spring to late summer: Giorgetti et al. 2000).

After cutting of a selected forage of the warm-season perennial grass, *Trichloris crinite,* to the same height, a study conducted monthly cuttings during a year. At each sampling date, the proportion of edible biomass with respect to total leaves (i.e., green + dry) was estimated. This proportion ranged between 90 (October) and 0% (September). Leaf crude protein varied from 20 to 9%, and neutral detergent fiber ranged from 64 to 68% during October and September, respectively (Avila et al. 2008).

Guevara et al. (1991) examined whether native rangeland grasses of the Mendoza plain could meet the protein and energy requirements of beef cows. They found that protein and energy levels of the main native grasses would allow a cow-calf operation without nutritional constraints, providing that calving should occur in November or December, and forage availability was great enough for selective grazing. Supplementation of the available forage, however, would be needed in some months. Major native grasses include the halophytic *Panicum urvilleanum, Aristida mendocina,* and *Pappophorum philippianum* (Guevara et al. 1991).

Mixed pastures were seeded in Mendoza: *Thinopyron elongatum* (10 kg ha^{-1}), *Melilotus albus* (5 kg ha^{-1}) and *Melilotus officinalis* (6 kg ha^{-1}). Their lowest dry matter production was greater than 4500 kg DM ha^{-1} yr^{-1}. This forage yield is much greater than the forage offer provided by rangeland vegetation. In addition, forage quality of the mixed pastures was much greater than that of the rangeland vegetation (Ochoa 1994).

Crude protein values might be overestimated in halophytes because of the presence of either nitrates or soluble, non-protein compounds that can be synthesis by osmoregulation (e.g., proline, glycinebetaine: Le Houérou 1992; Flowers and Colmer 2008). See also Masters, this volume.

Halophytic grasses differ from the Chenopodiaceae in the mechanism of osmotic adjustment. Tolerance to salinity in grasses is mostly based on maintaining a low leaf salt concentration. This is achieved by either excluding Na$^+$ and Cl$^-$ from the root surface area or secretion of these ions from the leaves. On the other hand, salinity tolerance in the Chenopodiaceae is based on a greater Na$^+$ and Cl$^-$ uptake, and the subsequent compartmentation of these ions in vacuoles, where they have a major role in osmotic adjustment (Flowers and Colmer 2008).

The lack of forage during the dry season is a major factor limiting livestock production in most of the semiarid western Argentina, where that is the only viable productive activity in the region (Morello and Saravia Toledo 1959; Abril et al. 2000). Therefore, any plant species like *Atriplex,* that can increase forage availability during the dry season, might allow a significant increase in productivity in areas chronically affected by rural poverty. Additionally, this shrub species has been recommended for the rehabilitation of eroded areas (Le Houérou et al. 1982). Introduction of *Atriplex* spp. might therefore be an appropriate alternative for rangeland rehabilitation in these eroded, degraded and salinized areas.

Prosopis is another genus of halophytes that includes about 45 species. The great morphological diversity of South American species of *Prosopis* suggests that the main center of irradiation is located in the Argentinean-Paraguayan Chaco (Burkart 1976). This region occupies the warm plains of Northern Argentina and Paraguay. The expansion of the genus towards colder and drier zones such as the Monte, Patagonia and Prepuna implies the acquisition and/or adjustment of morphological and physiological traits that allow plants to either avoid or tolerate the resource limitation and harsh environmental conditions of these regions (Villagra et al. 2010).

Cavagnaro and Passera (1993) found that *Prosopis flexuosa* showed greater water potentials than those determined on the C_4 *Atriplex lampa* during a whole vegetative cycle. *Prosopis* selections grew well on either nitrogen free media equivalent to 50% seawater (Felker et al. 1981) or nitrogen fixed media with xylem water potentials of −3.3 MPa (Felker and Clark 1982). Fruits of *P. chilensis* have 7–11% crude protein. Its young branches and leaves with a 13.5–18% crude protein are very palatable to livestock and wild herbivores.

Native plants of the arid west in Argentina have physiological characteristics which allow them to resist extreme arid conditions. The C_3 *Prosopis* spp. and the C_4 *Atriplex* spp. have particular characteristics that ensure their reproductive success and subsequent persistence under conditions of extreme aridity (Passera et al. 2010). These species are able to establish under conditions of extreme water stress which explain their presence mainly in arid and semiarid regions of the world. In addition, *Prosopis* spp. is able to compete by using superficial water or by the use phreatic water, even that coming from deep layers (Passera et al. 2010).

At the lowlands of valleys in Patagonia there are humid areas called "mallines". This is where the greatest stocking rate is concentrated. Total cover can diminish because of overgrazing and animal trampling. As a result of early erosion processes (on uncovered soil spaces), deterioration of the "mallines" is speeded up. Overgrazing of "mallines" has determined serious levels of salinization and eolic and hydric erosion. As a result, semideserts and extended zones of erosion have been generated (Movia 1972). The salty lowlands are characterized by the presence of saltgrass (*Distichlis* spp.), which can dominate the plant community because of the combined effects of overgrazing, aridity and salinity (Fiorio 2004). These are the edaphic communities defined as steppes of saltgrass (*D. scoparia* and *D. spicata*) which are in zones with a mean annual temperature lower than 13°C, and annual precipitation lower than 300 mm (Cabrera 1971).

Distichlis spp. is a short, very rhizomatous, grass. *Distichlis spicata* has a deep root system (Ansin et al. 1998) which helps determining its capacity to regulate the

(1) water stress because of summer drought, and (2) spring waterlogging (Sala et al. 1981). These plant traits constitute an advantage over *D. scoparia* (Gandullo 2004; Brizuela et al. 1990).

Distichlis spicata can have between 3 to 7% crude protein, and digestible dry matter is about 40% but as Masters (this volume) argues, the true value of many halophytes may be over estimated. Despite this, halophytes can contribute approximately 20% (up to 40% sometimes) of sheep and goat diets. Under good range conditions, the "mallines" can yield as much as 2000 kg ha^{-1} yr^{-1}. *Distichlis scoparia* is present in numerous communities. This suggests that the saline environments where it grows have a fluctuating source of sweet and phreatic water so that salinization and sodification are less serious (Conticello et al. 2002).

There are three species of saltgrass in Patagonia (i.e., *D. spicata*, *D. scoparia* and *D. australis*). These species can be found in saline lowlands which are mainly grazed by sheep. This is within a cold climate and under extreme conditions of aridity and salinity. Saltgrass is the unique C$_4$ species in the region.

Somlo et al. (1985) reported that consumption of saltgrass by sheep is low in Patagonia. However, Pelliza et al. (1997) found up to 30% saltgrass in sheep's diet. Saltgrass is an important part of cattle's diet during summer and winter in the Patagonia (Somlo et al. 1985) and in the Pampa Deprimida (Brizuela et al. 1990). It has also been found in the diet of horses and goats in Patagonia (Bonvissuto 2004), and of american camelids in the Puna (Aguilar and Newman 2005) and Patagonia (Pelliza et al. 2005). Pelliza et al. (2005) demonstrated its forage importance for sheep. Even though it was part of sheep's diet with rather low nutritional value, it represented more than 20% of the diet in some of the study cases. Utilization of saltgrass (*D. spicata* and *D. scoparia*) by cattle in the Pampa Deprimida comprised from 8 to 24% of the dry weight during summer months, when its quality is from medium to relatively low in the native communities developed on saline and alkaline soils (Brizuela et al. 1990). As a pioneer plant species, saltgrass can be useful in the initial steps of salinized farmland reclamation. Because of its poor ability to compete with other plant species when salinity diminishes, saltgrass can be easily replaced with other, more valuable crop species.

Saltmarshes are characterized by the abundant presence of terrestrial plants, although the sea covers them during several hours per day. Plants that can tolerate those extreme conditions are within the genus *Spartina* ("espartillos") and *Sarcocornia* (shrubs; "jumes") (Bortolus et al. 2009; Isacch et al. 2006). The "espartillos" are in the lower areas, while a more diverse amount of species with time occupies the upper areas at the same time that they get away from the coast towards more terrestrial conditions (Bortolus et al. 2009).

Along the Atlantic coast of South America, the northern saltmarshes (latitudes lower than 43° S) are dominated by *Spartina alterniflora* and *S. densiflora*, while the southern saltmarshes (latitudes greater than 43° S) are dominated by *Sarcocornia perennis* (Idaszkin and Bortolus 2010). The diverse plant species distribute in areas parallel to the coast as a result of differences in tolerance to immersion and salt concentration.

2.2 Chile

The total surface area of Chile is 75,700,000 ha. From this total surface, only 6.6% is available for tillage, and of that only 2.5% is cultivated land. The temperate, arid and semiarid region of Chile used for livestock production is greater than 6,000,000 ha. This region is between 29 and 34° of latitude south.

The main sylvopastoral systems in arid and semiarid zones of northern Chile, from north to south, are plantations with *Polylepis besseri* ("queñoa") in the precordillera (3000 to 4000 m.a.s.l.), and with *P. tarapacana* ("queñoa de altura") in the High Plateau (over 4000 m.a.s.l.). Tamarugo (*Prosopis tamarugo*) and algarrobo (*P. chilensis*) plantations are distributed in the Pampa del Tamarugal, in an extremely arid zone (2 mm yr^{-1} or less). In this zone, 20,500 ha have been planted with *Prosopis*, 88% of them with *P. tamarugo*. These plantations feed a variable herd of 7,000 to 9,000 sheep and goats per year. Production per tree varies from 20 to 70 kg yr^{-1} for 14- to 22-yr-old trees planted in densities of 100 trees ha^{-1} (Ormazábal 1991).

In the semiarid zone of the IV Region of Chile (Coquimbo), livestock production is the main economic activity. This area is characterized by an irregular amount and frequency of rainfall, and a lack of plant cover during many months. The result is that the main forage resource is constituted by annual herbaceous species, which makes a non-sustainable ecological system. In these lands, cover of trees and shrubs has been partially removed, and the desertification process has already started (Mulas and Mulas 2004).

The productive management of most of the farms consists on grazing livestock on each farm during winter and spring. During the dry period, animals are moved towards the cordillera or southern areas in the region (Meneses and Squella 1996). Since 1960, institutions such as the Corporación de Fomento de la Producción (CORFO), the University of Chile, the Servicio Agrícola y Ganadero (SAG), and the Instituto Nacional de Investigaciones Agropecuarias (INIA) have been working together to cope with this difficult situation. The region of Coquimbo is either dominated or it was in the past by abundant, unpalatable, shrubby vegetation. This suggested the possibility of introducing palatable species with similar morphophysiological characteristics to provide forage for livestock grazing. The main objective was to find species which were able to increase forage availability in arid areas affected by desertification. More than 48,000 hectares of Coquimbo are currently planted with forage shrubs of which *Atriplex nummularia* has been the most used species (more than 90% of the area), followed by *A. repanda* (Lailhacar 2000).

In the semiarid IV Region, *A. nummularia* is currently the most important species in reforestation and desertification control projects (Mulas and Mulas 2004). It produces from 10 to 20 kg DM ha^{-1} yr^{-1} per mm of rain (Le Houérou 1992). Its leaves and roots have responded positively to increasing NaCl concentrations up to 300 nM NaCl. Also, it is tolerant of grazing, has a fast regrowth, is resistant to diseases and propagates easily.

In some areas of Southern Coquimbo, where rainfall ranges from 100 to 220 mm yr^{-1}, yields of *A. nummularia* plantations have varied from 50 to 900 kg of dry matter ha^{-1} yr^{-1}, depending on age, field management and plant density. Average yields of 1,806 g plant^{-1} have been observed in areas with 143 mm yr^{-1} of rainfall (Soto 1996).

Planting density directly affects yields. For example, yields increased when *A. nummularia* planting densities increased from 625 to 10,000 plants ha^{-1}; however, leaf production decreased at planting densities over 2,500 plants ha^{-1} (Soto 1996).

Many trials have been conducted on the nutritive value of different species of *Atriplex* (Silva and Pereira 1976). In particular, the amino acid composition of *Atriplex* is in general well balanced, except for its low methionine content (Padilla 1986). *Atriplex repanda, A. undulata* and *A. clivicola* are shrubs palatable to livestock (Lailhacar 2000), and can markedly increase the productivity of lands previously comprised of wild herbaceous species (Concha 1975; Olivares and Gastò 1981).

In general, *A. nummularia* has similar crude protein content to the annual vegetation in spring, but that content is significantly higher than that in any other vegetation during the rest of the year (Benjamin 1980). Digestibility of its edible components, such as leaves and growing tips, can be as high as 70% in the spring, and as low as 40% during fall and winter. The poor expression of its high crude protein content on animal performance appears to be the ruminal degradation of amino acids to ammonia, which escapes absorption in the rumen (Norman et al., this volume). The palatability of most of the shrubs is low when green annual species are available. The high sodium content in the ash of the bush species increases the water requirements of animals which graze on those forage bushes.

Plantation with forage species allows the possibility of having green forage during the period of shortage (i.e., summer and fall) of rangeland vegetation, particularly during the prolonged droughts. It eases pressure to migrate to the Andes during summer. Reforestation (including agro-forestry) has diminished to date because the government stopped the planting of forest species. In addition, reforestation is less than expected from the private sector because of the low yields which are the result of mismanagement or natural causes (Ormazábal 1991).

Planting of herbaceous perennial species (i.e., mostly grasses) which had previously been eliminated because of overgrazing is critical. This is because forage shrubs need at least two years for establishment before they can be browsed. Forage shrubs must be managed appropriately to increase their positive effect on the herbaceous stratum. An excessive shrub diameter can limit the herbaceous production. This is because the shrub will physically interfere with the herbaceous plant development and it will reduce the incident radiation. This is especially true on shrubs (e.g., *A. atacamensis* and *A. deserticola*) where basal branches either touch or grow proximate to the soil surface. Production of an excessive amount of shrub litter should also be prevented. It can be a good source of organic matter and soil nutrients. However, seed germination of the herbaceous species could be favored by the first rains before the soil becomes wet enough as to allow penetration of the radicle. Substances which reduce the germinative process could also be provided from the shrub litter.

Most of the introduced shrubs evolved in the absence of a continuous grazing pressure. The absence of spines, the relatively free access of grazing animals to remove leaf primordia, and the inability of some bushes to produce viable, lateral buds make these species vulnerable to grazing animals. However, the low palatability of some of them offsets their low resistance to grazing. Nevertheless, that low palatability does not contribute to grazing resistance when alternative food sources are absent (Benjamin 1980).

Fruit yield and quality, seed viability and seed germination were compared among nine populations of *Atriplex repanda* in the arid region of Chile. At the population level, fruit yield was positively correlated with seed size, viability and germination. Seed viability was also positively correlated with seed size.

A recommendation on the use of *A. nummularia* is to associate this forage with *Opuntia ficus-indica* cladodes. The inclusion of young cladodes of cactus pear in the diet has increased milk production in goats (Azócar and Rojo 1992). This strong effect could be mainly attributed to the high water content of *Opuntia* cladodes which counterbalances the high salt content of *Atriplex* leaves.

Atriplex atacamensis and *A. halimus* did not show either visual symptoms of toxicity or reduced growth when growing in soils with high levels of available arsenic. Chlorophyll (and thiol) levels were not negatively affected in either species. Both species had the highest arsenic levels in roots. *Atriplex halimus* accumulated higher concentrations of As and B than *A. atacamensis* in the leaves. These plants resist contamination by arsenic, and can be used to generate plant cover in As-contaminated saline soils, thus preventing the dispersion of this metalloid (Tapia et al. 2013).

In Chile, yields of cladodes of *Opuntia ficus-indica* have ranged from 13 t DM ha^{-1} yr^{-1} in crops that only covered 30% of the land, to 40 t DM ha^{-1} yr^{-1} at higher plant densities, optimum watering and good fertilization. In non-irrigated lands of the central zone of Chile, yields of 8 t ha^{-1} yr^{-1} were obtained (Azócar 2001).

The effects of different levels of alfalfa hay and prickly pear cladodes (*Opuntia ficus-indica*) supplementation were determined on live weights and milk yields of goats. There were five treatments: (1) Control on rangeland; (2) rangeland and alfalfa hay *ad libitum*; (3) rangeland, and 84% alfalfa hay and 16% cladodes; (4) rangeland, and 79% alfalfa hay and 21% cladodes, and (5) rangeland, and 66% alfalfa hay and 34% cladodes (Azócar 2001). Results proved that the tested supplement may increase milk production, with an increment of 55% when only hay was provided. Hay replacement with 16, 21 and 34% cladodes favored even greater yields: 94, 104 and 125%, respectively. This was attributed to the high water content in the cactus tissues. For this reason the utilization of prickly pear cladodes as a summer supplement is of great importance to increase milk production in local goat herds.

2.3 Perú and Bolivia

Halophytic vegetation of the Andes and Perú has been reported by several authors (Brako and Zarucchi 1993; Galán de Mera et al. 2009, 2012; Montesinos Tubée 2012). There are communities of *Distichlis spicata* in the Peruvian shore and mountain ranges, at elevations up to 2500 m height. Presence of *D. spicata* is the most representative to identify halophytic communities. *Distichlis humilis* is frequent at heights greater than 3000 m. Communities of *D. humilis* and *Salicornia cuscoensis* have been reported in the Andean region of Cuzco (Müller and Gutte 1985). Halophytic communities have been described for the shore regions of Perú, where *Distichlis spicata*, *Sarcocornia neei* and other species of the Peruvian coast are dominant (Galán de Mera et al. 2009). *Distichlis spicata*, *Suaeda foliosa*, *Atriplex myriophylla*, and *Schoenoplectus americanus* were the most representative species in the three ecosystems of the Andes of southern Perú which had been exposed to either grazing or low-frequency fires or on abandoned fields (Montesinos Tubée 2012). Communities of *Muhlenbergia fastigata, Distichlis humilis, Atriplex nitrophiloides* and *Salicornia pulvinata* have been reported in the Andes of Bolivia under the presence of salty soils (Müller and Gutte 1985; Navarro 1993).

2.4 Brazil

Seeds of *P. juliflora* from Perú were introduced into Brazil in 1942 in Pernambuco. Thereafter, plantings extended to the states of the northeast region which are between the latitudes 1° and 18°30' S, and the longitudes 30°30' and 48°20' W, representing approximately 18.2% of the total surface area of the country. The semiarid areas occupy 75% of the region and 13% of Brazil. Soils are rather shallow, with rocky presence at the soil surface level, low capacity of soil moisture retention and reduced soil organic matter content. The region is covered by a xerophytic, tree-shrubby vegetation stratum called "caatinga", where legume species are dominant. Many of these species are forages, but they support very low stocking rates.

Prosopis juliflora, a nitrogen-fixing tree was planted to supply soil nutrient deficiencies. *Prosopis* species produce pods which help to supplement animals during dry periods. *Prosopis alpataco* is very useful for reforestation because of its (1) precocity, (2) resistance to drought conditions, (3) wood value, and (4) production of pods with a high palatability and nutritive value. It also has the (5) advantage of fructifying during dry periods (Ribaski 1987). New species of *Prosopis* were introduced in 1983 from three countries: *P. tamarugo, P. chilensis* and *P. alba* of the Pampa del Tamarugo from Chile; *P. velutina* and *P. glandulosa* from the United States, and *P. pallida* from Perú.

Most of the time, the protein level of the prairies and the grasses they support is not enough for either maintaining or gaining weight of the grazing animals during the dry period. Among these grasses, it is *Cenchrus ciliaris* which is well adapted to semiarid conditions. Attempts to grow herbaceous legumes are quite difficult because few species can resist the climatic conditions of northeastern Brazil (Ribaski 1987). Aware of this constraint, farmers planted *P. alpataco* to supplement animal nutrition mostly through its pod production. Also, associations of *Opuntia ficus-indica* with *P. juliflora* were made to shade *Opuntia*.

In Brazil, the most utilized forage halophyte is *Atriplex nummularia* because of its productivity, easiness of propagation and nutritional value. *Atriplex nummularia* showed 5.2, 6.1, 124.7, 19.3 and 149.4 g of Ca, Mg, K, Na y Cl, respectively, per kg leaves (Souza et al. 2011). Salt extraction on leaves and stems in a saline-sodic soil was 644, 758, 1059 and 1182 kg ha^{-1} at various levels of soil water availability: 35, 55, 75 and 95% of field capacity, respectively.

With the aim to reduce negative environmental impacts resulting from desalinization of brackish water in the semiarid Brazil, *A. nummularia* was cultivated during one year with four different volumes of aquaculture effluent (75, 150, 225 and 300 L of water per plant week^{-1}) generated from an intensive tilapia (*Oreochromis* sp.) raising system with a mean salinity of 8.29 dS m^{-1} (Porto et al.

2006). The highest dry matter production per unit of applied water was for the treatment of 75 L week^{-1}, which yielded 4.84 g L^{-1} of effluent. The same treatment removed 13.8% of the total salt incorporated into the soil (Porto et al. 2006). Dry matter yield at the different water levels ranged between 7.5 to 11.4 t ha^{-1} yr^{-1}. About 83 and 17% of it were forage (leaves 51.0%, stems 32%) and woody materials, respectively. The different levels of water had no effect on the chemical composition and the *in vitro* dry matter digestibility (IVDMD). Mean crude protein levels and IVDMD of the leaves were 15.1 and 67.9%, respectively (Barroso et al. 2006).

Moreno (2011) evaluated the performance, nutrient intake and digestibility, and meat quality of Santa Inés lambs fed with 30, 40, 50 and 60% of *A. nummularia* hay, using concentrated products for finishing those lambs. There was a linear reduction in dry matter intake as the level of *A. nummularia* hay increased in the diet. The intake of all digestible nutrients decreased with the hay inclusion, except for the crude protein intake, neutral and acid detergent fibers as well as nutrient digestibility. As consumption of mineral salt was reduced, that of *A. nummularia* increased. This showed a reduction in the daily lamb ingestion needs of commercial, mineral salt. Lamb meat was harder and the intramuscular fat decreased with the addition of *A. nummularia*. However, no differences were detected in sensory analysis. See also Abou El Ezz et al., this volume.

The effects of various levels of *A. nummularia* hay were determined on the intake of dry matter, crude protein and neutral detergent fiber in sheep. Five diets were evaluated using from 38.3 to 83.7% of *A. nummularia* associated with the forages watermelon (*Citrulus lanatus* cv. *citroides*) and shredded cassava root (*Manihot esculenta*) with 5% urea. Live weight gains were linearly reduced as hay levels increased in the diets (Rodrigues Souto et al. 2005).

Nutrient consumption and digestibility of a diet composed by 50% *A. nummularia* and 50% *Opuntia ficus-indica* was evaluated in goats and sheep (Alves et al. 2007). Diet composition was 7.1% crude protein, and 39.9% and 20.1% neutral and acid detergent fibers, respectively. Goats showed greater values than sheep for dry matter, mineral matter, and neutral and acid detergent fibers. Sheep had a greater diet acceptability than goats, but the latter showed a greater use of the diet nutrients. However, the diet of *A. nummularia* and *Opuntia ficus-indica* showed a low level of proteins and energy.

The effect of salinity on the growth and production of young cladode sprouts ("nopalitos") of *Opuntia ficus-indica* was studied. Irrigation levels with electrical conductivities of 2 to 21 dS m^{-1} were used (Murillo-Amador et al. 2001). All cladode variables considered (number of young cladodes, length and width of cladodes, dry weight, harvest index and relative growth rate) decreased with increasing salinity. As a result, these authors concluded that "nopalitos" of that species were sensitive to salt.

Cactus can be found in Northeastern Brazil. The most common, species studied were *Opuntia ficus-indica* and *Nopalea cochenillifera*. These species contributed greatly to sustainability and the socio-economic development of the semiarid region because they are able to survive large periods of low water levels (Santana 1992).

Chemical composition varies according to the species, plant age, timing of the year, and edaphic conditions. It is a food rich in water, minerals and carbohydrates. Those carbohydrates which have no fiber are an easily available energy source for microbial fermentation as long as they have low levels of neutral detergent fiber.

There are about 300 species of *Opuntia*. The most used are 12 species of the *Opuntia* genus and 1 species of *Nopalea* (*N. cochenillifera*) and are cultivated mostly in Pernambuco and Alagoas. There are 550,000 ha with plantations of cacti. Ferreira (2006) showed that they had low values of crude protein (4.81%), neutral detergent fiber (26.8%), acid detergent fiber (18.9%) and total carbohydrates (81.1%).

In Brazil, Santana (1992) reported a range of fresh weight yields from 107 to 205 t ha^{-1} yr^{-1} (approximately 16 to 31 t DM ha^{-1} yr^{-1}) in simulated conditions of high planting density, optimum watering and good fertilization.

Nutritional quality of *Opuntia* as a forage depends on plant species, variety, cladode age, crop management and growing season. It has a high biomass yield, high palatability and good nutritive value, soil adaptability and drought resistance (Le Houérou 1992). It has high water content (85 to 90%), low content of crude protein (3.5%), high ash content (260 g kg DM^{-1}) and neutral detergent fiber (185 g kg DM^{-1}). Crude protein decreases with increasing cladode age (1 to 5 years) from 5 to 3% DM, while fiber increases at the same time (from 9 to 20% DM). *Opuntia* cannot be fed alone due to its poor nitrogen content and needs to be supplemented with a cheaper N source, such as saltbush (*Atriplex* spp.) or non-protein nitrogen from treated straw. Watering animals during summer and drought periods is a serious problem in arid zones. Feeding with cactus cladodes reduce watering needs in dry areas. One of the major economical activities in northeastern Brazil is the production of cattle, sheep and goats. This livestock mostly feed on the native vegetation, with low productivity indexes.

3. Food and habitat use of some small mammals in South American drylands

Much of South America consists of diverse arid and semiarid regions characterized by high mammal endemism as a result of a complex interplay between place and lineage histories. Subsequently, we summarize information on food and habitat use of some small mammals in South America drylands. In the evaluation of food quality consumed by small herbivores, ingestion and digestibility appear as the most important factors. Food preference and digestibility decrease as the dietary fiber content increases (Bozinovic 1995).

The red vizcacha rat (*Tympanoctomys barrerae*) lives on halophytic habitats surrounding salt flats in the Monte desert of west-central Argentina. It is the first tetraploid mammal (Gallardo et al. 1999). The species has a reduced habitat and there is a patchy distribution surrounding salt basins. It occurs at low densities and is usually a solitary inhabitant of large, complex mounds. It specializes on leaves of halophytic vegetation (especially members of the family Chenopodiaceae: *Atriplex*, *Suaeda*, *Heterostachys*).

Mammalian hair is long and is known for its plasticity in evolving to serve various functions like defense, thermoregulation and crypsis.[1] The red vizcacha rat has a unique morphological feature: bundles of stiff hairs located behind the upper

---

[1] Crypsis is the ability of an organism to avoid observation or detection by other organisms.

incisors. They can function in place of teeth, and can remove the salty epidermis from the leaves of *Atriplex* before the edible parts of the leaves are consumed (Mares et al. 1997). These unique bristle bundles are not the only adaptation of the red vizcacha rat to a halophytic diet or to its specialized salt-flat habitat. The morphological and physiological adaptations of the red vizcacha rat to deserts show the overall similarity of these traits to those of unrelated rodents that inhabit similar habitats in other deserts of North America, Africa and Asia (see Degen and Kam, this volume, for results of a study on fat sand rats that rely exclusively on a single species of *Atriplex*). Similarities include diet composition and feeding behavior. It has a renal specialization, large medulla and long renal papilla for water balance and elimination of salt and urine concentrations, among other traits. The red vizcacha rat is more similar to these desert rodents than it is to con-familials that do not feed on halophytes.

The herbivorous rodent, the degu (*Octodon degus*) is an inhabitant of the semiarid environments of northern and central Chile. Degus show preference for food containing low fiber. But low fiber food is not available in the field during the dry season. In summer, grass consumption with a high percentage of fiber, nearly 60% is not a choice of the degus but a consequence of necessity. The low digestibility of high fiber food appears to be compensated for by increasing the volume of digesta in the alimentary canal. This is the result of changes in the rates of food intake, and so increases in turnover time of digesta. The digestive responses allowed them to increase the amount of energy obtained from fiber (Bozinovic 1995). Analysis of the whole organism of the degus show that *in vitro* analysis of enzymatic digestive activity and plant defenses cannot be used to explain and fully understand the physiological and behavioral effects of plant defenses on mammalian herbivores (Bozinovic et al. 1997).

The capybara (*Hydrochaeris hydrochaeris*) is an interesting case because it is the largest cecum fermenter and uses coprophagy as a part of its digestive strategy. The capybara hindgut content was heavier during the dry than wet season, but there were no significant seasonal differences between the stomach or small intestine and their contents. This suggests changes in the capacity of the hindgut, related to the seasonal variation in resource quality (Borges et al. 1996).

A small number of mammals [like the wild vicuña (*Vicugna vicugna*) and guanaco (*Lama guanicoe*)], and bird species depend upon the peatlands for grazing, nesting and water (Squeo et al. 2006b).

There is a group of alpine bofedales[2] in the arid grasslands of the central Andes. Peatlands in northern Chile are in the most arid part of it. Species of the Juncaceae are among the primary peat-forming plants. *Oxychloe andina* and *Patosia clandestine* (i.e., Juncaceae family) are the community dominants and primary peat formers, together with some Gramineae and other herbaceous species (Rutsatz 2000; Squeo et al. 2006b). Halophyte communities with *Atriplex atacamensis*, *Distichlis humilis*, *Muhlenbergia fastigiata*, *Senecio pampae*, *Suaeda foliosa* and *Tessaria absinthioides* occur around the salt areas. Water sources are fresh, and mildly saline ground waters originate from glaciers, rain and snowmelt. These peatlands are extremely fragile,

---

[2] The Bofedales are high altitude wetlands. They are considered to be the native grasslands of the Andean region.

depending on water, climate changes and human disturbances (like mining activity: Squeo et al. 2006a). The bofedales vegetation contrasts sharply with the surrounding terrestrial communities by having a plant cover usually greater than 70%, and a high plant productivity with biomasses over 1000 g m^{-2} (Squeo et al. 1993, 1994, 2006b).

More than 350 bofedales were identified in the Peruvian altiplano, covering a total surface area of 111,500 ha. In the bofedales at the Puna, crude protein ranged from 11.2 to 15.1% during the rainy season, and from 6.8 to 8.9% during the dry season. In bofedales of the humid Puna, cell wall ranged from 12.8 and 12.3% during the rainy season and 11.0 to 11.2% during the dry season (Siguayro Pascaja 2008). Dry matter availability at all locations and times during the year varied from 1000 to 1650 kg ha^{-1} (Siguayro Pascaja 2008).

4. Conclusions

South America is a huge continent of enormous geographic contrast ranging from high and dry mountains, alpine wetlands, coastal deserts, inland deserts and vast rangelands (pampas). In all of these regions there are salt tolerant and halophytic plants that provide sustenance for the vast inventories of cattle and other livestock as well as a myriad of specialized wildlife. Research has focused on how to mainstream the use of halophytic and salt tolerant plants into production systems.

References and Further Readings

Abou El Ezz, S., E.F. Zaki, N.H.I. Abou-Soliman, W.A. El Ghany and W. Ramadan (in press). Impact of halophytes and salt tolerant plant on livestock products. pp. 161–176, this volume.

Abril, A., M. Aiazzi, P. Torres and J. Argüello 2000. Nutritional value of *Atriplex cordobensis* grown in dry Chaco of Argentina. Rev. Arg. Prod. Anim. 20(3-4): 179–185.

Aguilar, M.G. and R. Newman. 2005. Junto a las vicuñas. http://www.FAO.org/Article/Agrippa/x9500S11.hm.

Alves, J.N., G.G.L. Araújo and E.R. Porto. 2007. Feno de erva-sal (*Atriplex nummularia*) e palma forrajeira (*Opuntia ficus-indica* Mill.) em dietas para caprinos e ovinos. Rev. Cient. Prod. Anim. 9(1): 43–52.

Ansin, O.A., E.M. Oyhamburu, E.A. Hoffmann M.C. Vecchi and M.C. Ferragine. 1998. Distribución de raíces en pastizales naturales y pasturas cultivadas de La Pampa Deprimida bonaerense y su relación con la biomasa forrajera. Rev. Fac. Agron. La Plata 103(2): 141–148.

Anon. 2008. Arable Cropping in a Changing Climate, 23–24 January 2008 http://www.hgca.com/events.past_events/32/32/Events/Events/Past%20Event%20Papers.mspx.

Avila, R., E. Quiroga, C. Ferrando and L. Blanco. 2008. Dinámica de la calidad a lo largo del año de dos gramíneas nativas de los Llanos de La Rioja. INTA EEA La Rioja.

Azócar, P. 2001. *Opuntia* as feeds for ruminants in Chile. *In*: Mondragon-Jacobo and Perez-Gonzalez (eds.). *Cactus (Opuntia* spp.) *as Forage*, FAO Plant production and protection papers N°169: 161p. FAO, Rome.

Azócar, P. and H. Rojo. 1992. Suplementación estival de cabras en lactancia con nummularia (*Atriplex nummularia*) y cladodios de tuna (*Opuntia ficus-indica*). Actas II Congreso Internacional de la Tuna y Cochinilla, 22–25 September 1992. Santiago, Chile, pp. 72–76.

Barroso, D.D., G.G.L. Araújo and E.R. Porto. 2006. Produtividade e valor nutritivo das frações forrageiras da erva-sal (*Atriplex nummularia*) irrigada com quatro diferentes volumes de efluentes da criação de tilapia em agua salobre. Agr. Tec. 27(1): 43–48.

Benjamin, R.W. 1980. The use of forage shrubs in the Norte Chico Region of Chile. pp. 299–302. *In*: H.N. Le Houérou (ed.). Browse in Africa: The Current State of Knowledge. ILCA (International Livestock Centre for Africa), Addis Ababa, Ethiopia.

Bonvissuto, G.L. 2004. Observaciones sobre la vegetación de Mamuel Choique. pp. 28–37. *In*: G.L. Bonvissuto (Coord.) (ed.). Elaboración de una propuesta integral para los sistemas de producción con mallines salino-sódicos de la cuenca del arroyo Mamuel Choique. Comunicación Técnica N° Area Rec. Nat. Pastizales Naturales INTA Series Comunicaciones Técnicas.

Borges, P.A., M.G. Domínguez-Bello and E.A. Herrera. 1996. Digestive physiology of wild capibara. J. Comp. Physiol. B. 166: 55–60.

Bortolus, A., E. Schwindt, P.J. Bouza and Y.L. Idaszkin. 2009. A characterization of Patagonian salt marshes. Wetlands 29(2): 772–780.

Bozinovic, F. 1995. Nutritional energetic and digestive responses of an herbivorous rodent (*Octogon degus*) to different levels of dietary fiber. J. Mammalogy 76(2): 627–637.

Bozinovic, F., F.F. Novoa and P. Sabat. 1997. Feeding and digesting fiber and tannins by an herbivorous rodent, *Octodon degus* (Rodentia: Caviomorpha). Comp. Biochem. Physiol. Part A: Physiol. 118(3): 625–630.

Brako, L. and J. Zarucchi. 1993. Catalogue of the flowering plants and gymnosperms of Peru. Monography Systematics Missouri Bot. Garden 45: 481–486, 923–925.

Brevedan, R.E., O.A. Fernández and C.B. Villamil. 1994. Halophytes as a resource for livestock husbandry in South America. pp. 175–199. *In*: V.R. Squires and A.T. Ayoub (eds.). Halophytes as a Resource for Livestock and for Rehabilitation of Degraded Lands. Kluwer Acad. Publ.

Brizuela, M.A., M.S. Cid, D.P. Miñón and R. Fernández Grecco. 1990. Seasonal utilization of saltgrass (*Distichlis* spp.) by cattle. Anim. Feed Sci. Technol. 30: 321–325.

Burkart, A. 1976. A monograph of the genus *Prosopis* (Leguminosae subfam. Mimosoideae). J. Arn. Arbor. 57: 219–249, 450–455.

Cabrera, A. 1971. Fitogeografía de la República Argentina. Bol. Soc. Arg. Bot. 1(1-2): 42.

Cavagnaro, J.B. and C.B. Passera. 1993. Water utilization by shrubs and grasses in the Monte ecosystems, Argentina. pp. 255–257. *In*: Proceedings IV International Rangeland Congress. Montpellier, France.

Cisneros, J.M., A. Degioanni, J.J. Cantero and A. Cantero. 2008. Caracterización y manejo de suelos salinos en el área Pampeana Central. pp. 17–46. *In*: E. Taleisnik, K. Grunberg and G. Santa María (eds.). La salinización de suelos en la Argentina: su impacto en la producción agropecuaria. Universidad Católica de Córdoba.

CONAF (Corporación Nacional Forestal). 1993. Plantaciones sp. Centraras, M., CONAF y privados en la IV Región. Programa Control Forestal IV Región. La Serena, Chile.

Concha, R. 1975. Consumo y ganancia de peso ovino durante el periodo primavera verano en una pradera natural biestratificada con *Atriplex repanda*. Tesis de MSc, Universidad de Chile, Chile.

Conticello, L., B. Cerazo and A. Bustamante. 2002. Dinámica de comunidades hidrohalófitas asociadas a canales de riego en el Alto Valle de Río Negro (Argentina). Gayana Bot. 59(1): 13–20.

Degen, A.A. and M. Kam. (in press). Energy and Nitrogen Requirements of the Fat Sand Rat (*Psammomys obesus*) when Consuming a Single Halophytic Chenopod. pp. 373–387, this volume.

Echeverría, I., J. Maddaloni and S.I. Alonso. 2001. Especies nativas e introducidas con valor forrajero. p. 522. *In*: J. Maddaloni and L. Ferrari (eds.). Forrajeras y pasturas del ecosistema templado húmedo de la Argentina (2ª ed.). INTA and Fac. de Cs. Agrarias, Univ. Lomas Zamora.

El Shaer, H. 2010. Halophytes and salt-tolerant plants as potential forage for ruminants in the Near East region. Small Ruminant Res. 91: 3–12.

Espino, L.M., M.A. Seveso and M.A. Sabatier. 1983. Mapa de suelos de la provincia de Santa Fé. Tomo II, p. 216.

Feldman, S.R. 2003. Ecología de *Spartina argentinensis* Parodi. Crecimiento y desarrollo de la planta y efecto del fuego sobre sus poblaciones y comunidades. Tesis doctorado. Univ. Nac. Córdoba, Fac. Cs. Exactas, Físicas y Naturales, p. 182.

Felker, P. and P.R. Clark. 1982. Position of mesquite (*Prosopis* spp.) nodulation and nitrogen fixation (acetylene reduction) in 3 m long phreatophytically simulated soils columns. Plant Soil 64: 297–305.

Felker, P., P.R. Clark, A.E. Laag and P.F. Pratt. 1981. Salinity tolerant of tree legumes; mesquite (*Prosopis glandulosa* var. *torreyana*, *P. velutina*, and *P. articulata*), algarrobo (*P. chilensis*), kiawe (*P. pallida*) and tamarugo (*P. tamarugo*) grown in sand culture on nitrogen-free media. Plant Soil 61: 311–317.

Ferreira, M. de A. 2006. Utilização da palma forrageira na alimentação de vacas leiteiras. *In*: 43°Reunião Anual da Sociedade Brasileira de Zootecnia 2006. João Pessoa, PB.

Fiorio, D. 2004. El recurso hídrico de la cuenca Mamuel Choique. pp. 25–27. *In*: G.L. Bonvissuto (Coord.) (ed.). Elaboración de una propuesta integral para los sistemas de producción con mallines salino-sódicos de la cuenca del arroyo Mamuel Choique. Comunicación Técnica N° Area Rec. Nat. Pastizales Naturales INTA Series Comunicaciones Técnicas.

Flowers, T.J. and T.D. Colmer. 2008. Salinity tolerance in halophytes. New Phytol. 179: 945–963.

Galán de Mera, A., E. Linares Perea, J. Campos de la Cruz and J.A. Vicente Orellana. 2009. Nuevas observaciones sobre la vegetación del sur del Perú. Del desierto Pacífico al altiplano. Acta Botánica Malacitana 34: 1–35.

Galán de Mera, A., J. Campos, C. Trujillo and E. Linares. 2012. Las comunidades vegetales relacionadas con los ambientes humanos en el sur del Perú. Phytocoenologia 41(4): 265–305.

Gallardo, M.H., J.W. Bickham, R.L. Honeycutt, R.A. Ojeda and N. Köhler. 1999. Discovery of tetraploidy in a mammal. Nature 401: 341.

Gandullo, R. 2004. New plant association of saline environments. Multequina 13: 33–37.

Garciandía, D.A. 2010. Desarrollo de subtropicales para bajos salino-sódicos. Megatérmicas en Carlos Tejedor, Bs. As. Sitio Arg. De Producción Animal. http://www.produccion-animal.com.ar.

Giorgetti, H.D., Z. Manuel, O.A. Montenegro, G.D. Rodríguez and C.A. Busso. 2000. Phenology of some herbaceous and woody species in central, semiarid Argentina. Phyton 69: 91–108.

Guevara, J.C., C.R. Stasi and O.R. Estevez. 1996. Effect of cattle grazing on range perennial grasses in the Mendoza plain, Argentina. J. Arid Environ. 34: 205–213.

Guevara, J.C., J.A. Paéz, R.F. Tanquilevich and O.R. Estevez. 1981. Economía de las explotaciones ganaderas, I. Tierras privadas del área centro este de la provincia de Mendoza. Cuaderno Técnico IADIZA 4: 1–39.

Guevara, J.C., J.B. Cavagnaro, O.R. Estevez, H.N. Le Houérou and C.R. Stasi. 1997. Productivity, management and development problems in the arid rangelands of the central Mendoza plains (Argentina). J. Arid Environ. 35: 575–600.

Guevara, J.C., L.I. Allegretti, O.R. Estevez, A.S. Monge, J.A. Paez and M.A. Cony. 2010. Diets of a domestic and wild herbivores grazing in common in a rangeland of Mendoza province, Argentina. *In*: B. Veress and J. Szigethy (eds.). Horizons in Earth Science Research. Vol. 1 Chapter 20: 463–477.

Guevara, J.C., O.R. Estevez, J.H. Silva and A. Marchi. 1991. Adequacy of native range grasses to meet protein and energy beef cow requirements in the plain of Mendoza, Argentina. IV Int. Rangeland Congress, France: 696–699.

Idaszkin, Y.L. and A. Bortolus. 2010. Does low temperature prevent *Spartina alterniflora* from expanding toward the austral-mos salt marshes? Plant Ecol. 212(4): 553–561.

Imaz, J.A., C. Antonelli and D.O. Gimenez. 2009. Implantación de forrajeras subtropicales en bajos salino-sódicos de la Cuenca del Salado. 1er Congreso de la Red Argentina de Salinidad, Córdoba.

Isacch, J.P., C.S.B. Costa, L. Rodríguez-Gallego, D. Conde, M. Escapa, D.A. Gagliardini and O.O. Iribarne. 2006. Distribution of saltmarsh plant communities associated with environmental factors along a latitudinal gradient on the south-west Atlantic coast. J. Biogeogr. 33: 888–900.

Lailhacar, S. 2000. Shrub introduction and management in South America. *In*: G. Gintzburger, M. Bounejmate and A. Nefzaoui (eds.). Fodder Shrub Development in Arid and Semi-arid Zones. Proceedings of the Workshop on Native and Exotic Fodder Shrubs in Arid and Semi-arid Zones, 27 October–2 November 1996, Hammamet, Tunisia. ICARDA, Aleppo (Syria) I: 77–100.

Lavado, R.S. 2008. Visión sintética de la distribución y magnitud de los suelos afectados por salinidad en la Argentina. pp. 11–15. *In*: E. Taleisnik, K. Grunberg and G. Santa María (eds.). La salinización de suelos en la Argentina: su impacto en la producción agropecuaria. Universidad Católica de Córdoba.

Le Houérou, H.N. 1982. Prediction of range production from weather records in Africa. Techn. Conf. on Climate in Africa. WMO, Geneva.

Le Houérou, H.N. 1992. The role of saltbushes (*Atriplex* spp.) in arid land rehabilitation in the Mediterranean Basin: a review. Agroforestru Syst. 18: 107–148.

Lewis, J.P., E.F. Pire, D.E. Prado, S.L. Stofella, E.A. Franceschi and N.J. Carnevale. 1990. Plant communities and phytogeographical position of a large depression in the great Chaco, Argentina. Vegetatio 86: 25–38.

Maddaloni, J. 1986. Forage production on saline and alkaline soils in the humid region of Argentina. Reclamation Reveget. Res. 5: 11–16.

Mares, M.A., R.A. Ojeda, C.E. Borghi, S.M. Giannoni, G.B. Diaz and J.K. Braun. 1997. How desert rodents overcome halophytic plant defenses. BioScience 47(10): 699–704.

Masters, D.G. (in press). Assessing the feeding value of halophytes. pp. 89–105, this volume.

Masters, D.G., S.E. Benes and H.C. Norman. 2007. Biosaline agriculture for forage and livestock production. Agric. Ecosys. Env. 119: 234–248.

Meneses, R.R. and N.F. Squella. 1996. Los arbustos forrajeros. pp. 150–170. *In*: I. Ruiz (ed.). Praderas para Chile. INIA, Santiago de Chile, Chile.

Montesinos-Tubée, D. 2012. Vegetación halófila de tres localidades andinas en la vertiente pacífica del sur de Perú. *Chloris Chilensis*: año 15 N° 2. URL://http:www.chlorischile.cl.

Morello, J. and C. Saravia Toledo. 1959. El bosque chaqueño II. Ganadería y el bosque en el oriente de Sala. Rev. Agron. Noroeste Arg. 3: 209–258.

Moreno, G.M.B. 2011. Feno de erva-sal (*Atriplex nummularia*) na terminação de cordeiros Santa Inés. Tesis Universidade Estadual Paulista. Brazil.

Mosconi, F.P., L.J.J. Priano and N.E. Hein. 1981. Mapa de suelos de la Provincia de Santa Fe. Tomo I, p. 246.

Movia, C. 1972. Formas de erosión eólica de la Patagonia. Photointerprétation, 6/3. Ediciones Technip, Paris.

Mulas, M. and G. Mulas. 2004. The strategic use of *Atriplex* and *Opuntia* to combat desertification. Tesis University of Sassari.

Müller, G. and P. Gutte. 1985. Salzpflanzengesellschaften bei Cusco/Perú. Wiss. Z. Karl-Marx-Univ. Leipzig, Math.-Naturwiss. R. 34: 402–409.

Murillo-Amador, B., A. Cortés-Avila, E. Troyo-Diéguez, A. Nieto-Garibay and H. Jones. 2001. Effects of NaCl salinity on growth and production of young cladodes of *Opuntia ficus-indica*. J. Agron. Crop Sci. 187: 269–279.

Navarro, G. 1993. Vegetación de Bolivia: el Altiplano meridional. Rivasgodaya 7: 69–98.

Norman, H.C.E. Hulm and M.G. Wilmot. (in press). Improving the feeding value of old man saltbush for saline production systems in Australia, pp. 79–87, this volume.

Norman, H.C., D.G. Masters and E.G. Barret-Lennard. 2013. Halophytes as forages in saline landscapes: Interactions between plant genotype and environment change their feeding value to ruminants. Env. Exp. Bot. 92: 96–109.

Ochoa, M.A. 1994. Producción de forraje en suelos salinos. Sitio argentino de Producción Animal. INTA EEA Rama Caída. http://www.produccion-animal.com.ar.

Olivares, A. and J. Gasto. 1981. *Atriplex repanda*: Organización y manejo de ecosistemas con arbustos forrajeros. Facultad de Ciencias Agrarias, Veterinarias y Forestales. Universidad de Chile, Santiago, p. 300.

Ormazábal, C.S. 1991. Silvopastoral systems in arid and semiarid zones of northern Chile. Agroforestry Systems 14: 207–217.

Otondo, J.J. 2011. Efectos de la introducción de especies megatérmicas sobre características agronómicas y edáficas de un ambiente halomórfico de la pampa inundable. Tesis. Bs. As., Fac. Agronomía, p. 58.

Padilla, J.F. 1986. Selección de procedencias de sereno (*Atriplex repanda* Phil.) en el secano costero de la Provincia de Elqui, IV Región. Tesis de Ingeniero Agrónomo. Facultad de Ciencias Agrarias y Forestales de la Universidad de Chile, p. 144.

Passera, C., J.B. Cavagnaro and C. Sartor. 2010. Plantas C3, C4 y CAM nativas del monte árido argentino. Adaptaciones y potencial biológico. En: C4 y CAM. Características generales y uso en programas de desarrollo de tierras áridas y semiáridas. Edit. CSIC, Madrid, Spain: 165–176.

Pelliza, A., L. Borrelli and G.L. Bonvissuto. 2005. El pasto salado (*Distichlis* spp.) en la Patagonia: una forrajera adaptada a la aridez y a la salinidad. Rev. Cient. Ccias Agrop. 9(2): 119–131.

Pelliza, A., P. Willems, V. Nakamatsu and A. Manro. 1997. Atlas dietario de herbívoros patagónicos. Proyecto de prevención y control de la desertificación en la Patagonia. INTA GTZ/FAO, p. 109.

Pensiero, J.F. 2009. Gramíneas nativas con potencialidades de cultivo com forrajeras para ambientes salinos. 1er Congreso de la Red Argentina de Salinidad. Córdoba, Argentina.

Porto, E.R., M.C.C. Amorim and M.T.D. Dutra. 2006. Rendimento da *Atriplex nummularia* irrigada com efluentes de criação de tilapia em rejeito de dessalinização de agua. Rev. Bras. Eng. Agr. Amb. 10(5): 97–103.

Ribaski, J. 1987. Comportamento da algaroba (*Prosopis juliflora* (SW) DC) e do capim-bufel (*Cenchrus ciliaris*) em plantio consociado na região de Petrolina, PE. Rev. Ass. Brasileira Algaroba 1(2): 171–225.

Rodrigues Souto, J.C.R., G.G.L. Araújo and D.S. Silva, E.R. Porto, S.H.N. Turco and A.N. Medeiros. 2005. Desempenho produtivo de ovinos alimentados com dietas contendo niveis crescentes de feno de erva-sal (*Atriplex nummularia* Lindl). Rev. Cien. Agron. 36(3): 376–381.

Rutsatz, B. 2000. Die Hartpolstermoore der Hochanden und ihre. Artenviefalt. Ver. D. Reinh.—Tuxen-ges. 12: 351–371.

Sala, O., A. Soriano and S. Perelman. 1981. Relaciones hídricas de algunos componentes de un pastizal de la Depresión del Salado. Rev. Fac. Agron. UBA 2(1): 1–10.

Santana, O.P. 1992. Punas forrajeras (*Opuntia ficus-indica* y *Nopalea cochenillifera*) en el noreste brasileño: una revisión. Actas II Congreso Internacional de tuna y cochinilla, 22–25 septiembre 1992. Santiago, Chile: 126–142.

Scherr, S. 1979. Soil degradation. A threat to developing-country food security by 2020? U.S.A. Food, Agriculture and the Environment, Discussion Paper 27. Food Policy Res. Institute. Washington, p. 71.

Siguayro Pascaja, R. 2008. Evaluación agrostológica y capacidad receptiva estacional en bofedales de Puna seca y húmeda del altiplano de Puna. Tesis Universidad Nacional del Altiplano, Puna, Perú.

Silva, E. and C. Pereira. 1976. Aislación y composición de las proteinas de hojas de *Atriplex nummularia* y *A. repanda*. Ccia. Inv. Agr. 3(4): 169–174.

Somlo, R., C. Durañoña and R. Ortiz. 1985. Valor nutritivo de especies forrajeras patagónicas. Rev. Arg. Prod. Anim. 5(9-10): 589–605.

Soto, G. 1996. Una especie pionera para las zonas aridas de Chile. Programa Conjunto FAO/PNUMA de Control de la Desertificacion en America Latina y el Caribe. Coquimbo (Chile).

Souza, E.R., M.B.G.S. Freire and C.W.A. Nascimento. 2011. Fitoextracão de sais pela *Atriplex nummularia* Lindl. sob estresse hídrico em solo salino sódico. Rev. Bras. Eng. Agr. Amb. 15(5): 477–483.

Squeo, F.A., B.G. Warner, R. Aravena and D. Espinoza. 2006a. Bofedales: turberas de alta montaña de los Andes centrales. Rev. Chil. Ha. Natural 79: 245–255.

Squeo, F.A., E. Ibacache, B.G. Warner, D. Espinoza, R. Aravena and J.R. Gutiérrez. 2006b. Productividad y diversidad florística de la Vega Los Tambos, Cordillera de Doña Ana: variabilidad interanual, herbivoría y nivel freático. pp. 333–362. *In*: J. Cepeda (ed.). Geoecología de la Alta Montaña del Valle del Elqui. Ediciones Universidad de La Serena. La Serena, Chile.

Squeo, F.A., H. Veit, G. Arancio, J.R. Gutierrez, M.T.K. Arroyo and N. Olivares. 1993. Spatial heterogeneity of high mountain vegetation in the Andean desert zone of Chile (30°S). Mountain Res. Develop 13: 203–209.

Squeo, F.A., R. Osorio and G. Arancio. 1994. Flora de los Andes de Coquimbo: Cordillera de doña Ana. Edic. Univ. La Serena, La Serena, Chile, 176 pp.

Tapia, Y., O. Díaz, C. Pizarro, R. Segura, M. Vines, G. Zúñiga and E. Moreno-Jiménez. 2013. *Atriplex atacamensis* and *Atriplex halimus* resist As contamination in Pre-Andean soils (northern Chile). Sci. Total Env. 450-451: 188–196.

Villagra, P.E., A. Vilela, C. Giordano and J. Alvarez. 2010. Ecophysiology of *Prosopis* species from the arid lands of Argentina: What do we know about adaptation to stressful environments? pp. 321–340. *In*: K.G. Ramawal (ed.). Desert Plants. Springer-Verlag. Berlin. Heidelberg.

Wainstein, P. and S. González. 1963. Valor forrajero de 13 especies de *Stipa* en Mendoza. Rev. Fac. Cs. Agrarias 9: 3–18.

Wainstein, P. and S. González. 1969. Valor nutritivo de plantas forrajeras del Este de la provincia de Mendoza. Reserva Reserva Forestal de Ñancuñán. Rev. Fac. Cs. Agrarias 15: 133–142.

Wainstein, P. and S. González. 1971a. Valor nutritivo de plantas forrajeras del Este de la provincia de Mendoza (Reserva Ecológica de Ñancuñán, I.) Deserta 67–75.

Wainstein, P. and S. González. 1971b. Valor nutritivo de plantas forrajeras del este de la provincia de Mendoza (Reserva Ecológica de Ñancuñán, II) Deserta 2: 77–85.

Wainstein, P., S. González and E. Rey. 1979. Valor nutritivo de plantas forrajeras de la Provincia de Mendoza, III Cuaderno Técnico 1: 97–108.

Zabala, J., J. Pensiero, J. Giavedoni, P. Tomas, H. Gutierrez, P. Widerhorn, M. Delbino, M. Ballesteros, M. Aracne, P. Imhoff and G. Schrauf. 2009. Colección y caracterización de recursos forrajeros nativos y naturalizados para ambientes salinos. 1er Congreso de la Red Argentina de Salinidad, Córdoba.

Improving the Feeding Value of Old Man Saltbush for Saline Production Systems in Australia

H.C. Norman,* E. Hulm and *M.G. Wilmot*

SYNOPSIS

This case study describes a research project investigating opportunities to improve old man saltbush (*Atriplex nummularia* Lindl.) for small ruminants, through simple plant selection. At the time of writing, much of this work had not yet been published in the scientific literature. The aim of this case study is to provide an overview of the processes, research iterations and achievements, leading to the release of the first commercial cultivar.

Dryland salinity is a major constraint to crop and livestock production in southern Australia. After many decades of on-farm research with large numbers of species, old man saltbush had become the leading species to improve profitability of saline land in low to medium rainfall areas; however feeding value was a major constraint. Whole-farm economic modelling suggested that the plant's energy value was the key objective to improve. A collection of germplasm from across the native range in Australia was conducted and genotypes from 27 populations and 2 subspecies were compared at 3 research sites. A number of methodological constraints were identified and overcome to aid plant selection. Old man saltbush *in vivo* digestibility standards were developed using metabolism crates and these were combined with laboratory analyses to develop a near

CSIRO Agriculture, Wembley, Western Australia.
* Corresponding author: Hayley.Norman@csiro.au

infrared spectroscopy prediction tool. This has allowed rapid and cost effective screening of large numbers of shrubs. In a novel approach to plant genetic improvement, sheep were used during all stages of the project to identify plants that are consistently preferred. Over 8 years of on-farm experiments, the team identified a cohort of old man saltbush genotypes with 20% higher organic matter digestibility, greater acceptability to sheep and 8 times more edible biomass production, when compared with the mean of the collection. In 2015, the first clonal cultivar was commercialised (cv. Anameka). There are several other clonal cultivars that are ready for release if Anameka does not perform across niches and work continues to develop a high-value seed line. Whole-farm economic analysis suggests that Anameka can double the profitability of saltbush plantations on farms. If improved profitability results in greater rates of adoption, the environmental health of agricultural landscapes in southern Australia will be improved.

Keywords: halophyte, shrub, diet selection, plant improvement, nutritional value, farming systems, salt affected, whole-farm economic modelling.

1. Introduction: climate and farming systems in south-western Australia

The south-western part of Australia is characterised by a Mediterranean-type climate with warm-to-hot, dry summers and mild, wet winters. Southern Australia differs from the eastern Mediterranean in terms of significant (but unpredictable) summer rainfall as monsoonal troughs occasionally descend from the tropics, bringing summer storms (di Castri 1981). The soils in this region are predominantly old, highly weathered and infertile. The mixed crop and livestock agro-ecological zone covers approximately 20 million ha and rainfall varies from 300 to 650 mm annually. Crops, predominantly wheat, barley and canola, are the most profitable enterprises although many farmers maintain sheep within their system. The sheep flock is primarily based on Merino ewes for wool production although rams of other breeds are used to produce lambs for domestic slaughter and live export. Arable land is rarely planted to perennial pastures they struggle to persist through the hot dry summers and they reduce flexibility in the cropping enterprise. Perennials are often planted on areas that are unsuited to cropping, including saline, waterlogged, acidic, sandy, stony and otherwise unproductive land (Lawes et al. 2014).

There are a number of significant challenges to farming systems in this region. It is estimated that 1.1 m ha of agricultural land in Western Australia is severely salt affected and a further 1.7–3.4 m ha is at risk (George et al. 2008). Other challenges include soil acidity, herbicide resistance within weed populations and a predicted increase in temperatures (Dolling and Porter 1994; Walsh and Powles 2014; Henry et al. 2012). Small ruminant production is characterised by constant changes in quantity and quality of the feedbase associated with seasons. In a typical system, sheep are moved between annual pastures and crop residues. In late summer and autumn, feed quality is poor and sheep tend to be supplemented with grain. This is the time of year when perennial plants are most valuable as a source of energy, protein and antioxidants to complement dry crop and pasture residues. Labour costs are high and it is not uncommon for individual farmers to manage more than 2000 breeding ewes as well as a large cropping program. Supplementary feeding regimes for sheep must therefore be simple and farmers seek to implement systems where sheep move between crop resides and pastures, selecting a diet that optimises their needs. For this reason, factors influencing voluntary intake and diet selection are of interest to livestock researchers.

2. Old man saltbush in southern Australia

Old man saltbush is a woody perennial shrub that occurs as a dominant species in widespread communities over a 4000 km range in arid and semi-arid zones of Australia (Parr-Smith 1982). As it is adapted to an arid climate it persists comfortably in agricultural zones of the Australian wheatbelt where rainfall is higher and more regular. The taxon is generally dioecious and octoploid (Nobs 1980). The two dominant subspecies are geographically separated with subsp. *nummularia* found in central and eastern Australia, while subsp. *spathulata* is restricted to Western Australia (Parr-Smith 1982).

Very few forage species tolerate the combined stresses of aridity and salinity in the saline valley floors of southern Australia. The drought tolerance mechanisms of old man saltbush include deep roots (can be > 4 m), osmotic control and slow growth when water is scarce (Barrett-Lennard 2003). In summer and autumn the salinity of the soil solution in the root-zone is often in excess of seawater concentrations (Barrett-Lennard 2003).

Being both perennial and in active growth through the summer and autumn period, saltbushes have the potential to reduce leakage of rainwater to water tables thus potentially reducing the effects of dryland salinity (Barrett-Lennard et al. 2005). At Yealering, Western Australia, data indicates that saltbushes play a role in reducing the impact of dryland salinity and salt export from saline areas (Bennett et al. 2012). Associated research at the same site has demonstrated that the addition of saltbush and annual understorey has quadrupled sheep grazing days in autumn and lead to improved biodiversity (Norman, unpublished). This is an example of how integrating livestock production in farming systems can lead to better environmental outcomes (Masters et al. 2010).

The feeding value of old man saltbush as a sole diet for sheep is generally poor, due to a combination of (1) low to moderate biomass production, (2) low to moderate digestibility of the organic matter, (3) excessive salt and/or sulphur accumulation and (4) excessive plant secondary compounds such as oxalate (Norman et al. 2004; Masters et al. 2007; Norman et al. 2010; Al Daini et al. 2013). It is therefore recommended that farmers supplement livestock grazing saltbush with crop stubbles, hay or grain (Norman et al. 2008). Despite these limitations, saltbush is a valuable source of crude protein, sulfur, vitamin E and minerals within meat and wool production systems (Ben Salem et al. 2010; Pearce et al. 2010; Fancote et al. 2013).

3. Old man saltbush improvement project

There has been little systematic effort to domesticate this shrub species and the majority of commercial plantations in Australia are derived from seed collected from native stands. The project was initiated with comprehensive whole-farm economic modelling studies to assess the limitations and opportunities provided by saltbush within the farming system. The modelling based on an average farm in several rainfall zones indicated that improving the energy value of old man saltbush would substantially increase farm profitability (O'Connell et al. 2006; Monjardino et al. 2010). Sensitivity analysis predicted that improving shrub digestibility by 10% would increase profits by three times the increment associated with increasing biomass production by 10%, or reducing the cost of establishment by 10% (O'Connell et al. 2006). The first selection criterion was therefore digestibility of the organic matter (OMD). The economic benefits rely on sheep choosing to incorporate the plants into their diets. Other studies had shown that sheep preferentially graze some individual shrubs before others, and some shrubs are not eaten (Norman et al. 2004). These differences in relative preference by sheep were likely to be associated with both nutritive factors and the presence of plant secondary compounds with antinutritional characteristics, and excessive salt and sulphur (Norman 2004, 2009, 2011). A secondary aim was therefore to select a cultivar with higher preference to sheep. The final selection criterion was improved production of 'edible' dry matter production and survival.

In 2006, seed was collected from 27 populations (referred to as provenances). Seed was grown in a shade house for 6 months and 60000 seedlings were planted in replicated blocks (n = 9) in three experimental sites, across three states of southern Australia (Hobbs and Bennell 2008). All plants were assessed for a range of agronomic traits and nutritive value was investigated at the provenance level.

In 2008, the three nursery sites were grazed with Merino sheep to assess relative palatability. This is perhaps one of the first times that sheep have been used in the initial stages of a plant improvement programme to identify superior genotypes. Across sites, sheep demonstrated with consistent preference for specific provenances and a consistent dislike of subspp. *spathulata* (Fig. 5.1).

Assessing the energy value (or OMD) of the saltbush was complicated by high levels of soluble salt and uncertainty about the suitability of traditional laboratory methods for (Masters et al. 2007; Norman et al. 2010; Masters, this volume). To generate samples with known *in vivo* organic matter digestibility, a series of metabolism crate feeding experiments were undertaken. Fourteen saltbush calibrations samples, with OMD ranging from 47 to 69% were generated. These samples were subject to *in vitro* enzymatic digestion and a linear relationship was generated to predict *in vivo* OMD from pepsin-cellulase digestion. Near infrared spectroscopy (NIRS) calibration equations were generated to predict *in vitro* OMD, crude protein, ash, neutral detergent fibre and acid detergent fibre (Norman and Masters 2010). Use of NIRS allowed for rapid and inexpensive screening of large numbers of plant samples.

There was significant variation between the leaves of provenances in OMD (Fig. 5.2), salt, crude protein and biomass growth. Generally, sheep preferentially grazed provenances with higher OMD and lower salt.

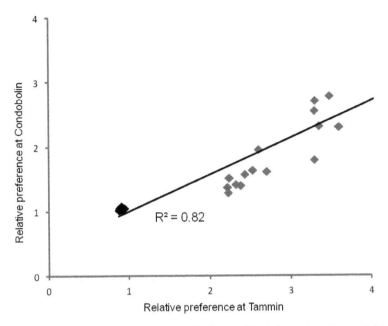

Fig. 5.1. Grazing preference of 27 provenances of old man saltbush from subspecies *spathulata* (black symbols) and subspp. *nummularia* (grey symbols) at two geographically distant sites (4000 km apart). A higher preference score indicates higher grazing preference (data previously published in Norman et al. 2011).

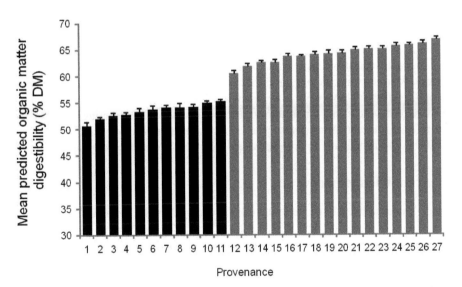

Fig. 5.2. Predicted organic matter digestibility of 27 provenances of old man saltbush (Mean ± SE) from subspecies *spathulata* (black columns) and subspp. *nummularia* (grey columns). Digestibility estimated using a pepsin cellulase assay, adjusted for *in vivo* digestibility using calibration samples.

In the second phase of the programme, 90 of the 60000 shrubs (the best 30 from each nursery site) were selected and vegetatively cloned as cuttings to ensure genetic purity. When compared to the average of the original collection, these plants had 20% higher digestibility, were consistently preferred by sheep and produced 8 times more biomass than the mean of the collection. These cuttings were planted into replicated blocks at 3 experimental sites across Australia and assessed for agronomic and nutritional traits over 3 years. Again, sheep were used to rank the genotypes for relative palatability and this was included as a selection criterion.

In the final phase of the work, the best 12 genotypes were identified, vegetatively cloned and planted in replicated block experiments at 13 sites across southern Australia. These sites, planted in 2012, varied in geographic location, salinity, soil type and rainfall and represented the majority of niches where saltbush is grown. Genotypes were again assessed for survival, productivity and nutritional value. The most promising 4 genotypes and an industry standard were harvested and fed to sheep in a final energy and nitrogen metabolism experiment. The saltbush was fed with cereal hay as 50% of the diet as sheep rarely select a diet of 100% saltbush.

The first cultivar, named Anameka after the private farm in Tammin where much of the work was conducted, was commercialised in 2014. Anameka must be propagated by vegetative cuttings as the breeding system of saltbush means that plants grown from seed may not be phenotypically similar to the parents. In the first year, over 250000 Anameka plants have been sold. There are several other clonal cultivars that are ready for release if Anameka does not perform across all niches. Our future goal is to reduce establishment costs and the project team are now undertaking work to generate seed from crosses of the best genotypes and to determine the feeding value of the F1 generation. A seed line would lead to reduced costs in commercial nurseries and allow for the possibility of direct seeding in paddocks. This should increase the scale of adoption and will leverage the associated environmental benefits.

We anticipate that Anameka and associated releases from this project will improve livestock production outcomes from farms with large areas of land that is marginal for cropping. This saltbush work is an example of utilising a multidisciplinary approach in forage development for livestock. By understanding the farming systems and identifying the economic drivers for livestock production, we have selected a cultivar that should substantially improve profitability of the livestock enterprise.

4. Acknowledgements

The old man saltbush improvement work, funded by the CRC Future Farm Industries, has involved partners from CSIRO, New South Wales Department of Primary Industries, Western Australian Department of Parks and Wildlife, Department of Water Land and Biodiversity Conservation in South Australia and the South Australian Research and Development Institute. Contributors include Peter Jessop, Mike Bennell, Richard Mazanec, Wayne O'Sullivan, Ram Nair, Greg Sweeney, Steve Hughes, Alan Humphries, David McKenna, Trevor Hobbs, Dean Revell, Ed Barrett-Lennard, Paul Young and Alan Rintoul. We are very grateful to all of our host producers for input and support, particularly Tony, Donna and Simon York from Tammin. We also acknowledge support (at various stages) by; Meat and Livestock Australia, Australian Wool Innovation, South Australian Government, Gillamii Centre, Western Murray Land Improvement Group and a large cohort of regional producer and land management groups.

References

Al Daini, H., H.C. Norman, P. Young and E.G. Barrett-Lennard. 2013. The source of nitrogen (NH4+ or NO2–) affects the concentration of oxalate in the shoots and the growth of *Atriplex nummularia* (old man saltbush). Functional Plant Biology 40: 1057–1064.

Barrett-Lennard, E.G. 2003. The interaction between waterlogging and salinity in higher plants: causes, consequences and implications. Plant and Soil 253: 35–54.

Barrett-Lennard, E.G., R.J. George, G. Hamilton, H. Norman and D. Masters. 2005. Multi-disciplinary approaches suggest profitable and sustainable farming systems for valley floors at risk of salinity Australian Journal of Experimental Agriculture 45: 1415–1424.

Bennett, D., R. George and R. Silberstein. 2012. Changes in run-off and groundwater under saltbush grazing systems: preliminary results of a paired catchment study. Department of Agriculture and Food Western Australia Resource Management Technical Report 381. ISSN 1039–7205.

di Castri, F. 1981. Mediterranean type shrublands of the world. *In:* F. di Castri, D.W. Goodall and R.L. Specht (eds.). Ecosystems of the World 11: Mediterranean-Type Shrublands. Elsevier Scientific Publishing Company, Amsterdam.

Dolling, P.J. and W.M. Porter. 1994. Acidification rates in the central wheatbelt of Western Australia 1. On a deep yellow sand. Australian Journal of Experimental Agriculture 34: 1155–1164.

Fancote, C.R., P.E. Vercoe, K.L. Pearce, I.H. Williams and H.C. Norman. 2013. Backgrounding lambs on saltbush provides an effective source of Vitamin E that can prevent Vitamin E deficiency and reduce the incidence of subclinical nutritional myopathy during summer and autumn. Animal Production Science 53: 247–255.

George, R., J. Clarke and P. English. 2008. Modern and palaeogeographic trends in the salinisation of the Western Australian wheatbelt: a review. Soil Research 46: 751–767.

Henry, B., E. Charmley, R. Eckard, J.B. Gaughan and R. Hegarty. 2012. Livestock production in a changing climate: adaptation and mitigation research in Australia. Crop and Pasture Science 63: 191–202.

Hobbs, T.J. and M. Bennell. 2008. Agroforestry species profiles for lower rainfall regions of southeastern Australia FloraSearch. 1b. Report to the Joint Venture Agroforestry Program (JVAP) and Future Farm Industries CRC. Published by RIRDC, Canberra. Available at https://rirdc.infoservices.com.au/downloads/07-080.

Lawes, R.A., P.R. Ward and D. Ferris. 2014. Pasture cropping with C4 grasses in a barley–lupin rotation can increase production. Crop and Pasture Science 65: 1002–1015.

Masters, D.G. (in press). Assessing the feeding value of halophytes. pp. 89–105, this volume.

Masters, D.G., S. Benes and H.C. Norman. 2007. Biosaline agriculture for forage and livestock production. Agriculture Ecosystems & Environment 119: 234–248.

Masters, D.G., D.K. Revell and H.C. Norman. 2010. Managing livestock in degrading environments. pp. 255–267. *In:* N.E. Odongo, M. Garcia and G.J. Viljoen (eds.). Sustainable Improvement of Animal Production and Health. Food and Agriculture Organization of the United Nations (FAO), Rome, Italy.

Monjardino, M., D.K. Revell and D.J. Pannell. 2010. The potential contribution of forage shrubs to economic returns and environmental management in Australian dryland agricultural systems. Agricultural Systems 103: 187–197.

Nobs, M.A. 1980. Chromosome numbers in Australian species of *Atriplex*. Carnegie Institute Yearbook. 79: 164–169.

Norman, H.C., C. Freind, D.G. Masters, A.J. Rintoul, R.A. Dynes and I.H. Williams. 2004. Variation within and between two saltbush species in plant composition and subsequent selection by sheep. Australian Journal of Agricultural Research 55: 999–1007.

Norman, H.C., D.G. Masters, M.G. Wilmot and A.J. Rintoul. 2008. Effect of supplementation with grain, hay or straw on the performance of weaner Merino sheep grazing old man (*Atriplex nummularia*) or river (*Atriplex amnicola*) saltbush. Grass and Forage Science 63: 179–192.

Norman, H.C., D.K. Revell and D.G. Masters. 2009. Sheep avoid eating saltbushes with high sulphur concentrations. Proceedings of the XIth International Symposium on Ruminant Physiology (ISRP), Clermont-Ferrand, France, September 6 to 9, 20.

Norman, H.C., D.K. Revell, D. Mayberry, A.J. Rintoul, M.G. Wilmot and D.G. Masters. 2010. Comparison of *in vivo* organic matter digestion of native Australian shrubs by sheep to *in vitro* and *in sacco* predictions. Small Ruminant Research 91: 69–80.

Norman, H.C. and D.G. Masters. 2010. Predicting the nutritive value of saltbushes (*Atriplex* spp.) with near infrared reflectance spectroscopy. International Conference on Management of Soil and Groundwater Salinisation in Arid Regions, Sultan Qaboos University, Muscat, January 11–14, 2010.

Norman, H.C., P. Jessop and M.G. Wilmot. 2011. The role of sheep in saltbush domestication—what can they tell us? 8th International Symposium on the Nutrition of Herbivores. Aberystwyth, Wales 6–9 Sept. 2011.

O'Connell, M., J. Young and R. Kingwell. 2006. The economic value of saltland pastures in a mixed farming system in Western Australia. Agricultural Systems 89: 371–389.

Parr-Smith, G.A. 1982. Biogeography and evolution in the shrubby Australian species of *Atriplex* (Chenopodiaceae). pp. 291–299. *In*: W.R. Barker and P.J.M. Greenslade (eds.). Evolution of the Flora and Fauna of Arid Australia. Frewville, SA: Peacock Publications.

Pearce, K., H. Norman and D. Hopkins. 2010. The role of saltbush-based pasture systems for the production of high quality sheep and goat meat. Small Ruminant Research 91: 29–38.

Salem, H.B., H.C. Norman, A. Nefzaoui, D.E. Mayberry, K.L. Pearce, D.K. Revell and P. Morand-Fehr. 2010. Potential use of oldman saltbush (*Atriplex nummularia* Lindl.) in sheep and goat feeding. Small Ruminant Research 91: 13–28.

Walsh, M.J. and S.B. Powles. 2014. Management of herbicide resistance in wheat cropping systems: learning from the Australian experience. Pest Management Science 70: 1324–1328.

PART 2

Nutritional Aspects

Part 2 of 4 chapters focuses on nutritional aspects and assesses the peculiar situation where many halophytic and salt tolerant feedstuffs have a high level of anti-nutritional compounds that inhibit digestibility and feed value. Some contain toxins that create health hazards or otherwise restrict their usefulness in livestock rations. Detailed consideration is given to what is known about such phyto-toxins and we also explore the pitfalls involved in evaluating the true value of such feedstuffs.

Assessing the Feeding Value of Halophytes

David G. Masters

SYNOPSIS

Halophytes have been used for forage for thousands of years, although it is only in recent times with the increase in saline soils internationally, that improvement in feeding value has become a priority. For efficient use and targeted improvement of halophytes there must be accurate methods of assessing feeding value for grazing ruminants. Such methods are available for traditional pasture plants and these may be appropriate (with calibration) for halophytes that survive through exclusion or excretion of salt. They are not appropriate for halophytes that accumulate salt and other elements and compounds required for osmoregulation and management of reactive oxygen species. For these plants, either specific methods or standard methods modified to allow for the components within the halophytes that may influence digestion and metabolism are required.

Keywords: revegetated saline land, *in vitro* analysis, *in vivo* methods, livestock, ash content, mineral ions, chenopods, vacuoles, edible biomass, voluntary feed intake, energy, protein, antinutritional, chemometrics, microbial crude protein, secondary compounds.

1. Introduction

Grazing halophytes on revegetated saline land provides an opportunity for a productive solution to a widespread environmental problem. Consequently, there have been many experiments to evaluate the usefulness of halophytes for grazing,

The University of Western Australia, Western Australia, Australia.
Email: david.masters@uwa.edu.au

with considerable debate about the feeding value of these plants. Early publications included two consistent observations. Firstly, that grazing animals do not grow as well as predicted from the dry matter available and the measured *in vitro* analysis of the plants (Morecombe et al. 1996) and secondly, that livestock often (but not always) preferentially consume other plant species which are opportunistically growing among the halophytes even though they appear to have lower nutritive value and dry matter availability (Graetz and Wilson 1980; Norman et al. 2010b). Such observations have slowed the adoption of the technology available to plant and graze chenopods. It is likely that the methods used to assess the feeding value of these plants are not appropriate or accurate and that progress in the selection and breeding of halophytes for livestock has been held back by the uncertainty around the measurements made. This review addresses the methods available for the assessment of feeding value of halophytes for livestock.

For the purposes of the review, the definition of halophytes provided by Flowers and Colmer (2008) is used. In this publication, halophytes are defined as plants with the ability "to complete a life cycle in a salt concentration of at least 200 mM NaCl" (approximately 17 dS/m) "under conditions similar to those that might be encountered in a natural environment". On the basis of the mechanisms they use to overcome the toxic effects of high ion concentrations, these plants can be broadly classified as salt excluders, accumulators or excreters (Norman et al. 2013; Ozgur et al. 2013). The excluders and excreters are predominantly grasses and legumes while chenopods tolerate salinity through uptake and compartmentalisation of sodium and chloride ions into vacuoles where these ions play the major role in osmotic adjustment (Flowers and Colmer 2008).

The salt-accumulating plants usually have high ion concentrations in their leaves that, in some cases, may equal or exceed concentrations in sea water (Flowers et al. 1977). Consequently, ash content may represent up to half of shoot dry weight (Flowers and Colmer 2008; Norman et al. 2013). The ash is predominantly sodium, potassium and chloride but may also include elevated magnesium, calcium and other elements that are essential or toxic to animals (Beadle et al. 1957; Mayberry et al. 2010; Redondo-Gómez 2013). The accumulation of mineral ions is not the only factor influencing the feeding value of halophytes. As part of a process to maintain osmotic pressure, plants may also accumulate organic acids (e.g., oxalate, malate, citrate), soluble carbohydrates (e.g., sucrose), sugar alcohols (e.g., sorbitol), quarternary ammonium compounds (e.g., glycinebetaine, choline), amino acids (e.g., proline, aspartate, glutamate), nitrates and methylated sulphonio compounds (e.g., dimethylsulphonioproprionate) (Al Daini et al. 2013; Briens and Larher 1982; Flowers and Colmer 2008; Flowers et al. 1977; Gil et al. 2013; Masters et al. 2007; Norman et al. 2013; Osmond 1967; Trossat et al. 1998).

As well as the compounds for osmoregulation, halophytes may accumulate antioxidants including α-tocopherol, β-carotene, tannins and various other phenols, flavonoids and antioxidant enzymes. These are required to manage the elevated production of reactive oxygen species resulting from high salinity and other stresses normally associated with the saline environment (Buhmann and Papenbrock 2013; Norman et al. 2013; Ozgur et al. 2013).

This unusual and highly variable composition of halophytes when compared to conventional pasture plants has a significant influence on feeding value. Consequently, the traditional methods to predict livestock growth and production from forage consumption are often unsuitable. This is particularly the case for the chenopods, so this review focuses primarily on halophytes that accumulate ions and/or secondary compounds. Methods used for traditional forages are suitable for the halophytic grasses and legumes that do not reflect their environment through changes in the composition of dry matter; these methods are well described elsewhere and will not be repeated here (AFRC 1993; CSIRO 2007; National Research Council 2000).

2. Potential livestock production from halophytes

2.1 Definitions

Feeding value is often used loosely within the literature to define livestock responses to feeding or the animal production potential of a herbage. A useful approach to understanding and measuring feeding value is through the definition provided by Ulyatt (1973). In this review feeding value is defined as the product of intake and nutritive value:

$$Feeding\ value = f\,(intake \times nutritive\ value) \qquad (\text{Ulyatt 1973})$$

Nutritive value is described as the concentration of nutrients in a feed or animal response per unit intake. For halophytes the more appropriate definition is "animal response per unit intake" rather than "concentration of nutrients" because the potential interactions between nutrients and other components of the forage means that animal response per unit intake is not defined by nutrient intake alone.

Feed intake within this definition is the amount voluntarily consumed when supply is not limited. Feed intake may be influenced by a range of physical and chemical properties of the forage, including height, density and architecture of the forage, spatial distribution, sward structure, fibre content, resistance to degradation, digested particle size, energy density, protein content and the presence of toxins or unpalatable secondary compounds. Intake is also influenced by energy demand and therefore the physiological state of the animal, as well as climate, disease and most important for livestock grazing halophytes, the availability and quality of the water supply (CSIRO 2007; Ulyatt 1973; Weston 2002). Voluntary feed intake is therefore influenced by many factors that also contribute to nutritive value, meaning the two components of feeding value are interdependent.

While feeding value, as defined above provides a basis for the assessment of livestock production potential from halophytes, it is incomplete without some consideration of available biomass. Potential biomass of halophytes is highly variable and is a key driver in selection and cultivation of plants for agricultural use. The approach taken in this review is to include edible biomass as part of the assessment of the production potential of halophytes as follows:

$$Livestock\ production\ potential\ (per\ unit\ area) = f\,(feeding\ value \times edible\ biomass)$$

Within the production potential equation, edible biomass is weight of edible dry matter per unit area of production.

3. Assessment of edible biomass

Estimation of edible biomass for traditional pastures includes a range of destructive and non-destructive methods often based on many small sample areas across a uniform pasture sward (' t Mannetje 1978). Such techniques may be useful in assessing edible biomass is some of the halophytic grasses and legumes growing in uniform low to moderate salinity, but they will not be useful when the halophyte is a shrub. For shrubs, options include destructive methods where representative shrubs are harvested and edible material is removed, dried and weighed. Total edible dry matter can then be estimated through counting of total shrubs in a known area (' t Mannetje 1978). Alternatively simpler and less destructive techniques are available. The 'Adelaide' technique is widely used; this requires the selection of a branch of the species to be assessed (defined as a 'unit'). Representative shrubs are then scored for the number of equivalent 'units' growing on the plant. At the end of the period of measurement, the selected branch is stripped of edible material, dried and weighed. Production of edible dry matter per unit area is calculated by multiplying the estimated average edible dry matter per shrub by the number of shrubs per unit area (Andrew et al. 1979).

For traditional pastures, edible biomass is assumed to be equal to total above-ground biomass; again this assumption can be applied to most halophytic grasses and legumes. This is not the case with woody plants, from which grazing livestock will consume only leaves and small stems. In these situations, edible biomass is usually assumed to be all leaves plus stems up to a maximum defined diameter. This diameter may range from 1.5 mm up to 6 mm (' t Mannetje 1978; Atiq-ur-Rehman et al. 1999; Norman et al. 2008) and will be influenced by plant species and stage of maturity, feeding conditions and animal species. Within any assessment of halophytic shrubs, the maximum diameter of edible biomass should be determined through observation and defined within subsequent publication.

Many publications describing the value of halophytes for grazing only provide information on biomass production. This information is essential for the determination of production potential, however, on its own, it is not enough and can lead to inaccurate conclusions; edible biomass and nutritive value are also required (Masters and Norman 2010).

4. Determining the nutritive value of halophytes

The methods used for assessment of nutritive value are not geographically consistent. This review will not cover all methods used in different parts of the world but will focus on the more common and accepted options used for traditional pastures.

In addressing the nutritive value, it is worthwhile considering the major components of feeds that contribute to this value. These include energy and protein, minerals and vitamins. For halophytes understanding and, where possible, quantifying the role of a range of pro- and anti-nutritional factors that may interact to change the energy, protein, mineral and vitamin utilisation is also vital.

4.1 Energy

4.1.1 In vivo methods

The most accurate method of determining the animal response per unit of feed intake is to measure this in a feeding experiment. Such experiments are time-consuming, expensive and for these reasons, unsuitable for screening large numbers of plants. Many years of research have been required to develop other *in vivo* and *in vitro* methods to provide rapid screening alternatives (France et al. 2000; Kitessa et al. 1999; Minson 1990).

It should be stated at the start of this section that there are no accurate, indirect methods for assessing the nutritive value of halophytes. All have limitations, some are useful indicators and others are misleading and unhelpful (Masters et al. 2005a; Norman et al. 2013). Most of the methods for prediction of energy value for traditional feeds are based around measurement of the digestibility and the conversion of digestibility to digestible energy (DE). Digestibility is derived for the feed as:

$$Dry\ matter\ digestibility\ (DMD) = \frac{Feed\ DM - Faeces\ DM}{Feed\ DM}$$

Alternatively it can be based on the energy yielding fraction of the feed as either:

$$Organic\ matter\ digestibility\ (OMD) = \frac{Feed\ OM - Faeces\ OM}{Feed\ OM}$$

or as:

$$Digestible\ organic\ matter\ in\ the\ dry\ matter\ (DOMD) = \frac{Feed\ OM - Faeces\ OM}{Feed\ DM}$$

These digestibility measurements are usually made with penned, mature livestock fed only the feed of interest and at the maintenance level of feeding. Forage is provided through cut and carry. Cut and carry can mean that newly harvested forages are fed fresh or that they are conserved through oven or air drying before feeding. Either process will result in some changes in nutritive value. Under best practice, feed of known digestibility is also used as a standard reference. From the measurement of digestibility and gross energy, DE can be calculated. In practice, the calculation of DE is bypassed and metabolisable energy (ME) is predicted directly from digestibility. Metabolisable energy is the energy from the diet available for utilisation after allowance for losses in faeces, urine and methane. It is generally accepted that:

$$Metabolizable\ energy\ (ME) = 0.81 \times Digestible\ energy\ (DE)*$$

Metabolizable energy is the measurement of most use for halophytes because it provides an indication of the energy available for maintenance, growth and production. It is expressed as MJ/Kg DM. There are a number of different equations used to convert digestibility to ME. These are available for various diets types (AFRC 1993; CSIRO 2007; Minson 1990).

---

* See (CSIRO 2007) for full discussion of the derivation of DE.

The application of digestibility measurements to halophytes and their use for prediction of ME is potentially very misleading. Given that the halophytes may have a mineral content between 6 and 50% (Gihad and El Shaer 1994) and that minerals are often absorbed and therefore appear to be digested but have no energy value, DMD alone cannot be used to predict ME without the measurement of gross energy in both feed and faeces. OMD does provide a measure of utilisation of the organic matter in the feed but conversion of digestibility expressed per unit of organic matter to energy expressed per unit of dry matter is best addressed using DOMD. DOMD does provide an indication of the amount of digestible organic matter per unit of feed dry matter. This is useful for preliminary assessment and comparison of the nutritive value of halophytes. Prediction equations for conversion of DOMD to ME have not yet been developed for halophytes, for roughages, the following equation has been proposed:

$$Metabolizable\ energy = 0.194DOMD - 2.577 \quad \text{(CSIRO 2007)}$$

However, while DOMD and the ME predicted from it are useful, the variable composition of halophytes still means that this may provide an overestimation of available energy. For example, betaine, choline, oxalates, tannins and nitrates may account for a significant proportion of the organic matter in the dry matter (as defined by loss during ashing), there are reports that glycine betaine alone may constitute up to 7% of shoot dry weight (Storey et al. 1977). These compounds are likely to make little or no contribution to the energy value of the diet but will still contribute to DOMD.

Similar comments apply to the use of *in sacco* or *in situ* techniques for halophytes. This methodology is used to predict the digestibility of feedstuffs through the measurement of disappearance of feed material from nylon bags suspended in the rumen (Quin et al. 1938). The technique allows for the comparative digestion of a number of different feedstuffs within the same rumen. Conditions within the bag or within the rumen can be changed in order to determine their impact on digestion. A further limitation when the technique is applied to novel feeds that contain anti- or pro-nutritional factors, or other components that will influence digestive kinetics (e.g., salt). These components will have little of no impact on apparent *in sacco* digestion because of a dilution effect within the rumen (Mould 2003). The digestion of the test feedstuff is also very sensitive to the composition of the basal diet and results may be erroneous if the basal diet is not similar to the test diet (Kempton 1980).

An alternative to the estimation of ME from digestibility of organic matter in the diet is to subtract the energy in urine, faeces and methane from the gross energy of the diet (Arieli et al. 1989). While the method is sound and avoids many of the errors caused by the unusual composition of halophytes, it is not a practical option for large scale screening and selection of forage options.

4.1.2 Chemical and in vitro methods

The requirement to rapidly assess and compare the nutritive value of feeds subsequently led to the prediction of metabolisable or net energy from methods based on proximate analysis, chemical analysis of fibre (Van Soest 1963), chemical analysis of fibre in combination with other chemical components (Lofgreen 1953; Maynard 1953; Weiss et al. 1992) or on simulated digestion in either rumen fluid (Czerkawski and Breckenridge

1977; Tilley and Terry 1963) or in a combination of digestive enzymes (Clarke et al. 1982; Newman 1972) or gas production (Busquet et al. 2006; Theodorou et al. 1994). These methods require, at a minimum, testing and calibration against *in vivo* methods carried out on the same type of feed material.

As a general principle, application of these methods to halophytes should focus on identification and measurement of feed components that are both digestible and provide energy. As with *in vivo* techniques, methods based on the digestion of inorganic matter should be avoided. Caution is also required when digestible fibre is determined in the laboratory. In the automation of fibre analysis, the step of differentiating between fibre and minerals through ashing the extracted sample (Faichney and White 1983; Goering and Van Soest 1970) is not always included. Unless this is performed as an additional step insoluble ash may be misinterpreted as fibre. Preliminary estimates of nutritive value should be based on the digestibility of organic matter or on measurement of the products of fermentation (e.g., gas production).

All *in vitro* methods involve the use of assumptions on how the fibre and other organic components of the diet will be digested in the rumen (and elsewhere in the digestive tract). These assumptions may apply to the halophytes that grow in low or moderate salinity, exclude or excrete salt and do not accumulate secondary compounds, but are unlikely to apply to salt accumulating halophytes. For example, a high concentration of salt in the diet significantly changes the rate of passage of nutrients through the rumen, and as a result the rate and extent of fibre digestion is changed (Hemsley et al. 1975; Masters et al. 2005b; Thomas et al. 2007). Salt is not the only factor to influence efficiency of digestion. Weston et al. (1970) fed *Atriplex nummularia* to sheep and reported that while the organic matter digestibility of saltbush was within the range reported for pasture grasses and legumes, the rumen played a less important role in the digestion of this organic matter and ruminal absorption of volatile fatty acids was impaired. Similarly, Mayberry et al. (2009) reported that feeding *Atriplex nummularia* to sheep reduced rumen efficiency and that this reduction in efficiency was not due to high salt concentrations. These are two of the very few investigations into the mechanisms of halophyte digestion in the rumen. The conclusions from these publications are that salt and other compounds influence fibre digestion in the rumen.

In any *in vitro* system, rate of passage must be simulated and manipulated by the user and the influence of other secondary compounds that alter rumen efficiency may or may not also be expressed *in vitro*. There appear to only be a few studies in which *in vivo* and *in vitro* digestibility of organic matter have been compared in diets that are 100% halophyte. Norman et al. (2010a) compared *in vivo*, *in vitro* and *in sacco* methods on eleven Australian native shrubs. A good relationship was established between *in vivo* OMD and *in vitro* OMD (measured using a pepsin-cellulase method) but the *in vitro* method still over-predicted OMD by 11% even after application of a correction factor to account for the effect of salt on digestibility. Conclusions were that this method could be used to rank shrubs within groups but was not appropriate for predicting livestock production potential of the plant. Of all the methods used, *in vitro* gas production was the most highly correlated with *in vivo* OMD. The study derived a series of correction or calibration equations for the compared methods. The calibration equations were first derived from the relationship between industry standard samples (of known digestibility) and the methods used. An additional correction factor was also applied to allow for the *in vivo* influence of salt in the plant material on digestibility. These salt correction equations were relevant because of the variable salt content both within and between halophytes species: a single correction factor cannot be applied when salt content is highly variable across samples. The correction equation was derived using data from two other experiments (Masters et al. 2005b; Thomas et al. 2007). These data were reanalysed to provide an estimate of the reduction in OMD caused by high salt intakes (Fig. 6.1). The correction equation applied to the *in vitro* methods was:

Reduction in OMD (%) = $-0.0002x^2 + 0.023x$ (where x = 0.793 x soluble ash in g/kg DM)[1]

What was clearly demonstrated in this research was that calibration of *in vitro* methods using commercial and verified internal non-halophyte standards is not sufficient to provide livestock feeding potential for halophytes. The approach requires internal calibration with plant material of similar composition with known digestibility and nutritive value properties derived from *in vivo* feeding trials.

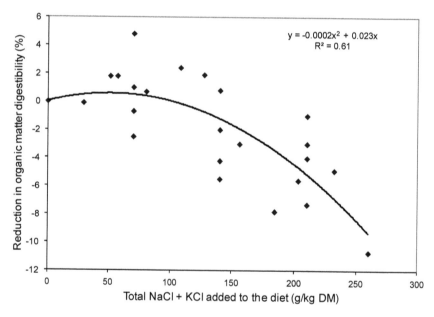

Fig. 6.1. Reduction in organic matter digestibility from adding salt (NaCl + KCl) to the diet. Data sources from Masters et al. (2005b) and Thomas et al. (2007).

4.1.3 Near infrared reflectance spectroscopy

Near infrared reflectance spectroscopy (NIRS) is a method that is fast, accurate if correctly calibrated and non-destructive. The method is now used widely to predict

---

[1] The figure of 0.793 x soluble ash was used as it represented the sum of NaCl and KCl in the samples.

the chemical composition and nutritive value of traditional forages and grains. The procedure is based on differential absorption of light by chemical moieties in the feed within the spectral range of 1100 and 2500 nm. The spectral data are processed using "chemometrics": a modelling procedure that relates spectral data to the measured characteristic of the feed (Givens et al. 1997; Kitessa et al. 1999; Landau et al. 2006). The method is rapid and capable of use for the prediction of many nutritive value characteristics including dry matter and organic matter digestibility, fibre, protein and minerals (Givens et al. 1991; Givens et al. 1997). It has also been shown that NIRS can be used for the prediction of feed intake (Coleman and Henry 2002). NIRS is still an indirect measure of nutritive value so calibration with appropriate plant species with measured *in vivo* and *in vitro* characteristics is required.

The methodology has been applied to shrubs, Meuret et al. (1993) determined organic matter, nitrogen, ADF and NDF using NIRS in 25 different Mediterranean tree and shrub species and concluded the accuracy of the NIRS prediction was similar to that obtained through classical analysis procedures. Similar conclusions were reached by Garcia Ciudad et al. (2004) when NIRS was used to assess forage quality of the leguminous shrub *Cytisus multiflorus*. Flinn et al. (1996) successfully used NIRS to predict nutritive value and the concentration of a range of minerals in the forage shrub tagasaste (*Chamaecytisus proliferus*). Problems with the procedure have been identified when large amounts of inorganic minerals are added to diets (Givens et al. 1997) but this has not been seen as a problem with halophytes and some progress has been made in the development of predictive equations for a range of saltbushes (*Atriplex* spp.) (Norman and Masters 2010). Esfahan et al. (2010) have reported on the use of NIRS to successfully predict DMD, water-soluble carbohydrates, crude protein (CP), crude fibre, ME and ADF in two salt accumulating halophytes (*Atriplex leucoclada* and *Sueda vermiculata*), however, it is not clear from the methods cited in this publication, if calibration equations were derived from relevant plant species.

Given the significant progress and value of NIRS in the prediction of nutritive value in traditional pasture plants, the methodology is an obvious long-term option for evaluation of halophytes. This option will require the development of multispecies calibrations for the prediction of livestock production potential and revegetation design within halophyte systems and single-species calibrations for the selection and breeding of improved cultivars for livestock within a single species. The most efficient way to approach this would be through the development and application of prediction equations at a small number of specialised laboratories. These laboratories would then offer the service of processing submitted samples or of applying calibrations to NIRS scans from uncalibrated sources. The integration of nutritive value, feed intake and biomass then provides a direction for both future research and commercial use of halophytes.

As with other predictive methods, NIRS does not directly take account of the impact of toxic components of the diet or any deficiency in specific nutrients. However, the technique may allow for measurement of known secondary compounds following future characterisation and calibration.

4.2 Protein

With very few exceptions, publications describing the protein value for halophytes use CP as the basis for evaluation. Crude protein is usually derived when nitrogen content is multiplied by 6.25. This is based on the assumption that protein contains 16% nitrogen (CSIRO 2007). Because ruminants have the ability to both degrade and construct protein within the rumen, CP alone is not an indicator of protein or amino acids available for absorption by the animal. The ability of forage to provide protein to the ruminant depends on the following factors (CSIRO 2007):

- The supply of nitrogen together with other nutrients in amounts sufficient to promote optimal rates of fermentation and growth of rumen microorganisms. Nitrogen supply alone is not an indicator of microbial crude protein (MCP) production. It is nitrogen in combination with other nutrients and, in particular, nutrients that supply rumen microbes with energy that determine MCP production.
- The amount of protein and amino acids in the feed that escape degradation in the rumen (undegraded dietary protein—UDP) and, together with MCP leaving the rumen, provide a supply of protein to be digested and absorbed by the animal.

The combination of MCP and UDP is termed truly digestible protein leaving the stomach (DPLS). It is this characteristic that best describes protein value of feed. This is unfortunately not a parameter that can be easily measured.

The evaluation of protein supply from halophytes is more complex than for traditional feeds. Firstly halophytes potentially contain high concentrations of non-protein nitrogen (NPN) compounds (e.g., amino acids, quarternary ammonium compounds, and nitrates) as part of the osmoregulation mechanism. For example, betaine concentrations that contributes up to 6 g N/kg DM are equivalent to about 4% CP (Storey et al. 1977). NPN as high as 42% of total nitrogen has been reported in *Atriplex barclayana* (Benjamin et al. 1992). Many of these NPN compounds are partially or fully degraded in the rumen and are therefore a source of nitrogen for MCP production (Mitchell et al. 1979) but as NPN are unable to contribute to the UDP component of DPLS. In addition, some of the true protein in halophytes would also be degraded in the rumen (RDP) and this would also contribute to the pool of nitrogen available for synthesis into MCP. The extent to which this NPN and RDP can be converted to MCP is highly dependent on the DE supply available for microbial fermentation. Various relationships have been published to predict the energy and nitrogen utilisation in the rumen (AFRC 1993; CSIRO 2007). For the purposes of this review, one of the relationships cited in CSIRO (2007) is used as described:

$$MCP\ (g) = 182 \times Digestible\ organic\ matter\ intake\ (DOMI)\ (kg)$$

By applying this equation to an example of a sheep consuming 1 kg of a typical *Atriplex* (saltbush) forage (DOMD = 45%) (Norman et al. 2010a), potential MCP production is 82 g. To supply the required nitrogen, the saltbush consumed would also need to have an RDP content of 82 g/kg DM intake or 8.2% RDP. Any amount of RDP above 8.2% would be unusable because of an energy deficiency unless it was both true protein and capable of escaping degradation in the rumen (UDP). This interpretation is supported by running a simulation using the livestock nutrition decision support package GrazFeed (Freer et al. 1997) with the saltbush described above (DOMD = 45% and ME = 6.0 MJ/kg DM) together with the estimates of 69% nitrogen digestibility with 60% of this as RDP and 28% as UDP reported by others (Arieli et al. 1989). With *ad libitum* feeding, a 50 kg sheep was simulated to be losing more than 100 g live weight per day due to inadequate energy supply but, with 9% CP in the diet, had more RDP and UDP than could be utilised. High CP was no value when ME was low.

Assessment of the protein value of halophytes is therefore best based on the ability to provide sufficient soluble nitrogen to support optimum MCP production. The contribution to UDP remains to be determined but is likely to be low. For this reason, CP analysis is useful but should not be used in the context of being an indicator of DPLS.

4.3 Minerals and secondary compounds

4.3.1 Minerals

The halophytes that do not exclude or excrete salt have high concentration of a range of essential minerals (Gihad and El Shaer 1994; Norman et al. 2002). Some of these are present at concentrations that exceed the Maximum Tolerable Level for ruminant consumption (National Research Council 2005). This does not necessarily mean they will cause clinical toxicity but may result in a depression in intake and therefore limit the ability of the animal to utilise the energy and protein available in the feed. For example, livestock decrease voluntary feed intake when sodium chloride in the diet exceeds 5% (Fig. 6.2) (Masters et al. 2005b; Thomas et al. 2007). Although there is

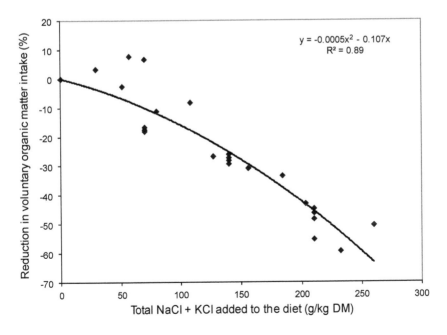

Fig. 6.2. Reduction in voluntary organic matter intake from adding salt (NaCl + KCl) to the diet. Data sources from Masters et al. (2005b) and Thomas et al. (2007).

now a good understanding of the effects of high sodium and potassium intakes from halophytes on grazing ruminants the same cannot be said for other minerals, for example, Mayberry et al. (2010) reported a large negative apparent absorption for magnesium, calcium and phosphorus in sheep fed *Atriplex nummularia*, even though the plants contained high concentrations of these minerals. There is therefore, still much to learn about the effects of high mineral intakes from halophytes on livestock health and production. For this reason, measurement of mineral composition is an integral part of the assessment of feeding value of halophytes.

4.3.2 Secondary compounds

Within this review, the term 'secondary compounds' is used loosely to include elements or compounds that have been reported in concentrations much higher than in traditional forage plants. They can be allocated into 3 broad categories:

- Osmolytes or ionic compounds that play a role in maintaining cell volume and fluid balance. Common examples listed earlier in this review include organic acids (e.g., oxalate, malate, citrate), soluble carbohydrates (e.g., sucrose), sugar alcohols (e.g., sorbitol), quarternary ammonium compounds (e.g., glycinebetaine, choline), amino acids (e.g., proline, aspartate, glutamate), nitrates and methylated sulphonio compounds (e.g., dimethylsulphonioproprinate).
- Antioxidants that are required to protect against cellular damage which will result from the high production of reactive oxygen species in halophytes under stress. Common examples include α-tocopherol, β-carotene, tannins and various other phenols and flavonoids.
- Anti-nutritional compounds to discourage damage to the plant from herbivores. Hundreds or even thousands of years of opportunistic and uncontrolled grazing have resulted in higher survival of unpalatable, low nutritive value plants with poor livestock production as an inevitable consequence. The compounds responsible for discouraging grazing are much less well characterised and may include alkaloids and saponins (Gihad and El Shaer 1994). The group may also include compounds that perform dual roles, oxalate, for example, appears to have a role in both osmoregulation and grazing deterrence.

Many of these compounds have potentially profound effects on grazing ruminants, but, most importantly these effects are not always negative. Some of the functional compounds may have a positive influence on livestock production (Attia-Ismail, this volume). Betaine may increase the ability of the animal to deal with osmotic stress and decrease intramuscular fat (Fernández et al. 1998). The α-tocopherol, β-carotene in halophytes can be converted to vitamins E and A respectively, and these vitamins play an essential role as antioxidants in animals as they do in plants. Providing access by sheep to *Atriplex nummularia* and *Atriplex amnicola* significantly increased vitamin E status and, as a consequence, storage quality of meat in comparison to sheep fed dry crop residues (Pearce et al. 2005). Condensed tannins at low concentration will improve protein utilisation but at high concentrations will depress feed intake and reduce growth (Waghorn et al. 2002). Sugars, amino acids and some organic acids may provide energy to livestock or, at a minimum, be readily digested with no negative impact.

Other compounds such as oxalates, nitrates, saponins and alkaloids are toxic at high concentrations and present a risk to health and production (Attia-Ismail, this volume).

With the high number of secondary compounds contained in halophytes along with high variation both between and within plant species, it is an impossible task to identify and measure all of these and then select for beneficial concentrations of each. Routine laboratory-based screening is either not available or not practical given the range of candidate compounds. Such an approach may, in any case be a flawed selection pathway and may just lead to plants that cannot cope with the environmental conditions in which they are planted. An alternative is the use of livestock in the selection process to screen for palatability or preference (Gihad and El Shaer 1994; Masters and Norman 2014). Through a range of pre- and post-ingestive sensory signals, ruminants make dietary choices (Provenza 1995), in some cases integrating signals to attain a minimal total discomfort (Forbes and Mayes 2002). This livestock preference can be incorporated into a halophyte selection program (Masters and Norman 2014). While the approach is useful, it should not be seen as infallible; livestock do avoid plants simply due to a lack of familiarity and show different preferences within a mixed sward (Provenza et al. 1998). This may lead to incorrect short-term conclusions; these can be managed through appropriate experimental design. The use of livestock in halophyte selection should be seen as an evolutionary step—traditional selection of pasture plants is based on a prescriptive approach where the scientist makes the decision on the appropriate plant characteristics. This alone is unlikely to be successful when the plants involved have a complex composition, partly developed to discourage grazing and are growing in hostile and uncontrolled environments.

The challenge for livestock and plant scientists in designing halophyte-based feeding systems is to find plants with the required internal mechanisms and therefore composition to allow healthy plant growth in a harsh environment but, with a set of compounds used in these mechanisms that can either be metabolised by the animal at minimal metabolic cost or provide the animal with production and/or health benefits. Identifying and understanding the role of secondary compounds with halophytes will result in systems that will benefit both the local environment and the grazing animal.

5. Conclusions

Livestock production from saline land and the profitability of halophyte systems can be increased with improved nutritive value, increased voluntary feed intake and increased biomass production from halophytes. These improvements can only be achieved if changes in potential livestock production can be measured and/or predicted. Measuring livestock production directly is accurate and informative but is not a practical approach for the screening and development of new plants and grazing systems. Higher throughput *in vivo, in sacco, in vitro* and laboratory methods are available for traditional pasture plants but these require selective use with modification and calibration for application to halophytes. The following parameters are suggested for routine description of halophytes for livestock:

1. *Edible biomass.* For halophyte shrubs, this has been estimated as all leaves and any stems below a minimum diameter. The minimum diameter may range from

1.5–6.0 mm and may be dependant of type of plant and stage of maturity. For halophytic grasses and legumes all dry matter is assumed edible.

2. *Digestible organic matter in the dry matter (DOMD)*. This can be estimated from *in vivo* or calibrated *in vitro* measurement and provides an indication of ME available. For halophytes containing high concentrations of minerals and secondary compounds DOMD is likely to overestimate the ME value.

3. *Crude protein.* Provides an indication of whether there is sufficient RDP and UDP to complement the ME provided by the plant. If ME is low, CP is of secondary importance.

4. *Minerals*. Ash and if possible macro and micro elements are required to assess potential toxicities and imbalances.

5. *Secondary compounds*. Key secondary compounds should be described and quantified where possible.

The longer term approach to improving halophytes for grazing ruminants will be best served through the application of methods such as NIRS that are low cost and high throughput for the prediction of nutritive value. These can be used in combination with grazing experiments to screen out unpalatable species. Integration of knowledge on plant physiology and animal nutrition will facilitate the selection of plants that use pro- rather than anti-nutritional compounds as part of the homeostatic control required to thrive in a harsh environment.

References

' t Mannetje, L. 1978. Measuring quantity of grassland vegetation. pp. 63–95. *In:* L. ' t Mannetje (ed.). Measurement of Grassland Vegetation and Animal Production. CAB International.

AFRC. 1993. Energy and Protein Requirements of Ruminants. CABI International, Wallingford, UK.

Al Daini, H., H.C. Norman, P. Young and E.G. Barrett Lennard. 2013. The source of nitrogen (NH_4^+ or NO_3^-) affects the concentration of oxalate in the shoots and the growth of *Atriplex nummularia* (oldman saltbush). Functional Plant Biology 40: 1057–1064.

Andrew, M.H., I.R. Noble and R.T. Lange. 1979. A non-destructive method for estimating the weight of forage on shrubs. Australian Rangeland Journal 1: 225–231.

Arieli, A., E. Naim, R.W. Benjamin and D. Pasternak. 1989. The effect of feeding saltbush and sodium chloride on energy metabolism in sheep. Animal Production 49: 451–457.

Atiq-ur-Rehman, J.B. Mackintosh, B.E. Warren and D.R. Lindsay. 1999. Revegetated saline pastures as a forage reserve for sheep: I. Effects of season and grazing on morphology and nutritive value of saltbush. Rangeland Journal 21: 3–12.

Attia-Ismail, S.A. (in press). Plant secondary metabolites of halophytes and salt tolerant plants. pp. 127–142, this volume.

Beadle, N.C.W., R.D.B. Whalley and J.B. Gibson. 1957. Studies in halophytes II. Analytic data on the mineral constituent of three species of *Atriplex* and their accompanying soils in Australia. Ecology 38: 340–344.

Benjamin, R.W., E. Oren, E. Katz and K. Becker. 1992. The apparent digestibility of *Atriplex barclayana* and its effect on nitrogen balance in sheep. Animal Production 54: 259–264.

Briens, M. and F. Larher. 1982. Osmoregulation in halophytic higher plants: a comparative study of soluble carbohydrates, polyols, betaines and free proline. Plant, Cell and Environment 5: 287–292.

Buhmann, A. and J. Papenbrock. 2013. An economic point of view of secondary compounds in halophytes. Functional Plant Biology 40: 952–967.

Busquet, M., S. Calsamiglia, A. Ferret and C. Kamel. 2006. Plant extracts affect *in vitro* rumen microbial fermentation. Journal of Dairy Science 89: 761–771.

Clarke, T., P.C. Flinn and A.A. McGowan. 1982. Low-cost pepsin-cellulase assays for prediction of digestibility of herbage. Grass and Forage Science 37: 147–150.

Coleman, S.W. and D.A. Henry. 2002. Nutritive value of herbage. pp. 1–26. *In*: M. Freer and H. Dove (eds.). Sheep Nutrition. CSIRO Publishing, Collingwood.

CSIRO. 2007. Nutrient Requirements of Domesticated Livestock. p. 270. *In*: M. Freer, H. Dove and J.V. Nolan (eds.). CSIRO Publications, East Melbourne.

Czerkawski, J.W. and G. Breckenridge. 1977. Design and development of a long-term rumen simulation technique (Rusitec). The British Journal of Nutrition 38: 371–384.

Faichney, G.J. and G.A. White. 1983. Methods for the Analysis of Feeds Eaten by Ruminants. CSIRO, Melbourne.

Fernández, C., L. Gallego and C.J. Lopez-Bote. 1998. Effect of betaine on fat content in growing lambs. Animal Feed Science and Technology 73: 329–338.

Flinn, P.C., N.J. Edwards, C.M. Oldham and D.M. McNeill. 1996. Near infrared analysis of the fodder shrub tagasaste (*Chamaecytisus proliferus*) for nutritive value and anti-nutritive factors. pp. 576–580. *In*: A.M.C. Davies and P. Williams (eds.). Near Infrared Spectroscopy: The Future Waves. NIR Publications, Chichester.

Esfahan, E.Z., M.H. Assarch, M. Jafari, A.A. Jafari, S.A. Javadi and G. Karimi. 2010. Phenological effects on forage quality of two halphyte species *Atriplex leucoclada* and *Sueda vermiculata* in four saline rangelands of Iran. Journal of Food, Agriculture & Environment 8: 999–1003.

Flowers, T.J. and T.D. Colmer. 2008. Salinity tolerance in halophytes. New Phytologist 179: 945–963.

Flowers, T.J., P.F. Troke and A.R. Yeo. 1977. The mechanism of salt tolerance in halophytes. Annual Reviews of Plant Physiology 28: 89–121.

Forbes, J.M. and R.W. Mayes. 2002. Food choice. pp. 51–69. *In*: M. Freer and H. Dove (eds.). Sheep Nutrition. CSIRO Publishing, Collingwood.

France, J., M.K. Theodorou, R.S. Lowman and D.E. Beever. 2000. Feed evaluation for animal production. pp. 1–9. *In*: M.K. Theodorou and J. France (eds.). Feeding Systems and Feed Evaluation Models. CABI Publishing.

Freer, M., A.D. Moore and J.R. Donnelly. 1997. GRAZPLAN: Decision support systems for Australian enterprises - II. The animal biology model for feed intake, production and reproduction and the GrazFeed DSS. Agricultural Systems 54: 77–126.

Garcia Ciudad, A., B. Fernández Santos, B.R. Vázquez de Aldana, I. Zabalgogeazcoa, M.Y. Gutiérrez and B. Garcia Criado. 2004. Use of near infrared reflectance spectroscopy to assess forage quality of a mediterranean shrub. Communications in Soil Science and Plant Analysis 35: 665–678.

Gihad, E.A. and H.M. El Shaer. 1994. Utilization of halophytes by livestock on rangelands. Problems and prospects. pp. 77–96. *In*: V. Squires and A.T. Ayoub (eds.). Halophytes as a Resource for Livestock and for Rehabilitation of Degraded Land. Kluwer Academic Publishers, Dordrecht, Netherlands.

Gil, R., M. Boscaiu, C. Lull, I. Bautista, A. Lidón and O. Vicente. 2013. Are soluble carbohydrates ecologically relevant for salt tolerance in halophytes? Functional Plant Biology 40: 805–818.

Givens, D.I., C.W. Baker, A.R. Moss and A.H. Adamson. 1991. A comparison of near-infrared reflectance spectroscopy with three *in vitro* techniques to predict the digestibility *in vivo* of untreated and ammonia-treated cereal straws. Animal Feed Science and Technology 35: 83–94.

Givens, D.I., J.L. De Boever and E.R. Deaville. 1997. The principles, practices and some future applications of near infrared spectroscopy for predicting the nutritive value of foods for animals and humans. Nutrition Research Reviews 10: 83–114.

Goering, H.K. and P.J. Van Soest. 1970. Forage fibre analyses Agricultural Handbook No. 379. Jacket No: 387–598. United States Department of Agriculture.

Graetz, R.D. and A.D. Wilson. 1980. Comparison of the diets of sheep and cattle grazing a semi-arid chenopod shrubland. Australian Rangeland Journal 2: 67–75.

Hemsley, J.A., J.P. Hogan and R.H. Weston. 1975. Effect of high intake of sodium chloride on the utilisation of a protein concentrate by sheep. II Digestion and absorption of organic matter and electrolytes. Australian Journal of Agricultural Research 26: 715–727.

Kempton, T.J. 1980. The use of nylon bags to characterise the potential degradation of feeds for ruminants. Tropical Animal Production 5: 107–116.

Kitessa, S., P.C. Flinn and G.G. Irish. 1999. Comparison of methods used to predict the *in vivo* digestibility of feeds in ruminants. Australian Journal of Agricultural Research 50: 825–841.

Landau, S., T. Glasser and L. Dvash. 2006. Monitoring nutrition in small ruminants with the aid of near infrared reflectance spectroscopy (NIRS) technology: A review. Small Ruminant Research 61: 1–11.

Lofgreen, G.P. 1953. The estimation of total digestible nutrients from digestible organic matter. Journal of Animal Science 12: 359–365.

Masters, D. and H. Norman. 2010. Salt tolerant plants for livestock production. pp. 23–31. *In*: International Conference on Soils and Groundwater Salinization in Arid Environments, Sultan Qaboos University, Muscat, Oman.

Masters, D.G., S.E. Benes and H.C. Norman. 2007. Biosaline agriculture for forage and livestock production. Agriculture, Ecosystems and Environment 119: 234–248.

Masters, D.G. and H.C. Norman. 2014. Genetic and environmental management of halophytes for improved livestock production. *In*: M.O. Muhammad Ajmal Khan, Bilquees Gul and Muhammad Zaheer Ahmed (eds.). Halophytes for Food Security in Dry Lands. In Press, Doha, Qatar.

Masters, D.G., H.C. Norman and E.G. Barrett-Lennard. 2005a. Agricultural systems for saline soil: the potential role of livestock. Asian-Australasian Journal of Animal Science 18: 296–300.

Masters, D.G., A.J. Rintoul, R.A. Dynes, K.L. Pearce and H.C. Norman. 2005b. Feed intake and production in sheep fed diets high in sodium and potassium. Australian Journal of Agricultural Research 56: 427–434.

Mayberry, D., D. Masters and P. Vercoe. 2010. Mineral metabolism of sheep fed saltbush or a formulated high-salt diet. Small Ruminant Research 91: 81–86.

Mayberry, D.E., D.G. Masters and P.E. Vercoe. 2009. Saltbush (*Atriplex nummularia*) reduces efficiency of rumen fermentation in sheep. Options Méditerranéennes 85: 245–250.

Maynard, L.A. 1953. Total digestible nutrients as a measure of feed energy. Journal of Nutrition 51: 15–21.

Meuret, M., P. Dardenne, R. Biston and O. Poty. 1993. The use of NIR in predicting nutritive value of mediterranean tree and shrub foliages. Journal of Near Infrared Spectroscopy: 45–54.

Minson, D.J. 1990. Forage in Ruminant Nutrition. Academic Press Inc., New York.

Mitchell, A.D., A. Chappell and K.L. Knox. 1979. Metabolism of betaine in the ruminant. Journal of Animal Science 49: 764–773.

Morecombe, P.W., G.E. Young and K.A. Boase. 1996. Grazing a saltbush (*Atriplex-Maireana*) stand by Merino wethers to fill the 'autumn feed-gap' experienced in the Western Australian wheat belt. Australian Journal of Experimental Agriculture 36: 641–647.

Mould, F.L. 2003. Predicting feed quality—chemical analysis and *in vitro* evaluation. Field Crops Research 84: 31–44.

National Research Council. 2000. Nutrient Requirements of Beef Cattle. 7th ed. The National Academies Press, Washington.

National Research Council. 2005. Mineral Tolerance of Animals. The National Academies Press, Washington.

Newman, D.M.R. 1972. A modified procedure for large scale pasture evaluation by digestibility *in vitro*. Journal of the Australian Institute of Agricultural Science 38: 212–213.

Norman, H.C., R.A. Dynes and D.G. Masters. 2002. Nutritive value of plants growing on saline land. pp. 59–69. *In*: 8th National Conference on the Productive Use and Rehabilitation of Saline Land, Fremantle, Australia.

Norman, H.C. and D.G. Masters. 2010. Predicting the nutritive value of saltbushes (*Atriplex* spp.) with near infrared reflectance spectroscopy. p. 124. *In*: Proceedings of the International Conference on Soils and Groundwater Salinization in Arid Regions, Sultan Qaboos University, Muscat, Oman.

Norman, H.C., D.G. Masters and E.G. Barrett Lennard. 2013. Halophytes as forages in saline landscapes: Interaction between genotype and environment change their feeding value to ruminants. Environmental and Experimental Botany 92: 96–109.

Norman, H.C., D.G. Masters, M.G. Wilmot and A.J. Rintoul. 2008. Effect of supplementation with grain, hay or straw on the performance of weaner Merino sheep grazing oldman (*Atriplex nummularia*) or river (*Atriplex amnicola*) saltbush. Grass and Forage Science 63: 179–192.

Norman, H.C., D.K. Revell, D.E. Mayberry, A.J. Rintoul, M.G. Wilmot and D.G. Masters. 2010a. Comparison of *in vivo* organic matter digestibility of native Australian shrubs to *in vitro* and *in sacco* predictions. Small Ruminant Research 91: 69–80.

Norman, H.C., M.G. Wilmot, D.T. Thomas, E.G. Barrett-Lennard and D.G. Masters. 2010b. Sheep production, plant growth and nutritive value of a saltbush-based pasture system subject to rotational grazing or set stocking. Small Ruminant Research 91: 103–109.

Osmond, C.B. 1967. Acid metabolism in *Atriplex* 1. Regulation of oxalate synthesis by the apparent excess cation absorption in leaf tissue. Australian Journal of Biological Sciences 20: 575–587.

Ozgur, R., B. Uzilday, A.H. Sekmen and I. Turkan. 2013. Reactive oxygen species regulation and antioxidant defence in halophytes. Functional Plant Biology 40: 832–847.

Pearce, K.L., D.G. Masters, G.M. Smith, R.H. Jacob and D.W. Pethick. 2005. Plasma and tissue α-tocopherol concentrations and meat colour stability in sheep grazing saltbush (*Atriplex* spp.). Australian Journal of Agricultural Research 56: 663–672.

Provenza, F.D. 1995. Postingestive feedback as an elementary determinant of food preference and intake in ruminants. Journal of Range Management 48: 2–17.

Provenza, F.D., J.J. Villalba, C.D. Cheney and S.J. Werner. 1998. Self-organisation of foraging behaviour: From simplicity to complexity without goals. Nutrition Research Reviews 11: 199–222.

Quin, J.J., J.G. van der Wath and S. Myburgh. 1938. Studies on the alimentary tract of merino sheep in South Africa. IV. Description of experimental technique. Onderstepoort Journal of Veterinary Science and Animal Industry 11: 341–360.

Redondo-Gómez, S. 2013. Bioaccumulation of heavy metals in Spartina. Functional Plant Biology 40: 913–921.

Storey, R., N. Ahmad and R.G. Wyn Jones. 1977. Taxonomic and ecological aspects of the distribution of glycinebetaine and related compounds in plants. Oecologia (Berlin) 27: 319–332.

Theodorou, M.K., B.A. Williams, M.S. Dhanoa, A.B. McAllan and J. France. 1994. A simple gas production method using a pressure transducer to determine the fermentation kinetics of ruminant feeds. Animal Feed Science and Technology 48: 185–197.

Thomas, D.T., A.J. Rintoul and D.G. Masters. 2007. Increasing dietary sodium chloride increases wool growth but decreases *in vivo* organic matter digestibility in sheep across a range of diets. Australian Journal of Agricultural Research 58: 1023–1030.

Tilley, J.M.A. and R.A. Terry. 1963. A two-stage technique for *in vitro* digestion of forage crops. Journal of the British Grasslands Society 18: 104–111.

Trossat, C., B. Rathinasabapathi, E.A. Weretilnyk, T.L. Shen, Z.H. Huang, D.A. Gage and D. Hanson. 1998. Salinity promotes accumulation of 3-dimethylsulfoniopropionate and its precursor S-methylmethionine in chloroplasts. Plant Physiology 116: 165–171.

Ulyatt, M.J. 1973. The feeding value of herbage. pp. 131–178. *In*: G.W. Butler and R.W. Bailey (eds.). Chemistry and Biochemistry of Herbage. Academic Press, London.

Van Soest, P.J. 1963. Use of detergents in the analysis of fibrous feeds. I. Preparation of fibre residues of low fibre content Journal of the Association of Official Agricultural Chemists 46: 835–829.

Waghorn, G.C., N.R. Adams and D.R. Woodfield. 2002. Nutritive value of herbage. pp. 333–356. *In*: M. Freer and H. Dove (eds.). Sheep Nutrition. CSIRO Publishing, Collingwood.

Weiss, W.P., H.R. Conrad and N.R. St. Pierre. 1992. A theoretically-based model for predicting total digestible nutrient values of forage and concentrates. Animal Feed Science and Technology 39: 95–110.

Weston, R.H. 2002. Constraints on feed intake by grazing sheep. pp. 27–49. *In*: M. Freer and H. Dove (eds.). Sheep Nutrition. CSIRO Publishing, Collingwood.

Weston, R.H., J.P. Hogan and J.A. Hemsley. 1970. Some aspects of the digestion of *Atriplex nummularia* (saltbush) by sheep. Proceedings of the Australian Society of Animal Production 8: 517–521.

7

Nutritional and Feed Value of Halophytes and Salt Tolerant Plants

Salah A. Attia-Ismail

SYNOPSIS

This chapter deals with different aspects of utilizing halophytes as animal feed components. The first point to be examined is the utility of laboratory-based methods to assess digestibility and feed value of salt tolerant and halophytic feedstuffs. The outcomes of animal feeding trials and the use of *in vivo* methods are analyzed. Attention is given to the vexing notions of palatability to the animals and preference shown toward certain feedstuffs and how these vary with time of year, the phenological stage of the plants, the form in which the feedstuff is presented (green vs. dry, chopped vs. whole), etc. Finally, factors affecting performance of animals when fed on saline diets are assessed.

Keywords: protein, energy, digestibility, palatability, preference, *in vitro, in vivo* ruminants, non ruminants, *Atriplex,* non protein nitrogen, laboratory-based methods, crude fiber, ensiling process, rumen, chemical composition.

1. Introduction

The human requirements for animal protein makes it increasingly important to utilize all available pastureland, which to a large extent lies in climatically unfavorable

Desert Research Center, Cairo, Egypt.
Email: saai54@hotmail.com

regions. In these areas, the animals often have to consume the only available feed resource which is halophytic plants (Attia-Ismail et al. 2009). However, feed resources deficiency is considered one of the basic constraints to improving animal productivity in arid and semi-arid regions. Improving the nutritional status of desert grazing livestock (sheep, camels and goats) particularly during the prolonged dry seasons would increase the average annual animal production by approximately 27%. Attention has been directed towards the necessity of utilizing the marginal resources, i.e., saline soils and underground water for producing unconventional animal feed ingredients. The native natural rangelands constitute the principal feed resources in the Egyptian deserts. The biomass production and quality of the natural rangelands in such areas vary considerably from season to season and from area to area depending on several factors, mainly environmental factors. They are widely distributed throughout several regions of Egypt due to the presence of numerous salines along the Mediterranean Sea and Red Sea shores and inlands (littoral salt marshes and inland salt marshes) as reported by Batanouny (1993); El Shaer and Attia-Ismail (this volume).

2. Green fodder production of halophytes

Saline soils and saline irrigations represent a serious production problem for crops because they suppress plant growth and, hence, yield. The yield potential of a halophyte plant depends on the plant species and the salt concentration. However, most studies on the biomass production of halophytes depend mainly on laboratory studies as there are fewer field studies. *Atriplex* spp., such as *A. nummularia*, *A. griffithii* and *A. hortensis*, could survive under high saline conditions; with optimal growth occurring at 5 to 10 g/l⁻¹ NaCl (Ramos et al. 2004). *Atriplex leucoclada* sample in an experimental site produced 3735 kg fresh weight and 2058 kg dry weight forage and 4015 kg seed in the high salinity location (Amouei 2013). Aronson (1985) calculated the presence of 1.26–2.09 kg/m² dry matter, and 15.5–39.5% fiber, and 10.2–19.5% crude protein in some species of *Atriplex*. The oilseed halophyte, *Salicornia bigelovii*, yields 2 ton ha⁻¹ of seed containing 28% oil and 31% protein, similar to soybean yield and seed quality (Glenn et al. 1999).

Suyama et al. (2007) compared alfalfa to *Thinopyrum ponticum* and found that this halophytic plant showed higher (85%) relative yield than conventional forages (43%) when irrigated with saline water. The most productive species yield 10 to 20 t ha⁻¹ of biomass on seawater irrigation, equivalent to conventional crops (Koyro et al. 2011).

Dagar (1995) found that *Kochia indica* produced fresh biomass of 8.5 kg per bush from March through August in India. Kallar grass (*Leptochloa fusca*) has been found to yield about 46 t/ha of green matter after five years when planted in extreme alkali soil in Pakistan (Malik et al. 1986). Ventura and Sagi (2013) collected information about the yield of some halophytic plants under several saline conditions (Table 7.1).

Collectively, the average biomass production from halophytes reaches the level of 4–5 billion tons calculated on the basis of FAO averages of 450 million hectares in the world. The development of salt tolerant fodder crops may help address the scarcity of good quality water in many arid agricultural regions of the world. Halophytes can, play an important role in the welfare of people in such regions. The greatest utilization of halophytes is their use as forage and fodder (Khan et al. 2006). Moinuddin et al.

Table 7.1. Yields obtained from halophyte crops grown under field conditions.*

Plant species	Salt concentration	Yield (Kg m^{-2} year^{-1})
Aster tripolium	40 mM	14.0 (fresh weight basis)
Atriplex lentiformis	500 mM	1.8 (dry weight basis)
Atriplex triangularis	150 mM	21.3 (fresh weight basis)
Batis maritima	500 mM	1.7 (dry weight basis)
Salicornia europaea	500 mM	1.5 (dry weight basis)
Salicornia persica	100 mM	15.0 (fresh weight basis)
Sarcocornia fruticosa	100 mM	28.0 (fresh weight basis)

*Source: Ventura and Sagi (2013).

(2012) mentioned that naturally occurring inland and coastal salt tolerant grasses have considerable potential as low cost non-conventional fodder crop. A number of saltbush species provide adequate nutrition with considerably preferred browse for domestic livestock.

3. Quality of halophytes and salt tolerant plants as animal feed components

The quality of a halophytic or salt tolerant plant as forage might be the extent to which it has the potentiality to get the required animal response. The quality of halophytes as forage varies greatly among and within each crop. In order to determine forage quality, different issues have to be taken into consideration.

Analyzing forages for nutrient content (chemical compositions) can be used to determine the quality of forage and if it is adequate to meet the animal requirements and to be used for proper ration supplementation. However, as Masters (this volume), points out there are some problems in accounting for the non-protein nitrogen and other non-nutritional components. Regardless of the chemical composition of the halophytes, an important factor, i.e., palatability of the plants has to be addressed first. Palatability will determine whether the halophyte would be consumed by the animal or not and to what extent. One may, therefore, look for other factors of quality assessment of halophytes as animal fed components. The factors that affect forage quality include palatability, nutrient contents (chemical compositions), plant secondary metabolites, feeding value (voluntary animal intake, nutrient digestibility), and eventually animal performance.

3.1 Palatability and feeding values of halophytes

The presence of different salts in halophytic and salt tolerant plants represents a major constraint to their palatability. Palatability is the first constraint to the consumption of these plant species. Palatability is defined as the ratio between the amount of feed ingested by animals and the amount offered for a given period of time. The palatability and feeding values of individual halophytes or other rangeland plants vary greatly from virtually zero to very high. In almost any forage populations, either natural or planted, of a given plant species of browse, there are various degrees of palatability from one plant to the other, ranging from highly palatable to poorly palatable. Palatability also depends on the relative abundance of the species under consideration on the rangeland; all other conditions being equal the palatability of a given taxon is inversely related to its abundance on the range. In addition to the internal plant factors, there are also internal animal factors such as: species of livestock, age, physiological and health status, feeding habits, nutritional status. The feeding value of any halophytic forage could also affect its palatability for the same animal. For instance: the content of crude fiber in forage plays an important role in its selection by livestock. Forages with high fiber content are usually better accepted by cattle than by sheep and goats; but this, in turn, depends on the proportions of the various components of fiber: cellulose, hemicelluloses, acid detergent fiber (ligno-cellulose), neutral detergent fiber (cell walls), etc. Mineral content may, also, be an important factor either limiting or favoring (usually limiting in low rainfall areas and favorable in high rainfall areas, when silica-free minerals are concerned). Table 7.2 shows the palatability of different halophytic plants as tested by different species of animals.

3.1.1 *Preference* versus *palatability*

Preference is the order in which a plant species is selected by animals within a given plant community or population, or at a grazing site, at a given time (Glenn et al. 1999). The word palatability should be replaced by preference. The former must be cancelled for the plant that may be palatable but not nutritive. Therefore, the word palatability is not objective. On the other hand, the ability of animals to select their feeds in the rangelands is, therefore, called preference for the animals are capable of selecting non harmful feeds. Ruminant animals are capable of selecting diets that lessens the hazards of potentially harmful feeds (Kyriazakis et al. 1998). Animals may select less or even unpalatable plants if they are nutritious and non harmful and then feed on them gradually until their rumen environment is adapted to these plants (Attia-Ismail, this volume). Therefore, the word palatability makes no sense in use whereas the word preference is preferred as it is more applied and subjective. An experiment was conducted to determine whether a secondary plant metabolite (oxalic acid) would influence the diet selection by animals (Duncan et al. 2000). They found that the rate of oxalic acid degradation in the rumen environment may influence diet selection. When animals were fed diets containing oxalic acid and tannins, rumen environment took time to adapt to these compounds depending on concentrations and prior nutritional background (Farid, M., personal communications). Microbial populations change gradually with prolonged increasing exposure to toxic substances and such gradual adaptation allows ruminants to increase their tolerance of some poisonous plants. Therefore, time is a crucial factor in adaptability to the plant secondary metabolites. This shows the ability of ruminal bacteria to adapt to some plant secondary metabolites. It might be safe to conclude that animals might be able to adapt themselves to the available feedstuffs in their local environment.

Table 7.2. Palatability of halophytic plants for different animal species.*

Plant species	Animal species		
	Goats	Camels	Sheep
Acacia albida	PP	HP	PP
Acacia elbaica	PP	HP	PP
Acacia mellifera	FP	HP	PP
Acacia reddiana	PP	HP	FP
Acacia tortilis	FP	HP	NP
Arocnemom glaucum	--	HP	NP
Astragalus eremophilus	--	HP	NP
Avicennia marina	HP	HP	NP
Blepharic ciliaris	--	HP	HP
Cadaba farinose	PP	HP	HP
Cadaba oblonifia	HP	PP	HP
Calligonum comosum	HP	HP	NP
Convolvulus hvstrix	PP	HP	HP
Halopeplis prefaliala	HP	HP	HP
Heliotropium leuteum	HP	HP	HP
Indigofera argente	--	PP	FP
Indigofera spinosa	PP	HP	NP
Leptadenia pyrotechnica	HP	FP	PP
Lycium shawii	HP	HP	PP
Maerua crassifolia	HP	HP	HP
Ochradenus baccatus	HP	HP	HP
Panicum turgidum	HP	HP	HP
Pergularia tomentosa	HP	HP	FP
Plantago ciliata	FP	HP	HP
Salsola baryosma	FP	HP	FP
Salsola tetrandra	--	PP	FP
Stipagrostis ciliata	PP	HP	--
Leptadenia pyrotechnica	FP	PP	--
Suaeda monaica	HP	HP	PP
Taverniera aegyptiaca	PP	FP	HP
Trichodesma ehrenbergu	HP	HP	--
Zygophyllum coccineum	PP	HP	HP

HP = highly palatable, FP = fairly palatable, PP = poorly palatable, NP = Not palatable.
*Source: El Shaer et al. (1997).

3.2 Chemical composition of halophytes

Halophytic plant species vary considerably in their chemical composition, nutritive value and palatability. The chemical composition of any animal feed is the first indicator of its nutritional value to the animals. Nutritive value is first determined by nutrient concentration. Nutrient contents are determined by the chemical composition. The nutrient content of plants varies from field to field, and from year to year, as a result of differences in plant species and varieties, soil fertility (if any, considering the saline soils), climatic factors (rain, wind, sunlight, temperature, etc.), stage of maturity, etc. Therefore, the only way to know the quality of forage is to have it analyzed. Chemical compositions are determined by traditional methods of feed analysis.

3.2.1 Ash and mineral composition

Mineral contents differ in halophytes from those of regular forages. The variations in mineral contents of halophytes depend on several factors (Attia-Ismail 2008). Among the many factors that cause variations are the plant species, stage of growth, season, soil and irrigation water salinity, etc. The aspects of ash contents and mineral compositions of halophytes are discussed in detail by Attia-Ismail (this volume). Upon the investigation of some mineral compositions of some halophytes in Australia, Albert and Marianne (1977) found a specific frequent pattern of ion content within specific taxon. They found that dicotyledons had an extraordinary accumulation of sodium (mainly Cl and less frequently SO_4 salts) and normal free oxalate concentrations of Chenopodiaceae and Caryophyllaceae, while other dicotyledons showed moderate salt concentrations. They also found K:Na ratios to be less than unity. They mentioned that monocotyledons were contrary to dicotyledons, where low salt contents are a characteristic of species like Poaceae, Cyperaceae and Juncaceae (where the K:Na ratios happen to be more than unity). Similar findings were detected by Gorham et al. (1980) in Wales. The impact of minerals on halophyte feeding to ruminants is discussed by Norman et al. (this volume). Data in Table 7.3 indicated that all plant species contained sufficient levels of major and trace elements. The levels of mineral concentrations appeared to cover the requirements of livestock except for P and S contents according to the international recommendation of mineral requirements (Kearl 1982) for livestock.

Therefore, P and S supplementation is necessary for grazing sheep and goats in particular, during the critical stages of animal productivity. Although some halophytic shrubs attained higher levels of some trace or major elements, their mineral

Table 7.3. Overall average values of some mineral composition in halophytic plants (DM basis) grown in Sinai and the North Western coast of Egypt.*

	Ca %	P %	Na %	K %	Mg %	S %	Zn Ppm	Cu Ppm	Fe Ppm	Mn Ppm
Nitraria retusa	1.96	0.22	5.35	0.66	0.36	0.14	32	11	567	62
Atriplex halimus	1.69	0.32	3.91	0.57	0.32	0.17	64	10	503	51
Tamarix aphylla	3.73	0.16	2.75	0.78	0.43	0.12	38	13	274	60
Zygophyllum album	2.26	0.14	2.89	1.14	0.64	0.09	41	7.78	393	52
Suaeda fruticosa	2.11	0.41	4.06	1.29	0.30	0.20	55	13	674	88
Salsola tetrandra	3.98	0.16	5.65	1.45	0.59	0.12	44	8.88	664	79
Tamarix mannifera	3.01	0.01	2.70	0.91	0.46	0.09	45	16	291	52
Haloxylom salicornicum	4.00	0.15	5.01	1.74	0.33	0.07	73	18	603	80
Halocnemon strobilaceum	4.5	0.14	5.21	2.01	0.39	0.10	91	17	621	87

*Adapted from: Raef (2012); El Shaer and Attia-Ismail (2002); Attia-Ismail (2008).

concentrations are still within the normal ranges and thus without any harmful effect on livestock production. However, diets for farm animals contain a mineral/trace element/vitamin supplement in order to overcome possible mineral deficiency that may occur when feeding rations including halophytes. It is, sometimes, necessary to include extra supplies of some minerals.

3.2.2 Protein contents and amino acid pools of halophytes and salt tolerant plants

It has been long recognized that environmental conditions play a major role in determining the quantity and quality of nutrients produced by halophytes. Proteins in halophytes play an intrinsic role in plant physiology. It is reported that protein level decreased under salinity is due to low uptake of nitrate ions (Agastian et al. 2000) and due to other factors.

Proteins play a major role in salt stress acclimation and plant cellular adjustment since proteins are involved in a wide range of cellular processes associated with salt acclimation (Kosová et al. 2013). In halophytic plants the excess salt concentrations seem to affect biosynthesis (Wang et al. 2014) of different nutrients within plant tissues including proteins (Hall and Flowers 1973). Salinity decreases protein synthesis and increases its hydrolysis in some plants, resulting in production of amino acids because salts have antagonistic effects on proteins (Kozlowski 1997). The response to salinity, also, is reflected in the amino acid pool in halophytes (Joseph et al. 2013). Salt stress regulation genes lead to changes in the protein profile that help plants to adapt to salt accumulation (Parker et al. 2006) and this may affect the amino acid profiles of proteins. It was found that some amino acids (e.g., aspartate and glutamate) play an important role in halophyte adaptation to salt stress. Rani (2007) found that the increase in salinity levels increased the concentrations of aspartate, glutamate, glycine, histidine, lysine, and arginine. Joseph et al. (2013) characterized the amino acid pool in some halophytic plants (Table 7.4) and found that lysine and threonine, essential amino acids which are absent in cereals, constitute a major part of the total essential amino acid pool in halophytes and that all essential amino acids were significantly higher in concentrations, when compared with the reference pattern of FAO/WHO (1990).

Within the salt tolerant sorghum types, protein content decreases as the salinity increases leading to the increase of non protein-nitrogen (Al-Khalasi et al. 2010). Also, Amouei (2013) found that the protein percent of *Atriplex leucoclada* decreased from low salinity to high salinity area. Meychik et al. (2013) found that upon an increase in salinity level, the amount of nitrogen per unit of protein mass increases in halophytes and concluded that at high salinity, the ability to synthesize non-protein nitrogen-containing compounds increases in the halophyte. They found also that upon a decrease or an increase in NaCl concentration (from 250 to 0.5 or from 250 to 750 mm), the amount of protein in root and leaf tissues of *Suaeda altissima* drastically increases. Naturally occurring inland (*Paspalum paspaloides* and *Paspalidium geminatum*) and coastal (*Aeluropus lagopoides* and *Sporobolus tremulus*) salt tolerant grasses in Pakistan were evaluated under optimal non-saline conditions and it was found that the protein percent differed between plant species (Moinuddin et al. 2012). It seems, therefore, that with a decrease in soil salinity, the available nitrogen increases

Table 7.4. Amino acid pools in some halophytes.**

	Atriplex nummularia leaves	*Salicornia brachiata*	*Arthrocnemum indicum*	*Suaeda maritima*	*Suaeda monoica*	*Sesuvium portulacastrum*
	mg/Kg	mg/100 g protein				
Histidine	2.78	9.62	3.44	3.21	3.44	2.53
Arginine	3.54	49.0	14.3	13.5	12.4	8.96
Threonine	3.06	33.9	10.1	10.5	9.4	6.22
Valine	4.15	51.5	14.0	13.1	12.4	8.18
Methionine	0.09	11.6	3.16	3.04	3.24	2.09
Isoluecine	3.32	46.0	11.2	10.5	9.56	6.34
Leucine	5.35	70.0	17.4	16.4	15.6	10.2
Phenylalanine	3.53	30.5	7.92	7.49	7.21	5.44
Lysine	2.84	89.9	27.6	26.6	25.6	24.6
Aspartate	7.11	86.3	23.8	22.1	21.9	15.6
Glutamine	7.72	124.2	32.1	30.2	29.6	19.7
Serine	3.23	29.9	8.75	8.04	8.49	6.42
Glycine	4.40	19.4	5.34	5.0	5.14	4.26
Alanine	4.26	44.2	11.4	9.0	9.45	7.28
Proline	3.45	14.1	3.67	3.56	3.49	3.63
Tyrosine	1.25	17.2	5.05	4.49	4.14	2.32
Cystine	0.45	3.25	0.97	0.75	0.88	1.25

**Adopted from Joseph et al. (2013).
* adopted from Abd El-Galil et al. (2014).

significantly. The increased leaf succulence of halophytes *Atriplex nummularia* is likely because the salt concentration gradient of cell cytosol and organelles became steeper and made the cell cytosol adjust osmotically (Silveira et al. 2003). Most of the halophytes that are utilized as animal feed components contain moderate amounts of crude protein (CP) which seems to cover the nitrogen requirements of grazing animals. Although, the protein content seems to be relatively high, a large proportion of such protein is present in the form of non protein nitrogen (NPN). Benjamin et al. (1992) reported that 42% of the nitrogen in *Atriplex barclayana* was non-protein nitrogen. Therefore, available sources of energy should be supplemented to animals for better utilization and efficient digestion of nitrogen (Masters, this volume). The intervention of different processes (digestive and synthetic) in the rumen with amino acid supply from the diets makes it more difficult to relate amino acid supply to tissue requirements. Amino acids—of halophytes and other feed plants—are rapidly degraded in the rumen. All essential amino acids can be synthesized by the rumen microorganisms but the supply of amino acids from microbial protein is limiting in quantity and quality for maximum rates of growth in young animals and for maximum milk production.

Table 7.5, compiled from several research papers, shows nitrogen and crude protein percent (nitrogen x 6.25). Values of protein contents varied according to plant maturity stage and plant part. For instance crude protein is 20.7% for leaves and twigs of *Atriplex repanda* while it is half this value (10.3%) in the fruits. *Suaeda foliosa* has 16.7% crude protein in leaves while the corresponding value of stem is 16.8%. *Pappophorum caespitosum* showed close values of protein contents in early fruitification (10.6%) and post fruitification (9.1%). The seeds of halophytes have high protein contents than other parts of the plant since they do not accumulate salts (O'Leary et al. 1985). It seems that the chenopods had higher CP than halophytic grasses.

However, proteins are an important nutrient for all animal species as they are a primary constituent of all body tissues and fluids such as hair, skin, fur, muscles, bones vital organs, blood and other body fluids. Unlike energy and fats, proteins cannot be stored in body as depots for future use. Therefore, proteins are required on a daily basis by animals for maintenance and reconstruction of worn out tissues, rather than for productive purposes of animals. It is, then, very essential to supplement animals

Table 7.5. Nitrogen and crude protein contents of different parts of some world halophytes.*

Halophytic plants	Plant part and/or maturity stage	N, %	Protein %
Acacia saligna	Whole	2.21	13.8
Anabasis brevifolia	Whole/Flowering	1.52	9.5
Atriplex halimus	Whole plant	2.11	13.1875
Atriplex nummularia	Whole plant	2.03	12.6875
Atriplex repanda	Leaves and twigs	3.31	20.6875
	Fruits	1.65	10.3125
Haloxylon ammodendron	Whole/Flowering	2.28	14.25
Halostachys caspica	Whole/Flowering	1.83	11.4375
Haloxylon salicornicum	Whole plant	2.8	17.5
Halocnemum strobilaceum	Whole plant	1.11	6.9375
Kalidium foliatum	Whole/Fruiting	3.03	18.9375
Karelinia caspica	Whole/Flowering	1.85	11.5625
Lyceum chilense	Fruitification/shoots	2.21	13.8125
Pappophorum caespitosum	Early fruitification	1.70	10.625
	Past fruitification	1.45	9.0625
Phragmites communis	Whole/Flowering	1.92	12
Reaumuria soongorica	Whole/Fruiting	1.64	10.25
Salsola tetranda	Whole plant	1.08	6.75
Suaeda foliosa	Leaves	2.67	16.6875
	Stem	2.69	16.8125
Suaeda fruticosa	Whole plant	1.94	12.125
Tamarix mannifera	Whole plant	1.22	7.625
Trichloris crinita	Early fruitification	1.54	9.625
	Past fruitification	1.51	9.4375
Zygophyllum album	Whole plant	1.05	6.5625

*Attia-Ismail et al. (2003; Gabr 2002; Shawkat et al. 2001; Kraidees et al. 1998; Warren et al. 1990; Vercoe 1986; El-Bassosy 1983).

with predetermined daily doses of proteins in their diets. Proteins contribute energy, and provide essential amino acids for rumen microbes as well as the animal itself. In small ruminants, the amount of protein is more important than the quality of protein.

Minimum protein requirements for adult sheep or cattle that are not reproducing or growing are 7 to 9%, and this increases to approximately 14 to 18% for growing or lactating animals (SCA 2007). Therefore, the amounts of crude protein in halophytes barely meet the maintenance, but not productive, requirements.

In evaluating proteins present as dietary nutrient in halophytes, one should take several issues into consideration. First is the high percent of non protein nitrogen portion of the crude protein content. The effect of high salinity levels on protein synthesis was discussed earlier. Second: is the high solubility of halophytic proteins. This high solubility stems from being present as leaf protein (which is usually characterized by high solubility) and because of the different mechanisms used by the plant to overcome the high salinity (which increase the portion of non protein nitrogen) as discussed earlier. The proteins in halophytes are mostly present in their leaves and stored later, after complete maturation, in the seeds (Table 7.5). Third is the high degradability of soluble proteins in the rumen by the microorganisms. Once these soluble proteins are ingested, the microbial community in the rumen acts upon them to degrade these proteins in order to build their own body proteins (microbial proteins). The lack of readily available carbohydrates together with the high protein degradation in the rumen causes the proteins of halophytes to be wasted and the animals could not profit from them. Examination of the literature revealed that the halophytes do have low digestible ether extract and soluble carbohydrate. The microbial protein supply is not sufficient to meet animal protein requirements at even the maintenance state (which explains the body weight loss of animals when fed halophytes alone). In order to increase the microbial protein supply to the ruminants, a readily available carbohydrate source should be included in the diet. The release of energy from these supplements coincides with the degradation of halophytic proteins creating the circumstances suitable to make more microbial protein. On the other hand, the low protein percent present in these cereal supplements may not be degradable and, thus, providing the animal with true protected proteins to the animals. This explains the positive response of animals fed halophytes when supplemented with energy concentrate. It is, then, necessary or perhaps vital to supplement animals fed on halophytes with cereal grain or other energy supplement.

3.3 Crude fiber contents of halophytes and salt tolerant plants

Plant cell wall consists of lignin, cellulose, and hemicellulose. Crude fiber is the residue remaining after the acid/alkaline hydrolysis of feed material. This insoluble residue is composed of polysaccharides such as cellulose, hemicellulose, and lignin. The Acid detergent fiber (ADF) is the extract of acidic detergent hydrolysis. It includes cellulose, lignin, acid detergent insoluble nitrogen (ADIN), and acid-insoluble ash (AIA). Therefore, the more the plant matures the more ADF content accrues. The lower the ADF the more energy is available to the animal. The Neutral detergent fiber (NDF) contains the fibers in ADF and hemicellulose. The NDF means how bulky the feed is.

Rations with low NDF are always preferred. The prime quality forage contains less than 31% ADF and less than 40% NDF on dry matter basis. Table (7.6) shows the crude fiber contents and their fractions of some halophytes. However, lignification of halophytic plants seems to get adversely affected by the salinity level. Lignification of the stems of the halophyte *Suaeda maritima* has been shown to be negatively correlated with salinity (Hagege et al. 1988). Upon an increase in salinity level over the optimum, lignin content in shoot tissues of *Suaeda altissima* significantly decreases (Meychik et al. 2013). A decrease in NDF as the salinity of irrigation water for Italian ryegrass (*Lolium multiflorum*) increases was found by Ben-Ghedalia et al. (2001). Fiber fraction (NDF, ADF and ADL) of *Panicum turgidum* recorded higher contents in dry season than in wet season (Badawy 2005). Also, Ewing (1981) reported that addition of NaCl to the nutrient solution caused a marked increase in the width of the cortex, as well as in the diameter of cortical cells in the stems of *Atriplex prostrata*. Some of the peroxidase isozymes were reported to be specific for lignin synthesis (Sato et al. 1993); these specific isozymes may decrease at high salinity. The high levels of fiber and ash contents could limit intake and digestibility of halophytic forages.

Table 7.6. Crude fiber and their fraction contents of different parts of some halophytes.*

	CF	NDF	ADF*	ADL	Cellulose	Hemi cellulose	Reference
Atriplex nummularia		60.5	33.1	10.3	22.8	27.4	Fahmy (1998)
Atriplex lentiformis		48.9	23.4	7.5	--	--	Donia (2002)
Atriplex halimus	42.1*						
Acacia saligna		58.72	35.31	18.4	16.91	23.41	Allam et al. (2006)
Tamarix mannifera	16.1*	49.0	33..3	12.2	21.0	15.7	Donia (2002)
Halocenum strobilaceum	14.6*	41.8	22.8	11.2	11.6	19.0	Donia (2002)
Kochia indica		59.5	39.5	10.9	28.4	20.2	El-shereef (2007)
Zygophyllum album	14.6*	44.9	31.4	13.2	18.2	13.5	Fahmy (1998)
Pearl millet#		74.3	43.9	5.6	--	--	Fahmy (1998)
Panicum turgidum	29.21	61.69	41.96	10.97	--	--	Badawy (2005)

#*Pennisetum glauca.*
The lower the ADF value, the higher energy in feeds.
* El Shaer and Attia-Ismail (2002).

3.4 Energy contents of halophytes

The amount of energy in feed can be expressed in the form of gross energy (total heat of combustion of a feed substance measured in calories or joules per unit weight of dry matter, GE), digestible energy (GE—energy lost in feaces, DE), metabolizable energy (DE—energy lost in urine and gases, ME) or net energy (ME—heat loss, NE) for maintenance and production. However, the commonly used energy term to express energy contents of halophytes is the metabolizable energy.

Most of the literature reporting/citing metabolizable energy values of halophytes depend mostly on estimates determined *in vitro*, which to some extent may produce unreliable values of energy contents. There is actually some relation between the *in vivo* and *in vitro* estimates, yet still the *in vitro* values do not represent the real values of energy contents of halophytes.

In their review, McEvoy and Jolly (2006) cited a table from Masters (2006) showing the discrepancy of the *in vitro* results compared with those obtained *in vivo* (Table 7.7).

However, the nutritive value of halophyte species such as metabolizable energy (ME) appears to depend strongly on plant maturity. Table 7.6 reveals that the examined halophytes have a potential of being a source of energy. Table 7.8 shows a comparison between halophyte forages and traditional ones. The values of both types of forages seem to be close and do not differ significantly which raises a question about the efficiency by which the energy in halophytes is utilized. The examination of literature showed contradicting results. Meneses et al. (2007) compared *Atriplex nummularia* hay with Alfalfa hay and found that ME intakes were not significantly different. Moinuddin et al. (2012) concluded that ME values in coastal grasses (such as *Aeluropus lagopoides* and *Sporobolus tremulus*) appeared to be sufficient for maintenance (not for production) of beef cattle. Thomas et al. (2007) also concluded that animals grazing on *A. nummularia* require energy supplementation more than any other kind of supplementation. El Shaer et al. (1990) found that the dry matter intake and, therefore, energy intake from *Zygophyllum album* to be zero. The low energy contents as indexed by nitrogen free extract of salt plants might be attributable for the decreased nutrient digestion and, therefore, utilization (Shawkat et al. 2001).

Table 7.7. Examples of *in vitro* and *in vivo* estimates of DOMD (adapted from Masters 2006 as cited by McEvoy and Jolly 2006).

	In vivo	Pepsin-cellulase	Pepsin-cellulase corrected*	NIRS
Sample 1	58			76
Sample 2	52	77	70	
Sample 3	45	77	71	

*corrected with non-halophyte calibration.

Table 7.8. Estimates of DE and ME of some halophytes compared to some traditional forages.

Halophytes	ME (Mcal/kg)	Traditional feeds	ME (Mcal/kg)
Aeluropus lagopoides[1]	2.30	*Medicago sativa*[3]	2.20
Sporobolus tremulus[1]	2.38	*Cynodon dactylon*[3]	2.49
Paspalum paspaloides[1]	2.53	*Sorghum vulgare*[3]	1.75
Paspalidium geminatum[1]	2.33	*Zea mays*[3]	2.97
Atriplex nummularia[2]	2.82	*Trifolium alexandrinum*[3]	1.99
Salsola tragus[2]	2.56	*Lolium multiflorum*[3]	2.50

1. Adapted from Moinuddin et al. (2012).
2. Adapted from McEvoy and Jolly (2006).
3. Adapted from Arab and Middle East Tables of feed composition (1979).

Regardless of how much energy is present in the feed or how much energy intake is ingested by the animal, the efficiency by which this energy is utilized matters. The feed conversion ratio is a good indicator of the efficiency by which energy is utilized. Energy efficiency is affected by several factors (http://cdn.intechopen.com/pdfs-wm/39472.pdf) such as dry matter intake (through energy transformation mechanisms from gross to net energy) and production (the increased animal production means high efficiency utilization). The efficiency of ME utilization is higher for maintenance than for production. The animals in arid and semi arid zones strive for living and therefore, high production is not of concern (in case of the dependence on halophytes only). Most of the chenopod shrubs contain high concentrations of salt and therefore, much of the apparently digestible material in these plants is salt and has no energy value.

The energy source should support animal performance (maintenance and production). The efficiency of the utilization of energy present in halophytes seems to be low enough to support animal production and sometimes maintenance. Konig et al. (2006) found that non-supplemented lactating ewes grazing *A. nummularia* lost 120 g/day while their lambs gained only 30 g/day. It was shown that (Abou El-Nasr et al. 1996) sheep fed saltbush hay; fresh *Acacia* and *Acacia* hay were not able to meet their maintenance requirements which contributed mainly to live body weight loss (see also Khamis 1988; El Shaer et al. 1991). Animals grazing/feeding on halophytes need a substantial amount of a high energy supplement, such as barley, in order not to lose weight and condition (Nizar et al., this volume). Al-Khalasi et al. (2010) found that as the salinity increased, the feed conversion efficiency decreased indicating clear effect of salinity on the energy utilization. This was translated into lower total and daily body weight gain of animals fed sorghum hay than those fed Rhodes grass (*Chloris gayana*) hay at the highest level of salinity. On the other hand (Al-Shorepy and Alhadrami 2008), found that the feed conversion ratio improved more by feeding a high level of halophyte *Distichlis* grass hay than when animals fed Rhodes grass. Thomas et al. (2007) concluded that animals grazing *A. nummularia* require energy supplementation more than any other kind of supplementation. The point of energy contents of halophytes seems to be related to the different mechanisms of salt tolerance. The halophytes seem to expend energy on these mechanisms to adapt to high salinity conditions and therefore, not much energy is left in these plants for the use as energy source for animals. However, more research, in this area is required to elucidate this ambiguous energy utilization of halophytes.

4. Feeding and nutritional value of halophytes and salt tolerant plants

4.1 Voluntary dry matter intake

Voluntary dry matter intake is considered one of the two major components (intake and digestibility) reflecting upon feeding and, hence, nutritional value of feeds for ruminants. However, there is an interrelationship between these two parameters. If feeds and their indigestible residues are retained in the digestive tract of the animal for long periods because of not being highly digestible the daily intake will be reduced.

Therefore, this relationship is positive. Actually, dry matter intake is more closely related to the rate of digestion of diets than to digestibility per se. The primary chemical composition of feeds, in general, that determines the rate of digestion is neutral detergent fiber (NDF). There is a negative relationship between the NDF content of feeds and the rate at which they are digested. Palatability of halophytes also affects the extent to which ruminants consume them. Warren et al. (1992) recorded dry matter intakes of 400–800 g/day for four *Atriplex* species (*A. undulata, A. lentiformis, A. amnicola* and *A. cinerea*) of different palatability indices fed to sheep.

When *Sorghum vulgare*, a salt tolerant plant, was irrigated with three levels of saline water and fed to male lambs; there were no treatment effects on total feed intake (Al-Khalasi et al. 2010). Dry matter intake was higher for lambs fed diets containing halophyte forages than for lambs fed on the *Cynodon dactylon*, grass control diet (Swingle et al. 1996). Daily dry matter (DM), organic matter (OM) intakes have been significantly increased when the lambs were fed 150 g/d, 300 g/d and 450 g/d seed of *Suaeda glauca* in diet (Sun and Zhou 2010).

4.1.1 Effect of season on feed intake of halophytes

Dry matter intakes (DMI) of halophytic forages vary according to the grazing season. It was found to be higher in wet season than in drought season in both sheep and goats (Badawy 2005). Rams consumed less DM of halophytic forages than bucks in drought season (34.4 vs. 44.1 gm/kg$^{0.75}$). Dry matter intake of *Panicum turgidum* was drastically reduced from 880.2 g/h/d in winter season to 358.3 g/h/d in dry season (Raef 2012).

4.1.2 Effect of chemical compositions of halophytes on feed intake

Table 7.9 shows that most of the halophytes contain high values of NDF (more than 40%) which makes them not of prime quality. They could range from good to poor quality forages. The higher the NDF, the lower the DMI of any forage. The high salt

Table 7.9. Dry matter intake (g/Kg W$^{0.75}$) of halophytic species by sheep and goats.*

Halophytes	Sheep	Goats
Fresh state:		
A. halimus	19.4	15.8
H. strobilaceum	18.2	14.2
T. mannifera	10.9	10.3
Z. album	2.12	2.94
Air-dried state:		
A. halimus	13.2	21.0
H. strobilaceum	9.45	8.23
T. mannifera	5.12	4.75
Z. album	0.00	0.00

*Adapted from Khamis (1988).

contents affect the DMI of animals. The presence of plant secondary metabolites also has its impact on DMI from halophytes. The low to moderate crude protein contents may affect the intake from halophytes.

4.1.3 Effect of processing of halophytes on feed intake

Most of the investigations carried out on halophytes indicated that the DMI of halophytes is poor when they are fed in a fresh or air dry form. In its air-dry state *Z. album* was not consumed and the DMI from *A. halimus* and *H. strobilaceum* were 13.2 and 9.45 g DMKgW$^{0.75}$ for sheep and 21.0 and 8.23 gKgW$^{0.75}$ for bucks, respectively (Table 7.9). The poor intake of the fresh and air-dried halophytic species could be attributed to three main factors: 1—high Na, Cl and silica (Attia-Ismail, this volume), 2—higher levels of ADL and NDF (Table 7.6) and 3—many shrubs contained higher levels of plant secondary metabolites (Attia-Ismail 2005).

Physical form of a halophyte feed affects intake. Grinding, pelleting and chopping of halophyte forages partially smash the structural organization of cell walls, thereby accelerating their breakdown in the rumen and increasing dry matter intake (El Shaer, this volume). Several other treatments may also affect the dry matter intake of halophytes (e.g., chemical treatments ensilage, etc.).

The ensiling process improved the quality of halophytes, enhanced their acceptability for sheep and goats (Table 7.10). Also, voluntary intake increased by ensiling a mixture of halophytic species (i.e., *A. halimus, H. strobilaceum, T. mannifera* and *Z. album*), with ground date seeds. Also, ensiling of less or unpalatable halophytes with other feed ingredients would improve forage acceptability; therefore, forage intake would be increased (El Shaer et al. 2002). It seems, therefore, that animals were only able to meet their maintenance requirements.

Table 7.10. Intake of halophytic silages by sheep and goats.

Items	T. mannifera		A. halimus		Z. album		H. strobilaceum	
	Goats	Sheep	Goats	Sheep	Goats	Sheep	Goats	Sheep
DMI (g/kgw$^{0.75}$/day)	42.15	44.17	59.69	59.87	22.53	24.54	59.41	58.21

Adapted from (El Shaer et al. 2002).

4.1.4 Effect of supplementation of halophytes on feed intake

The nutritive value of halophyte is characterized by moderate digestible crude protein and high mineral content, but low in digestible ether extract and soluble carbohydrate (Attia-Ismail et al. 2003). Such low level of readily available carbohydrates together with the rapid fermentation of its crude protein in the rumen may be responsible for the poor utilization of halophyte proteins. El Shaer (1997) reported that on poor quality pasture, energy supplementation could stimulate feed intake of two types of *Atriplex* (Table 7.11). The common effective energy supplementation source which is mainly

Table 7.11. Average values of saltbush intake (g/day/KgW$^{0.75}$) of during the wet season.*

Atriplex species		Sheep	Goat
A. nummularia: Solely fresh		40	-
	+ 150 g barley grains	59.8	58.0
	+ 250 g barley grains	63.2	60.0
A. halimus:	Solely fresh	19.4	15.8
	Solely fresh	13.4	14.8
	Solely dry	13.1	21.0
	+ 100 g barley grains	26.4	27.6
	+ 150 g barley grains	41.1	38.2

*Adapted from El Shaer (1997).

used for improving the intake and nutritive value of halophytes are barley grains. Also, Hossain et al. (2003) showed that feeding grazing goats with increased levels of dietary energy supplementation significantly increased total intake of dry matter.

Fedele et al. (1993) reported that the herbage intake level was influenced by (among other factors) the amount of supplementation. They found that with a high level of concentrate supplementation the grazing goats reduced herbage intake. Supplementing *Atriplex halimus* and *Acacia saligna* with either crushed barley grains or crushed date seeds as energy sources (Shawket et al. 2010) increased DMI than that from control diet (Berseem hay *ad lib.* plus crushed barley grains). They found that a crushed date seeds supplement was inferior to barley grain supplement.

4.1.5 Nutrient digestibility

Nutrient digestibility is an important indicator of the nutritive value of animal feeds. The process of aging and maturation of the ranges was found to be associated with a decline in digestibility and consequently nutritive value (El-Bassosy 1983). Dry matter digestibility of halophytic forages was higher in grazing season than in drought season with both sheep and goats (El Shaer et al. 1991; El Shaer 2010). Since halophytic plants are different from traditional feedstuffs, their nutrient digestion coefficients are affected by their nature and chemical and biochemical composition. High salt content is perhaps one of the major factors which affect digestion coefficients negatively through shortening the rumen turnover time (Warren and Casson 1992). High content of silica could be a reason for reduced nutrient digestibility of salt plants (Youssef et al. 2002).

Dry matter digestibility of halophytic plants varies greatly from 34.4% for *Juncus acutus* to 70.5% for *Nitraria retusa* (El Shaer and Zaharn 2002). El Shaer and Attia-Ismail (2002) mentioned that all chenopod shrubs contain high concentrations of digestible crude protein, perhaps because their crude protein included high percentage of non protein nitrogen.

Type of energy supplement could exert variable effects on nutrient digestibility. Supplementation of *Atriplex nummularia, Acacia saligna* and *Atriplex semibaccata* with 50% of maintenance requirements with barley grains increased digestion coefficients of DM, OM, CF and NFE but not that of CP and EE (El Shaer 1997). Increasing the supplementation level to 75% increased these values further. However, nitrogen intake increased with supplementation and, thus, the nitrogen balance. Attia-Ismail et al. (2003) supplemented dry salt plants with monensin. Monensin did not affect nitrogen intake nor balance positively. They also found that most of the nutrient digestion coefficients were not different except for NFE as a result of adding monensin. Feeding values (TDN and DCP) were also not affected by the inclusion of monensin.

Treatments like ensilage could also improve the digestibility and, hence, the nutritive value of halophytic plants. Different energy sources were also compared along with salt plant silage. When date seeds were used in partial replacement of the concentrate feed mixture (CFM), nutrient digestibility decreased (Shawket 1999).

5. Growth and performance of animals

The ultimate goal of evaluating feeds for ruminants is the animal performance represented by its productivity. In Southern Sinai, Hassan et al. (1980) indicated that sheep lost weight, but losses were minimal in the spring season (26 g a day as compared to 134 g a day in summer season). Also, sheep and goats were not able to maintain their live weights even when the pasture reached its best condition (Hassan et al. 1980). Kids raised on the salty pasture without any supplementation gained 0.36 g/kg $W^{0.75}$ on the average. Meanwhile, lambs lost 2.82 g/kgW$^{0.75}$ and 5.23% of their initial weights on the pasture (El Shaer 1997, 2010). Youssef et al. (2002) found that air dried salt plants covered only 35% of goats maintenance requirements for energy. Shawket et al. (2001) supplemented *Acacia saligna* and *Atriplex halimus* with barley grains at different levels and obtained comparable body weight gain to that of berseem hay. Gabr (2002) fed *Atriplex nummularia* and covered 50% of the maintenance requirements for energy from barley grains. He obtained 116.2 g/h/d as growth rate. Rams lost weight in drought season (–47.8 g a day) but gained weight (+24 g a day) in grazing season, whereas, bucks gained weight in both drought and grazing seasons (22.8 and 98.1 g a day, respectively).

Table 7.12. Performance of sheep fed fresh or ensiled *Atriplex* spp.*

Items	A. halimus		A. nummularia	
	Fresh	**Silage**	**Fresh**	**Silage**
Body weight changes, g/d	–31.0	35.5	6.66	91.1
DMI, g/d/KgW$^{0.75}$	33.4	59.9	59.8	69.7
Total water intake, ml/d/Kg W$^{0.75}$	246	208	207	245
DMD, %	58.1	63.0	62.6	65.4
TDN, g/d/Kg W$^{0.75}$	20.4	30.2	26.6	32.1
DCP, g/d/KgW$^{0.75}$	3.15	6.37	2.24	4.55

*Sources: El Shaer et al. (1990).

6. Summary and Conclusions

The utility of laboratory-based methods to assess digestibility and feed value of salt tolerant and halophytic feedstuffs has come into question. However, as understanding of the chemistry of the feedstuffs and the digestion processes within the animal becomes clearer the spectrum of proximal analysis might be refined and some less meaningful analyses may be dropped. Feeding trials with various salt tolerant and halophytic feedstuffs have many limitations when offered as the sole diet but merit of feeding them in mixed rations and augmenting with energy-rich supplements is well established.

References and Further Readings

Abd El-Galil, K., R.E. Khidr, S.E.M. El-Sheikh, A.A Salama, Henda A. Mahmoud, Mona M. Hassan, A.A. Abd El-Dayem and Fayza M. Salem. 2014. Utilization of atriplex leaves meal as a nontraditional feedstuff by local laying hens under desert conditions. Egypt. Poult. Sci. 34(II): 363–380.
Abou El Nasr, H.M., H.M. Kandil, D.A. El-Kerdawy, H.S. Khamis and H.M. El Shaer. 1996. Value of processed saltbush and acacia shrubs as sheep fodder under arid conditions of Egypt. Small Ruminants Res. 24: 15–20.
Agastian, P., S.J. Kingsley and M. Vivekanandan. 2000. Effect of salinity on photosynthesis and biochemical characteristics in mulberry genotypes. Photosynthetica 38: 287–290.
Albert, R. and P. Marianne. 1977. Chemical compositions of halophytes from Neusiedler lake region in Australia. Oecologia (Berl.) 27: 157–170.
Al-Khalasi, S.S. Osman Mahgoub, Isam T. Kadim, Waleed Al-Marzooqi and Salim A. Al-Rawahi. 2010. Salt Tolerant Fodder for Omani Sheep (Effects of Salt Tolerant Sorghum on Performance, Carcass, Meat Quality and Health of Omani Sheep). A Monograph on Management of Salt-Affected Soils and Water for Sustainable Agriculture, pp. 67–81.
Allam, S.M., K.M. Youssef, M.A. Ali and S.Y. AboBakr. 2006. Using some fodder shrubs and industrial by products in different forms for feeding goats in Sinai. J. Agric. Sci. Mansoura 31(3): 1371–1385.
Al-Shorepy, S.A. and G.A. Alhadrami. 2008. The effect of dietary inclusion of halophyte *Distichlis* grass hay *Distichlis spicata* (L.) on growth performance and body composition of Emirati goats. Emir. J. Food Agric. 20(2): 18–27.
Amouei, Alimohammad. 2013. Effect of saline soil levels stresses on agronomic parameters and fodder value of the halophyte *Atriplex leucoclada* L. (Chenopodiaceae). African Journal of Agricultural Research 8(23): 3007–3012.
Arab and Middle East Tables of fed composition. 1979. International feedstuffs Institute, Utah Ag. Exp. Stn., Logan, Utah.
Aronson, J.A. 1985. Economic halophytes—a global review. pp. 177–88. *In*: G.E. Wickens, J.R. Goodin and D.V. Field (eds.). Plants for Arid Lands. George Allen and Unwin, London.
Attia-Ismail, S.A. 2008. Role of minerals in halophyte feeding to ruminants. *In*: M.N.V. Prasad (ed.). Trace Elements: Nutritional Benefits, Environmental Contamination, and Health Implications.
Attia-Ismail, S.A. (in press). Mineral balance in animals as affected by halophyte and salt tolerance plant feed. pp. 143–160, this volume.
Attia-Ismail, S.A., A.M. Fayed and A.A. Fahmy. 2003. Some mineral and nitrogen utilization of sheep fed salt plant and monensin. Egypt. J. Nutr. and Feeds 6: 151–161.
Attia-Ismail, S.A., H.M. Elsayed, A.R. Asker and E.A. Zaki. 2009. Effect of different buffers on rumen kinetics of sheep fed halophyte plants. Journal of Environmental Science 19(1): 89–106.
Badawy, H.S. 2005. Nutritional Studies on Camels Grazing the Natural Ranges of Halaib-Shalateen Triangle Region. Ph.D. Thesis. Faculty of Agriculture Cairo University.
Batanouny, K.H. 1993. Ecophysiology of halophytes and their traditional use in the Arab World. Advanced course on Halophytes Utilization in Agriculture. 1223 Sept. 1993, Aghadir, Morocco.
Ben-Ghedalia, D., R. Solomon, J. Miron, E. Yosef, Z. Zomberg, E. Zukerman, A. Greenberg and T. Kipnis. 2001. Effect of water salinity on the composition and *in vitro* digestibility of winter-annual ryegrass grown in the Arava desert. Anim. Feed Sci. Technol. 91: 139–147.
Benjamin, R.W., E. Oren, E. Katz and K. Becker. 1992. The apparent digestibility of *Atriplex barclayana* and its effect on nitrogen balance in sheep. Anim. Prod. 54: 259–264.
Dagar, J.C. 1995. Characteristics of halophytic vegetation in India. pp. 55–76. *In*: M.A. Khan and I.A. Ungar (eds.). Biology of Salt Tolerant Plants. Karachi: University of Karachi.
Donia, A.M.A. 2002. Effect of salinity and drought on the nutritive value of *Atriplex lentiformis* and *Prosopis chilensis*. M.Sc. Fac. Sci., Ain Shams Univ.

Duncan, A.J., P. Frutos and S.A. Young. 2000. The effect of rumen adaptation to oxalic acid on selection of oxalic-acid-rich plants by goats. Br. J. Nutr. 83(1): 59–65.

El-Bassosy, A.A. 1983. Study of nutritive value of some range plants from Saloom to Mersa Matroh. Ph.D. Faculty of Agric., Ain Shams Univ. Shobra El-Khema, Egypt.

El Shaer, H.M. 2010. Halophytes and salt-tolerant plants as potential forage for ruminants in the Near East region. Small Ruminant Research 91(1): 3–12.

El Shaer, H.M. 1997. Practical Approaches for Improving Utilization of Feed Resources Under Extensive Production System In Sinai. Proc. Inter. Symp. on Systems of sheep and goat production, Bella, Italy, 25–27 Oct. 1997.

El Shaer. (in press). Halophytes and salt tolerant forages processing as animal feeds at farm level: Basic Guidelines, pp. 388–405, this volume.

El Shaer, H.M. and S.A. Attia-Ismail. 2002. Halophytes as animal feeds: potentiality, constraints and prospects. Int'l Symp. Optimum Utilization in Salt Affected Ecosystems in Arid and Semi Arid Regions, Cairo, Egypt, April 8–11, 2002, pp. 411–418.

El Shaer. H.M. and Attia-Ismail, S.A. (in press) Halophytic and salt tolerant feedstuffs in the Mediterranean Basin and the Arab region: an overview pp. 21–36, this volume.

El Shaer, H.M., H.M. El-Sayed, A.R. Askar, E.E. El-Hassanein and H.S. Soliman. 2002. Effect of feeding a combination of ensiled salty natural and cultivated shrubs on the digestion and animal performance of growing sheep in southern Sinai. Int'l Symp. Optimum Utilization in Salt Affected Ecosystems in Arid and Semi Arid Regions, Cairo, Egypt, April 8–11, 2002, pp. 462–472.

El Shaer, H.M., H.M. Kandil, H.M. Abou El-Naser and H.S. Khamis. 1997. Features and constraints of animal resources development in Shalaten-Halaib region. Egyptian Journal of Nutrition and Feeds 1: 121–128.

El Shaer, H.M., H.M. Kandil and H.S. Khamis. 1991. Salt marsh plants ensiled with dried broiler litter as a feedstuff for sheep and goats. J. Agri. Sci., Mansoura Univ. 1524–1534.

El Shaer, H.M., O.A. Salem, H.S. Khamis, A.S. Shalaby and M.F.A. Farid. 1990. Nutritional comparison studies on goats and sheep fed broiler litter ensiled with desert shrubs in Sinai. Proc. Of the Inte'l Goat Prod. Symp., Oct. 22–26, 1990, Tallahassee, F, USA, pp. 70–74.

El Shaer, H.M. and M.A. Zahran. 2002. Utilization of halophytes in Egypt: an overview. Proc. of the Inter. Conf. on Halophyte utilization and Regional Sustainable Development of Agriculture, Huanghua, Shijiazhang, China, 14–20 Sept. 2001.

Ewing, W.S.B. 1981. The effects of salinity on the morphological and anatomical characteristics of *Atriplex triangularis* Willd. Master's thesis, Department of Botany. Ohio University, Athens, OH.

Fahmy, A.A. 1998. Nutritional studies on halophytes and agricultural wastes as feed supplements for small ruminants in Sinai. Ph.D. Thesis, Fac. Agric., Cairo Univ.

FAO/WHO. 1990. Protein quality evaluation in Report of Joint FAO/WHO expert consultation; Food and Agricultural Organization of the United Nations: Rome, 23.

Fedele, V., M. Pizzillo, S. Claps, P. Morand-Fehr and R. Rubino. 1993. Grazing behavior and diet selection of goats on native pasture in Southern Italy. Small Rumin. Res. 11(4): 305–322.

Gabr, M.G. 2002. First experience of Matrouh Resource Management Project in salt bush utilization for animal feeding. Int'l Symp. Optimum Utilization in Salt Affected Ecosystems in Arid and Semi Arid Regions, Cairo, Egypt, April 8–11, 2002, pp. 419–425.

Glenn, E.P., J. Brown and E. Blumwald. 1999. Salt tolerance and crop potential of halophytes. Crit. Rev. Plant Sci. 18: 227–255.

Gorham, J., Ll. Hughes and R.G. Wyn Jones. 1980. Chemical composition of salt marsh plants from Ynys Mom (Anglesey): the concept of physiotypes. Plant, Cell and Environment 3: 309–318.

Hagege, D., C. Kevers, J. Boucaud and T. Gaspar. 1988. Activite's peroxydasiques, production d'e'thyle'ne, lignification et limitation de croissance chez Suaeda maritima cultive' en l'absence de NaCl. Plant Physiology and Biochemistry 26: 609–614.

Hall, J.L. and T.J. Flowers. 1973. The effect of salt on protein synthesis in the halophyte Suaeda maritime. Panta (Berl.) 110: 361–368.

Hassan, N.I., A.M. El-Serafy and M.F.A. Farid. 1980. Studies on pastures indigenous to Southern Sinai. I. Chemical composition and *in vitro* digestibility. J. Anim. Sci. 51 Suppl. 1, 239.

Hossain, M.E., M. Shahjalal, M.J. Khan and M.S. Hasanat. 2003. Effect of dietary energy supplementation on feed intake, growth and reproductive performance of goats under grazing condition. Pakistan Journal of Nutrition 2(3): 159–163.

Joseph, D., Kajal Chakraborty, C.S. Subin and Koyadan Kizhakedath Vijayan. 2013. Halophytes of Chenopodiaceae and Aizoaceae from South-East Coast of India as potential sources of essential nutrients and Antioxidants. Journal of Food and Nutrition Research 1(5): 97–107.

Kearl, L.C. 1982. Nutrient requirements of ruminants in developing countries. International Feedstuffs Institute. Utah Agric. Exp. Stn., Utah State University, Logan, Utah, USA.

Khamis, H.S. 1988. Nutritional studies on some Agricultural by products and some natural pasture plants in arid and semi-arid areas using sheep and goats. Ph.D. Thesis, Fac. Of Agric., Cairo Univ.

Khan, M.A. et al. 2006. Crop diversification through halophyte production on salt-prone land resources. CAB Reviews: Perspectives in Agriculture, Veterinary Science, Nutrition and Natural Resources, 1, No. 048.

Khan, M.A., I.A. Ungar and A.M. Showalter. 2005. Salt stimulation and tolerance in inter-tidal stem-succulent halophyte. J. Plant Nut. 28: 1365–1374.

Konig, R.C.W., K. Becker, R.W. Benjamin and H. Soller. 2006. Performance of sheep and goats with offspring on semi-arid saltbush (*Atriplex nummularia*) grassland ranges in the early dry season [Online, accessed July 12, 2006]. URL:http://www.ilri.cgiar.org/InfoServ/Webpub/Fulldocs/X5519b/x5519b0d.htm.

Kosová, K., Ilja T. Prášil and Pavel Vítámvás. 2013. Protein contribution to plant salinity response and tolerance acquisition. Int. J. Mol. Sci. 14: 6757–6789.

Koyro, H.W., M. Ajmal Khan and Helmut Lieth. 2011. Halophytic crops: A resource for the future to reduce the water crisis?. Emir. J. Food Agric. 23(1): 001–016.

Kozlowski, T.T. 1997. Responses of woody plants to flooding and salinity. Tree Physiology Monograph No. 1.

Kraidees, M.S, M.A. Abouheif, M.Y. Al-Saiady, A. Tag-Eldin and H. Metwally. 1998. The effect of dietary inclusion of halophyte *Salicornia bigelovii* Torr. on growth performance and carcass characteristics of lambs. Anim. Feed Sci. Technol. 76: 149–159.

Kyriazakis, I., D.H. Anderson and A.J. Duncan. 1998. Conditioned flavour aversions in sheep: the relationship between the dose rate of a secondary plant compound and the acquisition and persistence of aversions. British Journal of Nutrition 79: 55–62.

Malik, K.A., Z. Aslam and M. Naqvi. 1986. Kallar grass: a plant for saline land. Nuclear Institute for Agriculture and Biology, Faisalabad (Pakistan).

Masters, D.G. (in Press). Assessing the feeding value of halophytes. pp. 89–105, this volume.

Masters, D.G. 2006. Establishing the metabolisable energy value of halophytic shrubs *in vitro* —problems and possibilities, confidential report.

Masters, D.G., H.C. Norman and R.A. Dynes. 2001. Opportunities and limitations for animal production from saline land. Asian–Aust. J. Anim. Sci. 14: 199–211 (Special issue).

McEvoy, J.F. and S. Jolly. 2006. The nutritive value of rangelands plants of Southern and Western Australia: A review of the literature and a scoping study to demonstrate the inconsistencies between the current system of nutritive analysis and animal performance. AWI Project No: EC786. Productive Nutrition Pty Ltd.

Meneses, R., Gabriel Varela and Hugo Flores. 2012. Evaluating the use of *Atriplex nummularia* hay on feed intake, growth, and carcass characteristics of creole kids. Chilean Journal of Agricultural Research 72(1): 74–79.

Meychik, N.R., Yuliya I. Nikolaeva and Igor P. Yermakov. 2013. Physiological response of halophyte (*Suaeda altissima* (L.) Pall.) and glycophyte (*Spinacia oleracea* L.) to salinity. American Journal of Plant Sciences 4: 427–435.

Moinuddin, M., S. Gulzar, I. Aziz, A.A. Alatar, A.K. Hegazy and M. Ajmal Khan. 2012. Evaluation of forage quality among coastal and inland grasses from Karachi. Pak. J. Bot. 44(2): 573–577.

Nizar, M., G. Hajer and H. Kamel. (in press). Potential use of halophytes and salt tolerant plants in ruminant feeding: A Tunisian case study, pp. 37–59, this volume.

Norman, H.C., E. Hulm and M.G. Wilmot. (in press). Improving the feeding value of old man saltbush for saline production systems in Australia. pp. 79–88, this volume.

O'Leary, J.W., E.P. Glenn and M.C. Watson. 1985. Agricultural production of halophytes irrigated with seawater. Plant and Soil 89: 311–321.

Parker, R., T.J. Flowers, A.L. Moore and N.V. Harpham. 2006. An accurate and reproducible method for proteome profiling of the effects of salt stress in the rice leaf lamina. Journal of Experimental Botany 57: 1109–1118.

Raef, O. 2012. Nutritive evaluation of natural ranges in the south eastern corner of Egypt. M.Sc. Fac. Agric., Cairo Univ.

Ramos, J., M.J. Lopez and M. Benlloch. 2004. Effect of NaCl and KCl salts on the growth and solute accumulation of the halophyte *Atriplex nummularia*. Plant Soil 259: 163–168.

Rani, G. 2007. Changes in protein profile and amino acids in *Cladophora vagabunda* (Chlorophyceae) in response to salinity stress. Journal of Applied Phycology 19: 803–807.

Sato, Y., M. Sugiyama, R.J. GO´ recki, H. Fukuda and A. Kokmamine. 1993. Interrelationship between lignin deposition and the activities of peroxidase isoenzymes in differentiating tracheary elements of Zinnia. Planta 189: 584–589.

SCA. 2007. Standing Committee on Agriculture's Nutrient Requirements of Domesticated Ruminants. CSIRO 10 Publications, Melbourne, Australia.

Shawket, Safinaz M. 1999. Effect of energy level supplementation on the utilization of some pasture plants by goats. J. Agric. Sci. Mansoura Univ. 24: 4565–4573.

Shawket, Safinaz. M., I.M. Khatab, B.E. Borhami and K.A. El-Shazly. 2001. Performance of growing goats fed halophytic pasture with different energy sources. Egy. J. Nutr. and Feeds 4: 251–26.

Shawket, S.M., M.H. Ahmed and M.A. Ibrahim. 2010. Impact of feeding *Atriplex halimus* and *Acacia saligna* with different sources of energy on lambs performance. 3rd International Scientific Conference on Small Ruminant Development, Hurghada, Egypt, 12–15 April, pp. 191–208.

Silveira, J.A.G., R.A. Viégas, I.M.A. Rocha, A.C.O.M. Moreira, R.A. Moreira and J.T.A. Oliveira. 2003. Proline accumulation and glutamine synthetase activity are increased by salt induced proteolysis in cashew leaves. Journal of Plant Physiology, Cordoba 160(2): 115–123.

Sun, H.X. and D.W. Zhou. 2010. Effect of dietary supplement of seed of a halophyte (*Suaeda glauca*) on feed and water intake, diet digestibility, animal performance and serum biochemistry in lambs. Livestock Science 128: 133–139.

Suyama, H., S.E. Benes, P.H. Robinson, S.R. Grattan, C.M. Grieve and G. Getachew. 2007. Forage yield and quality under irrigation with saline-sodic drainage water: Greenhouse evaluation. Agricultural Water Management 88: 159–172.

Swingle, R.S., E.P. Glenn and V. Squires. 1996. Growth performance of lambs fed mixed diets containing halophyte ingredients. Anim. Feed Sci. Technol. 63: 137e148.

Thomas, D.T., A.J. Rintoul and D.G. Masters. 2007. Sheep select combinations of high and low sodium chloride, energy and crude protein feed that improve their diet. Applied Animal Behaviour Science 105: 140–153.

Ventura, Y. and M. Sagi. 2013. Halophyte crop cultivation: the case for *Salicornia* and *Sarcocornia*, Environmental and Experimental Botany 92: 144–153.

Vercoe, T.K. 1986. Fodder potential of selected Australian tree species. pp. 95–100. *In*: J.W. Turnbull (ed.). Australian *Acacias* in Developing Countries. Proc. of an International Workshop Held at the Forestry Training Center Gympie, Qld, Australia, 4–7 August, 1986 ACIAR Proceedings no. 16, Australian Center for International Research, Canberra.

Wang, J., Yaxiong Meng, Baochun Li, Xiaole Ma, Yong Lai, Erjing Si, Ke Yang, Xianliang Xu, Xunwu Shang, Huajun Wang and Di Wang. 2014. Physiological and proteomic analyses of salt stress response in the halophyte Halogeton glomeratus. Plant, Cell and Environment, 1: 15. published by JohnWiley & Sons Ltd.

Warren, B.E. and T. Casson. 1992. Performance of sheep grazing salt tolerant forages on revegetated salt land. Aust. Soc. Anim. Prod. 19: 237–241.

Warren, B.E., C.I. Bunny and E.R. Bryant. 1990. A preliminary examination of the nutritive value of four saltbush (*Atriplex*) species. Proc. Austr.

Youssef, K.M., H.M. Khattab, H.M. Kandil and S.M. Abdelmawla. 2002. Studies on fattening goats fed halophytic shrubs under arid conditions in Sinai. Int'l Symp. Optimum Utilization in Salt Affected Ecosystems in Arid and Semi Arid Regions, Cairo, Egypt, April 8–11, 2002, pp. 410–449.

Plant Secondary Metabolites of Halophytes and Salt Tolerant Plants

Salah A. Attia-Ismail

SYNOPSIS

Most halophytes produce secondary metabolites that affect animal performance when consumed, even when palatable to the animals. This chapter reviews the literature and provides a synthesis of current thinking. Some of these plant secondary metabolites (previously called anti-nutritional factors) may result in toxicity. The resulting adverse effects constitute the major constraint to the use of halophytes as livestock feed ingredient. Some examples of these plant secondary metabolites (PSM) are tannins, glucosides, flavonoids, alkaloids, terpenoids, cyanides, coumarin, nitrate, oxalate and organic acids, etc. PSM's are substances that when present in animal feed reduce the availability of one or more nutrients. These constituents have different effects on animal performance. Some are beneficial while others are deleterious including loss of appetite and reductions of dry matter intake and nutrients digestibility. It is estimated that almost 80,000 PSM compounds are found in plants and occur naturally. They are produced in plants for protective purposes for the plants themselves and to adapt to environmental stresses. Some of which are beneficial, some of which may be nutritionally valuable but many have no nutritional value or nutritionally detrimental effects.

Desert Research Institute, Cairo, Egypt.
Email: saai54@hotmail.com

Keywords: antinutritional, toxicity, secondary metabolites, tannins, glucosides, flavonoids, alkaloids, terpenoids, cyanides, coumarin, nitrates, nitrites, oxalates, organic acids, phenolic compounds, environmental stress, phytochemicals, detoxification, appetite, monensin.

1. Introduction

The future prosperity of feed resources in the countries located in the arid and semi-arid regions relies on the economically feasible use of marginal and long-neglected resources (El Shaer and Attia-Ismail 2002). The increasing demand for animal feed in most developing countries imposes certain pressures. In less developed countries, there are huge amounts of different natural resources that are not utilized or at least are inefficiently utilized. There should be more emphasis on alternative concepts of productivity and more efficient utilization of local resources. Also, recent shifts in both technological and practical potentialities in animal feeding have been strongly reinforced by economic pressure. The achievements, hence, are assessed as much on reduced or lowered cost inputs as on greater output. This may suggest that more attention is to be given to the most economical feeding while maintaining animal production at optimal levels. It is, then, that the use of non conventional feed resources is of more importance. Halophytes can, then, play a significant role in the well being of different peoples.

A number of halophytes have been used for different purposes. Aronson (1989) enumerated 1560 halophyte species which are already in use. There are, however, several broad situations in which halophytes are used by livestock. Utilization of the marginal resources such as saline soils, underground water and saline as well as salt tolerant plants to produce feed for animals (fodder crops) becomes necessary in order to improve nutritional status of these livestock.

The less and unpalatable plant species of halophytic fodders are dominant while the palatable and good quality species are always deteriorated because of over grazing (Attia-Ismail 2008). Most of these halophytes produce secondary metabolites that affect animal performance when consumed even when palatable to the animals. Some of these plant secondary metabolites (previously called anti-nutritional factors) may result in toxicity. The adverse effects of the plant secondary metabolites constitute the major constraint to the use of halophytes as livestock feed ingredient. Some examples of these plant secondary metabolites (PSM) are tannins, glucosides, flavonoids, alkaloids, terpenoids, cyanides, coumarin, nitrate, oxalate and organic acids, etc. These constituents have different effects on animal performance. Some are beneficial while others are deleterious including loss of appetite and reductions of dry matter intake and nutrients digestibility.

The term plant secondary metabolite is used widely to refer to a broad spectrum of plant metabolically produced compounds. PSM's are an extremely large group of compounds referred to also as phytochemicals (Acamovic and Brooke 2005). They are of small molecular weights and are widely distributed in plants (Edreva et al. 2008). They are produced in plants for protective purposes for the plants themselves and to adapt to environmental stresses and defined as those substances (Smitha-Patel et al. 2013) generated in natural feed stuffs by the normal metabolism of species and by different mechanisms, for, e.g., inactivation of some nutrients, diminution of the digestive process or metabolic utilization of feed which exert effects contrary to optimum nutrition (Kumar 1991).

Wink (1999a,b) estimated that almost 80,000 PSM compounds are found in plants and occur naturally. Some of which are beneficial, some of which may be nutritionally valuable but many have no nutritional value or nutritionally detrimental effects (Bento et al. 2005a). Toxicity of most poisonous plants is associated with stage of growth, temperature, site, rainfall precipitation, light, soils, weather conditions and kind of animal. However, there are striking differences among animal species with respect to the tolerance of plant toxicants. Ruminants are more tolerant to poisonous plants than non ruminants. Even within ruminant species there are great differences in tolerance of PSM's. For instance sheep are more resistant to certain type of halophytic plants than are cattle. There are also some variations among different animal species and their ages and sometimes their physiological status (Afifi 2004, unpublished data). However, Cheek and Shull (1985) found that it is possible for animal species to adapt to some toxic plants if they are allowed to be exposed for a period of time. For example, ruminants adapted to oxalate-containing plants such as *Halogeton glomeratus*; can tolerate concentrations that are lethal to non adapted animals. The animal's tolerance or intolerance to PSM's is subjected to considerable changes. Previous feeding experience of ruminants for example may affect tolerance of poisonous plants (Smith 1992). The same author pointed out that the PSM's present in halophyte plants may be subjected to chemical transformation not only by rumen microbes but also by enzymatic changes in gut mucosa, liver and other tissues. Some chemical changes increase toxicity of plant compounds and some cause detoxification. Nevertheless, feeding management or treatment protocols that alter rumen microbial could cause toxification and detoxification of PSM's. Some authors classified toxic compounds (Roige and Tapia 1996) into compounds that may be detoxified in the rumen, toxic compounds produced in the rumen and compounds which affect rumen metabolism.

PSM's are substances that when present in animal feed reduce the availability of one or more nutrients. They interfere with the intake, availability, or metabolism of nutrients in the animals (Attia-Ismail 2005) or they can be toxic. Their effects range from mild reduction in animal performance to death, even at relatively small intakes. Harmful effects of plant secondary metabolites cause great economic losses to livestock producers. However, PSM's can inhibit the growth of microbes and fungi and they are often referred to as defenses (Forbey et al. 2009).

Toxicity is rather affected by numerous factors such as rate of ingestion, types and rates of microbial transformations in the rumen, rates of gastro-intestinal absorption, rates and pathways of biotransformation in gut tissues, liver and kidney and effect of enzyme induction or inhibition. The PSM's may cause liver damage, renal failure, Anoxia, pancreatic hypertrophy, hypoglycemia, death and other pathological conditions in animals.

2. Divisions of plant secondary metabolites

Most of the secondary metabolites are classified based on their biosynthetic origin. Although this classification does not take into consideration all the groups of plant secondary metabolites, it gives a list of those frequently found in animal feeds. Plant secondary metabolites can, therefore, be divided into five major groups:

i) *The phenolic compounds*:
   1.1 Tannins
   1.2 Gossypols

ii) *Glycosides*:
   2.1 Saponins
   2.2 Cyanogens

iii) Alkaloids
iv) Nitrates
v) *Others*: Oxalates and Lectins (Haemagglutinins)

    Table 8.1 shows endogenous plant secondary metabolites present in some halophytic fodder crops. Table 8.2 shows the PSM's detected in some *Acacia* spp.

Table 8.1. Endogenous plant secondary metabolites present in halophytic fodder crops.

Fodder crops	Anti-nutritional factors
Atriplex nummularia	Saponin, Alkaloids, Tannins, Nitrate
Atriplex leucoclada	Saponin, Alkaloids, Tannins
Atriplex halimus	Saponin, flavonoids, Alkaloids, Tannins, nitrate
Diplache fusca	Flavonoids, Alkaloids
Halocnemum strobilecum	Saponin, flavonoids, alkaloids, tannins, nitrate
Haloxylon salicornicum	Saponin, flavonoids, alkaloids, tannins
Kochia eriophora	Alkaloids, Tannins
Juncus acutus	Flavonoids, Alkaloids, Tannins, nitrate
J. arabicus	Alkaloids, Tannins
J. subulatus	Alkaloids, Tannins, flavonoids
Limonium pruinosum	Saponin, alkaloids, Tannins
Nitraria retusa	Saponin, Tannins
Salsola glauca	Saponin, flavonoids, alkaloids
Suaeda fruticosa	Alkaloids, Tannins, nitrate
Tamarix aphylla	Saponin, Tannins
Salsola tetrandra	Nitrate
Tamarix mannifera	Saponin, tannins
Zygophyllum album	Saponin, flavonoids, alkaloids, tannins, nitrate
Sesbania sesban	Saponin, alkaloids.

Adapted from Fahmy 2004, unpublished data.

Table 8.2. Toxic PSM compounds in some *Acacia* species for ruminants.

Species	Part of plant	Toxin	Reference
A. aneura	phyllode	oxalate	Gartner and Hurwood 1976
A. aneura	phyllode	tannin	
A. burrowii	flowers	hydrogen cyanide	Cunningham et al. 1981
A. cambagei	phyllode	hydrogen cyanide	
A. cambagei	timber, bark	oxalate	
A. cana	browse	selenium	
A. deanei	browse	hydrogen cyanide	
A. decora	browse	abortive agent	
A. doratoxylon	browse	cyanogenic glycoside	
A. georgina	browse	hyrolytic enzyme only	Hall 1972
A. georgina	seeds/pods	fluoroacetate	Everist 1969
A. longifolia	browse	hydrogen cyanide	Cunningham et al. 1981

Adapted from Dynes and Schlink (2002).

3. The phenolic compounds

Hydrobenzoic acid (example of phenolic acids)

Phenolic compounds vary in their structure from simple phenols such as the derivatives of hydrobenzoic acid (Hydrolysable tannins) to condensed tannins (Tania et al. 2012). The two types differ in their effects on animals both on nutritional and toxic levels. The condensed tannins may affect the animals more than hydrolysable tannins with respect to nutrients digestibility. Hydrolysable tannins, however, may cause varied toxic manifestations due to hydrolysis in rumen (Smitha Patel et al. 2013). Among the most important are flavonoids, phenolic acids, coumarins and isoflavones which are widespread in vegetable crops such as fruits, vegetables, herbs, grains and seeds (Miniati 2007). A phenol is a phenyl ($-C_6H_5$) bonded to a hydroxyl (-OH) group. Tannins are responsible for the astringent taste of some plants.

Polyphenolic compounds are produced naturally in some plants in which these compounds are the group of tannins. Tannins are distinguished from other polyphenolic compounds by their ability to precipitate proteins (Haslam 1989; Silanikove et al. 2001). For instance, the net tannin percent showed in the *Sorghum* tannins may bind and precipitate at least 12 times their weight of protein (Jansman 1993). They also have the ability to complex with minerals (Reed 1995). Tannins are widely found in most plants, especially in trees, shrubs and herbaceous leguminous plants. Tannin contents differ from one plant to another. Table 8.3 shows the variations in tannin percent in some tree leaves (Rana et al. 2006). They also differ within the same plant from season to season. The concentration of tannins was almost two-fold higher in

Table 8.3. Tannin content of some tree leaves (% DM).

Tree	Total phenols	Net tannin	Condensed tannins
Acacia nilotica	16.2	14.6	1.1
Toona cililate	3.8	2.3	0.9
Bauhinia variegate	4.8	3.7	3.4
Phoenix acaulis	5.8	4.8	4.3
Anogeissus latifolia	17.4	15.9	0.4
Carrisa spinarum	6.6	4.5	4.6
Ougenia oojeiuealis	4.2	2.9	2.6
Leucaena leucocephala	4.9	2.1	0.8

Adapted from Rana et al. (2006).

the dry compared to the wet season of *Albizia procera* (Alama et al. 2005). There are reports of low digestibility for some accessions of *Calliandra calothyrsus*, a shrub legume that occurs throughout the tropics and sub-tropics, and this has been related to the variable concentrations of tannins (Salawu et al. 1999; Mupangwa et al. 2000).

3.1 Tannins

Simple tannins

Condensed tannins

Tannins are the second most abundant group of plant phenolics after lignin. They are water soluble phenolic compounds with a molecular weight greater than 500 (Smitha-Patel et al. 2013).

Tannins are classified into two main groups, hydrolysable and condensed tannins. The most important nutritional and toxicological aspect of tannins and their effects is forming strong complexes with proteins (Hagerman and Butler 1981). They form complexes also with carbohydrates in the feeds, and with digestive enzymes. Tannins could be beneficial or harmful to ruminant. It is said that tannin concentration from 2–4% in rations help protein to escape rumen degradation and increases the absorption of essential amino acids whereas 4–10% depresses feed intake and can be harmful or toxic.

General effect of tannins on animals: Benefits reported include increases in wool production, milk protein secretion, ovulation rate and the development of more nutritionally based and ecologically sustainable systems for disease control in grazing animals (Ben Salem et al. 2005). Soluble dietary proteins stabilize foam (Nguyena et al. 2005) in the rumen that can entrap gas bubbles and cause bloat (McLeod 1974). Therefore, tannins may be involved in bloat prevention (Waghorn 1990). Ram´ırez-Restrepo et al. (2005) fed diet containing 18–29 g condensed tannins/kg DM for *Lotus corniculatus*. They found that under commercial dryland farming conditions, the use of *L. corniculatus* during the mating season in late summer/autumn can be used to increase reproductive efficiency and wool production. They attributed these findings to the higher digestibility and metabolizable energy of *L. corniculatus* than pasture, and to the effect of condensed tannins having improved both protein digestion and absorption.

Effects of tannins on poultry: Monogastric animals are highly susceptible to tannins, more so than ruminant animals. Tannins cause leg abnormalities in chicks which are due to defective formation of bone matrix (Armstrong et al. 1974). They attributed that effect to the tannins which may cause depression in feed intake. The reduced feed intake may affect growth as a result. The complex interactions of tannins with proteins reduce not only intake but also the protein digestibility and, therefore, increased the fecal output of nitrogen. Alledredge (1994) has suggested that there is a considerable evidence to suggest that enzymatic proteins, as well as other endogenous proteins, comprise a considerable portion of excreted nitrogen. This also may result in a deficiency of one or more of the essential amino acids leading to reduced growth. A strong relationship between concentration of tannins in feeds and their palatability has been found (Mangan 1988). Adverse effects of tannins on feed palatability and consumption have been repeatedly reported (Makkar 2003; Hassan et al. 2003; Kim and Miller 2005). Tannins can increase the size of the parotid glands and damage the mucosal lining of the gastrointestinal tract of chickens, but to a lesser extent in the laboratory rat (Oritz et al. 1994). Chicks fed diets high in condensed tannins (faba beans hulls) had poor digestibility of amino acids (Longstaff and McNab 1991). They explained that the low digestibility may be due to an increased excretion of inactivated enzymes. Studies on the effects of condensed tannins have given similar results. Negative effect of tannins on starch digestibility in three week old chicks was detected (Flores et al. 1994). The extent of the depression depended on the quantity of tannins ingested.

Sorghum is a feed that is commonly used as poultry feed. It has high content of tannins. All sorghum varieties contain phenolic compounds, which can influence the color, appearance and nutritive value of the grain. Tannin contents in sorghum grains can vary considerably among different varieties (Gu et al. 2004). In the US the grain breeders have developed a new sorghum variety that has no tannins. This has its implications on the feeding value of sorghum. When new grain sorghum varieties were compared to other cereal grains using broilers, layers, and mature leghorn roosters, Huang et al. (2006) found that crude protein digestibility of sorghum versus corn in all three classes of birds was similar between the grain sources. Similar work using broilers by Ravindran et al. (2005) found that the digestibility of crude protein was higher for sorghum compared to corn (99 vs. 81%).

Effects of tannins on ruminants: Ruminant animals are more tolerant to tannins than monogastric animals. The degradation products of hydrolysable tannins within the gastrointestinal tract can be absorbed and cause toxicity (Acamovic and Brooker 2005). Tannins have negative effects (Barry and Manley 1986) on protein metabolism and decrease palatability of feeds (because of the astringency through tannin–salivary protein complex formation in the mouth) at very high levels (> 60 g/kg DM) and high levels (> 50 g/kg DM). At lower levels, however (10–30 g/kg DM), or trace levels (< 10 g/kg DM) tannins are beneficial (Balogun et al. 1998). Therefore, it is safe to say that condensed tannins can have both beneficial and detrimental effects on ruminants. Positive effects on ruminants include prevention of bloating and anathematic effects. Among some of the beneficial effects are that condensed tannins complex with soluble proteins in the rumen slow the rapid microbial degradation, and, therefore, increasing ruminal escape protein (Waghorn et al. 1999), which in turn leads to increased amino acids absorption in the lower gut (Barry and Manley 1986). Condensed tannins may also contribute to animal health by reducing the detrimental effects of internal parasites in sheep and the risk of bloat in cattle (Niezen et al. 1998). They also have beneficial antibiotic effect on animals (Aengwanich et al. 2009). The tannins, however, have variable effects on rumen microorganisms. The results of Bento et al. (2005a,b) indicated that there is considerable interaction between tannins, microbes and non-starch-polysaccharides (NSP) in animal feeds and that these interactions may influence the functional ability of microbes in the gastrointestinal tract of animals.

In some ruminants, particularly goats and camels, tannin-resistant rumen microbial populations have been described (Brooker et al. 1994). Browse species with high tannin content had inhibitory effects on rumen microbial fermentation (Osuga et al. 2005). Tannins also inhibit abomasal and intestinal structure and function (Robins and Brooker 2005). Tannin protein complexes in the rumen are considered stable (Andrabi et al. 2005). These complexes can dissociate in the region of post-rumen in response to the low pH that occur there (McNabb et al. 1996). The low pH in the abomasum as well as the high pH in the small intestine can stimulate dissociation. However, Rakhmani et al. (2005) investigated this relation where they fractionated condensed tannins into monomeric, oligomeric and polymeric components. A negative correlation between oligomers, flavonols and flavonol glycosides and DM digestibility *in vitro* was detected whereas a positive correlation between the polymeric proanthocyanidins and DM digestibility *in vitro* was observed. However, some rumen bacteria species that tolerate or degrade tannins have now been identified (McSweeney et al. 2001) in animals that are used to consuming tannin containing diets.

3.1.1 Methods to overcome tannin's effects

Several methods were used to overcome the adverse effects of tannins such as, alkali treatments including ferrous sulfate (Smitha Patel et al. 2013) and calcium hydroxide (Alama et al. 2005), polyethylene glycol (PEG) (Barry et al. 2001). Physical methods like soaking and drying (Reddy 2001) and heat treatment before feeding (Vitti et al.

2005) of forage may reduce the toxic level of tannin (Nuttaporn and Naiyatat 2009). Potential methods for reducing the effects of tannin include drying, chemical agents like urea, wilting and wetting with chemical agents, as well as gelatin high in proline content (Rusdi 2004) and ensiling (Attia-Ismail 2005).

Polyethylene glycol is commonly used as an additive to improve intake, digestibility, and live weight gain and wool growth in sheep and goats (Palmer and Jones 2000; Barry et al. 2001). Polyethylene glycol (molecular weight 4000 or 6000) suppresses negative effects of tannins. Polyethylene glycol-4000 prevents formation of complexes between tannic acid and protein and helps in the breakdown of already formed complexes thus liberating protein (Reddy 2001). PEG preferentially binds with tannins (Yildiz et al. 2005). They found that the addition of PEG seems unnecessary as it did not improve crude protein digestibility, nitrogen retention and body weight. The amount of PEG used has a large impact on the utilization of tannin-rich shrub foliage (Makkar 2003).

Chopping, water sprinkling, storage under aerobic and anaerobic conditions, urea, wood ash, activated charcoal and PEG 4000 treatments were evaluated for their efficiency in deactivating tannins in the shrub *Acacia cyanophylla* foliage (Ben Salem et al. 2005). They found that chopping, water soaking or storage under anaerobic conditions are efficient techniques and further improvement could be achieved when two or more of these techniques are combined. Economic considerations were an important barrier to the use of PEG and activated charcoal that proved effective in tannin deactivation in contrast to wood ash. Vitti et al. (2005) studied the effects of oven, sun and shade drying and of urea treatment. Urea treatment for 30 days reduced extractable tannins. The urea treatment was also most effective at reducing the *in vitro* effects of tannins compared to the other drying treatments. Drying has no negative effect on the biological activity of the tannins examined (Muetzel and Becker 2005). Hydrochloric acidic and calcium hydroxide solutions were used among attempts to deactivate tannins (Wina et al. 2005a). Soaking in calcium hydroxide solution, hydrochloric acid or water, removed 41–76% of tannin and all of the phenolics were removed from the recovered leaves. Soaking of the leaves of acacia also removed fermentable materials. Alkali treatment (either calcium hydroxide or potassium carbonate) reduced the concentrations of extractable tannin by as much as 92% of *Albizia procera*. Calcium hydroxide alone did not improve the feeding value of *Albizia*. It was concluded that calcium hydroxide does not deactivate the tannins in *Albizia* (Alama et al. 2005).

Rusdi (2004) explained that gelatin high in proline content forms an imide instead of amide bond with tannins, which cannot be hydrolysed by endogenous enzymes in mammals. They concluded that gelatine may improve digestibility of nutrients in livestock given a tannin containing diet. Abd El Halim (2003) and Abd El Rahman (1996) found that ensiling mixture of halophyte plants increased animal acceptability and feed intake by sheep and goats. The ensiling process sharply depressed the presence of some plant secondary metabolites (saponin, alkaloids, coumarin, volatile oils, flavonoids and tannins).

Another route of ameliorating the nutritive value of halophytes in the use of feed additives like monensin (Jones and Hegarty 1984). Fahmy et al. (2003) added monensin to a mixture of halophytes and obtained 14% increment in body weight gain of growing lambs. Supplementation with energy and protein supplements plays an important role supporting animal performance. Shawket et al. (2001) concluded that fresh saltbushes (*Atriplex nummularia*) supplemented with 50% ground barley and 50% ground date seeds may be a good source of energy for growing goat kids.

4. Nitrates

Nitrates are compounds that exist naturally in plants. Nitrates are used as fertilizers. Although nitrate poisoning in cattle occurred long before the use of nitrogen as fertilizer, they can be fed if properly managed (Smitha Patel et al. 2013). The poisoning from nitrate results from the conversion of nitrate into nitrite in the rumen which in turn, absorbed into the blood via rumen wall, converts hemoglobin (the oxygen carrying molecule) in the blood to methemoglobin, which cannot carry oxygen. The blood turns to a chocolate brown color rather than the usual bright red (Benjamin 2006).

Nitrate concentrations in feeds for livestock depend on plant species and environmental conditions. Nitrate itself is relatively nontoxic, but its metabolites, i.e., nitrite, nitric oxide, etc. are poisonous. Rumen microorganisms can incorporate nitrates into microbial protein by converting the nitrates to ammonia. However, if the amount ingested is large enough, nitrates can not be converted completely to ammonia and toxic levels of nitrite are absorbed. Benjamin (2006) reported that sheep and cattle fed poor diets seem to be more susceptible to nitrate poisoning; a case that may happen under arid conditions and a diet of halophytes.

Recommended levels of nitrate nitrogen (feed total dry matter basis) of animal feeds as stated by (Andrae 2008):

< 1000 ppm	level is safe to feed under all conditions
1000–1500 ppm	level is safe for non-pregnant animals under all conditions. It may be best to limit its use to pregnant animals to 50% of the total ration on a dry basis
1500–2000 ppm	level is safe if limited to 50% of ration's total dry matter
2000–3500 ppm	feeds should be limited to 35 to 40% of total dry matter in the ration
above 2000 ppm	nitrate nitrogen should not be used for pregnant animals
3500–4000 ppm	feeds should be limited to 25% of total dry matter in ration (Do not use for pregnant animals), and
> 4000 ppm	feeds containing over 4000 ppm nitrate nitrogen are potentially toxic. Do not feed.

5. Others

5.1 Oxalates

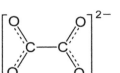

Oxalate is a common constituent of plants. Oxalates are poisonous when animals consume large quantities of oxalate containing plants without adaptation. If oxalates combine with calcium, they become insoluble (Gartner and Hurwood 1976). If this type of combination occurs in the rumen, insoluble oxalate may be excreted in the faeces. If it occurs in the blood, the crystals may precipitate in the kidneys and can cause kidney failure (Lincoln and Black 1980). Soluble oxalate may be degraded by rumen microflora (Allison et al. 1977). The occurrence of insoluble oxalates leads to problems in calcium and phosphorus metabolism. The content of oxalate in forage can be controlled by certain agricultural practices (fertilizer application). For example nitrate application and increased rate of potassium application resulted in higher contents of oxalates, while increased rate of calcium application, soluble oxalate content showed a decreasing trend and insoluble oxalate content showed a reverse trend (Rahman and Kawamura 2011).

5.2 *Lectins* (Hemagglutinins)

Lectins are glycoproteins that have the ability to bind to carbohydrate containing molecules which cause the agglutination of red blood cells as well as reduced growth, diarrhea, and interference with nutrient absorption. Lectin content in grain legumes in beans ranges from 1 to 3 percent. In general, the lectins in common beans are highly toxic, while lectins in peas and faba beans appear to be the least toxic. Moist heat treatment will destroy much of the lectin present, while dry heat treatment does not affect lectins that much (Anderson-Hafferman et al. 1992).

6. Summary of methods of alleviating the effects of PSM

The increasing demand for feed in most developing countries imposes certain pressures on animal production enterprises. Utilization of the marginal resources such as saline soils and underground water for producing feeds for animals (fodder crops) becomes necessary in order to improve nutritional status of livestock. Some of these fodder crops are toxic. Toxicity may result from the presence of a number of secondary metabolites. The consequences of toxic compounds in plants used as feedstuffs are not only their direct effects on animals, but also the economic loss resulting from poisoning the animals and the cost of treatments used to decrease its concentration in the feed.

The point sometimes is that ruminants, for instance, may convert a toxic compound to another toxic (cyanide to thiocyanate, which is goitrogenic) (Jones et al. 1997). The latter is less toxic, yet still toxic! A single plant may contain two or more toxic compounds which add to the difficulties of detoxification (Soetan and Oyewole 2009). Some detoxification methods may result in losses of nutrients which in turn lower the nutritive or feeding value of feeds. Cooking and/or germination treatments caused significant decreases in fat, total ash, carbohydrate fractions, anti-nutritional factors, minerals and B-vitamins of chickpeas (*Cicer arietinum*) (El-Adawy 2002). Yet, the germination treatment was less effective than cooking treatment in reducing trypsin inhibitor, hemagglutinin activity, tannins and saponins. However, it was more effective in reducing phytic acid, stachyose and raffinose. On the other hand, some detoxification treatments may improve the accessibility of nutrients by animals. Steam treatment, caused the PSM's to at least partially break down and some nutrients, such as fats, become better and therefore the nutritional value of the final animal feed is improved (Van Bruggen et al. 1993).

References and Further Readings

Abd El Rahman, H.H. 1996. Constraints and Possibilities for their alleviation to improve utilization of desert natural range plants for grazing ruminants. Ph.D. Thesis, Fac. of Agric. Cairo Univ., Egypt.

Abd El-Halim, A.M. 2003. Studies of some anti-nutritional factors affecting forage utilization by ruminants. Ph.D. Thesis, Fac. of Sci. Ain Shams Univ., Egypt.

Acamovic, T. and J.D. Brooker. 2005. Biochemistry of plant secondary metabolites and their effects in animals. Symposium on Plants as animal foods. Proceedings of the Nutrition Society 64: 403–412.

Aengwanich, W., M. Suttajit, T. Srikhun and T. Boonsorn. 2009. Antibiotic Effect of Polyphenolic Compound Extracted from Tamarind (*Tamarindus indica* L.) Seed Coat on Productive Performance of Broilers International Journal of Poultry Science 8(8): 749–751.

Alama, M.R., A.K.M.A. Kabir, M.R. Amin and D.M. McNeill. 2005. The effect of calcium hydroxide treatment on the nutritive and feeding value of *Albizia procera* for growing goats. Animal Feed Science and Technology 122: 135–148.

Alledrege, J. 1994. Effects of condensed tannins on browsers and grazers: Qualitative and Quantitative defense. Colorado State University, Fort Collins, Colorado, 805–823.

Allison, M.J., E.T. Littledike and L.F. James. 1977. Changes in ruminal oxalate degradation rates associated with adaptation of oxalate ingestion. J. Anim. Sci. 45: 1173–1179.

Andrae, J. and B. Pinkerton. 2008. Preventing prussic acid poisoning. Forage Leaflet, 13, Clemson University.

Anderson-Hafermann, J.C., Y. Zhang and C.M. Parsons. 1992. Effect of heating on nutritional quality of conventional and Kunitz trypsin inhibitor-free soybeans. Poul. Sci. 71: 1700–1709.

Andrabi, S.M., M.M. Ritchie, C. Stimson, A. Horadagoda, M. Hyde and D.M. McNeill. 2005. *In vivo* assessment of the ability of condensed tannins to interfere with the digestibility of plant protein in sheep. Animal Feed Science and Technology 122: 13–27.

Armstrong, W.D., W.R. Featherston and J.C. Rogler. 1974. Effects of bird resistant sorghum grain and various commercial tannins on chick performance. Poult. Sci. 53: 2137–2142.

Aronson, J. 1989. HALOPH: a database of salt tolerant plants of the world. Office of Arid Land Studies, Univ. Arizona, Tucson, Az., pp. 77.

Attia-Ismail, S.A. 2005. Factors limiting and methods of improving nutritive and feeding values of halophytes in arid, semi arid and coastal areas. International Conference On Biosaline Agriculture & High Salinity Tolerance, Mugla, Turkey, 9–14 Jan, pp. 91–99.

Attia-Ismail, S.A. 2008. Role of minerals in halophyte feeding to ruminants, Trace Elements: Nutritional Benefits, Environmental Contamination, and Health Implications. M.N.V. Prasad (ed.). Copyright # 2008 John Wiley & Sons, Inc., pp. 701–720.

Balogun, R.O. R.J. Jones and J.H.G. Holmes. 1998. Digestibility of some tropical browse species varying in tannin content. Anim. Feed Sci. Technol. 76(1-2): 77–88.

Bandyopadhyay, R., D.E. Frederickson, N.W. McLaren, G.N. Odvody and M.J. Ryley. 1998. Ergot: A new disease threat to sorghum in the Americas and Australia. Plant Dis. 82: 356–367.

Barry, T.N. and T.R. Manley. 1986. Interrelationships between the concentrations of total condensed tannins, free condensed tannins and lignin in *Lotus* sp. and their possible consequences in ruminant nutrition. Journal of Science and Food in Agriculture 37: 248–254.

Barry, T.N., D.M. McNeill and W.C. McNabb. 2001. Plant secondary compounds; their impact forage nutritive value and upon animal production. pp. 445–452. *In*: J.A. Gomide, W.R.S. Mattos and S.C. da Silva (eds.). Proceedings of the XIX International Grassland Congress. Sao Paulo, Brazil, 2001/02.

Benjamin David Neale. 2006. Effects of fertilizer application and cutting interval on nitrate accumulation in bermudagrass. M.Sc. (Agri.), Thesis: Univ.Tennessee, Martin.

Ben Salem, H., L. Saghrouni and A. Nefzaoui. 2005. Attempts to deactivate tannins in fodder shrubs with physical and chemical treatments. Animal Feed Science and Technology 122: 109–121.

Bento, M.H.L., T. Acamovic and H.P.S. Makkar. 2005a. The influence of tannin, pectin and polyethylene glycol on attachment of 15N-labelled rumen microorganisms to cellulose. Animal Feed Science and Technology, 19 August 2005 122(1–2): 41–57.

Bento, M.H.L., H.P.S. Makkar and T. Acamovic. 2005b. Effect of mimosa tannin and pectin on microbial protein synthesis and gas production during *in vitro* fermentation of 15N-labelled maize shoots. Animal feed science and technology, volumes 123–124, part 1, 30 September 2005, pp. 365–377.

Brooker, J.D., I. Skene, K. Clarke, L. Blackall, P. Muslera and L. O'Donovan. 1994. *Streptococcus caprinu* ssp. nov. A tannin resistant ruminal bacterium from feral goats. Lett. Appl. Microbiol. 18: 313–318.

Bush, L. and F.F. Fannin. 2009. Alkaloids. pp. 229–249. *In*: H.A. Fribourg, D.B. Hannaway and C.P. West (eds.). Tall Fescue for the Twenty-first Century. ASA, CSSA, and SSSA. Madison, WI.

Cunningham, G.M., W.E. Mulham, P.L. Milthorpe and J.H. Leigh. 1981. Plants of Western New South Wales. Soil Conservation Service of N.S.W. and N.S.W. Government Printing Office [no loc.].

Das, T.K., D. Banerjee, D. Chakraborty, M.C. Pakhira, B. Shrivastava and R.C. Kuhad. 2012. Saponin: Role in Animal system. Vet. World 5(4): 248–254.

Dynes, R.A. and A.C. Schlink. 2002. Livestock potential of Australian species of *Acacia*. Conservation Science W. Aust. 4(3): 117–124.

Edreva, A., V. Velikova, T. Tsonev, S. Dagnon, A. Gürel, L. Aktaş and E. Gesheva. 2008. Stress-protective role of secondary metabolites: diversity of functions and mechanisms. Gen. Appl. Plant Physiology, 2008, Special Issue 34(1-2): 67–78.

EFSA. 2011. Scientific Opinion on Pyrrolizidine alkaloids in food and feed. EFSA Journal 9(11): 2406.

EFSA. 2007. Cyanogenic Compounds as undesirable substance in animal feed Scientific Opinion of the Panel on Contaminants in the Food Chain. The EFSA Journal 434: 1–67.

EFSA. 2008. Gossypol as undesirable substance in animal feed Scientific Opinion of the Panel on Contaminants in the Food Chain. The EFSA Journal 908: 1–55.

EFSA. 2012. Scientific Opinion on Ergot alkaloids in food and feed. EFSA Journal 10(7): 2798.

EFSA. 2013. Scientific Opinion on Tropane alkaloids in food and feed Scientific Opinion on Tropane alkaloids in food and feed EFSA Journal 11(10): 3386.

El-Adawy, T.A. 2002. Nutritional composition and anti-nutritional factors of chickpeas (*Cicer arietinum* L.) undergoing different cooking methods and germination. Plant Foods for Human Nutrition (formerly Qualitas Plantarum) 57(1): 83–87.

El Shaer, H.M. and S.A. Attia-Ismail. 2002. Halophytes as animal feeds: potentiality, constraints and prodspects. Int'l Symp. Optimum Utilization in Salt Affected Ecosystems in Arid and Semi Arid Regions, Cairo, Egypt, April 8–11, 2002. pp. 411–418.

Everist, S.L. 1969. Use of Fodder Trees and Shrubs. Queensland Department of Primary Industries Advisory Leaflet no. 1024.

Ewing, W.N. 1998. The Feeds Directory Commodity Products Guide. Context, Ashby de la Zouch, UK.

Fahmy, A.A., S.A. Attia-Ismail and M.F. Afaf. 2003. Effect of monensin on salt plant utilization and sheep performance. Egy. J. Feeds and Nutr. 4: 581.

Flores, M.P., J.I. Castanan and J.M. Mcnabb. 1994. The effect of tannin on starch digestibility and TMEN of triticale and semi purified starches form triticale, and field beans. British Poultry Science 34: 281–286.

Forbey, J.S., A.L. Harvey, M.A. Huffman, F.D. Provenza, R. Sullivan and D. Tasdemir. 2009. Exploitation of secondary metabolites by animals: A response to homeostatic challenges. Integrative and Comparative Biology 49: 314–328.

Gardner, D.R., M.S. Thorne, R.J. Molyneux, J.A. Pfister and A.A. Seawright. 2006. Pyrrolizidine alkaloids in *Senecio madagascariensis* from Australia and Hawaii and assessment of possible livestock poisoning. Biochemical Systematics and Ecology 34: 736–744.

Gartner, R.J.W. and I.S. Hurwood. 1976. The tannin and oxalic acid content of *Acacia aneura* (mulga) and their possible effects on sulphur and calcium availability. Australian Veterinary Journal 52: 194–195.

Gu, L., Kelm Ma, J.F. Hammerstone, G. Beecher, J. Holden, D. Haytowitz, S. Gebhardt and Prior Rl. 2004. Concentrations of proanthocyanidins in common foods and estimations of normal consumption. J. Nutr. 134: 613–617.

Hagerman, A.E. and L.G. Butler. 1981. The specificity of proanthocyanidin-protein interactions. J. Biol. Chem. 256: 4494.

Hall, N. 1972. The Use of Trees and Shrubs in the Dry Country of Australia. Australian Government Printer, Canberra.

Hannaway, D.B., C. Daly, M.D. Halbleib, D. James, C.P. West, J.J. Volenec, D. Chapman, X. Li, W. Cao, J. Shen, X. Shi and S. Johnson. 2009. Development of suitability maps with examples for the United States and China. pp. 33–47. *In*: H.A. Fribourg, D.B. Hannaway and C.P. West (eds.). Tall Fescue for the Twenty-first Century. ASA, CSSA, and SSSA, Madison, WI.

Haslam, E. 1989. Plant Polyphenols-Vegetable Tannins Revisited. Cambridge University Press, Cambridge, U.K.

Hassan, I.A., E.A. Elzuber and H.A. Tinay. 2003. Growth and apparent absorption of minerals in broiler chicks fed diets with low or high tannin contents. Tropical Animal Health Production 35: 189–196.

Huang, K.H., X. Li, V. Ravindran and W.L. Bryden. 2006. Comparison of apparent ileal amino acid digestibility of feed ingredients measured with broilers, layers, and roosters. Poultry Sci. 85: 625–634.

Jansman, A.J.M. 1993. Tannins in feedstuffs for simple stomached animals. Nutrition Research Review 6: 209–236.

Jenkins, K.J. and A.S. Atwal. 1994. Effects dietary saponins on fecal bile acids and neutral sterols, and availability of vitamins A and E in the chick. J. Nutri. Biochem. 5: 134–138.

Jones, T.C., R.C. Hunt and N.W. King. 1997. Veterinary Pathology, 6th Edition.

Jones, R.J. and M.P. Hegarty. 1984. The effect of different proportions of *Leucaena Leucocephala* in the diet of cattle on growth, feed intake, thyroid function and urinary excretion of 3-hydroxy-4 LH. Pyridone. Aust. J. Agric. Res. 35: 317.

Kumar, R. 1991. Anti-nutritional factors the potential risks of toxicity and methods to alleviate them. Animal Feed Sci. and Tech. 30: 145–160.

Kim, H.S. and D.D. Miller. 2005. Proline rich in proteins moderate the inhibitory effect of tea on iron absorption in rats. Journal of Nutrition 135: 532–537.

Lincoln, S.D. and B. Black. 1980. Halogeton poisoning in range cattle. J. Am. Vet. Med. Assoc. 176: 717–718.

List, G.R. and G.F. Spencer. 1976. Fate of jimson weed seed alkaloids in soybean processing. Journal of the American Oil Chemists Society 53: 535–536.

Longstaff, M.A. and J.M. McNab. 1991. The inhibitory effects of hull Polysaccharides and tannins of field beans (*Vicia faba* L.) on the digestion of amino acids, starch and lipid and on digestive enzyme activities in young chicks. British Journal of Nutrition 65: 199–216.

Makkar, H.P.S. 2003. Effects and fate of tannins in ruminant animals, adaptation to tannins and strategies to overcome detrimental effects of feeding tannin rich feeds. Small Ruminant Research 49: 241–256.

Mangan, J.L. 1988. Nutritional effects of tannins in animal feeds. Nutrition Research Reviews 1: 209–231.

Masters, D.G. (in press). Assessing the feeding value of halophytes. pp. 89–105, this volume.

McLeod, M.N. 1974. Plant tannins, their role in forage quality. Min B.R. and Hart S.P. Tannins for suppression of internal parasites. Nutr. Abstr. Rev. 44: 803–815.

McNabb, W.C., G.C. Waghorn, J.S. Peters and T.N. Barry. 1996. The effect of condensed tannins in *Lotus pedunculatus* on the solubilization and degradation of ribulose-1 5-bisphosphate carboxylase (EC 4. 1. 1. 39; Rubisco) protein in the rumen and the sites of Rubisco digestion. Br. J. Nutr. 76: 535–549.

McSweeney, C.S., B. Palmer, D.M. McNeill and D.O. Krause. 2001. Microbial interactions with tannins: nutritional consequences for ruminants. Animal Feed Science and Technology 91: 83–93.

Miniati, E. 2007. Assessment of phenolic compounds in biological samples. Ann. Ist Super Sanità 2007 43(4): 362–368.

Muetzel, S. and K. Becker. 2005. Extractability and biological activity of tannins from various tree leaves determined by chemical and biological assays as affected by drying procedure. Animal Feed Science and Technology 125: 139–149.

Mulder, P.P.J., B. Beumer, E. Oosterink and J. de Jong. 2009. Dutch survey pyrrolizidine alkaloids in animal forage. Analytical Services & Development, Veterinary Drugs, Institute of Food Safety Wageningen University & Research centre.

Mupangwa, J.F., T. Acamovic, J.H. Topps, N.T. Ngongoni and H. Hamudikuwanda. 2000. Content of soluble and bound condensed tannins of three tropical herbaceous forage legumes. Animal Feed Science and Technology 83: 139–144.

Neale, B.N. 2006. Effects of fertilizer application and cutting interval on nitrate accumulation in bermudagrass. M.Sc. (Agri.), Thesis: Univ. Tennessee, Martin.

Nguyena, T.M., D.V. Binh and E.R. Ørskov. 2005. Effect of foliages containing condensed tannins and on gastrointestinal parasites. Animal Feed Science and Technology 121: 77–87.

Niezen, J.H., G.C. Waghorn and W.A.G. Charleston. 1998. Establishment and fecundity of *Ostertagia circumcincta* and *Trichostrongylus colubriformis* in lambs fed lotus (*Lotus pedunculatus*) or perennial ryegrass (*Lolium perenne*). Veterinary Parasitology 78: 13–21.

Nuttaporn Chanchay and Naiyatat Poosaran. 2009. The reduction of mimosine and tannin contents in leaves of *Leucaena leucocephala*. As. J. Food Ag-Ind. 2009, Special Issue, S137–S144.

Oda, K., H. Matsuda, T. Murakami, S. Katayama, T. Ohgitani and M. Yoshikawa. 2003. Relationship between adjuvant activity and amphipathic structure of soya saponins. Vaccine 21(17-18): 2145–2151. [doi:10.1016/S0264-410X (02)00739-9] Plohmann, B., Bader.

Oritz, L.T., C. Aizueta, J. Tervino and M. Castano. 1994. Effect of faba bean tannins on the growth and histological structure of the intestinal tract and liver of chicks and rats. British Poultry Science 35: 743–754.

Osuga, I.M., S.A. Abdulrazak1, T. Ichinohe and T. Fujihara. 2005. Chemical composition, degradation characteristics and effect of tannin on digestibility of some browse species from Kenya harvested during the wet season. Asian-Aust. J. Anim. Sci. 18(1): 54–60.

Palmer, B. and R.J. Jones. 2000. *In vitro* digestion studies using 14C-labelled polyethylene glycol (PEG): the effect of sample pretreatment on dry matter and nitrogen digestibility as well as PEG binding of *Calliandra calothyrsus*. Anim. Feed Sci. Technol. 86: 149–155.

Panter, K.E. 2005. Natural toxins of plant origin. pp. 11–63. *In*: W.M. Dabrowski and Z.E. Sikorski (eds.). Toxins in Food. CRC Press-Taylor and Francis Group, USA.

Potter, S.M., R. Jimenez-Flores, J. Pollack, T.A. Lone and M.D. Berber-Jimenez. 1993. Protein 45. Saponin interaction and its influence on blood lipids. J. Agri. Food Chem. 41: 1287–1291.

Rahman, M.M. and O. Kawamura. 2011. Oxalate accumulation in forage plants: some agronomic, climatic and genetic aspects. Asian-Aust. J. Anim. Sci. 24(3): 439–448.

Rajput, I.Z., Hu Song-hua, Xiao Chen-wen and G. Arijo Abdullah. 2007. Adjuvant effects of saponins on animal immune responses. J. Zhejiang Univ. Sci. B. 8(3): 153–161.

Rakhmani, S., J.D. Brooker, G.P. Jones and B. Palmer. 2005. Composition of condensed tannins from *Calliandra calothyrsus* and correlation with *in sacco* digestibility. Animal Feed Science and Technology 121(1–2): 109–124.

Ram´ırez-Restrepo, C.A., T.N. Barry, N. L´opez-Villalobos, P.D. Kempb and T.G. Harvey. 2005. Use of *Lotus corniculatus* containing condensed tannins to increase reproductive efficiency in ewes under commercial dryland farming conditions. Animal Feed Science and Technology 121: 23–43.

Rana, K.K., M. Wadhwa and M.P.S. Bakshi. 2006. Seasonal variations in tannin profile of tree leaves. Asian-Aust. J. Anim. Sci. 19(8): 1134–1138.

Ravindran, V.L.I. Hew, G. Ravindran and W.L. Bryden. 2005. Apparent ileal digestibility of amino acids in dietary ingredients for broiler chickens. An. Sci. 81: 85–87.

Reddy, D.V. 2001. Principles of Animal Nutrition and Feed Technology. Oxford IBM Publishing Company Pvt. Ltd., New Delhi.

Reed, J.D. 1995. Nutritional toxicology of tannins and related polyphenols in forage legumes. J. Anim. Sci. 73: 1516–1528.

Robert, M.F. and M. Wink. 1998. Alkaloids: biochemistry, Ecology and Medicinal applications. Plenum Press, New York.

Robins, C. and J.D. Brooker. 2005. The effects of *Acacia aneura* feeding on abomasal and intestinal structure and function in sheep. Animal Feed Science and Technology 121: 205–215.

Roige, M.B. and M.O. Tapia. 1996. The effect of plant and fungal toxins on rumen metabolism. Archivos-de-Medicina-Veterinaria. 28(1): 5–16, 108 ref.

Rosselli, M., K. Reinhart, B. Imthurn, P.J. Keller and R.K. Dubey. 2000. Cellular and biochemical mechanisms by which environmental oestrogens influence reproductive function, Human Reproduction Update 6(4): 332–350.

Rusdi. 2004. *In vitro* studies of the interaction between selected proteins and condensed tannins. Journal Agroland 11(1): 90–96.

Salawu, M., T. Acamovic, C.S. Stewart, T. Hvelplund and M.R. Weisbjerg. 1999. The disappearance of dry matter, amino acids and proanthocyanidins in the gastrointestinal tract from different fractions of *Calliandra* leaves. Animal Feed Science and Technology 79: 289–300.

Scanlan, N. and D.C. Skinner. 2002. Estradiol modulation of growth hormone secretion in the ewe: no growth hormone-releasing hormone neurons and few somatotropes express estradiol receptor α. Biology of Reproduction 66(5): 1267–1273.

Shawket, Safinaz M., I.M. Khatab, B.E. Borhami and K.A. El-Shazly. 2001. Performance of growing goats fed halophytic pasture with different energy sources. Egy. J. Nutr. and Feeds 4: 251–26.

Silanikove, N., A. Perevolotsky and F.D. Provenza. 2001. Use of tannin binding chemicals to assays for tannins and their negative postingestive effect in ruminants. Animal Feed Science and Technology 91: 69–81.

Smith, G.S. 1992. Toxification and detoxification of plant compounds by ruminants : an overview. J. Range Management 45: 25–30.

Smitha Patel, P.A., S.C. Alagundagi and S.R. Salakinkop. 2013. The anti-nutritional factors in forages—A review. Current Biotica 6(4): 516–526.

Soetan, K.O. and O.E. Oyewole. 2009. The need for adequate processing to reduce the antinutritional factors in plants used as human foods and animal feeds: A review. African Journal of Food Science 3(9): 223–232.

Strickland, J.R., M.L. Looper, J.C. Matthews, C.F.J. Rosenkrans, M.D. Flythe and K.R. Brown. 2011. Board invited review: St. Anthony's Fire in livestock: causes, mechanisms, and potential solutions. J.A.S. 89: 1603–1626.

Tânia da S. Agostini-Costa, R.F. Vieira, H.R. Bizzo, D. Silveira and M.A. Gimenes. 2012. Secondary Metabolites, Chromatography and Its Applications, Sasikumar Dhanarasu (Ed.), ISBN: 978-953-51-0357-8, InTech, Available from: http://www.intechopen.com/books/chromatography-and-itsapplications/secondary-metabolites.

Tewe, O.O., G. Gomez and J.H. Maner. 1980. Effect of extraneous linamarase on the hydrocyanic acid content of some tropical cassava varieties. Nigerian Journal of Nutrition Science 1: 27–32.

Van Bruggen, J., P. Veth and N.L. Sebastiaan. 1993. Method and device for reducing the amount of anti-nutritional factors in a mixture of raw material for animal feed. World Intellectual Property Organization (WO/1993/005664).

Velasquez-Pereira, J., C.A. Risco, L.R. McDowell, C.R. Staples, D. Prichard, P.J. Chenoweth, F.G. Martin, S.N. Williams, L.X. Rojas, M.C. Calhoun and N.S. Wilkinson. 1999. Long-term effects of feeding gossypol and vitamin E to dairy calves. J. Dairy Sci. 82(6): 1240–51.

Vitti, D.M.S.S., E.F. Nozella, A.L. Abdalla, I.C.S. Bueno, J.C. Silva Filho, C. Costa, M.S. Buenod, C. Longo, M.E.Q. Vieira, S.L.S. Cabral Filho, P.B. Godoy and I. Mueller-Harvey. 2005. The effect of drying and urea treatment on nutritional and anti-nutritional components of browses collected during wet and dry seasons. Animal Feed Science and Technology 122: 123–133.

Waghorn, G.C., J.D. Reed and L.R. Ndlovu. 1999. Condensed tannins and herbivore nutrition. In: Proceedings of the 18th International Grassl Congress, 111.

Waghorn, G.C. 1990. Effects of condensed tannins on protein digestion and nutritive value of fresh herbage. *In*: Proceedings of the Australian Society and Animal Production 18: 412–415.

WHO-IPCS (World Health Organisation-International Programme on Chemical Safety). 1988. Pyrrolizidine alkaloids. Environmental Health Criteria 80. WHO, Geneva, 1-345. Available from http://www.inchem.org/documents/ehc/ehc/ehc080.htm.

Wiedenfeld, H. and J. Edgar. 2011. Toxicity of pyrrolizidine alkaloids to humans and ruminants. Phytochemistry Reviews 10: 137–151.

Wina, E., B. Tangendjaja and I.W.R. Susana. 2005. Effects of chopping, and soaking in water, hydrochloric acidic and calcium hydroxide solutions on the nutritional value of *Acacia villosa* for goats. Animal Feed Science and Technology 122: 79–92.

Wina, E., S. Muetzel and K. Becker. 2005. The impact of saponins or saponin containing plant materials on ruminant production—A Review. J. Agric. Food Chem. 53: 8093–8105.

Wina, E., S. Muetzel and K. Becker. 2005a. The impact of saponins or saponin-containing plant materials on ruminant production-a review, Journal of Agricultural and Food Chemistry 53, 8093–8105.

Wina, E., S. Muetzel, E. Hoffman, H.P.S. Makkar and K. Becker. 2005b. Saponins containing methanol extract of Sapindus rarak affect munity structure *in vitro*, Animal Feed Science and Technology 121: 159–174.

Wina, E., B. Tangendjaja and I.W.R. Susana. 2005c. Effects of chopping, and soaking in water, hydrochloric acidic and calcium hydroxide solutions on the nutritional value of Acacia villosa for goats. Animal Feed Science and Technology, "Volume 122, Issues 1–2, 19 August 2005, pp. 79–92.

Wink, M. (ed.). 1999a. Introduction: Biochemistry, role and biotechnology of secondary metabolites. In Biochemistry of Plant Secondary Metabolism Annual Plant Reviews 2: 1–15. Boca Raton, FL: CRC Press.

Wink, M. (ed.). 1999b. Introduction: Biochemistry, role and biotechnology of secondary metabolites. In Biochemistry of Plant Secondary Metabolism Annual Plant Reviews 3: 1–16. Boca Raton, FL: CRC Press.

Woclawek-Potocka, I., C. Mannelli, D. Boruszewska, I. Kowalczyk-Zieba, T. WaVniewski and D.J. SkarHyNski. 2013. Diverse Effects of Phytoestrogens on the Reproductive Performance: Cow as a Model. International Journal of Endocrinology. Volume 2013. Article ID 650984, 15 pages http://dx.doi.org/10.1155/2013/650984.

Yildiz, S.I. Kaya, Y. Unal, D. Aksu Elmali, S. Kaya, M. Cenesiz, M. Kaya and A. Oncuer. 2005. Digestion and body weight change in Tuj lambs receiving oak (*Quercus hartwissiana*) leaves with and without PEG. Animal Feed Science and Technology 122: 159–172.

Mineral Balance in Animals as Affected by Halophyte and Salt Tolerant Plant Feeding

Salah A. Attia-Ismail

SYNOPSIS

Although salt tolerant plants, including true halophytes, can grow under adverse conditions in saline soils and produce biomass even when watered by seawater, their full potential as feedstuffs for livestock faces constraints. One of these constraints is the high content of salts and consequently the high mineral composition. In this chapter the type and attributes of the various minerals to be found in salt tolerant plants are examined. Measures to minimize the high mineral content are outlined.

Keywords: salt, ash, chenopods, saline soils, electrical conductivity, phytoextraction, bioavailability, mineral toxicity, mineral deficiency, pollution, toxicity.

1. Introduction

Up to 450 million hectares worldwide supports salt tolerant and halophytic plants (Glenn et al. 1999) that produce 4–5 billion tons of biomass each year (Squires et al., this volume). A significant proportion of this total biomass is potentially valuable fodder/forage for livestock. Halophytic plants have many uses: They can be used as animal feeds, vegetables, and drugs, etc. As an animal feed component, halophytes are promising because they have the potentiality of being a good feed resource. The

Desert Research Center, Cairo, Egypt.
Email: saai54@hotmail.com

potentialities of halophytes as animal feed components were recognized as early as the 1880s (Hutchings 1965). Yet, this potentiality does not go far because of several constraints that are limiting for which solutions have to be worked out. One of these constraints is the high contents of salts and consequently the high mineral composition. Attia-Ismail (2008) has reviewed in detail the role of minerals in halophyte feeding to ruminants.

2. The ash in halophytes and salt tolerant plants

The high content of ash is a typical characteristic of halophytes. Mineral contents differ in halophytes from those of regular forages. This has led to scientific concerns with regard as to what extent this may affect their feeding value to animals. Also, the mineral compositions of the high ash contents of halophytes have been controversial: Do they support animals' requirements of certain minerals; do they exceed the requirements to the extent that they may represent a poisoning threat to animals? However, mineral contents of halophytes vary considerably due to several factors (Attia-Ismail 2008), such as the plant species, stage of growth, season, etc.

Table 9.1 shows the ash and mineral contents of different parts of some halophytes around the world. Halophytes may represent a good source of minerals to ruminant animals. High ash contents of halophytes may compensate for mineral deficiency usually seen in animals in areas depending mainly on rangeland grazing as well as desert and coastal areas. Table 9.1 shows that ash contents of halophytes vary for whole plants, maturity stage, and plant parts. It runs up to more than 40% of plant materials. Plant parts differ also in their ash contents. Leaves and twigs of *Atriplex repanda*, for instance, have 23.1% whereas the fruits contain 32.65% (Table 9.1) representing 41% increment in ash content. *Salicornia bigelovii* spikes have 66% more ash than the stem. The stem of *Suaeda foliosa* has 117% more ash than leaves. Unless associated with non-digestible components such as silica, they may act as a source of minerals for animals.

Stage of plant maturity affects ash contents as well. *Pappophorum caespitosum* is an example. Its ash contents decreased after fructification by 98.5%. *Trichloris crinita* is a contrast case where ash contents increased after fructification by 152%. The variability in ash contents might be due to the physiological distribution and pools of minerals in different plant parts. Physiological and biochemical processes that take place in plants vary according to the plant species and this may explain the contrast in the percentage of ash content increment or decrement due to the process of fructification between both of *Pappophorum caespitosum* and *Trichloris crinita*.

2.1 Effect of saline water irrigation on ash contents

Halophytes grow naturally in saline environments, such as salt marshes, salt pans and salt deserts. Halophytic plant species and the salt tolerant plant species differ with regards to the degrees of tolerance to salinity. Halophytes also vary in accumulation of different ions. Chenopods, for example, are salt accumulators and have high Na^+ and Cl^- contents (Gul et al. 2000). The concentrations of Na^+ increased significantly

Table 9.1. Ash and mineral contents of some halophytic plants from different parts of the world.

Halophytic plants	Plant part and/or maturity stage	Ash %	Ca, %	P, %	Na, %	K, %	Mg, %	N, %	S, %	Zn, ppm	Cu, ppm	Fe, mg/Kg	Mn, ppm
Acacia aneura	Leaves	5.15	1.39	0.07	0.01		0.26		0.17	76			863
Acacia bidwilli	Leaves		2.43	0.09	0.06		0.24		0.21	37			49
Acacia concurrens	Leaves		0.97	0.05	0.25		0.26		0.18	47			917
Acacia deanei	Leaves		0.70	0.09	0.09		0.27		0.15	37			64
Acacia fimbriata	Leaves		0.30	0.06	0.13		0.25		0.12	33			83
Acacia glaucocarpa	Leaves		1.01	0.05	0.08		0.23		0.14	45			53
Acacia leptocarpa	Leaves		0.97	0.06	0.25		0.37		0.18	75			622
Acacia melanoxylon	Leaves		0.81	0.09	0.16		0.28		0.16	64			457
Acacia salicina	Leaves		3.52	0.11	0.02		0.29		1.13	55			30
Acacia saligna	Whole	8.83	3.75		1.15	1.05	6.14	2.21		140.5			
Acacia stenophylla	Leaves		2.08	0.10	0.16		0.35		0.58	123			42
Anabasis brevifolia	Whole/Flowering	17.5	3.19	0.08				1.52					
Aristida mendocina	Early fruitification	22.04	0.33	0.12				1.10					
Atriplex argentina	Leaves and twigs	6.95	0.67	0.08				2.71					
Atriplex atacamensis	Leaves and twigs	27	0.68	.11				2.55					
Atriplex coquimbensis	Leaves and twigs	30.56	0.87	0.10				2.69					
Atriplex halimus	Whole plant	29.20	1.69	0.32	3.91	0.57	0.32	2.11	0.17	64	10	503	51
Atriplex lampa	Twigs	30	0.15	0.01				1.80					
Ariplex nummularia	Whole plant	18.91	2.08	1.17	4.99	2.99	15.63	2.03		133.5	60.52		

Table 9.1. contd....

Table 9.1. contd.

Halophytic plants	Plant part and/or maturity stage	Ash %	Ca, %	P, %	Na, %	K, %	Mg, %	N, %	S, %	Zn, ppm	Cu, ppm	Fe, mg/Kg	Mn, ppm
Atriplex repanda	Leaves and twigs	23.1	1.9	0.12				3.31					
Atriplex repanda	Fruits	32.65	0.48	0.14				1.65					
Atriplex semibaccata	Leaves and twigs	26.6	0.57	0.09				3.00					
Cottea pappophoroides	Flowering	18-27	0.58	0.14				1.49					
Ephedra ochreata	Vegetative/twigs	25.8	1.94	0.20				0.99					
Glycyrrhiza inflate	Whole/Flowering	22.63	0.22	0.14				1.76					
Haloxylon ammodendron	Whole/Flowering	10.35	0.25	0.17				2.28					
Halostachys caspica	Whole/Flowering	17.9	0.52	0.18				1.83					
Haloxylom salicornicum	Whole plant	16.16	4.00	0.15	5.01	1.74	0.33	2.8	0.07	73	18	603	80
Halocnemum ssp. strobilaceum	Whole plant	18-36	2.45	0.14	7.00	1.65	0.39	1.11	0.10	99.7	17	621	87
Kalidium foliatum	Whole/Fruiting	30	0.48	0.08				3.03					
Karelinia caspica	Whole/Flowering	20.9	0.85	0.11				1.85					
Kochia Indica	Twigs, early bloom	17.2			2.52	1.87		2.13					
Larrea nitida	Vegetative/twigs	8.82	2.83	0.28				3.23					
Lyceum chilense	Fruitification/shoots	9.93	1.37	0.13				2.21					
Lyceum tenuispinosum	Budding	8.11	4.41	0.13				1.76					
Nitraria retusa	Whole plant	9.99	1.96	0.22	5.35	0.66	0.36		0.14	32	11	567	62
Pappophorum caespitosum	Early fruitification	35.93	0.28	0.14				1.70					
Pappophorum	Past fruitification	18.1	0.44	0.14				1.45					
Pappophorum philippianum	Past fruitification	40.1	0.38	0.11				1.37					

Phragmites communis	Whole/Flowering	43.41	0.24	0.06				1.92					
Reaumuria soongorica	Whole/Fruiting	10.89	0.65	0.12				1.64					
Salicornia bigelovii	Stem	25.1	1.01	0.08	6.44	0.75	0.40						
Salicornia bigelovii	Spikes	41.7	0.63	0.14	12.0	1.47	0.92						
Salsola tetrandra	Whole plant	12.90	3.98	0.16	5.65	1.45	0.59	1.08	0.12	44	8.88	6.64	79
Suaeda foliosa	Leaves	9.32	1.73	0.38				2.67					
Suaeda foliosa	stem	20.24	1.82	0.37				2.69					
Suaeda fruticosa	Whole plant	30.2	2.11	0.41	4.06	1.29	0.30	1.94	0.20	55	13	674	88
Suaeda physophora	Whole/Fruiting	8.79	0.41	0.16				2.88					
Tamarix aphylla	Whole plant	9.71	3.73	0.16	2.75	0.78	0.43	1.94	0.12	38	13	274	60
Tamarix mannifera	Whole plant	8.06	1.44	0.01	2.60	0.8	0.46	1.22	0.09	38.5	16	291	52
Tamarex ramosissima	Whole/Flowering	9.82	1.76	0.05				1.74					
Trichloris crinita	Early fruitification	14.32	0.27	0.13				1.54					
Trichloris crinita	Past fruitification	36.1	0.36	0.21				1.51					
Zygophyllum album	Whole plant	34.84	3.74	0.14	2.84	0.9	0.64	1.05	0.09	43.3	7.78	393	52

Adapted from: Abdelhameed et al. (2006); Attia-Ismail et al. (2003); Gabr (2002); Shawkat et al. (2001); Wilson (1974); Warren et al. (1990); Malcolm et al. (1989); Vercoe (1986); El-Bassosy (1983); Abiusso (1962); Wainstein et al. (1979); Kraidees et al. (1998) and Fahmy (2010).

at salinities greater than 300 mol.m^{-3} NaCl (Naidoo and Kift 2006). Saline water usually goes along with saline soils or in some cases independently of saline soils. Under the conditions of sea water irrigation the increase of salinity increased the ash content of plants both in the shoots and in the roots (Daoud et al. 2001) with some variations according to plant species and plant organ (Table 9.2). It appears from that table that the plant species, under test, had their ash content in shoots increased with increasing salinity of irrigation water except for a few where the increments in ash contents happen to occur mainly in the roots portion of the plants. However, the point of ash contents distribution among different physiological pools of the plant is not of concern to this discussion. Gul et al. (2010) reported *Kochia scoparia* to have increased ash contents in shoots more than in roots except for calcium and nitrate. The values of the mineral contents of *Kochia scoparia* shoots are presented in Table 9.3.

Table 9.2. Ash content (% of dry matter) in shoots and roots of plants watered with the five different dilutions of seawater.

	0% dilution		25% dilution		50% dilution		75% dilution		100% dilution	
	Shoots	Roots	Shoots	Roots	Shoots	Roots	Shoots	Roots	Shoots	Roots
A. tripolium	24.04	13.48	28.13	14.4	31.1	12.9	32	16.88	34.59	18.02
A. germinans	9.4	14.38	9.74	17.72	11.2	24.18	11.78	24.3	11.54	23.22
B. maritima	25.88	21.18	45.22	32.28	54.68	44.88	46.2	46.44	56.34	31.2
L. multiflorum	14.82	9.83	16.29	13.68	13.24	13.19	11.17	10.88	13.52	12.08
R. fasciculata	6.48	32.52	8.06	21.36	9.22	33.28	10.72	25.76	8.08	25.6
S. verrucosum	31.46	34.56	34.24	36.54	37.02	35.78	42.48	36.08	38.56	34.38
S. portulacastrum	29.34	22.52	38.88	21.36	39.56	20.86	40	14.58	34.48	13.98
S. alterniflora	9.82	15.04	13.52	17.31	13.94	21.64	16.9	32.66	18.12	38.16

Adapted from (Daoud et al. 2001).

Table 9.3. Change in cations and anions concentrations in the shoots of *Kochia scoparia* under various NaCl concentrations.

NaCl (mM)	Na$^+$ (mM)	K$^+$ (mM)	Ca^{++} (mM)	Mg^{++} (mM)	Cl$^-$ (mM)	SO$_4$ (mM)	NO$_3$ (mM)
0	960	17.0	0.7	66.6	973	30.09	0.04
300	1102	15.0	1.0	70.4	1162	68.7	0.19
600	1433	12.4	0.9	74.4	1503	43.6	0.17
900	1635	10.0	1.1	60.7	1702	37.2	0.39
1200	1788	11.8	0.6	57.9	1760	36.9	0.18
1500	1830	10.4	0.7	58.4	1835	27.7	0.56
1800	1890	10.5	0.7	60.4	1913	31.7	0.46

Adapted from Gul et al. (2010).

2.2 Effect of soil salinity on ash content

Soil salinity is a major constraint to crop production because it limits crop yield and restricts use of previously uncultivated land (Yokoi et al. 2002). Saline soils are the soils that have high percentages of salts. Salt-affected soils are distributed all over the world. Saline soils have an electrical conductivity at the saturation of soil extract of more than 4 dS/m (FAO 1988). Salt-affected soils are classified into two main classes according to the type of salts present in soils: saline soils (solonchak) and sodic soils (solonetz) (Szabolcs 1974). The most common salts present in saline soils are sodium, potassium, calcium and magnesium in the form of chlorides, sulfates or bicarbonates. Sodium and chloride are present in the highest percentage particularly in highly saline soils. The most common salts present in sodic soils are the carbonates (FAO 1988) particularly sodium.

Halophytes and salt tolerant plants grow in soils, where concentration of soil solution is about 5 g/l of total dissolved solids (85 mM NaCl or 7–8 dS/m) (Khan et al. 2006). They grow in habitats ranging from seacoast with salinity 0.5–1.0 M or higher (Ungar 1991).

The effect of saline soil level's stresses on the fodder value of *Atriplex* is determined by growing plants in 3 levels of salinity including low, medium and high levels of salinity (Amouei 2013). Ash, fat and calcium increased in the high salinity location while protein, phosphor and nutrient unit were reduced in the high salinity location (Table 9.4).

High accumulation of Na^+ and Cl^- accompanied by decreases in K^+, Ca^{2+} and Mg^{2+} with increasing salinity of soil in *Halopyrum mucronatum* were found by Khan et al. (1999).

Table 9.4. The mean comparison analysis related to the effect of saline soils levels on fodder value of *A. leucoclada.*

Salinity parameters	Ash, %	Fat %	Protein %	Ca (g/mg)	P (mg/kg)
High	13.5	0.271	9.32	0.57	0.220
Low	13.0	0.271	12.00	0.52	0.169
Medium	7.4	0.211	12.89	0.42	0.152

Adapted from Amouei (2013).

2.3 Effect of plant species on ash contents

Halophyte plant species differ in their ability to accumulate minerals (Reboreda and Cacador 2007). Table 9.5 shows the Cu, Cd and Pb partitioning in different tissues (roots + rhizomes), stems and leaves (mg m^{-2}) in areas colonized by *Halimione portulacoides* and *Spartina maritime* (Reboreda and Cacador 2007).

The ability of halophytes to accumulate minerals depends on the mechanism by which halophytes distribute minerals in different plant compartments. Minerals in halophytes are accumulated in different parts of the halophytes and the ratio of mineral distribution differs from one plant species to another. Some plant species accumulate salts in the underground parts while others accumulate them in the parts above the ground (e.g., shoots, stems, twigs). Halophytes have higher potentiality to phytoextract minerals from the soil than do glycophytes (Jordan et al. 2002). Legumes, generally, have a different mineral profile (being high in calcium, potassium, magnesium) than grasses which tend to be higher in manganese and molybdenum when grown under the same conditions.

Table 9.5. Cu, Cd and Pb partitioning (mean ± SD) in different tissues (roots + rhizomes), stems and leaves (mg m^{-2}) in areas colonized by *Halimione portulacoides* and *Spartina maritime.*

	Cu (mg/m^2)	Cd (mg/m^2)	Pb (mg/m^2)
S. maritime			
Roots	230.1 ± 40.5	19.3 ± 6.5	670 ± 235.4
Stems	1.8 ± 0.3	0.5 ± 0.1	7.9 ± 1.5
Leaves	0.8 ± 0.1	0.2 ± 0.1	3.0 ± 1.7
H. portulacoides			
Roots	660.9 ± 262.2	22.4 ± 8.6	1147.1 ± 483.3
Stems	3.9 ± 1.1	2.3 ± 0.3	22.5 ± 1.9
Leaves	1.8 ± 0.5	0.7 ± 0.1	7.7 ± 1.1

Adopted from (Reboreda and Cacador 2007).

2.4 Effect of season on ash contents

The season of growth affects not only ash contents of halophytes but also other growth parameters. Abbas (2005) analyzed *Zygophyllum qatarense* from saline and non-saline habitats in Bahrain Island for their ash content during four seasons: winter, spring, summer and fall. Ash content of leaves was higher than that in the roots from the two habitats and significant seasonal differences in the ash content in both roots and leaves in each of the two habitats existed. The variations in mineral contents of halophytes occur mainly in sodium and chloride ions. When these elements change due to any factor like season, soil salinity or any other, it may affect the concentrations of other elements. However, the results in this aspect are contradictory. Agha et al. (2009) compared their results with those obtained by (Epstein 1972) and (Cramer et al. 1985). Nutrient deficiencies can occur in plants when high concentrations of Na$^+$ in the soil reduce the amounts of available K$^+$, Mg^{2+} and Ca^{2+} or when Na$^+$ displaces membrane bound Ca^{2+}. On the other hand, their results revealed that Ca^{2+} and Mg^{2+} contents were reduced in shoots of *A. stocksii* when salinity increased.

3. The mineral contents in halophytes and salt tolerant plants

Since the salt marshes are areas for the accumulation of many pollutants, they are considered to be important sinks, especially for metal pollutants (Doyle and Otte 1997).

At high salinity, the growth of halophytes is presumably limited by many factors but mainly because of an imbalance in nutrient uptake, essentially K and Ca (Nasim et al. 2008). They found that salinity stress significantly decreased potassium concentration and K:Na ratio in leaves of *Eucalyptus camaldulensis* in Pakistan). The imbalance in nutrient uptake resulting from salinity stress may lead to excessive accumulation of Na$^+$ and Cl$^-$ in tissue (Tahir et al. 2006). The first adaptation mechanism to high salinity is the exclusion of Na$^+$ ion from sodium sensitive sites of the plants because the osmotic adjustment is usually achieved by the uptake of inorganic ions and this accumulation of ions is often accompanied by mineral toxicity and nutritional imbalance (Arzani 2008).

3.1 Mineral role in ruminant nutrition

Minerals, in general, whether major or minor, play a crucial role in the lives of animals and affect their production to a greater extent. Minerals have four vital functions in the bodies of animals. They have a structural function in bones and other structural tissues of the body, physiological function in body fluids as electrolytes, regulatory function as they regulate several metabolic processes in the body, and catalytic function when they enhance the enzyme activities (Underwood and Suttle 1999).

3.2 Bioavailability of mineral contents to animals

Excess or deficient mineral intake may adversely affect diet and mineral utilization. However, the degree of absorption (bioavailability) of minerals may affect the state of mineral utilization by animals. The bioavailability of minerals is an important factor that has to be taken into consideration. It varies, however, among different minerals because of different conditions. Haenlein and Ramirez (2007) found that the degree of absorption of minerals differs and that it was low for Cu, Mn, Mg, Zn, Ca and Fe.

4. Minerals deficiency in halophyte included diets

Ash contents of halophytes may limit animal feed intake and digestibility coefficients of ruminant diets. Therefore, some restriction is placed on the use of these plant species in livestock rations. Adequate trace mineral intake and absorption is required for a variety of metabolic functions including immunity, reproduction, etc. The deficiency results from either excessive sweating or through the feeding on halophyte plants either deficient in certain minerals or may have imbalanced mineral ratios. The opposite applies for mineral toxicity.

Animals inhabiting desert areas suffer from high environmental heat stress. A typical characteristic of animals in desert areas is to dissipate heat through several adaptive mechanisms. One of them is to sweat. Sweating in such animals (Farid 1989) is accompanied by the excretion of salts from the body. This author cited (from MacFarlane et al. 1963) that camels' sweat is rich in bicarbonates (pH 8.2–8.5) and is particularly high in potassium, four times as much as sodium. When animals are exposed to high environmental temperatures, the sweating rate increases as a result (about 0.5 Kg per day in sheep and goats). Thus, higher concentrations of urea, sodium, potassium and chloride might be found in the sweat of animals (Farid 1989). Sweat was found to contain different amounts of Mg, Na, K, Ca, and Cl in man (Verde et al. 1982).

Animals in these areas suffer from mineral deficiency or are exposed to high concentrations of various minerals in either feedstuffs or drinking water (if brackish water is used) or both (Squires, this volume). Walker (1980) found that the concentration of calcium and potassium is usually higher than that of other minerals; the average being 1 to 1.5% in Southern African browses.

Goats browsing on naturally growing halophytes in Mexico (Haenlein and Ramirez 2007) were deficient in minerals intake. They found significantly low supply of Mg, Cu, Mn, and Zn from the range plants browsed by goats compared to the requirements for these minerals. On the other hand, sheep selecting different range plants had deficiency in Ca, Mg, K, Cu, and Mn. Similarly, phosphorus deficiencies in acacias were reported by Vercoe (1986) and Craig et al. (1991) which lead to an imbalance in the Ca/P ratio in foliage. *Panicum turgidum* was fed to camels in south eastern region of Egypt (Badawy 2005). Mineral composition of this halophyte is presented in Table 9.6. The major problem encountered was the extreme low Ca/P ratio which reached 1:19 in wet season and 1:32 in dry season. This ratio is far below than that is generally recommended (2:1) for domestic animals. They found a similar trend in Mg concentrations. It tended to be higher in range during dry season than in wet season. The levels of Mg and K in *Panicum turgidum* were inadequate compared with the dietary requirements of ruminants recommended by NRC (1981).

Table 9.6. Mineral content in *Panicum turgidum* as basal diet for camels.

Mineral	Wet season	Dry season	DS + M
Ca %	1.69	1.62	1.60
P %	0.09	0.05	0.04
Mg %	0.23	0.27	0.26
Na %	0.35	0.42	0.40
K %	0.23	0.31	0.30

Badawy (2005).

4.1 Excessive minerals in livestock rations in dry areas

Some halophytes attain higher levels of some trace or major elements than other traditional feedstuffs (Attia-Ismail 2005). It seems that some of their mineral concentrations are within normal ranges to the extent that they may support normal animal physiological functions. On the other hand, excessive concentrations may reach toxic levels for the animals. Certain mineral ratios are skewed. For instance, calcium-phosphorus ratio present in most halophytes does not match that required by animals to support normal functions. In general, excessive sodium, chloride and potassium contents of halophytes may have several drawbacks to the animal health. The capacity of the animal body to store excess electrolytes or to excrete in the faeces is very little (Masters et al. 2007). Marai et al. (1995) found animals to have low rectal temperature and high water retention when exposed to higher levels of sodium.

5. Recommended mineral allowances

Minerals required by animals for proper functioning of the body and for the optimal production are either macro or micro-minerals. Macro-minerals are known to be Ca, P, Na, Cl, Mg, K, and S while the micro-minerals are Fe, I, Cu, Zn, Mo, Mn, Co, and Se.

Recent studies have revealed some of the requirements of goats, and sheep in dry areas. NRC (1981) has also recommended mineral requirements for goats, and for sheep (NRC 1985). Table 9.7 shows some recommended mineral requirements. Clear variations are present when matching mineral supply of halophytes to the requirements

Table 9.7. Recommended mineral requirements of sheep and goats.

Mineral	NRC	
	Lactating Goats, 1981	Sheep, 1985
Na, % of DM	0.09–0.18	0.09–0.18
Cl, % of DM	--	--
Ca, % of DM	0.20–0.80	0.20–0.82
P, % of DM	0.20–0.40	0.16–0.38
Mg, % of DM	0.12–0.18	0.12–0.18
K, % of DM	0.50–0.80	0.50–0.80
S, % of DM	0.14–0.26	0.14–0.26
I, PPM	0.10–0.80*	0.10–0.80
Fe, PPM	30–50*	30–50
Cu, PPM	10–20*	7–11
Mo, PPM	0.50–1.0*	0.5
Co, PPM	0.10–0.20*	0.1–0.2
Mn, PPM	20–40*	20–40
Zn, PPM	20–33*	20–33
Se, PPM	0.10–0.20*	0.1–0.2

*Rick Machen, Texas Ag. Ext. Serv.

of these animal species. Therefore depending on halophytes as the only mineral source for the animals may result in either mineral deficiency of some elements, toxicity of some others, or malabsorption of some other elements because of the presence of different proportions of minerals (e.g., Ca/P, K/Na, Ca/Na ratios). However, mineral requirements differ according to several conditions such as the season (summer versus winter), especially for the desert animal, and physiological state (pregnant, lactating, growing, etc.).

5.1 Effect of halophytes feeding on mineral utilization

Attia-Ismail et al. (2003) fed sheep a mixture of salt plants and found that Ca intake (mg/Kg BW$^{0.75}$) of salt plant (SP) fed animals was higher (Table 9.8) (P < 0.05) than that of the control group. In a previous study (Fahmy et al. 2001); animals consumed more salt plants than control (hay). Therefore, amount of Ca intake of SP group and, hence the balance, was higher (P < 0.05) than control. Yet, serum Ca concentrations (mg %) of SP group was almost similar to that of the control group. Sodium (Na) intake of salt plant fed groups was, logically, higher than the control group (Table 9.8).

Fahmy (1998) found Na intake to vary from 805 to 1507 mg/Kg BW$^{0.75}$ for silages made of a mixture of salt plants; a range close to that obtained in the work carried out by Attia-Ismail et al. (2003). The Na balance in the salt plant fed animals was; however, lower than that of the control group. The excreted Na in this experiment might be higher for the salt plant fed animals than the control group. Serum Na did not differ (Table 9.8). Potassium utilization followed the same trend as Na did except that serum K of control was higher than the salt plant fed animals.

Table 9.8. Mineral utilization of sheep fed a mixture of halophytic plants (Attia-Ismail et al. 2003).

	Halophyte fed sheep	Control group
Calcium utilization		
Ca intake (mg/BW$^{0.75}$)	914a	484b
Ca balance (mg/BW$^{0.75}$)	914a	341b
Serum Ca, mg%	8.2a	10.9a
Sodium utilization		
Na intake (mg/BW$^{0.75}$)	1546a	1068b
Na balance (mg/BW$^{0.75}$)	252a	316a
Serum Na, mg%	421a	431a
Potassium utilization		
K intake (mg/BW$^{0.75}$)	1138	1049
K balance (mg/BW$^{0.75}$)	8b	61a
Serum K, mg%	25.5a	61b
Zinc utilization		
Zn intake (mg/BW$^{0.75}$)	45.2a	31.2b
Zn balance (mg/BW$^{0.75}$)	40.4a	28.6b
Serum Zn, mg%	1.22a	0.85b

a and b: values bearing different superscripts at the same row differ significantly ($P < 0.05$).

Fahmy (2010) fed *Kochia indica* planted in salt affected soil and irrigated with saline water (8000 ppm) in South Sinai, Egypt. He fed the plant *ad libitum* to lambs for 90 days and obtained positive N, Na and K balances.

Camels fed on rations containing *Atriplex nummularia* showed a higher blood Na values (448.3 and 454.3 mg/100 ml), than those of other tested groups (Kewan 2003). Khattab (2007) had the same results when *Atriplex halimus* was fed to sheep (Table 9.9).

Table 9.9. Blood minerals content of Barki ewes fed the experimental diets during pregnancy and lactation periods (mean ± SE).

Items	Control	Mixture of *Atriplex halimus* and *Acacia saligna*	
		Fresh	Silage
Na (mmol/l)	148.68	173.15	167.85
K (mmol/l)	4.68	5.30	167.85
Ca (mg/dl)	11.57	9.70	9.01
Zn (µg/dl)	96.33	78.65	77.03
Mg (mg/dl)	2.53	1.53	1.26
Cu (µg/dl)	117.5	53.3	47.5
Se (µg/dl)	11.58	22.60	19.87

Adapted from Khattab (2007).

6. Conclusions

Mineral composition of feedstuffs, including those from salt tolerant and halophytic plants, is an important contributor to nutrition but mineral imbalance can lead to problems, including toxicity. Some halophytes are phytoaccumulators and there may be toxic quantities of some elements in their tissues. When saline diets are offered there may be need to supply additional mineral supplements, e.g., calcium, to balance the intake and prevent unwanted side effects. Better understanding of the important role of mineral elements in animal nutrition has been summarized here and further reading is provided in the references.

References and Further Readings

Abbas, J.A. 2005. Seasonal Variations of Ash Content of the Halophyte *Zygophyllum qatarense* Hadidi from Saline and Non-Saline Habitats In Bahrain. Pak. J. Bot. 37(4): 853–858.

Abdelhameed Afaf, A.E., S.M. Shawkat and I.M. Hafez. 2006. Physiological studies on the effect of feeding salt plants in ewes under semi arid conditions. 4th Sci. Conf. Physiol. Applic. For Anim. Wealth Dev., Cairo, Egypt, July 29–30, pp. 113–132.

Abiusso, N.G. 1962. Composicion quimica y valor alimenticio de algunas plantas indigenas y cultivadas en la Republica Argentina. Rev. Investgaciones Agricolas. 16: 193–247.

Agha, F., B. Gul and M.A. Khan. 2009. Seasonal variation in productivity of *Atriplex stocksii* from a coastal marsh along the Arabian sea coast. Pak. J. Bot. 41(3): 1053–1068.

Ahmad, K., Z.I. Khan, M. Ashraf, E.E. Valeem, Z.A. Shah and L.R. Mcdowell. Determination of forage concentrations of lead, nickel and chromium in relation to the requirements of grazing ruminants in the salt range, Pakistan. Pak. J. Bot. 41(1): 61–65.

Amouei, Alimohammad. 2013. Effect of saline soil levels stresses on agronomic parameters and fodder value of the halophyte *Atriplex leucoclada* L. (Chenopodiaceae). African Journal of Agricultural Research 8(23): 3007–3012.

Arzani, A. 2008. Improving salinity tolerance in crop plants: a biotechnological view. *In Vitro* Cell. Dev. Biol.-Plant 44: 373–383.

Attia-Ismail, S.A. 2008. Role of Minerals in Halophyte Feeding to Ruminants. Trace Elements as Contaminants and Nutrients: Consequences in Ecosystems and Human Health, Edited by M.N.V. Prasad, pp. 701–720.

Attia-Ismail, S.A., A.M. Fayed and A.A. Fahmy. 2003. Some mineral and nitrogen utilization of sheep fed salt plant and monensin. Egyptian J. Nutr. Feeds 6: 151–161.

Attia-Ismail, S.A. 2005. Factors limiting and methods of improving nutritive and feeding values of halophytes in arid, semi-arid and coastal areas. Conference on Biosaline Agriculture & High Salinity Tolerance, January 9–14, 2005, Mugla, Turkey.

Attia-Ismail, S.A., H.M. Elsayed, A.R. Asker and E.A. Zaki. 2009. The effect of different buffers on rumen kinitics of sheep fed halophyte plants. J. Environ. Sci. 19: 89–106.

Badawy, H.S. 2005. Nutritional studies on camels fed on natural ranges in Shalatin-Halaib Triangle rewgion. Ph.D. Cairo University.

Barrio, J.P., S.T. Bapat and J.M. Forbes. 1991. The effect of drinking water on food intake responses to manipulations of rumen osmolality in sheep, proceeding of the nutrition society, 50, 98A.

Bennink, M.R., T.R. Tyler, G.M. Ward and D.E. Johnson. 1978. Ionic milieu of bovine and ovine rumen as affected by diet. J. Dairy Sci. 61: 315–323.

Craig, G.F., D.T. Bell and C.A. Atkins. 1991. Nutritional characteristics of selected species of *Acacia* growing in naturally saline areas of Western Australia. Australian Journal of Experimental Agriculture 31: 341–345.

Cramer, G.R., A. Lauchli and V.S. Polito. 1985. Displacement of Ca2+ by Na+ from the plasmalemma of root cells. Plant Physiol. 79: 207–211.

Daoud, S., M.C. Harrouni and R. Bengueddour. 2001. Biomass production and ion composition of some halophytes irrigated with different seawater dilutions. First International Conference on Saltwater Intrusion and Coastal Aquifers-Monitoring, Modeling, and Management. Essaouira, Morocco, April 23–25, 2001.

Dobson, A., A.F. Sellers and V.H. Gatewood. 1976. Absorption and exchange of water across rumen epithelium. American Journal of Physiology 231: 1588–1594.

Doyle, M.O. and M.L. Otte. 1997. Organism-induced accumulation of Fe, Zn and As in wetland soils. Environ. Pollut. 96(1): 1e11.

El-Bassosy, A.A. 1983. A study of the nutritive value of some range plants from El-Saloom and Mersa Mattroh. Ph.D. Thesis, Faculty of Agriculture, Ain Shams University, Egypt.

Epstein, E. 1972. Mineral Nutrition of Plants: Principles and Perspectives. John Wiley and Sons, Inc., New York, London, Sydney, Toronto.

Fahmy, A.A., S.A. Attia-Ismail and A.M. Fayed. 2001. Effect of Monensin on salt plant utilization and sheep performance. Egypt. J. Nutr. and Feeds 4: 581–590.

Fahmy, A.A. 1998. Nutritional studies on halophytes and agricultural wastes as feed supplements for small ruminants in Sinai. Ph.D. Thesis, Cairo Univ., Egypt.

Fahmy, A.A. 2010. Productive performance of lambs fed *Kochia indica* shrubs under desert conditions of Sinai. Egypt. J. appl. Sci. 25: 17–28.

FAO. 1988. Salt-affected soils and their management. By Abrol IP, Yadav JSP, and Massoud FI. FAO Soils Bull. 39.

Farid, M.F.A. 1989. Water and minerals problems of the dromedary camel (an overview). Options Méditerranéennes—Série Séminaires 2: 111–124.

Gabr, M.G. 2002. First experience of Matrouh Resource Management Project in salt bush utilization for animal feeding. International Symposium on Optimum Utilization in Salt Affected Ecosystems in Arid and Semi Arid Regions, Cairo, Egypt, April 8–11, 2002, pp. 419–425.

Glenn, E.P., J.J. Brown and E. Blumwald. 1999. Salt tolerance and crop potential of halophytes. Crit. Rev. Plant Sci. 18: 227–255.

Gul, B., D.J. Weber and M.A. Khan. 2000. Effect of salinity and planting density on physiological responses of *Allenrolfea occidentalis*. West. North Am. Nat. 60: 186–197.

Gul, B., R. Ansari, I. Aziz and M.A. Khan. 2010. Salt tolerance of *Kochia scoparia*: a new fodder Crop for highly saline arid regions. Pak. J. Bot. 42(4): 2479–2487.

Haenlein, G.F.W. and R.G. Ramirez. 2007. Potential mineral deficiencies on arid rangelands for small ruminants with special reference to Mexico. Small Ruminant Research 68: 35–41.

Harrison, D.J., D.E. Beever, D.J. Thomson and D.F. Osbourn. 1975. Manipulation of rumen fermentation in sheep by increasing the rate of flow of water from the rumen. Journal of Agricultural Science, Cambridge 85: 93–101.

Hutchings, S.S. 1965. Grazing management of salt-desert shrub ranges in the Western United States. Proc. 9th Int. Grass Cong. 2: 1619–1625.

Jordan, F.L., M.M. Robin-Abbott, M. Raina and P. Glenn. 2002. A comparison of chelator-facilitated metal uptake by a halophyte and a glycophyte. Environ. Toxicol. Chem. 21: 2698–2704.

Kewan, K.Z. 2003. Studies on Camel nutrition Ph.D. Thesis, Fac. Agric., Alexandria Univ., Egypt.

Khan, M.A., R. Ansari, B. Gul and M. Qadir. 2006. Crop diversification through halophyte production on salt-prone land resources. CAB Reviews: Perspectives in Agriculture, Veterinary Science, Nutrition and Natural Resources 1–9, No. 048.

Khan, M.A., R. Ansari, H. Ali, B. Gul and B.L. Nielson. 2009. *Panicum turgidum*, a potentially sustainable cattle feed alternative to maize for saline areas. Agriculture, Ecosystems and Environment 129: 542–548.

Khan, M.A., I.A. Ungar and A.M. Showalter. 1999. Effects of salinity on growth, ion content, and osmotic relations in *Halopyrum mucronatum* (L.) Stapf. Journal of Plant Nutrition 22: 191–204.

Khattab, I.M.A. 2007. Studies on halophytic forages as sheep fodder under arid and semi arid conditions in Egypt. Ph. D. Thesis. Alexandria University.

Kraidees, M.S., M.A. Abouheif, M.Y. Al-Saiady, A. Tag-Eldin and H. Metwally. 1998. The effect of dietary inclusion of halophyte *Salicornia bigelovii* Torr. on growth performance and carcass characteristics of lambs. Anim. Feed Sci. Technol. 76: 149–159.

Lopez, S., B. Hovell and N.A. Macleod. 1994. Osmotic pressure, water kinetics and volatile fatty acid absorption in the rumen of sheep sustained by intragastric infusion, British Journal of Nutrition 71: 153–168.

MacFarlane, W.V., R.H.J. Morris and B. Howard. 1963. Turnover and distribution of water in desert camels, sheep, cattle and kangaroo. Nature, London 197: 270–271.

Malcolm, C.V. and T.C. Swaan. 1989. Screening shrubs for establishment and survival on salt-affected soils in southwestern Australia. Department of Agriculture Western Australia Technical Bulletin 81, 35p.

Marai, I.F.M., A.A. Habeeb and T.H. Kamal. 1995. Response of livestock to excess sodium intake. pp. 173–180. *In*: C.J.C. Phillips and P.C. Chiy (eds.). Sodium in Agriculture. Chalcombe Publications, Canterbury.

Masters, D.G., S.E. Benes and H.C. Norman. 2007. Biosaline agriculture for forage and livestock production. Agriculture, Ecosystems and Environment 119(2007): 234–248.

Moinuddin, M., S. Gulzar, I. Aziz, A.A. Alatar, A.K. Hegazy and M. Ajmal Khan. 2012. Evaluation of forage quality among coastal and inland grasses from Karachi. Pak. J. Bot. 44(2): 573–577.

Naidoo, G. and J. Kift. 2006. Responses of the saltmarsh rush *Juncus kraussii* to salinity and waterlogging. Aquat. Bot. 84: 217–225.

Nasim, M., M. Saqib, T. Aziz, S. Nawaz, J. Akhtar, M. Anwar-ul-Haq and S.T. Sah. 2008. Changes in growth and ionic composition of *Eucalyptus camaldulensis* under salinity and waterlogging stress; a lysimeter study. Soil & Environ. 27(1): 92–97.

NRC. 1981. Nutrient Requirements of Goats. Washington, DC: National Research Council, National Academic Press.

NRC. 1985. Nutrient Requirements of Sheep. Washington, DC: National Research Council, National Academic Press.

Reboreda, R. and I. Cacador. 2007. Halophyte vegetation influences in salt marsh retention capacity for heavy metals. Environ. Pollut. 146: 147–154.

Shawket, S.M., I.M. Khatab, B.E. Borhami and K.A. El-Shazly. 2001. Performance of growing goats fed halophytic pasture with different energy sources. Egyptian J. Nutr. Feeds 4: 215–264.

Squires, V.R. (in press). Water requirements of livestock fed on halophytes and salt tolerant forage and fodders. pp. 287–302, this volume.

Szabolcs, I. 1974. Salt Affected Soils in Europe. The Hague: Martinus Nijhoff.

Tahir, M.A., Rahmatullah, T. Aziz, M. Ashraf, S. Kanwal and M.A. Maqsood. 2006. Beneficial effects of silicon in wheat (*Triticum aestivum* L.) under salinity stress. Pakistan Journal of Botany 38(5): 1715–1722.

Tokalioglu, S., S. Kartal and A.A. Gunis. 2000. Determination of heavy metals in soil extracts and plant tissues at around of zinc smelter. Intl. J. Environ. Analytical Chem. 80: 210–217.

Underwood, E.J. and N.F. Suttle. 1999. The Mineral Nutrition of Livestock. New York: CABI Publishing.

Ungar, I.A. 1991. Ecophysiology of Vascular Halophytes. CRC Press, Boca Raton.

Vercoe, T.K. 1986. Fodder potential of selected Australian tree species. pp. 95–100. *In*: J.W. Turnbull (ed.). Australian *Acacias* in Developing Countries, Proc. of an international workshop held at the Forestry Training Center Gympie, Qld, Australia, 4–7 August 1986. ACIAR Proceedings no. 16, Australian Center for International Research, Canberra.

Verde, T., R.J. Shephard, P. Corey and R. Moore. 1982. Sweat composition in exercise and in heat. J. Appl. Physiol. 53: 1540–1545.

Wainstein, P., S. Gonzalez and E. Rey. 1979. Valor nutritivo de plantas forrajeras de la provincea de mendoza. III. Cuaderno Teccnico (IADIZA) 1: 97–108.

Walker, B.H. 1980. A review of browse and role in livestock production in Southern Africa. *In*: H.N. Le Houerou (ed.). Browse in Africa, the Current State of Knowledge. ILCA Addis Ababa, Ethiopia.

Warner, A.C.I. and B.D. Stacy. 1965. Solutes in the rumen of the sheep. Quart. J. Exp. Physiol. 50: 169–184.

Warren, B.E., C.I. Bunny and E.R. Bryant. 1990. A preliminary examination of the nutritive value of four saltbush (*Atriplex*) species. Proc. Austr. Soc. Anim. Prod. 18: 424–427.

Wilson, A.D. 1974. Water requirements and water turnover of sheep grazing semi-arid pasture communities in New South Wales. Aust. J. Agric. Res. 25: 339.

Yokoi, S., F.J. Quintero, B. Cubero, M.T. Ruiz, R.A. Bressan, P.M. Hasegawa and J.M. Pardo. 2002. Differential expression and function of *Arabidopsis thaliana* NHX Na+/H+ antiporters in salt stress response. Plant Journal 30: 529–539.

PART 3

Experience with Halophyte Feeding

Part 3 of 4 chapters reviews experiences on using halophytic and salt tolerant feedstuffs from around the world. Feeding trials have been conducted on a wide range of feedstuffs (trees, shrubs, grasses and forbs—even sea weed and date seeds) when they were fed to small ruminants (sheep and goats), cattle, camels and in various forms (fresh, dry or ensiled).

Impact of Halophytes and Salt Tolerant Plants on Livestock Products

S. Abou El Ezz,* Engy F. Zaki, Nagwa H.I. Abou-Soliman, W. Abd El Ghany and W. Ramadan

SYNOPSIS

This chapter focuses on the livestock products in terms of meat, milk and wool, as affected by feeding halophytes and salt-tolerant fodder crops. It covers the impact of feeding such feed materials on meat and its products, meat quality and composition, carcass traits, meat processing and nutritional quality. Also, it reviews the influence of feeding such fodders on productive performance of dairy animals, quality of milk and dairy products in terms of physicochemical and organoleptic properties. The impact on skin and coat traits, i.e., chemical and physical characteristics of fibers and skin characteristics are also discussed.

Keywords: meat, milk, dairy products, skin, coat traits, desert, salinity, handicrafts, wool.

1. Introduction

Food security is becoming an important issue throughout the world especially in the developing countries. Meat and milk are the main sources of animal protein being rich in essential amino acids and other nutrients required for normal growth and human health. Dairy and meat production is constrained by many factors including mainly

Desert Research Center, Cairo, Egypt.
*Corresponding author: samirabouelezz@hotmail.com

the genetic and environmental factors such as the scarcity and poor quality of feed resources and adverse climatic conditions which lead to frequent and extended drought periods. Halophytic plant and salt-tolerant fodder species can offer a range of nutrients for animals; provide a valuable reserve feed for grazing animals particularly under drought conditions or fill regular gaps in feed supply caused by seasonal conditions. In arid and semi-arid countries, sheep, goats and camels are generally accepted as valuable meat and milk-producing animals in the marginal areas. Their wool is used for making tents, rugs and other handmade products in most local communities, so producing good quality wool can improve the economic value of this raw material. The improvement in textiles and other handmade products leads in turn to higher income and better livelihoods.

Livestock production occupies approximately 30% of global arable land (FAO 2008a). With fast depletion of natural resources, ever-increasing population pressure and rising living standards, it has become extremely important to diversify the present-day animal agriculture to meet increasing demand for animal products. Food security is becoming a pressing issue throughout the world, particularly in the developing countries. Most of these countries have started to control their natural resources to attain this goal. Meat and milk are the main sources of animal protein being rich in essential amino acids and other nutrients required for normal growth and human health. Meat price is dramatically increasing in most developing countries mainly due to the devastating increase in human population and the relative decrease in animal population as well as low productive efficiency. Milk production in Africa and in other developing countries elsewhere is growing slowly; this growth is mostly due to the increase in numbers of producing animals rather than a rise in productivity per head (FAO 2012).

Wool is used for making tents, rugs and other handmade products in most local communities, so producing good quality wool could improve the economic value of this raw material used in handmade textile products which leads to increased income and better livelihoods. Generally, animal production is constrained by many factors such as scarcity, higher prices and poor-quality of feed resources, and adverse climatic conditions which lead to frequent and extended drought periods (FAO 2012). Frequent droughts exacerbate the shortage of feed of sufficient nutritional quality which is the reason for actively searching for alternative feedstuffs. Halophytic and salt-tolerant fodder species can offer a range of nutrients for animals; provide a valuable reserve feed for grazing animals particularly under drought conditions or fill feed gaps caused by seasonal conditions. In arid and semi-arid countries, sheep, goats and camels are generally accepted as valuable meat and milk-producing animals, especially in the marginal areas. These animal species can survive the prevailing harsh conditions in the desert regions and are capable of converting the poor and irregular growth of vegetation into meat and other products.

2. Impact of halophytes and salt tolerant fodders on meat quality

2.1 Effect on carcass traits

A number of studies have identified that ingesting or grazing saltbushes or a diet high in salt may result in carcasses with lower proportions of fat and higher proportions of lean compared to sheep grazing either a grain-hay based diet or a pasture-stubble paddock (Pearce et al. 2008a,b).

The effect of dietary inclusion of halophyte *D. spicata* grass hay on growth performance and body composition of Emirati goats was studied by Al-Shorepy and Alhadrami (2009). However, hot carcass weights and dressing percentage for the animals fed diets containing 100% *Distichlis* grass hay were significantly ($P < 0.05$) heavier than for the control animals.

Several studies were conducted to evaluate the impact of feeding saltbush with or without supplementation on carcass traits of animals. For examples, the effect on carcass traits of grazing lambs on either predominantly saltbush (*Atriplex nummularia*), supplemented with pasture or oat grain for 68 days, was compared to lambs grazed predominantly on lucerne (*Medicago sativa*) as reported in Hopkins and Nicholson (1999). They found that there was no significant ($P > 0.05$) difference between the treatments for live weight. On the other hand, there was a significant ($P < 0.05$) treatment effect on hot carcass weight for animals fed saltbush. That group of lambs was lighter than those from the lucerne group due to the high nutritional quality compared to that of saltbush. Another study was conducted by Shehata and Mokhtar (2005) on feeding fresh *Acacia saligna* or *Atriplex halimus* plus barley grains to sheep and goats (over a three month fattening period) to study their growth performance and carcass characteristics. Results showed that the halophytic forages supplemented with barley grains can be used successfully in fattening the growing lambs and kids with no adverse effects on their growth performance and meat characteristics.

The long term consumption of saltbush and barley prior to slaughter did not result in decreases in carcass weight and improvement in dressing percentages as demonstrated by Pearce et al. (2008a); it might be due to an improved hydration status at slaughter. The effect of feeding goats on air-dried halophytic plants (D2) compared to those fed berseem hay (as control group, D1) on carcass trait was studied by Youssef et al. (2002). The dressing percentage on basis of fattening weight of D1 and D2 were 46.9 and 44.5 (without offal) and 49.31 and 47.3 (with offal), respectively. Dressing percentage on an empty weight basis was not significantly different.

Working with camels, Shehata et al. (2004) evaluated the utilization of some non-conventional feedstuffs (fresh *Atriplex nummularia*, and/or *Acacia saligna*, and ensiled rice straw, as roughages), in a fattening trial using young male camels, and the impact on growth performance and carcass yield. The authors indicated that such halophytic plants supplemented with nontraditional concentrates (e.g., ground date seeds) can be successfully used in feeding growing camels for a period of eight months with no adverse effects on their wholesale cuts, physical components (lean meat, fat, bone and boneless meat) and fat deposits.

2.2 Effect on eating quality

Many consumers define meat eating quality in terms of tenderness and juiciness, with more tender and juicy meat having higher eating quality. However, in most instances, flavor of meat is also used to help determine acceptance. A trained taste panel could significantly differentiate meat from lambs grazing annual legumes, lucerne (*Medicago sativa*) and/or *Atriplex* sp. from the meat of control lambs fed grass, even when fed for short periods of time (Park et al. 1972). The intensity of the characteristic lucerne taint described as a sharp or pungent odor and a dirty or sticky flavor was found to increase with increasing length of grazing lucerne. In addition, Hopkins and Nicholson (1999) studied the meat quality of lambs grazed on either saltbush (*Atriplex nummularia*) plus supplements or lucerne (*Medicago sativa*) and found treatment had a significant effect on aroma strength ($P < 0.05$); samples from lambs in the saltbush/hay group (SH) and those in saltbush/grain (SG) group having a stronger aroma than those from lambs grazed on lucerne (L) alone. No treatment effect for liking of aroma was found. Flavor strength was not significantly ($P > 0.05$) stronger for samples from groups SH and SG than for samples from group L. There was no effect of treatment on tenderness or juiciness and overall panelists ranked the samples similarly for acceptability. Similarly, Pearce et al. (2008a,b) studied the effect of grazing a saltbush and barley ration on eating quality of sheep meat; no significant improvements or detriments for eating quality in terms of liking of flavor and aroma, tenderness, juiciness and overall acceptance were detected. On fattening goats, Youssef et al. (2002) studied the effect of air dried halophytic shrubs (D2) as goat fattening fodder compared to those fed berseem (*Trifolium alexandrinum*) hay as good quality roughage (D1) on physical properties of goat meat. They found that water holding capacity value of D2 was higher (7.1 cm^2) than that of D1 (6.8 cm^2) and differences were not significant. Mean value of tenderness indicated that D1 had higher value (3.7 cm^2) than D2 (3.4 cm^2) of goat meat. Differences between two groups were not significant, so it means that such halophytic fodders could be used safely for fattening goats.

On the other hand, Shehata (2005) recommended that the use of unconventional concentrate mixture (UCM) plus a mixture of halophytic shrubs in feeding growing camels for a period of eight months would improve economical efficiency of meat production under arid and semi-arid conditions with no adverse effects on the sensory and physical characteristics of camel meat. Similarly on camels, the effect of feeding a range of salty plants (*Atriplex* and *Acacia*) on the sensory characteristics of camel meat were evaluated (Zaki 2007). It was indicated that the means for any attribute of sensory test of all camel meat under study was not significantly different ($P > 0.05$) due to the type of ration for camels on different types of roughages. They also found that using the desert plants especially *Atriplex* and *Acacia* in camel feed can improve the physical properties of meat. Cooking loss of camel meat of *Atriplex* group was significant lower ($P > 0.05$) when compared with other groups. *Atriplex* group was better in water holding capacity and plasticity than the other feeding groups.

2.3 Effect on meat nutritional quality

The high vitamin E content of meat from sheep fed saltbush is an important selling point for saltbush meat because at 6.3 mg/kg of vitamin E, the level is equivalent to some vegetables, oils, nuts, green leafy vegetables, and fortified cereals which are common food sources of vitamin E (Traber 1999). Vitamin E (α-tocopherol) has been shown to be an effective antioxidant, especially when included in animal diets and is accepted by consumers on the basis that it is a naturally occurring substance. Vitamin E is a major, lipid-soluble antioxidant, and one of its primary functions is to maintain and protect biological membranes against lipid peroxidation (Jose et al. 2008). Vitamin E from *Atriplex* spp. slows the oxidation of lipids in meat and delays the oxidative change of oxymyoglobin to brown metmyoglobin, thus improving both the flavor and increasing the shelf-life of meat (Pearce et al. 2005 and 2010). Wulf et al. (1995) fed 500 mg/head/day of α-tocopherol acetate and observed α-tocopherol levels in the muscle of 5.9 mg/kg fresh muscle weight. Comparable figures were obtained by Jose et al. (2008) when fed 400 mg/head/day of α-tocopherol acetate and observed levels of 5.1 mg/kg fresh muscle weight. *Atriplex* species have been reported to contain α-tocopherol at concentrations of 116–139 mg/kg DM (Pearce et al. 2005). In addition, it appears that saltbush is a superior natural source of vitamin E compared with pellet–grain–hay-based diets.

Fatty acids are the most important lipid fraction of meat. Their role is very particular because they have a part in the immune function, prevention of inflammation and as energy source (Webb et al. 1994). In addition, fatty acids composition plays an important role in the definition of meat quality, as it is related to differences in organoleptic attributes, especially flavor and in nutritional value of fat for human consumption (Wood and Enser 1997). There is an increasing demand for 'healthier' meat with higher levels of unsaturated fats. If saltbush meat does contain lower levels of unsaturated fat in the depot fat, it may be branded as 'healthy meat' which will only add to its commercial acceptability. The effects of grazing on saltbush on the fatty acid composition of sheep meat have not been previously investigated too much; limited data are available. However, Pearce et al. (2008b) studied the effect of grazing a saltbush and barley ration on the fatty acid composition. There was a significant difference in the sum of unsaturated fatty acids between the treatments in both fat depots but no differences in the saturated fat content were observed between treatments. Moreover, Russo et al. (1999) investigated the fatty acid profile of tissues in lambs fed three different diets: (1) a concentrate with barley flakes (BC) and lucerne hay, (2) a concentrate with maize oil (MC) and lucerne hay and (3) only concentrate with maize oil and slaughtered at 105 days. The fatty acid profile was affected by the different diets, the intramuscular fat from lambs on diet 3 was more saturated (43.88%) and with less essential fatty acids (w-3: 0.9% and w-6: 10.15%) than the fat of the other two diets (diet 1: respectively, 42.9, 1.3 and 10.5%; diet 2: respectively, 43.2, 1.2 and 10.7%).

2.4 Effect on meat processing

Tenderness, taste and palatability of camel meat products could be increased by the processing of the meat produced from camels fed saltbushes (Ahmed 2001). It was recommended that camel meat could be processed in the way as beef meat with similar consumer products leading to increased cash flow for camel meat, which is cheaper than beef. On these trends, Zaki (2007) studied the effects of feeding a range of salty plants (*Atriplex* and *Acacia*) on the chemical, physical, microbial and sensory characteristics of camel meat and the quality of the basturma[1] prepared from

---

[1] A highly seasoned, air-dried cured beef of Anatolian origin, which is now part of the cuisines of the former Ottoman Empire.

(*Biceps femoris*) muscles. It was found that basturma produced from camels in *Acacia* group recorded the highest protein content and the lowest fat content. However, no significant differences were found in Thiobarbaturic acid values of fresh basturma processed from camel meat fed on different types of roughages. Basturma processed from *Atriplex* group had the highest sensory score for any attributes followed by basturma from control and *Acacia* groups.

Generally, using the desert plants especially *Atriplex* and *Acacia* in feeding animals can produce a good source of animal protein, improve physical properties of meat products and reduce the high cost of meat production.

3. Impact of feeding halophytes and salt-tolerant plants on milk production and quality characteristics of milk and dairy products

3.1 Impact on milk production

The influence of feeding halophytes and salt-tolerant plants on milk production of ruminants was investigated in many studies, which showed varied results. Some of these studies mentioned that feeding halophytes to ruminants has adverse effects on milk production. For instance, Shetaewi et al. (2001) found that feeding concentrate diet plus green *Acacia saligna* to Damascus goat over 10 weeks led to a decline in milk production of about 20% compared with those fed *Trifolium alexandrinum* hay.

On the other hand, increased milk yield in response to supplementation with halophytes was also confirmed. Shetaewi et al. (2001) stated that feeding concentrate diet plus green *Acacia saligna* or *Oryza sativa* straw to Damascus goat resulted in an increase of about 65% in milk production of group fed *Acacia* compared with that fed *Oryza* straw. A research study by Mekoya et al. (2009) showed that long-term supplementation of *Sesbania sesban* at 30% of the ration, resulted in 13% increase ($P < 0.05$) in milk production of ewes over those supplemented with concentrates (0% *S. sesban*). In addition, the peak lactation for *S. sesban*-supplemented ewes was higher and persisted longer than for concentrate-supplemented ewes. Similarly, studies on camels showed that long-term feeding fresh *Atriplex halimus* supplemented with yellow corn plus barley grains instead of berseem hay in the diet resulted in an increase ($P < 0.05$) in milk yield and camels fed *Atriplex* diet had a two-peak lactation curve at the 5th and 7th months after parturition, whereas a lactation curve of camels fed berseem hay diet had one peak at the 4th month after parturition (Shawket and Ibrahem 2012). Similar results were obtained by Ahmed et al. (2013) on she-camels fed either *Atriplex halimus* or camel thorn (*Acacia erioloba*) supplemented with concentrate feed mixture, but this increment was not significant ($P < 0.01$).

However, other studies declared that consumption of halophytes and salt-tolerant plants by lactating animals has no effect on milk yield. For instance, Meneses and Flores (1999) mentioned the use of *Acacia saligna* as a supplemental forage for grazing goats did not show positive effects on milk production. No significant differences ($P < 0.05$) in average milk yield among lactating goats fed rations consisting of 60% dried hays of clover, *Pennisetum typhoides* (pearl millet) or *Kochia indica* plus 40%

concentrate feed mixture (El-Shereef 2007). Recently, Abbeddou et al. (2011a,b) concluded that *Atriplex* leaf diet offered to Awassi ewes did not significantly affect ($P = 0.005$) milk yield.

There are several factors responsible for these varying results, from the chemical composition, the nutrients content of halophytes and salt tolerant fodders (El Shaer 2010). It is well known that good quality feed is essential for higher milk yield; if the feed is high in fiber and low in energy, milk production will diminish dramatically (Campbell and Lasley 1985). Halophytes are usually fairly poor in energy (El Shaer 2010); additional energy source supplements were needed to improve ($P < 0.05$) milk yield of dams grazing halophytic pastures during lactation stages (El Shaer 1981). The anti-nutritional substances found in most halophytes, could be another reason to be deleterious or beneficial for animal production depending mainly on their nature (Makkar 2003; Attia-Ismail, this volume), animal species and physiological stage (Ben Salem 2010) and concentration in the feed materials (Makkar 2003; Ben Salem 2010; El Shaer 2010). For example, low content of tannins in the forages increases milk yield and has positive effects on milk protein concentration and yield because of the protection of proteins from microbial degradation in the rumen and the higher availability of essential amino acids as reported by Vasta et al. (2008). However, Maamouri et al. (2011) mentioned the natural protection of protein from microbial degradation in the rumen by tannins of the *Acacia cyanophylla* did not improve ($P > 0.05$) sheep-milk production. On the other hand, high concentration of tannins in the forages reduces rumen microbial activity and feed amino acid digestion, in the intestine (Vasta et al. 2008). A study was performed by Shetaewi et al. (2001) showed that fresh acacia should not be fed to small ruminants as a sole diet for extended periods due to its tannin or other anti-nutritional substances and supplemental feeding is required to enhance the milk production of goats. Abu-Zanat and Tabbaa (2006) concluded that the inclusion of *Atriplex halimus* and *A. nummularia* foliage, compared to shredded barley straw, in the diets of lactating Awassi ewes did not result in significant ($P < 0.05$) differences in milk production of ewes. They attributed that to the negative effects of oxalates and tannins present in the browse of *Atriplex*. Many studies confirmed the addition of polyethylene glycol in the diet is efficient in deactivating tannins (Vasta et al. 2008; Ben Salem 2010; Meneses et al. 2012) and hence improving the efficiency of utilization of several shrubs containing tannins by animals (Makkar 2003); and consequently increases milk yield and maintains milk fat and protein concentration, counterbalancing the dilution effect caused by the higher milk yield (Vasta et al. 2008).

For the above mentioned reasons, these plants should not be used as sole diets (Shetaewi et al. 2001; El Shaer 2010) and appropriate mixing of these species, based on their complementary roles, could bring balance to the diet for nutrients and to minimize the adverse effects of secondary compounds and the excess of minerals (Ben Salem 2010; El Shaer 2010; Masters, this volume). There is a maximum level of mixing or inclusion of halophytes and salt-tolerant plants in the diet. If this level is exceeded, it impairs animal productive performance, including milk production. For example, Abu-Zanat and Tabbaa (2006) mentioned that substituting barley straw by *Atriplex* browse up to 50% significantly reduced the cost of roughage component of sheep diets but did not adversely affect milk yield. El-Shereef (2007) concluded that *Pennisetum typhoides* hay was good roughage that could be used instead of clover up to 60% in the ration without any adverse effect on performance of lactating goats. Meneses et al. (2012) recommended goat diets should not include more than 25% *Acacia* forage during pregnancy and lactation to avoid affecting milk production. Additionally, Raju et al. (2012) stated that *Acacia nilotica* pods may be incorporated up to 33% in concentrate mixture of lactating goats without affecting milk production and its composition and it may be increased up to 50% by treating with $Ca(OH)_2$ before use.

Ensiling and haylage treatments could reduce the concentrations of secondary metabolites, ash content and/or affect the lingo-cellulose bonds (El Shaer et al. 2005). In this respect, the effect of feeding ensiled halophytes on milk yield was investigated. For example, El-Shereef (2012) showed that utilization of *Kochia indica* silage with concentrate feed mixture could be used in feeding lactating ewes without any adverse effects on their total milk yield.

3.2 Impact on physicochemical properties of milk

Milk composition is affected by both genetic and environmental factors (Butler 2013). Diets can affect fat content, fatty acid profile and protein content of milk but changes in milk protein are not as dramatic as those observed for milk fat (Santos 2002; Butler 2013).

Based on ICARDA (1996), the milk of ewes grazing saltbush contained less fat than milk from ewes grazing native pasture. Complete substitution of barley straw, as control forage, by *Atriplex halimus* leaves in a diet composed of forage and concentrate in a ratio of 0.3:0.7 resulted in significant increase ($P < 0.01$) in milk solids but not fat percentage and casein number ($P < 0.05$) of ewes fed *Atriplex* diet (Abbeddou et al. 2011a,b). They returned the elevation of casein number to the highly digestible *Atriplex* leaves, which were helpful to promote casein formation. The substitution did not influence milk total solids and fat content (Abbeddou et al. 2011a,b). With regard to fatty acids, diet containing *Atriplex* leaves led to lower proportion of 18:0 in milk fat ($P = 0.018$) and promoted a higher proportion of the n-3 fatty acids. The proportion of 18:2 *c*9, *c*12 did not significantly differ and the proportion of medium-chain fatty acids was similar (Abbeddou et al. 2011a). In response to their particular composition, *Atriplex* leaf diet influenced milk fat properties in a favorable way. Abbeddou et al. (2011a) stated that antiradical activity of the milk resulting from the *Atriplex* leaf diet was 1.2 fold higher ($P < 0.001$) than that of milk produced from barley-straw diet. It could be attributed to the active secondary metabolites that exist in *Atriplex* leaves like phenols, which probably transferred from feed to milk and consequently improved antioxidative stability of the milk. The impact of feeding *Atriplex* on camel milk properties was also examined. Substitution of *Atriplex* for berseem hay in the diet caused marked increase ($P < 0.05$) in protein content of camel milk (Skawket and Ibrahem 2012). They ascribed this elevation to the higher content of crude protein in *Atriplex* than berseem hay. In a similar study, incorporation of fresh *Atriplex* or camel thorn in the diet of she-camels led to significant increase ($P < 0.01$) in cholesterol and insulin when compared to berseem hay-fed group. In addition, there were no significant differences among groups in other milk constituents (Ahmed et al. 2013).

Studies on the effect of *Acacia* on milk composition were undertaken. For example, milk total solids, fat and protein percentages decreased in does fed concentrate diet plus green *Acacia* in comparison to those fed concentrate diet plus berseem hay (Shetaewi et al. 2001). The differences in milk composition could be owing to the higher digestibility coefficients of the nutrients in berseem hay than those in *Acacia* leaves. On the other hands, Maamouri et al. (2011) indicated that milk fat and protein content did not differ ($P > 0.05$) among ewes grazing rye grass (*Lolium*) pasture and supplemented with concentrate or concentrates plus different proportions of *Acacia* (tannins). However, such findings are not compatible with those of Wang et al. (1996) for sheep and Woodward et al. (1999) for cows, who found that protein content of milk increased with tannin administration in the diet. Moreover, Maamouri et al. (2011) reported that *Acacia* supply decreased ($P < 0.05$) milk urea nitrogen. They concluded the lower values of milk urea nitrogen resulted maybe from better use of protein with tannin protection.

The effect of feeding halophytes on ash content and mineral composition of milk was investigated. Ewes grazing saltbush had a higher total mineral content in their milk than pasture-fed ewes (ICARDA 1996; Chadwick et al. 2009), with higher concentrations of K, Mn, B, P and Zn and lower Al and Fe concentrations (Chadwick et al. 2009). Contrary to expectations, feeding ewes on *Atriplex* leaves (rich in electrolytes) did not increase either milk sodium or potassium contents and did not affect the electrical conductivity compared with barley straw diet (Abbeddou et al. 2011b). Some studies also showed that inclusion of *Kochia* hay for does (El-Shereef 2007) or *Kochia* silage for ewes (El-Shereef 2012) in the diet increased ($P < 0.05$) milk ash percentage. This percentage for goat milk increased, without significant differences, as the proportion of *Kochia* silage in the ration increased (Ahmed et al. 2001). Concerning camel milk, inclusion of fresh *Atriplex* or camel thorn in the diet of she-camel led to significant increase ($P < 0.01$) in ash content when compared to berseem hay-fed group (Ahmed et al. 2013). Despite the fact that milk of camels fed *Atriplex* diet had higher ($P < 0.05$) concentration of Na, K and Ca than that of camels fed berseem hay diet, the electrical conductivity value of the milk from camels fed *Atriplex* diet was lower ($P < 0.05$) than those fed berseem hay diet (Shawket and Ibrahem 2012). They attributed this finding to the ability of oxalates and tannins, which exist in *Atriplex*, to bind minerals and form insoluble salts and thus lowering the electrical conductivity of milk.

3.3 Impact on gross composition and organoleptic properties of milk products

Information on the effect of feeding halophytes on quality of dairy products is scarce indeed. According to the annual report of ICARDA for 1996, the quality of the cheese and yoghurt made from milk of ewes grazing saltbush may be somewhat lower than the same products made from milk of ewes grazing native pastures. Yields of cheese solids and of total solids in yoghurt were somewhat lower when ewes grazed saltbush. Abbeddou et al. (2011b) reported that the yield of cheeses produced with milk of ewes fed *Atriplex* leaf diet was similar to those produced with milk of ewes fed barley straw ones. Also, gross composition of yoghurt and cheese was not significantly ($P < 0.05$) affected by *Atriplex* supplemented diet. A recent research study demonstrated that yoghurt made from fresh *Atriplex* diet milk (yoghurt manufactured from mixture of camel and buffalo milk at a ratio of 1:1) had significantly ($P < 0.01$) higher cholesterol and insulin, non-significantly higher pH, total solids, total protein and ash, lower acidity and fat than those of yoghurt made from berseem hay diet milk (Ahmed et al. 2013).

Regarding the sensory properties, it is known that type of fodder affects the taste and odor of milk. Grazing on *Atriplex halimus*, for example, gives a salty taste to the milk (Gast et al. 1969). The taste and odor of Jibneh Khadra, non-aged soft cheese in the Middle East, became better ($P < 0.05$) by feeding *Atriplex* leaf diet. In addition, this cheese was clearly firmer ($P < 0.05$) than cheeses produced from animals fed straw barley (Abbeddou et al. 2011b). They ascribed that to the high value of casein number in milk from which the cheese was made. With regard to yoghurt, the reports of ICARDA (1996) indicated that yoghurt made from milk of ewes grazing saltbush had lower scores for the taste and appearance than that produced from ewes grazing native pastures. In another study (Abbeddou et al. 2011b), *Atriplex*-leaf based yoghurt was softer ($P < 0.05$) than yoghurt produced with barley straw treatment (control); odor was higher with control, whereas taste scores did not differ between two diets and no off-flavors were observed as a result of changing diet. Ahmed et al. (2013) found that yoghurt made from milk of camels fed berseem hay (yoghurt manufactured from mixture of camel and buffalo milk at a ratio of 1:1) had the highest score of flavors, body and texture, appearance, acidity and total panel scores than those of yoghurt made from fresh *Atriplex* or fresh camel thorn diet milk.

In conclusion, it is important to identify and get enough information on the chemical composition and nutritive value of halophytes and salt tolerant fodders included in dairy animal diets and the responses of these animals to nutrients and anti-nutritional components. Such information will ensure the optimal use of these plants and consequently improve milk production and quality for high quality dairy products which could satisfy consumers' tastes. In addition, more studies are required on the impact of these plants on the technological characteristics of milk.

4. Impact of feeding halophytes and salt tolerant plants on skin and coat traits

4.1 Chemical and physical characteristics of fibers

There was improvement in wool growth efficiency and fiber diameter in sheep fed saltbush as reported by Hemsley et al. (1975). Feeding sheep on high salt levels of diet made up of 89% protein concentrate caused a high response of wool growth to protein available for absorption in the small intestine. This result may be a consequence of the higher water intake leading to an increased rate of passage of nutrients (Squires, this volume). Increased rate of passage will mean lower degradation of protein by microorganisms in the rumen and therefore more undegraded protein available for absorption in the small intestine (Hemsley et al. 1975). The amount of wool grown per Kg of organic matter intake increased by up to 50% when sodium and potassium chloride were included as a part of normal roughage based diet at 25% of the dry matter as demonstrated by Masters et al. (2005). From the point of view of a specialist wool growing enterprise, this result is highly significant. It indicates that total wool production per Kg of available organic matter could be increased by including plant material with a high salt concentration.

Performance of sheep fed silage mixture of *Tamarix mannifera* (50%), *Atriplex nummularia* (25%) and *Acacia saligna* (25%) supplemented with nitrogen bacteria and cellulytic bacteria and mixture of both supplements showed an increase in fiber diameter, staple length wool weight and yield compared with animals fed fresh halophytes as reported by Helal and Fayed (2013). The increase in wool production is a result of increasing both fiber diameter and staple length. Feeding sheep on fresh *Acacia* caused reduction in nitrogen digestibility and decreased wool growth and yield, as well as sulfur absorption as stated by Pritchard et al. (1988). A reduction in wool growth (8.9%), when *Acacia saligna* leaves were offered *ad libitum* to sheep fed on balanced diet of lupins (*Lupinis* sp.) and straw, was detected by Krebs et al. (2003). Moreover, Hunter et al. (1990) reported a reduction in fiber diameter of up to 40% in nutritional saline stressed animals.

Sheep fed on saltbush pastures had low wool strength and this might be due to the restriction effect of tannins on amino acids and sulfur availability for metabolisable protein synthesis (Gartner and Hurwood 1967; Morcombe et al. 1994). Similarly, Barry (1985) found a negative effect of tannins (especially condensed tannins) on wool growth. Treated halophytes fodder as a silage supplemented with nitrogen bacteria lead to improvement in the quality of wool represented in increasing fiber diameter and staple strength. When fiber diameter increases, staple strength tended to increase (Gartner and Hurwood 1967); this is accompanied with availability of sulfur amino acids to the follicles. It is worthy to note that processing of fresh halophytes as silage or haylage tended to decrease the concentration of anti-nutritional factors, e.g., tannins, oxalates, flavonoids and alkaloids that could affect the chemical and physical characteristics of fibers as reported by many workers (Abd El-Halim 2003; Khattab 2007; El Shaer 2010). For instance, wool production significantly ($P < 0.05$) declined in animals fed fresh halophytes compared with those fed silage. This decrement might be due to the decrease in both fiber diameter and staple length. Staple length decreased about 7% and 21% for haylage and hay, respectively compared with berseem hay as reported by Taha et al. (2009). Fiber diameter decreased by 2.34 μm (7%) and 3.47 μm (10%) in both haylage and hay salt tolerant plants, respectively compared with the control group. Both kemp and non-medullated fiber increased with salt tolerant plant groups compared with berseem hay group. Medullated fibers increased in the group fed on halophytes. Fine fibers take the opposite trend of medullated fibers with no significant differences between both treatments of salt tolerant plants. Haylage process could improve the quality of salt tolerant plants. That could explain the decrease in the production of wool, staple length, fiber diameter and staple strength in hay from salt tolerant plants group compared with haylage one. Most important industrial wool characteristics like staple strength, staple length, yield and wool production were higher in animals fed haylage compared to those fed hay (control group) as reported by Younis et al. (2012). Moreover, concentration of anti-nutritional substances such as saponins, tannins and alkaloids were higher in *Kochia* hay more than *Kochia* haylage; so, as mentioned earlier haylage processing could improve the nutritional value of salt tolerant plants and halophytes compared to hay or fresh state materials (El-Essawy et al. 2009). The decrease in both fiber diameter and staple length would lead to a decrease in the wool production, which reached 19 and 31% in salt tolerant plants groups compared with the berseem hay group (Younis et al. 2012). Yield also significantly ($P < 0.05$) decreased in both treatments containing salt tolerant plants, Staple strength was higher insignificantly and no significant differences in staple elongation were found between berseem hay group and salt tolerant plant group.

The salt concentration in saltbush forage may have beneficial effects on wool growth. Comparisons were made between wool growth in sheep fed a 60:40 "old man saltbush[2]/barley diet" and a "control diet" (composed of lupins, barley and oaten hay) of similar value in metabolisable energy and crude protein (Barrett-Lennard et al. 2003). Sheep on the old man saltbush/barley diet recorded 20% higher rates of wool growth than the control group. Sheep fed old man saltbush leaf, cocky chaff,[3] or a 3:2 mixture of cocky chaff/saltbush leaf; sheep fed the mixture or the saltbush diet had 25% higher wool growth rates than sheep fed cocky chaff alone.

Sulfur is essential for synthesis of structural proteins and is a component of three amino acids (Cystine, Cysteine and Methionine), several vitamins, Insulin and coenzyme (A) (McDonald et al. 2002). Sulfur deficiency limits live weight gain, and as wool contains about 4% sulfur, deficiency is especially problematic in wool production systems (SCA 2007). Grasses such as *Thinopyrum ponticum* and *Puccinellia ciliata* (0.13–0.16% DM) would be deficient in sulfur for sheep. The concentration of wool sulfur was higher significantly in the ensilage halophytes group than that of the berseem hay group and the fresh halophytes mixture groups (Taha et al. 2009).

Slight increase was found in all fiber amino acids contents for salt tolerant plants hay and haylage as compared with berseem hay group as reported by Abd El-Ghany et al. (2012). Results recorded marked differences in Cysteine, Alanine, Aspartic acid, Argnine, Thrionine and Methionine between all groups. Amino acids; Cysteine, Alanine, Glutamic acid, Proline and Methionine showed a significant increase in haylage group whereas only Methionine achieved a marked increase in the two treatment groups compared to berseem hay group. Methionine significantly increased in coarse and fine fibers of both salt tolerant groups, while Alanine tended to be higher in only fine fibers of haylage group as compared with berseem hay group. The authors reported that no significant effect of treatment was detected on wool characteristics, except cotting score[4] which was significantly increased in hay group and haylage group. Fiber cross sectional area decreased in both coarse and fine fibers of hay group and haylage group than berseem hay group; this decline was higher in haylage group than hay group. Feeding on haylage might cause an increase in fiber length and fiber fineness through decreasing fiber cross sectional area, and an increase of crimp frequency in fine fibers of haylage group than hay group and a partial decline in cotting score. On

---

[2] *Atriplex nummularia.*
[3] Cocky chaff The husks of grain separated by threshing or winnowing.
[4] Cotted Fleeces—Fleeces in which fibers are matted and felted. Most common in coarse wooled sheep and may be due to ill health or lack of sufficient yolk. Increase in oilage or waste as fibers are broken as fleece is torn apart.

the other hand, feeding salt tolerant plants mixed with molasses may cause an increase in fiber amino acids contents specially Cysteine, Alanine, Glutamic acid, Proline and Methionine which in turn make changes in physical characteristics of wool fibers viz.; an increase in fiber length and cotting score, and a decrease in fiber cross sectional area and crimp frequency (Abd El-Ghany et al. 2012).

4.2 Fiber follicles and skin characteristics

Performance of sheep fed silage mixture of *Tamarix mannifera* (50%), *Atriplex nummularia* (25%) and *Acacia saligna* (25%) supplemented with nitrogen bacteria and cellulytic bacteria and mixture of both showed that silage with both cellulytic and nitrogen bacteria group and nitrogen bacteria group had significantly higher values of external diameter, internal diameter, fiber diameter, wall thickness, medulla thickness, concentration of general proteins in primary and secondary follicle structures (outer root sheath, inner root sheath and fiber) in both primary and secondary follicles compared with feeding fresh halophytes (Helal and Fayed 2013). Dimensions of the primary and secondary wool follicles were lower significantly in animals fed on fresh halophytes mixture compared with those of the berseem hay group or ensiled halophytes mixture as indicated by Taha et al. (2009). The concentrations of general proteins in primary and secondary follicles structures showed a similar trend. Meanwhile, concentrations of general carbohydrates in primary and secondary follicles structures showed no significant differences between all studied groups. The researchers revealed the importance of ensilage processing of the halophytic plants to decrease the negative effects of the fresh form of these plants on wool production and characteristics (Taha et al. 2009; Helal and Fayed 2013). The undesirable effect of feeding halophytic forages on wool growth could be attributed to the lower availability of amino acids especially the sulfur amino acids to skin follicles due to the formation of tannin-protein complex. Ensilage processing of halophytes decreases the negative effect of the feeding halophytic plants on wool growth and characteristics due to decrease in the levels of the anti-nutritional factors contained in ensiled forages (Taha et al. 2009).

It could be concluded that feeding sheep on treated salt tolerant plants or halophytes (adding bacteria or in ensiled form) showed an increase in fiber diameter, staple length, wool weight, staple strength and yield compared with animals fed fresh salt tolerant plants or halophytes. Feeding salt tolerant plants mixed with molasses may cause an increase in fiber amino acids contents specially Cysteine, Alanine, Glutamic acid, Proline and Methionine which in turn make changes in physical characteristics of wool fibers. Follicle diameter, wall thickness, and concentration of general proteins in structure of follicles increased by adding different types of bacteria with halophytes as compared with animals fed on fresh halophytes.

Finally, using some halophytes and salt-tolerant plants in feeding animals can improve physical properties of meat products and reduce the high cost of meat production. Feeding of such fodders has positive or negative impacts on milk production and composition. This depends on many factors, including type, composition and mixing ratio of these plants in the diet. Therefore, it is necessary to identify and get enough information on these plants and the responses of lactating animals to such fodders. This information will ensure the optimal use of these plants and thus improve milk yield and quality for high quality products which could satisfy consumer needs. On the other hand, feeding sheep on treated salt-tolerant plants or halophytes shows an improvement in chemical and physical characteristics of fibers and skin characteristics compared with animals fed fresh salt-tolerant plants or halophytes.

References and Further Readings

Abbeddou, S., B. Rischkowsky, Mel-D. Hilali, H.D. Hess and M. Kreuzer. 2011b. Influence of feeding Mediterranean food industry by-products and forages to Awassi sheep on physicochemical properties of milk, yoghurt and cheese. J. Dairy Res. 78: 426–435.

Abbeddou, S., B. Rischkowsky, E.K. Richter, H.D. Hess and M. Kreuzer. 2011a. Modification of milk fatty acid composition by feeding forages and agro-industrial byproducts from dry areas to Awassi sheep. J. Dairy Sci. 94: 4657–4668.

Abd El-Ghany, W.H., F.E. Younis, W.A. Ramadan, N.H.M. Ibrahim and H.M. El Shaer. 2012. Effect of feeding on salt tolerant plants on physical and chemical properties of coat fibers in sheep. J. of Anim. And Poultry Prod. 3(11): 485–496.

Abd El-Halim, A.M. 2003. Studies of some anti-nutritional factors affected forage utilization by ruminants. Ph.D. Thesis, Faculty of Agriculture, Ain Shams Univ. Egypt.

Abu-Zanat, M.M.W. and M.J. Tabbaa. 2006. Effect of feeding *Atriplex* browse to lactating ewes on milk yield and growth rate of their lambs. Small Rumin. Res. 64: 152–161.

Ahmed, M.A., H.A. El-Metoly, A.A. Khir and T.A. Deraz. 2013. Effect of feeding she-camel by desert plants on the milk yield, composition and the probiotic yoghurt properties made from the milk. Egypt. J. of Appl. Sci. 28: 352–371.

Ahmed, M.E., A.M. Abd El-hamid, F.F. AbouAmmou, E.S. Soliman, N.M. El-kholy and E.L. Shehata. 2001. Response of milk production of Zaraibi goat to feeding silage containing different levels of Teosinte and *Kochia*. Egypt. J. Nutrition and feeds 4: 141–153.

Ahmed, S.A. 2001. Camel meat and customer's psychology. Camel Newsletter 18: 33–37.

Al-Shorepy, S.A. and G.A. Alhadrami. 2009. The effect of dietary inclusion of halophyte *Distichlis* grass hay on growth performance and body composition of Emirati goats. Emirates J. Food Agric. 20: 18–27.

Attia-Ismail, S.A. (in press). Plant secondary metabolites of halophytes and salt tolerant plants. pp. 127–142, this volume.

Barrett-Lennard, E.G., C.V. Malcolm and A. Bathgate. 2003. Saltland pastures in Australia: a practical guide (2nd Edition), Land Water and Wool Sustainable Grazing on Saline Lands Sub-program, September 2003.

Barry, T.N. 1985. The role of condensed tannins in the nutritional value of *Lotus pedunculatus* for sheep 3. Rates of body and wool growth. British Journal of Nutrition 54: 211–7.

Ben Salem, H. 2010. Nutritional management to improve sheep and goat performances in semiarid regions. Revista Brasileira de Zootecnia 39: 337–347 (supl. especial).

Butler, G. 2013. The effect of feeding and milking system on nutritionally relevant compounds in milk. Max Rubner Conference 2013. Health Aspects of Milk and Dairy Products. October 7–9, 2013, Karlsruhe, Germany, pp. 8.

Campbell, J.R. and J.F. Lasley. 1985. Animals that Serve Humanity. McGraw Hill, New York, USA.

Chadwick, M.A., I.H. Williams, P.E. Vercoe and D.K. Revell. 2009. Feeding pregnant ewes a high-salt diet or saltbush suppresses their offspring's postnatal renin activity. Animal 3: 972–979.

El Shaer, H.M. 1981. A comparative nutrition study on sheep and goats grazing Southern Sinai desert range with supplements. Ph.D. Thesis, Faculty of Agriculture, Ain Shams University, Egypt.

El Shaer, H.M. 2010. Halophytes and salt-tolerant plants as potential forage for ruminants in the Near East region. Small Rumin. Res. 91: 3–12.

El Shaer, H.M., F.T. Ali, Y.S. Nadia, S. Morcos, S.S. Emam and A.M. Essawy. 2005. Seasonal changes of some halophytic shrubs and the effect of processing treatments on their utilization by sheep under desert conditions of Egypt. Egyptian J. Nutr. Feeds 8: 417–431.

El-Essawy, A.M., K.M. Youssef, A.A. Fahmy and H.M. El Shaer. 2009. Effect of feeding Kochia and Pearl Millet as haylage or hay to sheep on the fate of anti-nutritional factors of the diet. Egyptian J. Nutrition and Feeds 12(3): 441–459.

El-Shereef, A. Afaf. 2007. Nutritional studies on the use of some fodder plants in feeding small ruminants under desert conditions. M.Sc. thesis, Faculty of Agriculture, Cairo University, Egypt.

El-Shereef, A. Afaf. 2012. Productive performance of Barki ewes fed on untraditional feed mixtures under desert conditions. Ph.D. thesis, Faculty of Agriculture, Cairo University, Egypt.

FAO. 2008a. Livestock, a major threat to environment. Rome, Italy. Available at: http://web.archive.org/web/20080328062709/http%3A//www.fao.org/newsroom/en/news/2006/1000448/index.html).

FAO. 2012. Food and Agriculture Organization of the United Nations. Accessed September, 2014.

Gartner, R.J.W. and I.S. Hurwood. 1967. Tannin and oxalic acid contents of *Acacia aneura* (Mulga) and their possible effect on sulphur and calcium availability. Australian Veterinary Journal 52: 194–6.

Gast, M., L. Mauboisj and J. Adda. 1969. Le laitet les produitslaitiers en Ahaggar. Centre Rech. Anthr. Prehist. Ethn.

Helal, A. and A. Fayed. 2013. Wool characteristics of sheep fed on halophytes plants ensiled by some biological treatments. Proceedings Book of 4th International Scientific Conference on Small Ruminant Development) (3–7 Sept., 2012, Sharm El Shiekh). Egyptian Journal of Sheep and Goat Sciences 8(1): 131–139.

Hemsley, J.A., J.P. Hogan and R.H. Weston. 1975. Effect of high intakes of sodium chloride on the utilization of a protein concentrate by sheep II. Digestion and absorption of organic matter and electrolytes. Australian Journal of Animal Science 26: 715–727.

Hopkins, D.L. and A. Nicholson. 1999. Meat quality of wether lambs grazed on either saltbush (*Atriplex nummularia*) or supplements of Lucerne (*Medicago satvia*). Meat Sci. 51: 91–95.

Hunter, L., J.B. Van Wyk, P.J. De Wet, P.D. Grobbelaar, P.S. Pretorius, J. de V. Morris and W. Leeuner. 1990. The effect of nutritional and lambing stress on wool fiber and processing characteristics. Wool Tech. Sheep Breed (Sep., 1990/Oct., 1990 & Dec., 1990/Jan., 1991), 89–91.

ICARDA. 1996. International Center for Agricultural Research in the Dry Areas (ICARDA) Annual Report.

Jose, C.G., D.W. Pethick, G.E. Gardner and R. Jacob. 2008. The colour stability of aged lamb benefits from Vitamin E supplementation. Proc. 54th Int. Cong. Meat Sci. Technol., Cape Town, South Africa.

Khattab, I.M. 2007. Studies on halophytic forages as sheep fodder under arid and semi arid conditions in Egypt. Ph.D. Thesis. Faculty of Agriculture. Alexandria University. Egypt.

Krebs, G.L., D.M. Howard, D. May and M. van Houtert. 2003. The value of *Acacia saligna* as a source of fodder for ruminants. A report for the rural Industries Research and Development Corporation. RIRDC Publication No 02/165.

Maamouri, O., N. Atti, K. Kraiem and M. Mahouachi. 2011. Effects of concentrate and *Acacia cyanophylla* foliage supplementation on nitrogen balance and milk production of grazing ewes. Livest. Sci. 139: 264–270.

Makkar, H.P.S. 2003. Effects and fate of tannins in ruminant animals, adaptation to tannins, and strategies to overcome detrimental effects of feeding tannin-rich feeds. Small Rumin. Res. 49: 241–256.

Masters, D.G. (in press). Assessing the feeding value of halophytes. pp. 89–105, this volume.

Masters, D.G., A.J. Rintoul, R.A. Dynes, K.L. Pearce and H.C. Norman. 2005. Feed intake and production in sheep fed diets high in sodium and potassium. Aust. J. Agric. Res. 56: 427–434.

McDonald, P., R.A. Edwards, J.F.D. Greenhalgh and C.A. Morgan. 2002. Animal Nutrition, 6th Edition. Pearson Educated Limited, Essex UK.

Mekoya, A., S.J. Oosting, S. Fernandez-Rivera, S. Tamminga and A.J. Van der Zijpp. 2009. Effect of supplementation of *Sesbania sesban* to lactating ewes on milk yield and growth rate of lambs. Livest. Sci. 121: 126–131.

Meneses, R. and y H. Flores. 1999. Evaluación de caprinos de reemplazo y adultos en el últimotercio de preñez y lactanciaalimentados con *Acacia saligna*. AgriculturaTécnica (Chile) 59: 26–24.

Meneses, R.R., Y.V. Olivares, M.S. Martinoli and H.P. Flores. 2012. Effect of feeding *Acacia saligna* (LABILL.) H.L. WENDL. on goats stabled during late Pregnancy and lactation. Chil. J. Agric. Res. 72: 550–555.

Morcombe, P.W., G. Young and K.A. Boase. 1994. Short term, high density grazing of a saltbush plantation reduces wool staple strengths in withers. Proceedings of the Australian Society of Animal Production 20: 313–316.

Park, R.J., J.L. Corbett and E.P. Furnival. 1972. Flavour differences in meat from lambs grazed on lucerne (*Medicago satvia*) or phalaris (*Phalaris tuberosa*) pastures. Journal of Agricultural Science (Cambridge) 78: 47–52.

Pearce, K.L., D.G. Masters, R.H. Jacob, G. Smith and D.W. Pethick. 2005. Plasma and tissue α-tocopherol concentrations and meat colour stability in sheep grazing saltbush (*Atriplex* spp.). Aust. J. Agric. Res. 56: 663–672.

Pearce, K.L., H.C. Norman and D.L. Hopkins. 2010. The role of saltbush-based pasture systems for the production of high quality sheep and goat meat. Small Ruminant Research 91: 29–38.

Pearce, K.L., H.C. Norman, M. Wilmot, D.W. Pethick and D.G. Masters. 2008a. The effect of grazing a saltbush and barley ration on the carcass and eating quality of sheep meat. Meat Sci. 79: 344–354.

Pearce, K.L., D.W. Pethick and D.G. Masters. 2008b. The effect of ingesting a saltbush and barley ration on the carcass and eating quality of sheep meat. Animal 2: 479–490.

Pritchard, D.A., D.C. Stocks, B.M. O'Sullivan, P.R. Martin, I.S. Hurwood and P.K. O'Rourke. 1988. The effect of polyethylene glycol (PEG) on wool growth and live weight of sheep consuming Mulga (*Acacia aneura*) diet. Proceeding of Australian Society of Animal Production 17: 290–293.

Raju, K., S.N. Rai and A.K. Singh. 2012. Effect of feeding Acacia nilotica pods on body weight, milk yield and milk composition in lactating goats. Indian J. Anim. Res. 46: 366.

Russo, C., G. Preziuso, L. Casarosa, G. Campodoni and D. Cianci. 1999. Effect of diet energy source on the chemical–physical characteristics of meat and depot fat of lambs carcasses. Small Ruminant Research 33: 77–85.

Santos, J.E.P. 2002. Feeding for milk composition. In Proc. VI Intern. Cong. on Bovine Medicine. Spanish Association of Specialists in Bovine Medicine (ANEMBE), Santiago de Compostela, Spain, pp. 163–172.

SCA. 2007. Standing Committee on Agriculture's Nutrient Requirements of Domesticated Ruminants. CSIRO Publications, Melbourne, Australia.

Shawket, Safinaz M. and A.H. Ibrahem. 2012. Proceedings of the 3rd Conference of the International Society of Camelid Research and Development (ISOCARD), 29th January-1st February, 2012, Muscat, Sultanate of Oman.

Shehata, M.F., K.Z. Kewan, Shawket, M. Safinaz and A.M. Nour. 2004. Feedlot performance and carcass traits of growing one-humped camels fed on non-conventional rations. J. Agric. Sci. Mansoura. Univ. 29: 6911–6923.

Shehata, M.F. 2005. Carcass traits and meat quality of one humped camels fed different halophytic forages: 2- physical, chemical and sensory characteristics of camel meat. J. Agric. Sci., Mansoura Univ. 30: 1943–1952.

Shehata, M.F. and M.M. Mokhtar. 2005. Growth performance and carcass characteristics of growing Barki lambs and Baladi goat kids fed halophyticforages. J. Agric. Sci. Mansoura Univ. 30: 1877–1886.

Shetaewi, M.M., A.M. Abdel-Samee and E.A. Bakr. 2001. Reproductive performance and milk production of Damascus goats fed acacia shrubs or berseem clover hay in North Sinai, Egypt. Trop. Anim. Health Pro. 33: 67–79.

Squires, V.R. (in press). Water requirements of livestock fed on halophytes and salt tolerant feeds and fodders. pp. 287–302, this volume.

Taha, E.A.T., A. Samia, Hekal and A.H. Mahmoud. 2009. Some wool characteristics of barki sheep fed on some halophytes under desert conditions, J. Agric. Sci. Mansoura Univ. 34(1): 151–166.

Traber, M.G. 1999. Vitamin E. *In*: M.E. Shils, J.A. Olson, M. Shike and A.C. Ross (eds.). Modern Nutrition in Health and Disease. Williams & Wilkins, Baltimore, USA.

Vasta, V.A. Nudda, A. Cannas, M. Lanza and A. Priolo. 2008. Alternative feed resources and their effects on the quality of meat and milk from small ruminants. Anim. Feed Sci. Tech. 147: 223–246.

Wang, Y., G.B. Douglas, G.C. Waghorn, T.N. Barry and A.G. Foote. 1996. Effect of condensed tannins in *Lotus corniculatus* upon lactation performance in ewes. J. Agric. Sci. Camb. 126: 353–362.

Webb, E.C., N.H. Casey and W.A. Van Niekerk. 1994. Fatty acids in the subcutaneous adipose tissue of intensively fed SA Mutton merino and Dorperwethers. Meat Science 38: 123–131.

Wood, J.D. and M. Enser. 1997. Factors influencing fatty acids in meat and the role of anti-oxidants in improving meat quality. Brit. J. Nutr. 78: S49–S60.

Woodward, S.L., M.J. Auldist, P.J. Laboyrie and E.B.L. Jansen. 1999. Effect of *Lotus corniculatus* and condensed tannins on milk yield and milk composition of dairy cows. Proceedings of the New Zealand Society of Animal Production. Proc. NZ Soc. Anim. Prod. 59: 152–155.

Wulf, D., J. Morgan, S. Sanders, J. Tatum, G. Smith and S. Williams. 1995. Effects of dietary supplementation of vitamin E on storage and case life properties of lamb retail cuts. J. Anim. Sci. 73: 399–405.

Younis, F.E., W.H. Abd El-Ghany, A. Helal and H.M. El Shaer. 2012. Study of hematological and biochemical parameters and some coat characteristics of sheep fed on salt tolerant plants. Egypt. J. Basic Appl. Physiol. 11: 371–387.

Youssef, K.M., H.M. Khattab, H.M. Kandil and S.M. Abdelmawla. 2002. Studies on fattening goats fed halophytes shrubs under Arid conditions in Sinai. Int. Symp. on Optimum Resources Utilization in Salt-Affected Ecosystems in Arid and Semi Arid Regions, Cairo. 8–11 Apr., pp. 440–449.

Zaki, E.F. 2007. Evaluation of one humped camel meat and its products as affected by feeding on different types of roughages. M.Sc. Thesis, Fac. Agric., Cairo Univ., Egypt.

Review of Halophyte Feeding Trials with Ruminants

James J. Riley

SYNOPSIS

The format of this literature review is aimed at assisting those interested in halophyte forage development and/or feeding halophytes to ruminants to determine the state of the research and to direct them to some of the appropriate sources. A Review of Reviews highlights recent reviews relevant to the researcher who wishes to become familiar with one or more fields related to feeding halophytes to ruminants. Halophyte forage research related to ruminant nutrition is briefly summarized. Sheep, Goat, Camel and Cattle, feeding related trials are summarized in separate parts. The articles reviewed were based on research in 19 countries; namely: USA, Australia, Pakistan, Syria, Jordan, Israel, Tunisia, Algeria, Oman, Iran, The Netherlands, Egypt, China, Japan, UAE, Uzbekistan, Eritrea, Kuwait and Saudi Arabia.

Keywords: *Atriplex, Salicornia, Rhagodia, Chamacecytisus, Geijera, Maireana, Gamanthus, Sesbania, Hordeum, Prosopis*, sheep, goats, camels, cattle, salinity, seawater irrigation, toxicity.

1. Introduction

Demands on freshwater are increasing in deserts mostly related to increases in population, changes in lifestyles, and global climate change. Water discharges from power and desalination installations have increased the preponderance of high and

---
Associate Professor (retired) Soil, Water and Environmental Science Department, College of Agriculture and Life Sciences, The University of Arizona, Tucson, Arizona 85721 USA.
Email: jrjayjay2@gmail.com

moderately saline water sources. Consequently, there are more efforts to tap native vegetation to replace conventional forage sources and to replace irrigation of forage crops requiring freshwater irrigation with saline water resources of varying degrees of salinity.

This chapter focuses on increasing the breadth of forage resources for animal production, and the response of ruminants to a range of feeding sources with halophytes. As more attention has been given to sheep and goat trials the preponderance of the papers reviewed related to one or both of these animals. Fewer studies were found on camels and cattle.

2. Review of reviews

There have been a number recent reviews on topics related to the feeding of halophytes to ruminants. The 91st volume of <u>Small Ruminant Research</u>, published in 2011, contains the highest density of papers relevant to the topic of feeding halophytes to ruminants. Articles from this source and others are highlighted and briefly described here as a guide to the reader and as a supplement to the write-ups that follow. Not all these review papers relate directly to the feeding of halophytes to ruminants, but nevertheless are pertinent to the management and development of arid land agriculture and husbandry.

Al-Shorepy and colleagues reported on numerous halophyte trials in the UAE at the UAE University's Zayed International Agricultural and Environmental Research Program and the International Center for Biosaline Agriculture which aims to "promote the use of sustainable agricultural systems that use saline water to grow forages, field crops, vegetables, fruits and trees" (Al-Shorepy et al. 2010b). This research has the potential for wider scale applications in other hot desert areas with similar shortages of potable water and a range of saline water resources in coastal and inland areas. Halophytes offer the potential to provide at least a portion of the diets for small ruminants. Five trials with goats and sheep fed a range of levels of halophytes are reported later in this chapter under the appropriate ruminant heading.

Ben Salem and others reviewed numerous relevant papers in a wide range of conditions and farming systems (Ben Salem et al. 2010). It is concluded that Oldman Saltbush (*Atriplex nummularia*) is a good supplement for sheep and goats, but cannot be fed alone as a feed since the plant is high in salt, has only moderate energy levels, and the plant is low in biomass production.

It grows well in marginal saline lands. "To cope with these harsh conditions, this species accumulates high levels of salt and oxalates on its leaves rendering them less palatable and decreasing their nutritive value."

However its high growth rate (C_4) complements conventional feeds in hot seasons. It is high in crude protein and vitamins E and S. The presence of vitamin E and betaine may improve meat quality. Its genetic diversity offers chances for selection of genotypes best suited for given sites.

El-Kebawy and others report on research at the Dubai Desert Conservation Reserve which includes several exclosures to camel grazing some of which include grazing by antelopes, Oryx, and gazelles, which along with camels are of historic and traditional interest to the region (El-Kebawy et al. 2009). Camel populations have increased significantly as the region has developed. Rangelands have been degraded as camel grazing is unrestricted. Data are presented on the competition for rangeland plant species by small antelopes and camels over several range substrates.

El Shaer has written this superb review paper on a wide range of halophytes or salt tolerant plants for feeding sheep, goats, and camels (El Shaer 2010). This should be considered required reading for those considering use of these plants for ruminant feeds.

"Many studies showed that these plants could be used advantageously as alternative feeds to replace totally or partially common feedstuffs, thus to alleviate feeding costs. However, the presence of high contents of ash, plant secondary metabolites and non-protein nitrogen (NPN) should be taken into consideration when formulating diets containing halophytes and or salt-tolerant forages for small ruminants."

"... provision of energy supplements... is necessary to overcome maintenance and or moderate production requirements of sheep and goats fed halophytes and or salt tolerant forage-based diets."

These plants can grow in saline habitats. A brief summary of some of the specific results are presented in Table 11.1 (El Shaer 2010):
Salsola kali and *Kochia indica* are highlighted in the following categories (Table 11.2) to call attention to their frequent possession of desirable traits for forage crops.

Table 11.1. Relative palatability of selected halophytes to livestock in Egypt (El Shaer 2010).

<u>Highly or Fairly Palatable</u> X= palatable O= not palatable

Plant sp.	Goat	Sheep	Camel
Tamarix mannifera	0	0	X
Salsola tetrandra	X	X	X
Nitraria retusa	X	X	X
Atriplex halimus	X	X	0
Suaeda fruiticosa	X	X	X

"Although this review paper focused on studies carried out in Near East region, mainly in Egypt, it seems that a wide range of halophytes (*Atriplex* spp., *Kochia* spp., etc.) and salt-tolerant grasses (Sorghum, Sudan grass, etc.) could be considered as promising feed resources for small ruminants raised in saline lands and or in arid and semi-arid regions."

This paper by Masters and colleagues has several tabular summaries of studies on species that have been evaluated for consumption by livestock as well as constituents that have nutritional or anti-nutritional compounds, plus a summary of sheep production on saline sites (Masters et al. 2007).

Table 11.2. Key attributes of *Salsola kali* and *Kochia indica* relative to common halophytes in Egypt (El Shaer 2010).

Plants with Highest Dry matter %	
Salsola kali	*89.8*
Limoniastrum monopetalum	*48.6*
Salsola vermiculata	*45.0*
Plants with Dry Matter Digestibility (Highest)%	
Suaeda fruticosa	*70.4*
Kochia indica	*67.9*
Tamarix mannifera	*65.3*
Plants with Crude Protein % (Highest)	
Atriplex leucoclada	*15.1*
Salsola kali	*15.0*
Haloxylon salicornicum	*14.8*
Plants with Lowest Ash Content %	
Salsola kali	*12.2*
Juncus acutus	*12.3*
Kochia indica	*15.1*

An article by Norman and others takes a broad overview of requirements for growing halophytes in saline environments and their potential as source of nutrition for ruminants (Norman et al. 2013). Figure 11.1 (Norman et al. 2013) provides a visual framework highlighting the interaction of key factors related to providing adequate feeds for ruminants in saline landscapes.

Irrigated biomass production is 10-times dryland production. Halophytes have lower metabolisable energy (ME) than traditional forages. Most do not have enough ME to maintain live weight energy requirements. Halophytes tend to have high concentrations of indigestible fiber, salt, plus minerals and toxins.

"The productivity of saline agriculture systems may be improved by increasing halophyte feeding value. Increasing metabolisable energy is considered the most important factor. Measurement of relative palatability by grazing animals may assist in identifying genotypes with higher NV (nutritive value)."

Halophytes meet dietary requirements of crude protein and potassium and greatly exceed the dietary requirements for sodium, chloride, and iron. Most halophytes studied exceed dietary requirements for sulfur, phosphorus, calcium, chloride, magnesium, zinc and manganese. Some may contain toxins on the other hand some may have the potential to provide supplemental amounts of nutrients often found deficient in human diets.

Pearce and colleagues thoroughly review the topic of ruminants consuming halophytes and include tabulations and plots of research results (Pearce et al. 2010). Some of the conclusions are given below:

Grazing saltbush has the potential to produce animals with less fat, boost muscle and vitamin E as well as improving meat color and stability. The biggest problem is the low to moderate energy content of halophytes (most require concentrate with

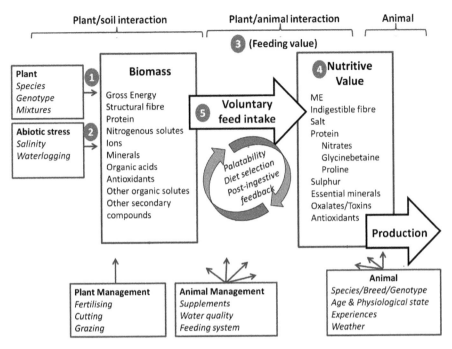

Fig. 11.1. Framework for thinking about the use of halophytes for livestock production. Livestock production depends on the utilization of biomass. The quantity and chemical composition of biomass is a function of plant genotype, abiotic environment and management. Biomass has no value unless it is eaten and its chemical composition influences both voluntary feed intake and nutritive value. Feedback loops between the stomach and brain regulate diet selection and intake. Livestock production is also influenced by animal genotype and husbandry factors. This review focuses on the impact of salinity on chemical composition of biomass, intake and nutritive value; the numbered circles indicate the different section numbers of this review (Figure 1 in Norman et al. 2013).

grain supplement or grazing time on high quality pasture). However, there is untapped potential for the goat meat industry via meat product from halophyte grazing.

"Goats may have an increased potential for liveweight gain and higher carcass lean and achieve a similar elevation in muscle vitamin E levels to sheep because of their browsing ability."

Waldron and others prepared a thorough review which has sections on

- Adaption to arid and semiarid rangelands
- Nutritional quality
- Stimulatory effects on forage intake and nutrient digestion
- Livestock performance

Kochia prostrate may be thought of as the alfalfa of the desert. It is valuable forage for sheep and goats as it is drought and salt tolerant (Waldron et al. 2010).

"Forage kochia has the potential to improve the sustainability of small and large ruminate production in areas that are threatened with extended drought and increasing salinity."

Gray, Muir, and Patricia noted that many of the halophytes being evaluated for ruminant feeding trials are not native to the lands where they are being evaluated (Gray et al. 2013). *Kochia prostrate* (forage kochia) has been widely proscribed in the USA as a means of re-establishing forests destroyed by fire. This study presents data showing that the plant has spread to unseeded areas on nearly 90% of the sites reviewed.

This should lead to caution in introducing exotic halophyte species.

Riley and associates observed that there is a growing source of saline water sources due to processes (cooling towers, desalination plant concentrate, etc.) which may present new opportunities for growing halophytes for ruminant feeds (Riley et al. 1997). This paper includes an Arizona, USA, case study highlighting the potential benefits of growing halophytes with the concentration from a membrane filtration desalination plant.

Toderich and colleagues published a report focused on Uzbekistan and Central Asia but the information on growth of a number of plant species over a wide range of salinity is potentially useful over a wider area, especially, figures and tables comparing magnitude of the growth of a number of halophytes growing over a wide range of soil salinity (Toderich et al. 2009). For example, the drainage water and groundwater at the Kyzylkesek experimental site have TDS values ranging from 2600–9000 mg/l.

It is noted that halophyte species with a minimum concentration of mineral ions are relatively more palatable and valuable for livestock production. Fresh biomass of forage species grown on highly saline soils sharply decreases with increasing salinity gradient. Wild halophytes with low salt content are better alternative forages. Use of strip alley cropping and windbreaks are good cropping systems to improve or maintain the quality of ranges.

Vasta and others prepared a review paper which analyzes the relationship between feeds and the quality of meat and milk with goats and sheep (Vasta et al. 2008). Although, the emphasis in not on halophytes the paper contains some data on halophytes. For example, they report that grazing saltbush (*Atriplex* spp.) preserves lamb meat color stability suggesting that high levels of Vitamin E in these shrubs protects myoglobin from oxidation.

Walker and associates report that *Atriplex halimus* (Mediterranean saltbush) is a halophytic shrub that grows in a wide range of conditions in North Africa and Saudi Arabia (Walker et al. 2014). "The importance of A. *halimus* in the functioning of ecosystems is reflected in its promotion of soil biota, while it also acts as a food plant for mammals and arthropods. ... The protein-rich shoot material ...makes it an important fodder species for livestock, particularly sheep and goats. However, its low energy value means that it should be supplemented with carbohydrate-rich material, such as cereal straw."

Masters and colleagues conducted an analysis of range management in Australia, but the principles are applicable to other areas with similar concerns or problems (Masters et al. 2006). For example, they note: "Management of dryland salinity in Australia will require changes in the design and utilisation of plant systems in agriculture... In areas already affected by salt a range of plants can be grown from feeding high value legumes with moderate salt tolerance through to high salt tolerant shrubs. A hectare of these plants may support between 500 to 2000 sheep grazing days per year."

3. Summary of halophyte forage studies

A summary of halophytes studied as potential forage crops is included in Table 11.3 which demonstrates the diversity and location of these trials.

Many halophyte studies have as at least part of their objective the production of halophytes for animal forage, but do not include actual feeding trials. These trials are summarized in this section of this chapter. Studies with an actual feeding or animal response component are not summarized here but in the appropriate ruminant category included in the trials (see below).

Atriplex species were included most frequently in the non-animal forage trials and among the *Atriplex* sp. *Atriplex nummularia* was the focus of most of the *Atriplex* trials, hence these studies are summarized in separate sections (see below). Where other species were an integral part of a given trial, their evaluation is included with that of the *Atriplex* sp.

Atriplex sp. *forage trials*

Glenn and others conducted a trial with *A. nummularia* which was irrigated with two levels of saline water to demonstrate the potential of utilizing the blowdown from a power plant in central Arizona, USA (Glenn et al. 1998). At the site there were two sources of non-potable water, stormwater runoff (TDS 1149 mg/l) and cooling tower blowdown (TDS 4100 mg/l). A. *nummularia* was found to have a higher water use efficiency, productivity and consumptive water used than conventional forage crops grown nearby. The nutritional content was judged satisfactory for ruminant forage and determined to be a good means of utilizing a source of saline water which would otherwise have been discharge to the local aquifer or evaporation pond.

Kumara Mahipala and associates reported on the chemical composition of browse in Western Australia including: *Atriplex amnicola* and *Atriplex nummularia*, halophyte species indigenous to Western Australia (Kumara Mahipala et al. 2009). They are high in ash and low in tannin. These and other existing or potential browse species were compared to the composition of *Avena sativa* (Oats) hay which is commonly used in this region as a ruminant feed. The *Atriplex* spp. had higher levels of crude protein and lower levels of neutral detergent fiber compared to Avena *sativa*. The atriplex browses were rich in calcium, magnesium and zinc which are deficient in A. *sativa*. The *Atriplex* halophytes plus *Rhagodia eremacea* (Goomalling) were low in phenolics and high in ash. It was concluded that all these species, except *Chamacecytisus palmensis* (Tree Lucerne), could supplement poor quality basal diets but since they are poor in digestibility and metabolisable energy content they would not be suitable as a sole diet for sheep.

Table 11.3. Forage Species Studied, Author and Country (H: Halophyte; C: Control).

Author	Popular Name	Scientific Name		Country
		Atriplex nummularia +		
Glenn 1998	Saltbush/Old Man Saltbush	*Atriplex nummularia*	H	USA
Mahipala 2009	Saltbush/Old Man Saltbush	*Atriplex nummularia*	H	Australia
Mahipala 2009	River Saltbush	*Atriplex amnicola*	H	Australia
Mahipala 2009	Goomalling	*Rhagodia eremacea*	H	Australia
Mahipala 2009	Oats	*Avena sativa*	C	Australia
Mahipala 2009	Tree Lucerne/Tagasate	*Chamacecytisus palmensis*	H	Australia
Mahipala 2009	Golden Wattle/Orange Wattle	*Acacia saligna*	H	Australia
Watson 1994	Saltbush/Old Man Saltbush	*Atriplex nummularia*	H	USA
Watson 1994	Atriplex	*Atriplex deserticola*	H	USA
Watson 1994	Wavy-leaved Saltbush	*Atriplex undulata*	H	USA
Watson 1994	Cattle Saltbush	*Atriplex polycarpa*	H	USA
Watson 1994	Four-winged Saltbush	*Atriplex canescens*	H	USA
Watson 1994	Four-winged Saltbush	*Atriplex canescens*	H	USA
Slavich 1999	Saltbush/Old Man Saltbush	*Atriplex nummularia*	H	Australia
Norman 2010	Golden Wattle/Orange Wattle	*Acacia saligna*	H	Australia
Norman 2010	River Saltbush	*Atriplex amnicola*	H	Australia
Norman 2010	Saltbush/Old Man Saltbush	*Atriplex nummularia*	H	Australia
Norman 2010	Creeping Saltbush	*Atriplex semicaccata*	H	Australia
Norman 2010	Small-leaved bluebush	*Maireana brevifolia*	H	Australia
Norman 2010	Alfalfa (Lucerne)	*Medicago sativa*	C	Australia
Norman 2010	TreeLucerne/Tagasate	*Chamacecytisus palmensis*	H	Australia
Norman 2010	Rhagodia	*Rhagodia preissii*	H	Australia
Norman 2010	Wilga	*Geijera parviflora*	H	Australia
		Other *Atriplex* sp. +		
Masters 2010a	River Saltbush	*Atriplex amnicola*	H	Australia
Riasi 2008	Kochia	*Kochia scoparia*	H	Pakistan
Riasi 2008	Atriplex	*Atriplex dimorphostegia*	H	Pakistan
Riasi 2008	Suaeda	*Suaeda arcuata*	H	Pakistan
Riasi 2008	None	*Gamanthus gamacarpus*	H	Pakistan
		Other Halophytes		
Kafi 2010	Kochia	*Kochia scoparia*	H	Iran
Sandhu 1981	Kallar Grass	*Diplachne fusca*	H	Pakistan
Alshour 1997	Kunth	*Leptochoa fusca*	H	Egypt
Alshour 1997	Aiton	*Spartina patens*	H	Egypt
Alshour 1997	Smyrna	*Sporobolus virginicus*	H	Egypt
Alshour 1997	Dixie-course	*Sporobolus virginicus*	H	Egypt
Alshour 1997	Alfalfa (Lucerne)	*Medicago sativa*	C	Egypt

Table 11.3. contd....

Table 11.3. contd.

Author	Popular Name	Scientific Name		Country
Khanum 2010	Hunnigreen	*Sorghum bicolor*	H	Pakistan
Khanum 2010	Jumbo	*Sorghum (bicolor* x *Sudanese* 1)	H	Pakistan
Khanum 2010	Revolution-BMR	*Sorghum (bicolor* x *Sudanese* 2)	H	Pakistan
Khanum 2010	Guinea Grass	*Panicum maximum*	H	Pakistan
Khanum 2010	Ruzi Grass	*Brachiaria ruzziensis*	H	Pakistan
Khanum 2010	Sesbania	*Sesbania aculeate*	H	Pakistan
Fahmy 2010	Sudan Grass	*Sorghum sudanese*	H	Egypt
Fahmy 2010	Pearl Millet	*Pennisetum glaucum*	H	Egypt
Fahmy 2010	Sorghum Grass	*Sorghum bicolor* x Johnson grass	H	Egypt
Sun 2012	Suaeda	*Suaeda glauca*	H	China
Ishikawa 2002	Saltwort	*Salicornia herbacea*	H	Japan
Riley 1994	Dwarf Saltwort	*Salicornia bigelovii*	H	Arabian P.
Khan 2009	Seepweed	*Suaeda fruticosa*	H	Pakistan
		REVIEW PAPERS		
Waldron 2010	Forage kochia	*Kochia prostrata*	H	USA
Gray 2013	Forage kochia	*Kochia prostrata*	H	USA
El Shaer 2010	Tamarisk tree	*Tamarix mannifera*	H	Egypt
El Shaer 2010	Salsola	*Salsola tetrandra*	H	Egypt
El Shaer 2010	None	*Nitrara retusa*	H	Egypt
El Shaer 2010	Sea Orache	*Atriplex halimus*	H	Egypt
El Shaer 2010	Seablite	*Suaeda fruticosa*	H	Egypt

Watson and associates reported on a study of five *Atriplex* species in addition to *A. nummularia* which were grown in the San Joaquin Valley of California, USA (Watson et al. 1994). They included single species of *Atriplex deserticola, Atriplex undulate, Atriplex polycarpa*, and two species of *Atriplex canescens*. The saline irrigation water (EC 17 dS/m) had substantial levels of Boron (5–10 mg/l) and Selenium (0.15 mg/l). The accumulation of Se and B were similar in all the *Atriplex* sp. Higher values were found in plants with the longest period of growth (27 months). It was concluded that the best use of the harvested atriplex species and was as a blend of up to one-third of the diets for cattle.

Slavich and others reported on trials in which *Atriplex nummularia* was planted in Western Australia for forage and depletion of shallow saline groundwater aquifers (Slavich et al. 1999). Consequently, water added was minimized and consisted of irrigation water drainage and rainfall runoff. During dry times plants were irrigated with saline groundwater. Water use was lower than expected due to limited leaf area and low stomatal conductance. Hence, the plantation had little hydrologic affect.

Norman and colleagues reported on eleven shrub accessions, native to Australia, which were tested for their organic matter digestibility (Norman et al. 2010a). They were: *Acacia saligna, Atriplex amnicola*, three accessions of *Atriplex nummularia, Atriplex semicaccata, Maireana brevifolia, Medicago sativa, Chamaecytisus palmensis, Rhagodia preissii*, and *Geijera paviflora*. The study compared three methods of determining digestibility of organic matter. The enzyme based pepsin-cellulose method was judged to be most suitable for ranking digestibility of shrubs. Detailed results are included in the summary for Sheep fed halophytes.

Other trials including Atriplex sp.

Masters and others conducted a greenhouse trial in Western Australia which was utilized to determine the nutritive value of River Saltbush (*Atriplex amnicola*) irrigated with a range of saline water levels (Masters et al. 2010). They conclude, "The practical consequences of the observed variation and interactions are that selection of genotypes for improved nutritive value is possible but that plants should be screened within the environment where they will be used." The nutritive value was influenced by NaCl level and genotype. IVOMD (*in vitro* organic matter digestibility) and Metabolisable energy were reduced by added salt (higher ash content). The treatment range was from 0 to 120 mM NaCl for the first 5 weeks and 0 to 400 mM NaCl for the second 5 weeks (for comparison seawater is about 600 mM). No difference was observed in crude protein levels between treatments. "neither ADF nor NDF should be used as guide to nutritive value or selection of superior saltbush genotypes." At 400 mM Na and K concentrations in the plants may depress feed intake. It was observed that K, Ca, Cu and Zn were decreased with increasing concentrations of NaCl. Cation-anion excess leads to a risk of Ca deficiency (especially during pregnancy).

Riasi and colleagues gathered four halophytes in the Iran desert which were analyzed for their chemical composition; namely: *Kochia scoparia, Atriplex dimorphyostegia, Suaeda arcuate* and *Gamanthus gamacarpus* (Riasi et al. 2008). "...the chemical composition of *Kochia* and *Atriplex* was notably different from those of *Suaeda* and *Gamanthus*. All of these halophytic plants had high concentrations of Na, K, Cl, Cu and Se, and low levels of Ca, P and Mg." Crude protein levels were lower for Kochia and *Atriplex* compared to the other halophytes. The *in situ* rumen observations from this trial are included in the summary for cattle in this chapter.

Other halophyte trials

Kafi and associates analyzed the yield potential of *Kochia scoparia* when given low levels of irrigation of highly saline waters at the Ferdowsi Research Station in Iran (Kafi et al. 2010). Sabzevar, a local cultivar of *Kochia* was studied in two trials.

In the first trial, the electrical conductivity, EC, of the irrigation water was increased in 5 dS/m steps from 5 to 20 dS/m. No significant decrease in biological yield was found up to 15 dS/m and only a slight yield decrease at an EC of 20 dS/m. The local cultivar, Sabzevar produced a yield four times that of the imported genotype,

Brujerd. The Sabzevar yields approximate those of conventionally grown forage crops such as alfalfa.

In the second trial, the irrigation water quality was maintained at an EC of 5 dS/m while the amount of applied water was varied from 100% of the optimal irrigation requirement to 40% of the optimal in 20% steps. The average dry weight yield of both *Kochia* cultivars was highest at the 100% irrigation level with a slight decrease down to 60% of the optimal level of irrigation and a much steeper yield reduction below that level. Seed yields decreased proportionally to the lowering of the amount of irrigation water applied while the protein content of the seeds increased. "The farmers along the desert areas ... are in extreme need of fodder for their animals, and the introduction of this species would provide an impressive economic benefit."

Sandhu and colleagues conducted salinity response studies in a greenhouse in Pakistan on Kallar grass (*Diplachne fusca* Beauv) (Sandhu et al. 1981). Control had an EC of 3 mmhos/cm composed of Hoagland's medium. Salinity treatments were 5, 10, 20, 30, and 40 mmhos/cm of Electrical Conductivity using NaCl, Na_2SO_4, $CaCl_2$, or $MgCl_2$. The ash content increased proportionally to the treatment level. The tissue water content was the same up to the 10 mmhos EC and then decreased with the level of treatment. The fresh plant tissue fresh weight and dry weight decreased with increasing level of salinity. The salt content of the leaves was found to be reduced when washed with fresh water. Glycinebetain and proline increased with treatment level as they apparently are related to osmotic adjustment. Kallar grass appears promising for exploitation of saline soils. The fodder looks to be palatable for sheep, goats, buffalo and cattle.

Ashour and associates conducted field trials at the Saline Agriculture Experimental Station on the Suez Gulf on sandy soil using irrigation with 12.5, 25.0, 37.5 and 50% seawater applied weekly in the hot seasons and bi-weekly in the cooler seasons (Ashour et al. 1997). "The tested grass species were *Leptochloa fusca* (L.) (Kunth) (local), *Spartina patens* (Aiton) Muhl., *Sporobolus virginicus* (L.), Kunth (Smyrna-smooth) and S. *virginicus* (Dixie-coarse) (exotic)." *Medicago sativa* (Alfalfa) was used as a control or to contrast with halophyte species.

- *Sporobolus virginicus* (Dixie) dry weight yield was greater at all levels of salinity, reaching a peak at 37.5% seawater and decreasing to its lowest level at 50% seawater dilution.
- *Medicago sativa* yield was approximately equal to the lowest yield of *Sporobolus virginicus* (Dixie) yield when irrigated with seawater diluted to 12.5% seawater and decreased linearly with increasing salinity.
- *Spartina patens* was lower than *Medicago sativa* at all levels except when salinity was 37.5% of seawater concentration.
- *Leptochloa fusca* and *Sporobolus virginicus* (Smyrna) had the lowest dry weight yields for all levels of salinity but showed little difference in dry weight yields vs. salinity.
- *Sporobolus virginicus* (Dixie) was the only test species whose dry weight exceeded the control (*Medicago sativa*) at all levels of salinity. At the highest level of salinity (50% seawater) it had the highest level of crude protein and ash content, but lowest fiber content of all the halophytes.

Khanum and colleagues conducted agronomic trials with halophytes grown on sandy loam saline soil (ECe = 2.5–3.1 ds/m) (Khanum et al. 2010). The results of the feeding trials are summarized under the summary of the sheep feeding trials.

These trials were conducted at a research station in Faisalabad, Pakistan on moderately saline soils. The forage species planted and yields are given in Table 11.4:

Table 11.4. Performance of selected salt tolerant fodder species in Pakistan (Khanum et al. 2010).

Forage	80 days	
	Dry Matter Yield t/ha	Crude Protein g/kg
• Hunnigreen (Sorghum bicolor)*	13.2a	83.0c
• Jumbo (Sorghum bicolor x Sorghum Sudanese)*	8.7b	71.3d
• Sesbania aculeate	7.7c	110.9a
• Revolution-BMR (Sorghum bicolor x Sorghum Sudanese)*	7.5c	102.8b
• Guinea grass (Panicum maximum)	4.9d	55.4e
• Ruzi grass (Brachiaria ruzziensis)	4.6d	74.7d

*the 3 sorghum varieties were cut 6cm from the ground 50 days after sowing. All the plants were grown to 80 days.

"…brown midrib (BMR) forage genotypes contain less lignin and may have altered lignin chemical composition with better digestibility than their normal cousins… This study was …planned to confirm potential for DM (dry matter) production and nutrient digestibility of three sorghum cultivars, two grass species and one legume on medium saline lands."

Not surprisingly, the legume (Sesbania) has the highest crude protein (CP) followed closely by Revolution-BMR sorghum. Looking at total CP produced calculated by the product of CP and DM, the highest value is Hunnigreen, then: Sesbania, Revolution-BMR, Jumbo, Ruzi grass, and Guinea grass.

Fahmy and others conducted halophyte trials in the Sinai Peninsula. The feeding trial summary is included in the sheep feeding trials in this chapter (Fahmy et al. 2010). This entry only addresses forage grasses grown in the Sinai Peninsula.

"Three annual salt-tolerant grasses: Sudan grass (*Sorghum Sudanese*), Pearl Millet grass (*Pennisetum americanum*), and Sorghum grass (*Sorghum* bicolor x Johnson grass) were cultivated on salt-affected soil of South Sinai Research Station. Two levels of underground water salinity were used for irrigation of crops: level 1 (L1, 4000 ppm) and level 2 (L2, 7000 ppm total salts)." Crops irrigated with Level 1 generally had higher yields than those irrigated with level 2 water. Most Level 2 crops had higher ash content and higher crude protein content.

Sun and Zhou report on halophyte trials conducted in China. This entry only covers agronomic factors of plant selection and growth (Sun and Zhou 2010). The results of the sheep feeding trials are in the summary of halophyte feeding trials with sheep.

Trials were conducted in Harbin, China where seeds of Suaeda (*Suaeda glauca*) were harvested from degraded grassland in Heilongjiang Province, China and a control hay of *Leymus chinensis* which dominates grassland in the western region of the Songnen plain. The area has become more saline over the last 50 years and the halophyte population has increased, hence the importance of the study. "*Suaeda glauca* and S. *glauca* hay samples… were harvested from native saline-alkali grassland at the Ecological Research Station for Grassland Farming…" Dry matter per unit area of harvested *Suaeda* peaked in August (Season: June–October). Crude protein (CP) and ash content were highest initially and decrease over the growing period. Seed formed in October.

Ishikawa and others conducted research on saltwort in Japan. This entry only includes information on agronomic aspects of this research (Ishikawa et al. 2002). The results on the feeding trials are found under the Goat Feeding trials summaries, later in this chapter.

"Saltwort (*Salicornia herbacea* L.), a kind of halophyte in Japan, is under investigation as a feed source for livestock as well as a potential oilseed crop in salt marshes and salt fields. The present study was undertaken to analyze the nutritive value of saltwort as a feed for ruminants… The results indicate that saltwort is rich in … minerals (mainly sodium chloride)."

"Analysis of the chemical composition of saltwort…the ash content of the plant was about 40%." Most of the ash can be accounted for by the Na and Cl content. Saltwort had a crude protein content of 11.7%.

Riley and colleagues report on agronomic trials with *Salicornia bigelovii* which demonstrated good yield potential in trials on the Arabian Peninsula when irrigated with seawater (Riley et al. 1994). It is viewed as a potential source of oilseed and forage with a biomass production of over 15 mt/ha over a growing period of about 10 months. The seed is about 10% of the total biomass and has a high oil and protein content with a low salt content. The vegetative portion, less the seed, has a low protein content (6%) and a high salt content (30%). The results of feeding trials with sheep and goats are given later in the chapter with similar trials findings.

Lymbery and others conducted trials with saltgrass. They observed that "…there is much interest in using saline groundwater to grow marine fish on salt-affected farmland, the disposal of nutrient enriched, saline aquaculture effluent is a major environmental problem (Lymbery et al. 2013)." This study investigated the potential for removing some of the nutrients using the halophyte, Salt grass (*Distichlis spicata*).

This study used samples of outflow water from four aquaculture ponds in Western Australia to evaluate the removal of nutrients by *Distichlis*. "…NyPa forage may remove more than 90% of nitrogen and phosphorus from effluent".

Karmiris and associates studied coastal grazing lands in Greece. They recorded the diet selection by coexisting cattle (*Bos taurus*), feral horses (*Equus cabalus*), European hares (*Lepus europaeus*) and white-fronted geese (*Anser albifrons*) was evaluated in the coastal grazing lands of the Evros Delta, Greece (Karmiris et al. 2011)."

About 20% of the study area was covered with water at high tide. Plants included grasses and halophytes. The latter composed 60% of the vegetation. Surveys of plants were made seasonally. "Halophytes were the dominant available forage category since they constituted almost 57% of the total available food categories in the study area." About 28% of the forage for cattle came from halophytes.

Khan and colleagues planted Desert bunchgrass (*Panicum turgidum*) on the Zia Model Farm in Hub Kund, Balochistan, Pakistan in plots between berms on saline soil (EC 10–15 mS/cm) which was flood irrigated with brackish groundwater (EC 10–12 mS/cm) at rates just sufficient to keep the root zone moist (Khan et al. 2009). "For the purpose of diminishing soil salt content, (*Suaeda fruiticosa*) Seepweeds was used as a salt scavenger intercrop. *Suaeda* seedlings (10–12 cm high) were planted 50 cm apart on bunds between plots of *P. turgidum* and irrigated as needed". Typically, the leaves of *S. fruiticosa*, as a salt accumulator, achieves an ash content of 35–50% of the dry matter content. "Locals burn leaves of this species to obtain soda ash which is then used to make soap." "The aerial part of the *Suaeda* plants … must be cut frequently… to prevent it from competing for light, water and nutrients with the main crop."

"Use of *S. fruiticosa*, a salt accumulator, in combination with *P. turgidum* has proven to be effective. Soil salinity has registered only a marginal increase during a full year of irrigation with brackish water."

When harvested the Desert bunchgrass plants had 13% ash, 18% crude fiber, 25% carbohydrates, and 13% crude protein. For comparison, Zea maize (corn) has 7% ash, 25% crude fiber, and 11% crude protein. *P. turgidum* is used as a forage crop for cattle. The results of one trial are given in the Cattle Feeding Trials Summary in this chapter.

4. Summary of halophyte sheep feeding trials

A summary of the reviewed trials in which sheep were fed halophytes is given in Table 11.5. The results summarized here are organized by types of sheep and primary halophyte feeds.

Trials with Awassi sheep

Abbeddou and colleagues conducted a trial based at ICARDA in Allepo, Syria. The only halophyte fed to Fat-tailed Awassi (castrated males) was *Atriplex halimus* leaves (Abbeddou et al. 2011). The control forage was *Hordeum vulgare* (Barley Straw). All diets were iso-nitrogenous and balanced with barley grain, wheat bran and cotton-seed meal. The diets were supplemented with an energy and vitamin mix.

The *Atriplex* leaves were high in Na and K with high quality organic matter, as a result animals fed *Atriplex* drank more than twice as much as those fed barley straw. Excessive Na intake via *Atriplex* leaves increased urinary N losses, but less than expected, but did cause diarrhea. *Atriplex* had the highest digestibility but low in metabolizable energy equal to that of barley straw with limited palatability which caused significant refusals. N intake was thus reduced but was compensated by lower fecal N losses. "*Atriplex* leaves, if fed restrictedly, could be used to replace barley straw when access to water is *ad libitum*." *Atriplex* leaves had higher crude protein than barley straw. It was judged that *Atriplex* leaves would provide good nutritional requirements during periods of feed shortage.

Table 11.5. Summary of Sheep fed halophytes trials.

Author	Location	Type of Sheep	Type of Sheep Scientific name	Halophytes fed Popular name	Halophytes fed Scientific name	Halophytes fed Parts fed	Irrigation water Salinity	Sheep Drinking Water
Abbeddou, 2011	Allepo, Syria	Fat-tailed Male Lambs	Awassi	Sea Orache	Atriplex halimus	Leaves	NA/wild growing	potable
Abu-Zanat, 2006	Mafraq, Jordan	ewes	Awassi	Sea Orache	Atriplex halimus	Leaves/twigs	NA/nursery clippings	potable
				Old Man Saltbush	Atriplex nummularia	Leaves/twigs	NA/nursery clippings	potable
Al-Shorepy, 2010	Al-Ain, Abu Dhabi, UAE	male & female idigenous lambs	Awassi	Sporobolus	Sporobolus virginicus	hay	20,000 ppm	potable
Al-Shorepy, 2010	Al-Ain, Abu Dhabi, UAE	Indigenous ewe lambs	Awassi	Sporobolus	Sporobolus virginicus	grass	25,000+ ppm	potable
				Rhodes Grass	Chloris gayana	hay	25,000+ ppm	potable
Al-Shorepy, 2010	Al Ain, Abu Dhabi, UAE	Indigenous sheep/lambs	Awassi	Sporobolus	Sporobolus virginicus	hay	20,000 ppm	potable
Degen, 2010	Beer Sheva, Israel	Fat-tailed sheep	Awassi	Four-winged Saltbush	Atriplex canescens	leaves	Range grown	potable
				Sea Orache	Atriplex halimus	leaves	Range grown	potable
				Old Man Saltbush	Atriplex nummularia	leaves	Range grown	potable
Obeidat, 2008	Irbid, Jordan	male lambs	Awassi	Mesquite	Prosopis juliflora	pods	nr Jordan Valley	potable
Chadwick, 2009	Perth, Australia	pregnant ewes & offspring	Merino	Saltbush	Atriplex nummularia	grazed	Range grown	potable
				salt pellets			Range grown	potable
Digby, 2010	Adelaide, Australia	lambs from ewes fed either high salt (13% NaCl) or conventional salt (0.5% NaCl)	Merino	conventional diet		mixed	Range grown	1.5% NaCl ad libitum
Mayberry, 2010	Crawley, W. Australia	wethers	Merino	Old Man Saltbush	Atriplex nummularia	salt pellets or OMS	Range grown	potable
Norman, 2010	Crawley, W. Australia	18 mo old wethers	Merino	Old Man Saltbush	Atriplex nummularia	grazing	Range grown	potable
				Wavey Leaf Saltbush	Atriplex undulata	grazing	Range grown	potable
Pearce, 2008	Western Australia	lamb wethers/hoggett wethers	Merino	Old Man Saltbush	Atriplex nummularia	grazing	76-80 mS/m	76-80 mS/m
				Wavey Leaf Saltbush	Atriplex undulata	grazing	76-80 mS/m	76-80 mS/m
				River Saltbush	Atriplex amnicola	grazing	76-80 mS/m	76-80 mS/m
Sun, 2010	Harbin China	male lambs	Merino	Suaeda	Suaeda glauca	Seeds & capsules	degraded grassland	potable
Norman, 2010	Australia	wethers some cannulated	Merino	Creeping Saltbush	Atriplex semicaccata			
				Blue-leaf Wattle	Acacia saligna			
				River Saltbush	Atriplex amnicola			
				Old Man Saltbush	Atriplex nummularia			
				Cotton Bush	Maireana brevifolia			
				Alfalfa (Lucerne)	Medicago sativa			
				Tagasaste	Chamaecytisus palmensis			
				Salt Bush	Rhagodia presissii			
				Australian Willow	Geijera paviflora			
Fahmy, 2010	South Sinai, Egypt	adult rams	Barki	Pearl Millet	Pennisetum americanum	fodder grass	4,000 ppm/7,000 ppm	potable
				Sudan grass	Sorghum sudanese	fodder grass	4,000 ppm/7,000 ppm	potable
				Sorghum grass	Sorghum bicolor	fodder grass	4,000 ppm/7,000 ppm	potable

Table 11.5. contd....

Table 11.5. contd.

Author	Location	Type of Sheep	Type of Sheep Scientific name	Halophytes fed Popular name	Halophytes fed Scientific name	Halophytes fed Parts fed	Irrigation water Salinity	Sheep Drinking Water
Helal, 2012	South Sinai, Egypt	adult rams	Barki	Highest ash silage	*Tamarix mannifera*	silage		potable
				Lowest ash silage	*Acacia saligna*	silage		potable
				Median ash silage	*Atriplex nummularia*	silage		potable
Sun, 2011	Harbin China	ram lambs	Ujmuqin	Suaeda	*Suaeda glauca*	hay	Saline/alkaline grassland	potable
Kanum, 2010	Faisalabad, Pakistan	ewes	Lohi	Hunnigreen	*Sorghum bicolor*		forage grown on	potable
				Jumbo Sorghum	*S. bicolor x S. sudanese*		medium saline lands	
				Revolution-BMR	*S. bicolor x S. sudanese*		rain, groundwater, canal	
				Guinea grass	*Panicum maximum*			
				Ruzi grass	*Brachiaria ruzziensis*			
				Sesbania	*Sesbania aculeate*			
Al Khalasi, 2010	Al-Batinah Coast, Oman	Male lambs	Omani	Sorghum (Super Dan)	*Sorghum vulgare*	chopped forage	3,6,9 dS/m	potable
Kraidees, 1998	Riyadh, Saudi Arabia	lambs	Najdi	Salicornia	*Salicornia bigelovii*	stems or spikes	Seawater from ArabianGulf	potable
Wang, 2009	Xilinhot, Mongolia, China	wethers	Mongolian	conventional diet		chemical additives	tropical grasslands	potable
Ktita, 2010	Ariana, Tunisia	lambs	Barbarine	Seaweed	*Ruppia maritima*	washed & dried	lake&lagoon	potable
				Seaweed	*Chaetomorpha linum*	washed & dried	lake&lagoon	potable
Swingle, 1996	Tucson, USA	wethers lambs	Suffolk x Rambouillet	Atriplex forage	*Atriplex barclayana*	forage	Seawater from Gulf of Calif.	potable
Swingle, 1996	Tucson, USA	wethers lambs	Rambouillet	Salicornia forage/meal	*Salicornia bigelovii*	hay or meal	Seawater from Gulf of Calif.	potable
				Suaeda forage	*Suaeda esteroa*	hay	Seawater from Gulf of Calif.	potable
Riley, 1993	Kuwait	wethers lambs	Australian dwarf	Salicornia	*Salicornia bigelovii*	hay/seed	Seawater from Arabian Gulf	potable
Riley, 1994	Sharjah, UAE	sheep	unspecified	Salicornia	*Salicornia bigelovii*	vegetative portion	Seawater from Arabian Sea	potable
Tesfa, 2011	Asmara, Eritrea	sheep	unspecified	Halophyte	Mixture	unspecified	Seawater from Red Sea	potable
Ben Salem, 2010 **Review Paper no trials**				Old Man Saltbush	*Atriplex nummularia*			

Abu-Zanat and Tabbaa conducted trials at the Kalideaya Nursery in Mafraq in the Jordan Valley (300 m below sea level) (Abu-Zanat and Tabbaa 2006). Awassi ewes, 3–5 years old were fed *Atriplex* browse composed of leaves and small twigs of *Atriplex halimus* and *Atriplex nummularia*. The treatments produced no significant effect on final body weight or milk production. The nutritive value (crude protein and metabolizable energy) of the *Atriplex* browse was better than that of the control, shredded barley straw and (tibin). It was concluded that substituting *Atriplex* browse for tibin substantially reduced the cost of the roughage component of the diet.

Al-Shorepy and others report on trials conducted at the Al Foah area of the UAE University in Al-Ain, Abu Dhabi, UAE (Al-Shorepy et al. 2010a). Twenty males and twenty females lambs were fed the halophyte, *Sporobolus virginicus* grass hay with Rhodes Grass (*Chloris gayana*) as the control hay diet. *Sporobolus* was grown on site and irrigated with 20,000 ppm groundwater. Male and female lambs were fed *Sporobolus* at rates of 100, 70, 30, or 0% replacement of Rhodes Grass. Feed and water were offered *ad libitum*. Feed intake was greater for halophyte treatments, but there were no significant differences in Feed Conversion Ratio or growth, indicating that *Sporobolus* could be substituted for Rhodes Grass with a net saving of potable irrigation water.

Al-Shorepy and colleagues conducted several trials, however, this summary includes only information on the three trials with Awassi sheep fed either Rhodes Grass (*Chloris gayana*) or sporobolus (*Sporobolus virginicus*) or a mixture of the two hays (Al-Shorepy et al. 2010). Forage was supplemented with a commercial nutrient concentrate in all trials.

Experiment 1. There were two treatments both using Awassi ewe lambs. Animals were fed either Rhodes Grass or sporobolus. The latter was grown in saline desert lands where it was irrigated with highly saline groundwater (Approx. 25,000 ppm). The dry matter intake of sporobolus was significantly higher than for the Rhodes Grass, but the animals fed sporobolus consumed 23% more potable water. There was no difference in final body weight.

Experiment 2. Rhodes grass was replaced by sporobolus at rates of 100, 66.6, 33.4, or 0%. The sporobolus was produced at the International Center for Biosaline Agriculture in the UAE utilizing saline groundwater with a salinity of 20,000 ppm. The weight gain varied with age, but the final body weights and average daily gain were not significantly different.

Experiment 3. Lambs were fed 0, 30, 70, or 100% of sporobolus with the remaining diet composed of Rhodes Grass. There was no significance in final weight, growth rate, or feed conversion ratio.

Degen and associates conducted these trials at Beer Sheva, Israel in the Negev desert with Awassi sheep (Degen et al. 2010). The trials at the same site with goats are included in the summary of goat research later in this chapter.

Four Awassi sheep were allowed to feed *ad libitum* in cafeteria trials on two trees (*Acacia salicina* and *Acacia saligna*); one leguminous shrub (*Cassia sturtii*); and three halophytic shrubs (*Atriplex canescens, Atriplex halimus*, and *Atriplex nummularia*). "*A. saligna* was the most preferred feed...(by Jacobs selectivity index) and the two *Acacia* species comprised more than ...70% of the dietary intakes in...sheep...."

Obeidat and others conducted these trials at the Agriculture Centre for Research and Production in Jordan at an elevation of 510 m above sea level (Obeidat et al. 2008). The basic (control) diet for the Awassi lambs contained 500 g/kg barley grain (*Hordeum vulgare*). The treatments given here were for replacing 100 g/kg or 200 g/kg of barley with Mesquite (*Prosopis juliflora*) pods. Seven male lambs were entered in each treatment. All diets were isonitrogenous/isocaloric with 17% crude protein on a dry matter basis. The cost of the diet was reduced the more barley was replaced by *Prosopis*. Both *Prosopis* treatments had higher feed conversion ratios than the control. At the end of the trial there was no significant differences in any of the measured parameters for the animals. Meat quality for all treatments was similar, except the water holding capacity of the *Prosopis* fed sheep was lower than the control.

Trials with Merino sheep

Chadwick and colleagues conducted two trials in Western Australia with Merino Lambs (Chadwick et al. 2009).

Trial 1. Offspring born to ewes that were fed a high-salt diet (14% NaCl) were compared to offspring borne to ewes receiving the control diet (2% NaCl). Also, offspring of sheep previously grazing saltbush (mainly *Atriplex nummularia*) were compared to sheep grazing a control diet. After 8 months, their preference for salt was tested by giving them high or low salt pellets. "The salt content of the diet of the ewe had no effect on their offspring's preference for salt (P > 0.05)."

"High-salt offspring consumed about 14% less food than control offspring (P < 0.05) and drank 35% less water than control offspring when consuming the low salt pellets, or when given a choice between low salt and the high-salt pellets (P < 0.01)." Saltbush offspring had consistently lower plasma rennin activity (PRA) than pasture offspring, regardless of the salt content of their diet.

Trial 2. "Sheep were given 25 g of NaCl as an oral dose ... administered by a syringe with a plastic tube placed in the mouth of the sheep and, four days later they were given 50 g of salt (in solution)." "Saltbush offspring had a lower rennin activity than the pasture offspring before the salt dose and it continued to be lower during the 25 g salt dose and during the first 4 h of the 50 g dose (P < 0.05)."

High-salt offspring excreted lower volume of urine than controls during the 30 hr period. They excreted less salt than the controls. No difference in water intake was observed. Saltbush offspring gained less weight than the pasture fed offspring.

"Saltbush offspring show salt handling adaptions that are conducive to them being better prepared to handle excess salt in their diet, such as faster rate of salt excretion. In contrast, high-salt offspring have a tendency to retain salt which may be detrimental to them when consuming excess salt in their diet."

Digby and others carried out this trial with Merino lambs at The University of Adelaide, Roseworthy campus, Australia (Digby et al. 2010). The working hypothesis was that salt-programmed lambs may not need to drink as much saline water as control lambs and that voluntary feed intake of salt-programmed lambs would be reduced (i.e., Fetal programming of lambs borne to ewes exposed to high salt during pregnancy).

"…exposure of pregnant ewes to high-salt affects the physiological responses of their offspring to the consumption of saline drinking water over a period of 2 days. The main differences are reflected in the temporal patterns of change in water intake, sodium excretion and aldosterone concentration following the consumption of saline drinking water… The capacity of the programmed lambs to deal with salt insult throughout altered thresholds could be further explored by exposing them to salty water for period greater than 2 days."

Mayberry and associates directed two trials in Western Australia to determine the effect of a high salt diet on the mineral metabolism of 18 month old Merino wethers (Mayberry et al. 2010). The treatments included adding salt pellets to the diet or leaves of Old Man Saltbush (*Atriplex nummularia*) at the same level of salinity.

Trial 1. The objective of this trial was to measure the concentration of Na and K in the rumen over a 24 hr period following feeding. Rumen cannulae were installed in 16 sheep. The saline diet was stepped up to full level (approximately 20% salt) or saltbush leaves over a two week period. Then, the full diet was maintained for 3 weeks. "…there was no relationship between rumen pH and the amount of salt ingested or the salinity of the rumen fluid. The average pH of rumen fluid from sheep fed the saltbush diet was higher than the pH of the rumen fluid from sheep fed any other diet ($P < 0.05$)."

Trial 2. The design was similar to Trial 1, except the full diet period was extended for 5 weeks. Faecal samples were collected the last 6 days of the experiment. "Apparent digestion of Na, K and Cl was high in all diets compared to the apparent digestion of Mg, Ca and P ($P < 0.05$). …There was a net loss of Mg, Ca and P from sheep fed saltbush…."

"…these preliminary results suggest strongly that saltbush may be unsuitable as forage for sheep with high nutritional demands, and mineral deficiencies may contribute to the poor performance of sheep grazing saltbush-based pastures."

Norman and colleagues implemented trials with Merino sheep grazed pastures primarily composed of *Atriplex nummularia* and *Atriplex undulate* (Norman et al. 2010a). They tested two hypotheses; namely:

Hypothesis 1: Rotational grazing of saltbush-based pastures will increase sheep productivity and reduce pattern of rapid live weight gain followed by liveweight loss. The differences between the treatments were small and therefore this hypothesis was rejected.

Hypothesis 2: "…the intake of saltbush, as a proportion of the total diet, will be negatively correlated to the digestibility of the understorey sward." The data supported this hypothesis.

Pearce and associates raised Merino sheep on either saltbush-based pasture on saline land or control of barley stubble (Pearce et al. 2008). Both were given barley grain supplement. The first experimental trials were conducted near Goomalling, Western Australia. The second trial was conducted in Wickenpin, Western Australia.

Experiment 1. 6 mo old wethers were give free access to water with a conductivity of about 78 mS/m. Oldman Saltbush (*Atriplex nummularia*) and Wavey Leaf Saltbush (*Atriplex undulate*) were grazed for 14 weeks. The saltbush provided about 80% of the total forage. There was no difference in the initial live weight at the start of the trial, but at the end the sheep fed the control diet of barley (*Hordeum vulgare*) were significantly heavier.

Experiment 2. 18 mo old hogget wethers were given free access to water with a conductivity of 78 mS/m. The sheep grazed on pastures of Old Man Saltbush and River Saltbush (*Atriplex amnicola*) with a control diet of barley for a period of 14 weeks. The saltbush provided about 80% of the total forage. Sheep grazing on saltbush had significantly higher live weight gain.

Sodium intake was significantly higher for the animals grazing on saltbush. While there was no significant difference in the treatment eating quality in both trials, the sheep fed saltbush had less fat. Grazing saltbush for finishing was judged not appropriate as the sheep had reduced carcass weights. Grazing saltbush increased urine weight but decreased urine concentration. "Despite the low growth rates...these experiments demonstrated that saltbush can still be effectively use as a maintenance ration without compromising carcass and eating quality."

Sun and Zhou carried out trials in Harbin, China with Northeast Merino male lambs fed seeds of Suaeda (*Suaeda glauca*) harvested from a degraded grassland in Heilongjiang Province, China with a control diet of *Leymus chinensis* hay plus a concentrate fed at a rate of 300 g/da (Sun and Zhou 2010). The *suaeda* seeds were offered at rates of 0, 150, 300, and 450 g/da. The *L. chinensis* and water were offered *ad libitum* to all animals.

The addition of *suaeda* did not negatively impact lamb performance in terms of daily weight gain and health status. However, dry matter, organic matter and water intake increased with the treatment levels. "This study shows that the seed of S. *glauca* might be used at a level of up to 450 g/d in lamb diets without compromising their feedlot performance..." The best level was found to be 300 g/da.

Norman and others evaluated eleven shrub accessions native to Australia for their digestibility when fed to sheep (Norman et al. 2010b). The accessions tested included: *Acacia saligna, Atriplex amnicola,* three accessions of *Atriplex nummularia, Atriplex semicaccata, Maireana brevifolia, Medicago sativa, Chamacecytisus palmensis, Rhagodia preissii, Geijera parviflora* in Crawley, Western Australia.

In vivo measurements of organic matter digestibility (OMD) were measured in three year old Merino wethers. A selection of the same animals were cannulated and used for *in sacco* observations and as a source of rumen fluid for the *in vitro* trials. Two of the selected shrubs were eliminated from the trial, namely: *Rhagodia preissii* and *Geijera paviflora*. "The relationship of *in vivo* OMD...was evaluated with 9 forage shrub accessions. Of all the methods used, gas production in *in vitro* batch fermentation showed the best relationship with *in vivo* OMD... (and) could be used to predict OMD or metabolisable energy supplied to livestock grazing or feeding on all the plants tested."

Trials with Barki Sheep

Fahmy and associates conducted trials at the South Sinai Research Station (about 200 km Southeast of Cairo) using adult Barki rams, which were fed three annual salt-tolerant grasses; namely: Sudan grass (*Sorghum Sudanese*), Pearl Millet grass (*Pennisetum americanum*) and Sorghum grass (*Sorghum bicolor*) which were cultivated on salt-affected soils (10 dS/m) (Fahmy et al. 2010). "Two levels of underground water salinity were used for irrigation of crops: level 1 (L1, 4000 ppm) and level 2 (L2, 7000 ppm total salts)." The grasses evaluated are not generally considered halophytes, but are tolerant to relatively high level of saline soil and saline water irrigation. The fodder production was greater at the lower salinity level of irrigation water.

The Barki rams growth and the digestion of dry matter were better for the grasses grown at level 1 irrigation. The rams fed grasses grown at level 2 irrigation had higher levels of digestible crude protein and they retained more N, Cu, Zn, and Co. Both were judged good quality summer fodders.

Helal and Fayed implemented trials at the Ras Sudr Research Station in South Sinai, Egypt (Helal and Fayed 2013). Twenty-four adult Barki rams were fed several silage mixtures for 60 days. Silage was prepared from three salt tolerant plants; namely: Manna (*Tamarix mannifera*), Blue leaf wattle (*Acacia saligna*) and Oldman Saltbush (*Atriplex nummularia*), mixed at rates of 50%, 25%, and 25%, respectively. Silage was prepared in several ways; Control was prepared without additives; the second batch had cellulytic bacteria added; the third batch had nitrogen bacterial added; and the fourth batch had cellulytic and nitrogen bacterial added. A nutritive concentrate was given to all animals. Silages were provided twice daily *ad libitum*. Potable water was provided continuously *ad libitum*. "Using halophytes in animal feeding significantly depressed the industrial characteristics of wool fibers." The wool weight, fiber diameter, and yield were significantly higher for the third and fourth treatments. There was no significant difference of staple length between all treatments.

Trials with Other Sheep

Sun and colleagues divided eighteen Ujmuqin ram lambs into three treatments of feeding *Suaeda glauca* hay in Changling, China (Sun et al. 2011). All animals were fed 500 g/kg of hay of either *Leymus chinensis*; a mixture of equal portions of *Leymus chinensis* and *Suaeda glauca*; or *Suaeda glauca* alone. All treatments included Maize flour, Soybean meal, Calcium phosphate, and a trace elements premix. Ten g/kg of salt was added to the control feed (w/o Suaeda).

"Supplementation at 25 or 50% (suaeda hay) significantly increased the digestible energy, DM, OM and ADF compared with the (treatment without suaeda hay). Water intake, urinary excretion, water content of fresh faeces and sodium excretion from urine linearly with increasing S. *glauca* content, but potassium excretion from urine decreased… This study indicates that S. *glauca* is a suitable hay when added as 25% of the lamb diet."

Khanum and associates fed eighteen 7 mo. old Lohi breed ewes *ad libitum* one of the following salt tolerant forage species: Hunnigreen (*Sorghum bicolor*), Jumbo (*Sorghum bicolor* x *Sorghum Sudanese*), Revolution-BMR (*Sorghum bicolor* x *Sorghum Sudanese*), Guinea grass (*Panicum maximum*), Ruzi grass (*Brachiaria ruzziensis*), and *Sesbania aculeate* for 21 days in trials at Faisalabad, Pakistan (Khanum et al. 2010). The diets were composed of 70% of the forage plus 30% wheat bran.

The forage digestibility was determined for forages cut at 50 and 80 days. Revolution-BMR had the highest level of digestibility in terms of dry matter, crude protein and crude fiber for all trials. All measures of digestibility decreased from forage cut at 50 days and that cut at 80 days. Guinea grass had the lowest level of digestibility by sheep as measured by dry matter and crude protein. Hunnigreen and Jumbo had lowest value of crude fiber digestibility when fed forage cut at 50 days. Jumbo and *Sesbania aculeate* digestibility as measured by crude fiber were lowest when fed forage cut at 80 days.

Al Khalasi and others conducted feeding trials at the Agriculture Research Station of Sultan Qaboos University in Oman (Al Khalasi et al. 2010). "Thirty-two, 3-month–old Omani male lambs were randomly distributed into four groups of eight lambs each. The first group was fed a control diet of Rhodes grass hay (RGH) plus a commercial concentrate. The other groups were given one of three sorghum hays irrigated with water containing one of the three different concentrations of salt plus the commercial concentrate." These were 3, 6, and 9 dS/m. "This study indicated that salinity-tolerant sorghum forage grown with irrigated water up to 9 dS/m salinity could be used as complete roughage source without affecting the health or performance of Omani sheep.

Kraides and colleagues reported that "Sixty-three Nadji ram lambs (22.8 kg) were used to evaluate the effect of dietary inclusion of *Salicornia bigelovii* by products on growth performance, carcass characteristics and mineral and water intake (Kraides et al. 1998)." The feeding trials were conducted at King Saud University, Riyadh, Saudi Arabia. The lambs were fed either dried stems or spikes (seed head less seeds) of Salicornia produced with seawater irrigation. Rhodes grass (*Chloris gayana*) hay was the control diet, which was substituted for by three levels of *salicornia* stems or spikes ranging from 100 to 300 g/kg. All diets included vitamin premix, maize, wheat bran, soybean meal, and alfalfa. All feed treatments and water were offered *ad libitum*.

Dry matter intake was not affected by treatments with salicornia stems, but decreased above the 100 g/kg level of salicornia spikes. Water intake increased proportionally with the level of salicornia stems or spikes. Inclusion of up to 300 g/kg of stems or 100 g/kg spikes did not adversely affect growth or carcass characteristics. salicornia did not affect empty body weight but did increase size of kidneys and hearts of rams. The warm carcass weight decreased linearly with increasing level of spikes included in the diet.

Wang and colleagues conducted a study with Mongolian wethers in Xilinhot, Mongolia, China (Wang et al. 2009). The animals were fed conventional diets with water available freely. "The basal diet consisted of 1.29 kg mixed hay and 0.43 kg concentrate mixture based on dry matter (DM)." Treatments consisted of the addition of control (no addition) or Flavomycin, or Ropadiar or Saponin.

"Results indicated that intake and digestibility were unaffected by treatments. The NH_3–N concentration of rumen liquor was lower for additive treatments versus the control treatment. Higher concentrations of volatile fatty acid (VFA) were observed in the saponin … and ropadiar … treatments." The average methane production was reduced by all of the treatments.

Rijba Ktita and others conducted these trials with Barbarine lambs in Ariana, Tunisia (Rjiba Ktita et al. 2010). Mineral and vitamin supplement and 300 g/da of oat hay was given to all. The control diet was composed of 80% barley and 17.5% soybean meal. Fresh water available at all times. About 20% of the treatment diets were seaweeds. The seaweeds were washed with fresh water to reduced salts and then air dried.

The total dry matter, DM, intake, final body weight, and average daily gain were lower for the animals fed 20% *Chaetomorpha linum*, but feed conversion ratio was highest for the animals fed the same treatment. "This study showed that the two seaweeds… could be associated to barley and soybean meal to make concentrates without causing any deleterious effect on growth performance of Barbarine lambs."

Swingle and others conducted trials in Tucson, Arizona, USA (Swingle et al. 1996). Halophytes, *Atriplex barclayana*, *Suaeda esteroa*, and *Salicornia bigelovii* were grown on seawater near Puerto Penasco, Sonora, Mexico. The straw of *Salicornia* (what is left after threshing and removal of seeds) and the meal of *Salicornia* (seed after oil extracted) were evaluated in the trials. Potable water was provided at all times in the trials.

Trial 1. Twenty-four Suffolk x Rambouilet wethers were selected for this trial from the University of Arizona sheep flock. Forages were 30% of the diet dry matter with Bermuda grass (*Cynodon dactyon* (L.) pers.) as the control forage diet. All diets had 70% concentrate with 12.5–15% protein. The treatment forages for this trial included: *Atriplex* forage with cottonseed meal; *Salicornia* forage with cottonseed meal; and *Cynodon* hay with *Salicornia* seed meal. The lower levels of protein in halophytes necessitated addition of meal to bring up level of protein so all were equal. Trial 1 results: "The highest dry matter intake was by lambs fed the diet containing Salicornia forage. Intakes by lambs fed diets containing *Cynodon* hay with either cottonseed meal or *Salicornia* meal were similar… Daily gains… did not differ among diets. Because of their higher dry matter intakes, lambs fed diets containing halophyte forage had poorer feed efficiencies than those fed *Cynodon* hay… Lambs fed diets containing halophyte forages consumed up to 110% more water per day and 50% more water per kilogram of dry matter intake than those fed the diets with *Cynodon* hay." The dressing percentage was slightly higher for lambs fed halophyte forages.

Trial 2. Twenty-four Rambouilet wethers were used for this trial which included the same control at Trial 1 with forages of *Salicornia*, *Suaeda*, and a synthetic forage in which NaCl was added to Bermuda grass to assess the effect of excess salt on feed intake and growth.

Trial 2 results: "Lambs fed natural halophyte forages had higher feed intake than lambs fed *Cynodon* hay… but animals fed the artificial halophyte had the lowest rates of feed intake and weight gain. The form in which NaCl is incorporated into a diet turned out to be important in determining its acceptability to the lambs. Lambs had similar feed efficiencies on all diets."

"It is concluded that halophyte forages and seed products could become important feed resources for small ruminate production wherever production of these unique plants can be justified agronomically…"

Riley and associates conducted feeding trials included goats and sheep (Riley et al. 1994). Only the latter results are presented here. The goat trials are included under the goat feeding trials, later in the chapter.

"A 97 day feeding trial with …12 sheep (unspecified breeds) was initiated in December 1988 in Kalba, Sharjah, United Arab Emirates…Washed *Salicornia* hay was substituted for Rhodes grass (*Chloris gayana*). In addition, differing amounts of *Salicornia* seed were added to the diets." See Table 11.6 for details of treatments:

Table 11.6. Treatments of sheep (unspecified breed) in feeding trial in Kalba, Sharjah, UAE, December 1988 (Riley et al. 1994).

Treatments	Salicornia hay	Salicornia seed	Rhodes grass	Yellow Corn Dates Wheat Bran
1	50%	25%	0	25%
2	60%	20%	0	20%
3	70%	15%	0	15%
4 (Control)	0	0	50%	50%

"…there were no significant differences in weight gain. The mean weight gain for all treatments was 70.8 g/head/da…"

Note: "The term "hay" is used to indicate dried vegetative material containing immature seeds. The term "straw" is used to describe dried vegetative material from which the seeds have been removed."

Riley and Abdal initiated in April 1989, a two-month sheep feeding trial at the Sulaibiya Research Site of the Kuwait Institute for Scientific Research using *Salicornia bigelovii* hay harvested in 1988 from the Kalba Agriculture farm in Sharjah, UAE (Riley and Abdal 1993). Fifty-four, seven month old Australian wether lambs were randomly distributed among 6 dietary treatments (Table 11.7); namely:

Table 11.7. Summary of results of sheep feeding trial in Kuwait 1989 (Riley and Abdal 1993).

Treatment Number	Dietary Treatments Components			Aggregate Crude Protein	Wt. Gain g/h/d	Feed Consp. g/h/d	FCR g/g
	Concentrate	Alfalfa	Salicornia hay				
	%	%	%	%			
1	0	100	0	10.6	93.7	1046	11.2
2	50	50	0	12.2	142.7	1040	7.3
3	50	37.5	12.5	11.7	154.7	1228	7.9
4	50	25	25	11.3	114.7	1076	9.4
5	50	0	50	10.4	45.4	895	18.2
6	0	0	100	7.0	−78.5	633	---

The weight gain is highly correlated with the crude protein content. Diets with less than 8.9% protein had negative growth rates. The growth of Treatment 3 with 12.5% salicornia hay; 37.5% alfalfa; and 50% of concentrate had the highest growth rate. Based on these data it was concluded that *Salicornia* hay can be incorporated in the diet of sheep at a rate up to 25%. In future trials, the high protein concentrate could be contributed by *Salicornia* meal.

Tesfa and Menari conducted a feeding trial at the Ghatelai Ram Farm, seventy kilometers North of Asmara, Eritrea at an elevation range of 280–450 meters above sea level (Tesfa and Mehari 2011). The trial included 80 goats and 80 sheep, with equal numbers of male and female animals. Animals were fed at 3% of body weight/ day from 0 to 42 days and 3.5% body weight/day from 43 to 84 days. Potable water was provided *adlibitum*. The materials included in the diets were:

1. Halophyte: mixture of Salicornia (*Salicornia bigelovii*), mangroves and algae with 7.65% Crude Protein (CP)
2. Sorghum (*Sorghum bicolor*) husks. 8.05% CP
3. Shishay: a local commercially produced complete feed at the Shishay Animal Feed Processing Plant in Eritrea with 16.3% CP
4. Mineral Salts (unspecified)

The feeding treatments for sheep and goats was:

Treatment 1 100% Halophyte: (ash 21.6%) (CP 7.6%)
Treatment 2 30% Halophyte + 70% Sorghum husks: (ash 10.9%) (CP 8.2%)
Treatment 3 30% Shishay + 70% Sorghum husks: (ash 6.3%) (CP 10.6%)
Treatment 4 100% Sorghum husks + Mineral salts: (ash 6.8%) (8.1%)

The results for sheep feeding is given in Table 11.8, below (results of goat feeding are given with other goat feeding trial results in this chapter):

Table 11.8. Summary of outcomes from a sheep feeding trial in Eritrea where halophyte feedstuffs were fed (Tesfa and Mehari 2011).

Treatment	Ave. Daily Gain (all sheep) g/day	Dry Matter Intake g/head/day	Feed Conversion Rate Column 3/Column 2
1	112.3	540	4.8
2	53.3	740	13.9
3	68.0	510	7.5
4	47.8	790	16.5

Note: A high Feed Conversion Ratio (FCR) means that an animal must eat more of the diet to gain a unit weight.

The sheep fed all halophyte diet (treatment 1) had the highest daily gain and the lowest FCR, indicating the sheep had to ingest a minimal amount of the feed to reach the same rate of gain as other animals. The FCR was highest for treatment 4 feed only Sorghum Husks. All the sheep in Treatment 1 had the highest final body weight and those fed Treatment 4 the lowest. In all treatments the male animals were heavier than females. The results demonstrate the benefit of feeding the all halophyte diet.

5. Summary of goat feeding trials with halophytes

Table 11.9 presents a tabular summary of all goat feeding trials with halophytes.

Al-Shorepy and colleagues conducted feeding trials with sheep and goats. This summary reports only on the feeding trials with goats (Al-Shorepy et al. 2010b). Other aspects of the paper are summarized in the Review of Reviews at the beginning of this document and in the sheep feeding trials summary in this chapter. The research was conducted at the United Arab Emirates University, Al Ain, Abu Dhabi, UAE.

Experiment 4

This experiment was performed to study the effect of inclusion of Distichlis grass (*Distichlis spicata*) hay as a source of forage in whole-mixed diets on growth, feed and voluntary water intake and carcass composition of Emirati goat kids. Eighteen male and 22 female local goat kids were used in this study. The kids were fed in groups of the same sex. The treatments were formulated to have 100, 66.7, 33.3, or 0% *Distichlis* hay as a forage in replacement of Rhodes grass (*Chloris gayana*) commonly used in the region. Commercial concentrate was offered equally to all groups to meet animal requirements. All goats had free access to potable water.

There was no significant differences in initial and final body weights for all the treatments. Kids fed 100% *Distichlis* grass had the highest feed intake and daily feed intake, while the control fed animals had the highest feed conversion ratio. The goats in the treatment with 33.4% *Distichlis* had the highest daily water intake. The animals fed 33.4 to 100% *Distichlis* had equal average daily gains, all higher than the control

Table 11.9. Summary Goat fed halophyte trials.

Author/date	Location	Type of Goats	Scientific Name	Halophytes Popular Name	Halophytes Scientific Name	Halophytes Part fed	Irrigation Water Salinity	Goats drinking water
Al-Shorepy, 2010	Al-Ain, Abu Dhabi UAE	goat kids	Emirati	Distichlis grass	Distichlis spicata	hay	25,000 ppm	potable
				Mediterranean Saltbush	Atriplex halimus	hay	saline	potable
				Brewers Saltbush	Atriplex lentiformis	hay	saline	potable
				Old Man Saltbush	Atriplex nummularia	hay	saline	potable
				Seashore Dropseed	Sporobolus virginicus	hay	25,000 ppm	potable
Glenn, 1992	Kalba, Sharjah UAE	male and female	Damascus	Dwarf Glasswort	Salicornia bigelovii	fodder	Seawater, Arabian Sea	potable
Riley, 1994	Kalba, Sharjah UAE	Kid goats	Damascus	Dwarf Glasswort	Salicornia bigelovii	fodder	Arabian Gulf	potable
Degen, 2010	Beer Sheva, Israel	local male goats	baladi or Negev	Four wing Saltbush	Atriplex canescens	fodder	Range grown	potable
				Mediterraneian Saltbush	Atriplex halimus	fodder	Range grown	potable
				Old Man Saltbush	Atriplex nummularia	fodder	Range grown	potable
Ishikawa, 2002	Ibaraki,Japan	castrated goats	Native goats	Saltwort	Salicornia herbacea	fodder	salt marshes or fields	potable
Nawaz 1994	Faisalbad, Pakistan	dwarf (teddy)	unspecified	River Saltbush	Atriplex amnicola	fodder	Range grown	potable
El-Rahman, 2014	Giza, Egypt	male goats	Baladi	Golden Wattle	Acacia saligna	leaves and twigs	Range grown	potable
				Black Mustard	Brassica nigra	leaves and twigs	Range grown	potable
Tesfa, 2011	Asmara, Eritrea	goats	unspecified	Halophyte	Mixture	unspecified	Seawater, Red Sea	potable
Abdou,2011	Sinai, Egypt	goat kids	Black Desert	Old Man Saltbush	Atriplex nummularia	leaves and twigs	Range grown	potable

diet. "It appears that the rate of passage of digesta might have accelerated in the lambs fed ...*Distichlis* grass hay... might ... account for the lower feed conversion efficiency and higher gain."

Experiment 5

"*Atriplex* and Sporobolus grass were grown at the International Center for Biosaline Agriculture. Twenty female Emirati goat kids were used in this study." "The halophytes produced at the ICBA were irrigated with saline groundwater with a salinity of about 25,000 ppm. Goats were fed a diet of *Atriplex* and or *Sporobolus*. Commercial concentrate was offered equally to all groups to meet animal requirements." All goats had free access to potable water. The *Atriplex* feed was a mixture of A. *halimus*, A. *lentiformis*, and A. *nummularia*. The treatments were: 1. Mixture 1:1 of *Atriplex* and Sporobolus; 2. Only *Atriplex*; 3. Only *Sporobolus*; and 4. The control Rhodes grass (*Chloris gayana*).

Animals on diets containing *Sporobolus*, alone or mixed with *Atriplex* had the highest daily feed intake while those fed a mixture of *Atriplex* and Sporobolus had the highest daily water intake. The final body weight was lowest for animals fed only *Atriplex* and the highest was the diet of 50% *Atriplex* plus 50% *Sporobolus*.

Degen and others conducted several feeding trials. Only the results of the goat feeding trials are included here (Degen et al. 2010). Sheep feeding trial results are in the summary of sheep feeding trials earlier in this chapter.

These trials were conducted at the Wadi Hashash research station, 20 km south of Beer Sheva, Israel. Local male goats of 1.5 to 1.0 years old were selected from Bedouin flocks. Leaves were collected from six mature plants for each of the following species: *Acacia salicina, Acacia saligna, Atriplex canescens, Atriplex halimus, Atriplex nummularia* and *Cassia sturtii*. Four local, male, crossbred goats known as baladi or Negev were in each treatment. "The animals were allowed to feed on the six fodders for 15 min after which time the refusals of each fodder species were recorded. This procedure was repeated for 5 consecutive days."

The *Atriplex* species had the highest levels of crude protein and ash. They had the lowest levels of gross energy and condensed tannins. The *Acacia* trees had high levels of *in vitro* metabolizable energy, condensed tannins, neutral detergent fibre, acid detergent fibre, and acid detergent lignin. *Cassia* had highest level of gross energy and lowest levels of *in vitro* metabolizable energy, crude protein, ash, and neutral detergent fibre. "The two *Acacia* species comprised more than... 86% of the DMI (dry matter intake) in ...goats ...."

Glenn and associates conducted a feeding trial in Kalba, Sharjah, UAE with 28 male and 27 female Damascus-type goats (Glenn et al. 1992). The test forage was *Salicornia bigelovii* grown near the shore of the Arabian Sea in Kalba and irrigated with seawater. For the feeding trials *Salicornia* straw (seeds removed) was utilized. The control forage was Rhodes grass (*Chloris gayana*). There were four treatments; namely: A. consisted of chopped Rhodes grass (Control); B. Chopped 50/50 mixture of chopped Rhodes grass and *Salicornia* straw; C. Chopped *Salicornia* washed with seawater to reduce salt content; D. Chopped Rhodes grass presented at a rate equal to the consumption of B. (The purpose of the latter was to provide a control for B. in which it was thought that the intake of B. might be restricted by the high salt content). A high protein concentrate (16.8% protein) was provided equally to all animals at a rate of 250 g/head/day, prepared by Sharjah Feed Mills.

Unwashed *Salicornia* hay had a NaCl content of approximately 30% which was reduced to 8–13% by washing with seawater and pressing. "Protein levels were approximately the same in washed and unwashed biomass, but crude fiber, neutral detergent fiber, acid detergent fiber and lignin increased on a percentage basis upon washing as a consequence of loss of cell soluble materials. ... which indicates that easily digested organic cell solubles (sugars, amino acids, soluble proteins) were removed as well as salt."

"Rhodes grass had nearly twice the protein content, much lower NaCl content, higher fiber content, but approximately the same metabolizable energy content as washed or unwashed S. *bigelovii* biomass."

The weight gains for the males was higher than the females in all treatments. There were no significant differences in weight gains of animals in all treatments when compared by sex. Animals in Diet C had the highest feed consumption and feed conversion ratio.

"...S. *bigelovii* biomass can be an effective forage material for goats in the Arabian Peninsula and similar coastal desert locations where the crop can be produced using saline water."

Ishikawa and colleagues conducted feeding trials in Ibaraki, Japan with 5 Japanese native castrated goats (Ishikawa et al. 2002). It was designed to determine whether Saltwort (*Salicornia herbacea*) could be a potential forage for ruminants in Japan. The control was alfalfa (*Medicago sativa*) (fed in cubes). The treatment feed was an alfalfa based diet supplemented by 15–20% of Saltwort. "Mean DM (dry matter) intake of the goats offered a mixed diet (alfalfa + saltwort) tended to be low compared with that of the animals offered a base diet...Goats consumed 70% of the Saltwort diets offered. Mean water consumption of the goats offered base and mixed diets was similar..."

"It was shown that the amount of digestible crude fiber, NDF and ADF contained in saltwort were fairly low as a result of the low digestibility of these components of the halophyte. On the other hand, it was demonstrated that the DCP (digestible crude protein) of the halophyte was relatively high."

"...we conclude that saltwort could be a suitable feeding resource for use as protein or mineral supplement for livestock reared around salinized areas where high quality feed is not available."

Riley and others completed a feeding trial with an unspecified breed of goats in 1988 in Kalba, Sharjah, UAE (Riley et al. 1994). (Sheep were included in the same trial, but only the goat trial results are included here.) The trials consisted of feeding mixtures of *Salicornia* hay with *Salicornia* seed, using Rhodes grass as forage control treatment. A protein supplement composed of yellow corn, dates, and wheat bran was added to all treatments.

Note: "The term "hay" is used to indicate dried vegetative material containing immature seeds. The term "straw" is used to describe dried vegetative material from which the seeds have been removed."

There were no significant differences determined due to the death of several animals from rinderpest and birthing. The average gain was 12.9 g/head/da.

Goats generally maintain good productivity when *Salicornia* forage was substituted for local forage up to a level of 100%.

Nawaz and colleagues conducted a feeding trial with 25 female dwarf (teddy) goats was conducted in Faisalabad, Pakistan (Nawaz et al. 1994). They were divided into five groups corresponding to 5 feeding regimes utilizing *Atriplex amnicola*, known as River Saltbush, and/or Sudex (Sorghum x Sudan grass hybrid). Sudex was substituted for by River Saltbush in increments of 25% from 0 to 100%. In general, the crude protein was higher and the ash content lower for the treatments with the most Sudex. The weight gain, feed intake and feed efficiency, were higher for treatments with more Sudex, while feed efficiency, and water intake were greater for higher levels of *Atriplex*. The lowest feed efficiency and digestibility were highest with 100% *Atriplex*. Carcass parameters and organoleptic evaluation of the meat were not significantly different. "Both hematological as well as biochemical values remained within the normal ranges... except for cholesterol and triglycerides-cholesterol was significantly decreased while triglycerides increased with *Atriplex* based regimes."

El-Rahman and associates carried out several feeding trials. Only the goat results are presented here (El-Rahman et al. 2014). The trials took place at the Nubaria Experimental Farm in Giza, Egypt.

Acacia saligna (Golden Wattle) and *Brassica nigra* (Black Mustard) hay (leaves and stems) were evaluated for their nutritive potential using 9 mature male Baladi goats. The animals were fed hay from Berseem (Clover: *Trifolium alexandrinum*) as the Control or *Acacia saligna* or *Brassica nigra* treatments. All were supplemented with Barley Grains which provided 50% of the animal maintenance requirement. The experimental treatments were offered *ad lib*. Golden Wattle had the highest level of Crude Protein, CP (13.5%), with Black Mustard (12.8%) and Berseem (11.8%) following. A. *saligna* had the highest content of anti-nutritional factors (saponins, alkaloids, flavonoids, tannins and courmarins). The average dry matter intake for the goats from *Acacia* or *Brassica* spp. was 20.8 g/kg/da compared to 32.0 g/kg/da for control treatment with Berseem. The weight gains during the trial were not reported.

"From these results, it is recommended that to improve the native range species as well as to cultivate some high potential shrubs, making hay of succulent parts of less palatable species with other feed ingredients will increase the quality and quantity of feed resources."

Abdou and others carried out feeding trials to "...investigate the effect of feeding goats on fresh *Atriplex nummularia*...when added with different sources of energy supplementation (concentrate feed mixture, CFM, ground barley grains or ground date stones..."

The trial was conducted at the Suder Research Station of the Desert Research Center in southern Sinai over a period of 105 days (Abdou et al. 2011).

The treatments were:

Treatment 1: Berseem hay (*Trifolium alexandrinum*) + concentrate feed mixture (CFM)
Treatment 2: *Atriplex nummularia* + Ground date stone (GDS)
Treatment 3: *Atriplex nummularia* + Barley grains (BG) (*Hordeum vulgare* L.)
Treatment 4: *Atriplex nummularia* + (50% barley grains + 50% ground date stone)
Twenty-eight black Desert goat kids were divided randomly into the four treatments.

Treatment 2 had the lowest average daily gain and digested crude protein as well as the highest total water intake. All other treatments were not significantly different from each other. The economic feed efficiency (ratio of price of total live body weight gain to the price of feed consumed to reach that gain) was highest for Treatment 2, followed by T4, T3 and T1.

The inclusion of economic feed efficiency should be made standard for all comparative feed experiments.

Tesfa and Mehari conducted a feeding trial at the Ghatelai Ram Farm, seventy kilometers North of Asmara, Eritrea at an elevation range of 280–450 meters above sea level (Tesfa and Mehari 2011). The trial included 80 goats and 80 sheep, with equal numbers of male and female animals. Animals were fed at 3% of body weight/day from 0 to 42 days and 3.5% body weight/day from 43 to 84 days. Potable water was provided *ad libitum*. The materials included in the diets were:

1. Halophyte: mixture of Salicornia (*Salicornia bigelovii*), mangroves and algae with 7.65% Crude Protein (CP)
2. Sorghum (*Sorghum bicolor*) husks. 8.05% CP
3. Shishay: a local commercially produced complete feed at the Shishay Animal Feed Processing Plant in Eritrea with 16.3% CP
4. Mineral Salts (unspecified)

The feeding treatments for sheep and goats was:

Treatment 1 100% Halophyte: (ash 21.6%) (CP 7.6%)
Treatment 2 30% Halophyte + 70% Sorghum husks: (ash 10.9%) (CP 8.2%)
Treatment 3 30% Shishay + 70% Sorghum husks: (ash 6.3%) (CP 10.6%)
Treatment 4 100% Sorghum husks + Mineral salts: (ash 6.8%) (8.1%)

The results for goat feeding is given in Table 11.10 below (results of sheep feeding are given with other sheep feeding trial results, earlier in the chapter):

Table 11.10. Summary of goat feeding trials fed on halophytic feedstuffs in Eritrea (Tesfa and Mehari 2011).

Treatment	Ave. Daily Gain (all goats) g/day	Dry Matter Intake g/head/day	Feed Conversion Rate Column 3/Column 2
1	53.0	590	11.1
2	26.3	640	24.0
3	40.9	560	13.7
4	44.9	750	16.7

Note: A high Feed Conversion Ratio (FCR) means that an animal must eat more of the diet to gain a unit weight.

The goats fed all halophyte diet (treatment 1) had the highest daily gain and the lowest FCR, indicating the goats had to ingest a minimal amount of the feed to reach the same rate of gain as other animals. The FCR was highest for treatment 2 fed 30% Halophyte + 70% Sorghum Husks. All the goats in Treatment 1 had the highest final body weight and those fed Treatment 2 the lowest. In all treatments the male animals were heavier than females. The results indicate the benefit of feeding a mixture of halophytes.

6. Interaction between halophytes and camels (*Camelus dromedaries*)

A tabular summary of camel trials and observations is given in Table 11.11.

El Shaer's excellent paper was summarized in the Review of Reviews section of this chapter, which includes a summary of plants that are considered palatable or not for camels (*Camelus dromedaries*) (El Shaer 2010).

El-Kebawy and others paper is summarized in the Review of Reviews in this chapter, which is more general than the prime focus of this chapter's section on camels consuming halophytes (El-Kebawy et al. 2009).

The paper has been authored by members of the Biology Department of the UAE University in Al-Ain, Abu Dhabi, UAE and the Terrestrial Environment Research Center, Environment Agency in Abu Dhabi, UAE. "Camel grazing affects over 90% of the Arabian Peninsula, of which 44% is severely or very severely degraded. ...the camel population density in the UAE is much higher (2.99 camels/km²), than other countries of the region, e.g., Saudi Arabia (0.12 camels/km²)...Camels generally graze on a broad spectrum of fodder plants, including thorny bushes, halophytes, and aromatic species..." which leads to a reduction in the population of many species. "...seven years after substituting small antelopes for camels in the AMR (Al Maha Resort), there was a marked recovery of perennial plants, including medium and large shrubs... most of the recovered plants are palatable, especially for camels."

While not specifically uniquely related to halophytes, the improved management of camel grazing would help improve stands of halophytes among other plants and perhaps open an opportunity for halophyte feeding trials with camels.

Table 11.11. Camel Feeding Trials Summary.

Author	Location	Type of Animals Popular Name	Type of Animals Scientific name	Halophytes fed Popular name	Halophytes fed Scientific name	Halophytes fed Parts fed	Irrigation water Salinity	Animal Drinking Water
El-Keblawy	Al-Ain, Abu Dhabi, UAE	Camels	Camelus dromedarius	Mixture range	unspecified	Mixture	Range fed	unspecified
Study of Camel habitat and affect of Small Antelope			Oryx leucoryx					
Haddi	Algerian Sahara	Camels	Camelus dromedarius	Soueda	Suaeda mollis	Browse	Range fed	potable
				Legtaf	Atriplex halimus	Browse	Range fed	potable
		Cows	Bos primigenius	Eldjel	Salsola vermiculata	Browse	Range fed	potable
				Ezzerita	Limoniastrum guyonianum	Browse	Range fed	potable
				Ghardes	Nitraria retusa	Browse	Range fed	potable
Laudadio	Southern Tunesian desert	Camels	Camelus dromedarius	Saltwort-leaved R.	Reaumuria vermiculata	Browse	Range fed	unspecified
				French Tamarix	Tamarix gallica	Browse	Range fed	unspecified
				Salt tree	Nitraria retusa	Browse	Range fed	unspecified
				?	Zygophyllum album	Browse	Range fed	unspecified
				Sea Orach	Atriplex halimus	Browse	Range fed	unspecified
				?	Salicornia arabica	Browse	Range fed	unspecified
				?	Salsola tetragona	Browse	Range fed	unspecified
				?	Salsola tetrandra	Browse	Range fed	unspecified
				?	Suaeda mollis	Browse	Range fed	unspecified
				Indian Walnut	Aeluropus littoralis	Browse	Range fed	unspecified
				Blady grass	Imperata cylindrical	Browse	Range fed	unspecified
				Field wormwood	Artemisai campestris	Browse	Range fed	unspecified
				Sea Heath	Frankenia thymifolia	Browse	Range fed	unspecified
				?	Limoniastrum guyonianum	Browse	Range fed	unspecified
Al-Qwaimer	Riyadh, Saudi Arabia	Camels	Camelus dromedarius	Salicornia	Salicornia bigelovii	Hay	42 ppt Arabian Gulf	potable
El Shaer	Mataria, Cairo, Egypt	REVIEW PAPER						

Haddi and others used slaughtered camel (*Camelus dromedaries*) rumen fluid and a fistulated cow (*Bos primigenius*) to access rumen to determine the *in vitro* fermentation of halophytes which occur in the Algerian Sahara (Haddi et al. 2003). See Table 11.12:

Table 11.12. The effect of stage of maturity on nutritive value of Algerian Sahara halophytic feedstuffs for camels (Haddi et al. 2003).

Halophytes analyzed		Early cuttings	Late cuttings
Scientific Name	Local Name	Highest readings	Highest readings
Suaeda mollis	Soueda	High in fiber (NDF and ADF) and CP	CP high
Atriplex halimus	Legtaf		High in fiber ADF and NDF + Lignin
Salsola vermiculata	Eldjel	Highest ash content ASHI	CP decreased with maturity High ash ASHI
Limoniastrum guyonianum	Ezzerita	Highest ADL	
Nitraria retusa	Ghardes	Hi in ash ASHI	Hi in ash ASHI
Medicago sativa (Control)	Alfalfa	Not reported	Not reported

Crude protein CP/Acid detergent fiber ADF/Acid detergent lignin ADL/Acid detergent insoluble ash ASHI/ Neutral detergent fiber (NDF).

The main difference between shrubs was the crude protein CP content. The dromedary inoculum produced higher fermentation per unit time and per unit of organic matter compared to the fistulated cow. *S. vermiculata* produced higher gas production with camel rumen and had the highest gas production in early season. In the late season, *A. halimus* in camel rumen had the highest gas production and *L. guyonianum* had the lowest gas production in camel rumen. In general, camel rumen is better adapted to ferment halophyte shrubs than cow rumen. "All shrubs had lower *in vitro* fermentation characteristics than *Medicago sativa* used as a standard hay".

Laudadio and associates report on a survey of the southern Tunisian desert to determine the key characteristics of halophytes typically eaten by camels (Laudadio et al. 2009). The survey was taken in spring. "… a survey was carried out to determined camels pasture quality, dietary preference and to characterize the chemical characteristics and nutritional value of different halophytes plants…"

Table 11.13 shows the range of concentration of selected constituents.

"…a better knowledge of the food consumed by camels in pasture would allow the formulation of suitable rationing plans for this species within the different physiologic situations and mainly would give to us the possibility to integrate conveniently the ration so that to reduce the impacts of eventual deficiencies that can represent a limiting factor to the production performance of these animals, which in North Africa constitute a fundamental subsidy to the human activity."

Table 11.13. Range of key characteristics of halophytes eaten by camels in Tunisia (Laudadio et al. 2009).

Surveyed plants and range of selected chemical composition	Dry Matter DM%	Crude Protein CP%	Ash %	Na	K	Ca
				Mg/100g DM		
Plants collected						
Reaumuria vermiculata						
Tamarix gallica			– 14.7			
Nitraria retusa						
Zygophyllum album	– 13.5					
Atriplex halimus					+1134	
Salicornia arabica						
Salsola tetragona				– 9		
Salsola tetrandra						
Suaeda mollis		+ 16.5		+ 9237		– 481
Aeluropus littoralis						
Imperata cylindrical		– 4.1				
Artemisia campestris						
Frankenia thymifolia	+ 40.1					
Limoniastrum guyonianum			+36.8		– 215	+ 7185

– Lowest   + Highest

Al-Qwaimer reports on a trial with eleven Najdi camels with two treatments (Al-Qwaimer 2000). The control diet was Rhodes grass and 5% wheat straw. Salicornia hay (plants with immature seed heads) replaced Rhodes grass in the second treatment diet. It is assumed that the *Salicornia* hay was grown on the Ras Al Zawr farm in Saudi Arabia on the Arabian Gulf where it was irrigated with water from the Gulf with a salinity of about 42 ppt. Some minor adjustments were made with other ingredients to attain an iso-nitrogenous diet with a crude protein level of 13.5% and 16.5% ash. The trial was conducted at King Saudi University in Riyadh, Saudi Arabia. The camels were offered the diets *ad libitum* for 120 days. The animals fasted for 24 hours before slaughter. Weights of body parts were recorded. There was no significant difference in the total or individual body weights of kidney, heart, and liver. The proximate chemical composition of the *M. longissimus* portions of the carcasses (percentage of moisture, protein, ash, and ether extract) were essentially the same for both diets, with the exception of Na, which was higher in the animals fed *Salicornia* hay (59.0% for camels fed Rhodes grass and 68.4% for those fed *Salicornia* hay (P < .05)).

7. Summary of trials feeding halophytes to cattle

A tabular summary of cattle feeding trials is given in Table 11.14.

Riasi and associates conducted cattle feeding trials. "The objective of this study was to evaluate four Iranian halophytic plants … for their chemical composition and *in situ*

Table 11.14. Cattle Feeding Trials Summary.

Author	Location	Type of Cattle	Type of Cattle Scientific name	Halophytes fed Popular name	Halophytes fed Scientific name	Halophytes fed Parts fed	Irrigation water Salinity	Sheep Drinking Water
Riasi 2008	Mashhad, Iran	Steers	Holstein	Summer cypress	*Kochia scoparia*	leaves and twigs	Range	potable
					Atriplex dimorphstegia	leaves and twigs	Range	potable
					Suaeda arcuate	leaves and twigs	Range	potable
					Gamanthus gamacarpus	leaves and seeds	Range	potable
Khan 2009								
	Balochistan, Pakistan	Calves	unspecified	Desert bunchgrass	*Panicum turgidum*	whole plants	EC 12mS/cm	EC 1.1 mS/cm
Nolte 2014	N. coast The Netherlands	unspecified	unspecified	Salt marshes	Salt marshes	Forage	Seawater/Rain	potable

ruminal and post ruminal disappearance of DM (dry matter) and CP (crude protein) (Riasi et al. 2008)." The halophytes collected in the central Iran desert were: *Kochia scoparia, Atriplex dimorphostegia, Suaeda arcuata,* and *Gamanthus gamacarpus.*

The feeding trials were conducted at Ferdowsi University of Mashhad, Mashhad, Iran. Four Holstein steers, 21–27 months of age with ruminal fistula and duodenal cannula were used. All were housed in concrete floor pens. Halophyte samples were introduced to the ventral sac of the rumen for 12 hours or into the small intestine at a rate of one bag per 30 minutes. *Gamanthus* had the highest DM dry matter and Ash content. *Kochia* had the highest CP crude protein content (P < 0.05). *Kochia* and *Suaeda* had the lowest DM; *Atriplex* had the lowest CP and *Kochia* had the lowest Ash content. *Gamanthus* had the highest effective degradability for DM and CP.

It was concluded that *Kochia* and *Atriplex* had the best potential for providing supplemental feed nutrients for cattle.

Khan and others conducted a trial in which Desert bunchgrass (*Panicum turgidum*) was planted on the Zia Model Farm in Hub Kund, Balochistan, Pakistan in plots between berms on saline soil (EC 10–15 mS/cm) and flood irrigated with brackish groundwater (EC 10–12 mS/cm) at rates to just keep the root zone moist (Khan et al. 2009). "For the purpose of diminishing soil salt content (*Suaeda fruticosa*), Seepweed was used as a salt scavenger intercrop. Suaeda seedlings (10–12 cm high) were planted 50 cm apart on bunds between plots of *P. turgidum* and irrigated as needed". When harvested the P. *turgidum* plants had 13% ash, 18% crude fiber, 25% carbohydrates, and 13% crude protein.

Four groups of four calves each were fed one of four diets twice daily. They were evaluated in two trials in succession each lasting two months. There were two control diets (neither with *P. turgidum*) one of these had no concentrates added. In the two treatment diets (with *P. turgidum*) one did not have Sweet corn or Maize (Zea mays). The feeds used in the trials included Wheat straw, Maize, *Panicum turgidum*, Wheat bran and Concentrates. All the diets were fed 12 kg/4 calves twice daily in the first trial and 16 kg/4 calves twice daily in the succeeding 3 month period. Fresh drinking water was provided as needed. The animals were slaughtered after 4 months.

The calves fed the control diet plus concentrate (without *P. turgidum*) had the greatest weight growth. The calves fed complete diet with maize (with *P. turgidum*) had the meat with the highest protein content. "... *P. turgidum*, ... has similar effectiveness to conventional feed such as maize for animal production."

Nolte and others conducted a study located in "Noord-Friesland Buitendjks" in the Netherlands (Nolte et al. 2014). It is a temperate salt march situated on the northern mainland coast. The site has typically been grazed by cattle and horses for the last 20 years. Treatment blocks were employed for each animal species. The height of the canopy in areas grazed by cattle was higher and had a lower range than the areas grazed by horses. Also, the mean range (patch size) was significantly lower for cattle. "... for increased vegetation-structure patchiness in salt marshes we would recommend the use of cattle rather than horses and low rather than high herbivore densities."

8. Conclusions

Most trials demonstrate that single halophytes, like other forages, are mostly insufficient to provide a total diet. A nutrient concentrate is almost always required for ruminants. Blending of feed sources shows the most promise, ideally with only halophytes, but including both halophytes and non-halophytes as required until investigations find the best combination. Potable water is the preferred source of drinking water for ruminants, so increasing animal production may require some additional sources of suitable water. However, production of halophytes can result in considerable saving of high quality water for crop production as they can be grown with brackish or saline water—even seawater. The latter reduction in potable water for irrigation is much greater than the increase in drinking water for ruminants, hence a significant saving of fresh water resources.

I would encourage that all future ruminant feeding trials include the economic feed efficiency (ratio of price of total live body weight gain to the price of feed consumed to reach that gain).

It should be noted that the seed from seed-bearing halophytes can be an important source of energy, protein, and nutrients for ruminant diets and thus reduce or eliminate the need for supplemental conventional feeds. A fully balanced halophyte diet for ruminants would not only be more economical, but also, greatly reduce the need for fresh water to irrigate or produce conventional dietary supplements.

Caution should be taken when introducing exotic halophytic species to new areas, lest they reduce or eliminate indigenous vegetation.

9. Acknowledgements

I wish to express my appreciation to the University of Arizona Libraries for their assistance in preparing this review. I wish to especially thank Ms. Jeanne L. Pfander and Ms. Elizabeth Kline, Associate Librarians, who reacquainted and guided me through the information management systems available in the libraries of The University of Arizona. Also, I wish to complement the staff who implement the Inter-Library Loan service. Having used this service, years ago, I was astounded at the speed with which publications from far away are made available in an easily used format.

References and Further Reading

Abbeddou, S., S. Rihawi, H.D. Hess, L. Iñiguez, A.C. Mayer and M. Kreuzer. 2011. Nutritional composition of lentil straw, vetch hay, olive leaves, and saltbush leaves and their digestibility as measured in fat-tailed sheep. Small Ruminant Research 96(2-3)(4): 126–35.

Abdou, A.R., E.Y. Eid, A.M. ElEssawy, A.M. Fayed, H.G. Helal and H.M. El Shaer. 2011. Effect of feeding different sources of energy on performance of goats fed saltbush in *Sinai*. The Journal of American Science 7(1): 1040,1050. 51 ref.

Abu-Zanat, M.M.W. and M.J. Tabbaa. 2006. Effect of feeding *Atriplex* browse to lactating ewes on milk yield and growth rate of their lambs. Small Ruminant Research 64(1-2)(7): 152–61.

Al Khalasi, S.S., O. Mahgoub, I.T. Kadim, W. Al-Marzouqi and S. Al-Rawahi. 2010. Health and performance of *Omani* sheep fed salt-tolerant sorghum (*Sorghum bicolor*) forage or *Rhodes* grass (*Chloris gayana*). Small Ruminant Research 91(1)(6): 93–102.

Al-Qwaimer, A.N. 2000. Effect of dietary halophyte *Salicornia bigelovii* Torr. on carcass characteristics, minerals, fatty acids and amino acids profile of camel meat. Journal of Applied Animal Research 18(2)(Dec): 185–92.

Al-Shorepy, S.A., G.A. Alhadrami and A.J. Al-Dakheel. 2010a. Growth performances and carcass characteristics of indigenous lambs fed halophyte *Sporobolus virginicus* grass hay. Asian-Australasian Journal of Animal Sciences 23(5)(May): 556–62.

Al-Shorepy, S.A., G.A. Alhadrami and A.I. El Awad. 2010b. Development of sheep and goat production system based on the use of salt-tolerant plants and marginal resources in the United Arab Emirates. Small Ruminant Research 91(1)(6): 39–46.

Ashour, N.I., M.S. Serag, A.K.A. El Haleem and B.B. Mekki. 1997. Forage production from three grass species under saline irrigation in Egypt. Journal of Arid Environments 37(2)(Oct 1997): 299–307.

Ben Salem, H., H.C. Norman, A. Nefzaoui, D.E. Mayberry, K.L. Pearce and D.K. Revell. 2010. Potential use of oldman saltbush (*Atriplex nummularia* Lindl.) in sheep and goat feeding. Small Ruminant Research 91(1)(6): 13–28.

Chadwick, M.A., P.E. Vercoe, I.H. Williams and D.K. Revell. 2009. Dietary exposure of pregnant ewes to salt dictates how their offspring respond to salt. Physiology & Behavior 97(3-4)(6/22): 437–45.

Degen, A. Allan, S. El-Meccawi and M. Kam. 2010. Cafeteria trials to determine relative preference of six desert trees and shrubs by sheep and goats. Livestock Science 132(1-3)(Aug): 19–25.

Digby, S.N., D. Blache, D.G. Masters and D.K. Revell. 2010. Responses to saline drinking water in offspring born to ewes fed high salt during pregnancy. Small Ruminant Research 91(1)(6): 87–92.

El Shaer, Hassan M. 2010. Halophytes and salt-tolerant plants as potential forage for ruminants in the Near East region. Small Ruminant Research 91(1)(Jun): 3–12.

El-Keblawy, A., T. Ksiksi and H. El Alqamy. 2009. Camel grazing affects species diversity and community structure in the deserts of the UAE. Journal of Arid Environments 73(3)(3): 347–54.

El Rahman, H.H.A., M.M. Shoukry, A.A. Abedo, Y.A. El Nomeary, F.M. Salman and M.I. Mohamed. 2014. Nutritional evaluation of some halophytic plants by range animals (sheep and goats). Global Veterinaria 13(1): 75,82. 32 ref.

Fahmy, A.A., K.M. Youssef and H.M. El Shaer. 2010. Intake and nutritive value of some salt-tolerant fodder grasses for sheep under saline conditions of south Sinai, Egypt. Small Ruminant Research 91(1)(6): 110–5.

Glenn, E., R. Tanner, S. Miyamoto, K. Fitzsimmons and J. Boyer. 1998. Water use, productivity and forage quality of the halophyte *Atriplex nummularia* grown on saline waste water in a desert environment. Journal of Arid Environments 38(1)(Jan): 45–62.

Glenn, Edward P., Wayne E. Coates, James J. Riley, Robert O. Kuehl and Roy S. Swingle. 1992. *Salicornia bigelovii* Torr.: A seawater-irrigated forage for goats. Animal Feed Science and Technology 40(1)(12): 21–30.

Gray, Erin C., Muir and Patricia S. 2013. Does *Kochia prostrata* spread from seeded sites? an evaluation from southwestern Idaho, USA. Rangeland Ecology & Management 66(2): 191–203.

Haddi, Mohamed-Laid, Stefano Filacorda, Khalid Meniai, Frédéric Rollin and Piero Susmel. 2003. *In vitro* fermentation kinetics of some halophyte shrubs sampled at three stages of maturity. Animal Feed Science and Technology 104(1-4)(2/20): 215–25.

Helal, A. and A.M. Fayed. 2013. Wool characteristics of sheep fed on halophyte plants ensiled by some biological treatments. Egyptian Journal of Sheep and Goat Sciences 8(1)(3–7 September, 2012): 131. 43 ref.

Ishikawa, N., K. Shimizu, T. Koizumi, T. Shimizu and O. Enishi. 2002. Nutrient value of saltwort (*Salicornia herbacea* L.) as feed for ruminants. Asian-Australasian Journal of Animal Sciences 15(7)(Jul): 998–1001.

Kafi, Mohammad, Hajar Asadi and Ali Ganjeali. 2010. Possible utilization of high-salinity waters and application of low amounts of water for production of the halophyte *Kochia scoparia* as alternative fodder in saline agroecosystems. Agricultural Water Management 97(1)(1): 139–47.

Karmiris, I., P.D. Platis, S. Kazantzidis and T.G. Papachristou. 2011. Diet selection by domestic and wild herbivore species in a coastal mediterranean wetland. Annales Zoologici Fennici 48(4): 233–42.

Khan, M. Ajmal, Raziuddin Ansari, Haibat Ali, Bilquees Gul and Brent L. Nielsen. 2009. *Panicum turgidum*, a potentially sustainable cattle feed alternative to maize for saline areas. Agriculture Ecosystems & Environment 129(4)(Feb): 542–6.

Khanum, S.A., H.N. Hussain, M. Hussain and M. Ishaq. 2010. Digestibility studies in sheep fed sorghum, sesbania and various grasses grown on medium saline lands. Small Ruminant Research 91(2010) 63–68.

Kraidees, M.S., M.A. Abouheif, M.Y. Al-Saiady, A. Tag-Eldin and H. Metwally. 1998. The effect of dietary inclusion of halophyte *Salicornia bigelovii* Torr. on growth performance and carcass characteristics of lambs. Animal Feed Science and Technology 76(1-2)(12/1): 149–59.

Kumara Mahipala, M.B.P., G.L. Krebs, P. McCafferty and L.H.P. Gunaratne. 2009. Chemical composition, biological effects of tannin and *in vitro* nutritive value of selected browse species grown in the West Australian mediterranean environment. Animal Feed Science and Technology 153: 203–215.

Laudadio, Vito, Vincenzo Tufarelli, Marco Dario, Mohamed Hammadi, Mabrouk Mouldi Seddik, Giovanni Michele Lacalandra and Cataldo Dario. 2009. A survey of chemical and nutritional characteristics of halophytes plants used by Camels in Southern Tunisia. Tropical Animal Health and Production 41(2)(Feb): 209–15.

Lymbery, Alan J., Gavin D. Kay, Robert G. Doupe, Gavin J. Partridge and Hayley C. Norman. 2013. The potential of a salt-tolerant plant (*Distichlis spicata* cv. NyPa forage) to treat effluent from inland saline aquaculture and provide livestock feed on salt-affected farmland. Science of the Total Environment 445(Feb 15): 192–201.

Masters, D., N. Edwards, M. Sillence, A. Avery, D. Revell, M. Friend, P. Sanford, G. Saul, C. Beverly and J. Young. 2006. The role of livestock in the management of dryland salinity. Australian Journal of Experimental Agriculture 46(6-7): 733–41.

Masters, D.G., S.E. Benes and H.C. Norman. 2007. Biosaline agriculture for forage and livestock production. Ecosystems and Environment 119: 234–48.

Masters, David, Mohamad Tiong, Philip Vercoe and Hayley Norman. 2010. The nutritive value of river saltbush (*Atriplex amnicola*) when grown in different concentrations of sodium chloride irrigation solution. Small Ruminant Research 91(1)(6): 56–62.

Mayberry, Dianne, David Masters and Philip Vercoe. 2010. Mineral metabolism of sheep fed saltbush or a formulated high-salt diet. Small Ruminant Research 91(1)(6): 81–6.

Nawaz, S., S.H. Hanjra and R.H. Qureshi. 1994. Effect of feeding *Atriplex amnicola* on growth and carcass quality of dwarf goats. Tasks for Vegetation Science. ed. V.R. Squires and A.T. Ayoub Vol. 32.

Nolte, Stefanie, Peter Esselink, Christian Smit and Jan P. Bakker. 2014. Herbivore species and density affect vegetation-structure patchiness in salt marshes. Agriculture, Ecosystems & Environment 185(0)(3/1): 41–7.

Norman, H.C., M.G. Wilmot, D.T. Thomas, E.G. Barrett-Lennard and D.G. Masters. 2010a. Sheep production, plant growth and nutritive value of a saltbush-based pasture system subject to rotational grazing or set stocking. Small Ruminant Research 91(1)(6): 103–9.

Norman, H.C., D.K. Revell, D.E. Mayberry, A.J. Rintoul, M.G. Wilmot and D.G. Masters. 2010b. Comparison of *in vivo* organic matter digestion of native Australian shrubs by sheep to *in vitro* and *in sacco* predictions. Small Ruminant Research 91(1)(6): 69–80.

Norman, Hayley C., David G. Masters and Edward G. Barrett-Lennard. 2013. Halophytes as forages in saline landscapes: Interactions between plant genotype and environment change their feeding value to ruminants. Environmental and Experimental Botany 92(Aug): 96–109.

Obeidat, Belal S., Abdullah Y. Abdullah and Fatima A. Al-Lataifeh. 2008. The effect of partial replacement of barley grains by *Prosopis juliflora* pods on growth performance, nutrient intake, digestibility, and carcass characteristics of Awassi lambs fed finishing diets. Animal Feed Science and Technology 146(1-2) (9/15): 42–54.

Pearce, K.L., H.C. Norman and D.L. Hopkins. 2010. The role of saltbush-based pasture systems for the production of high quality sheep and goat meat. Small Ruminant Research 91(1)(6): 29–38.

Pearce, K.L., H.C. Norman, M. Wilmot, A. Rintoul, D.W. Pethick and D.G. Masters. 2008. The effect of grazing saltbush with a barley supplement on the carcass and eating quality of sheepmeat. Meat Science 79(2)(6): 344–54.

Riasi, A., M. Danesh Mesgaran, M.D. Stern and M.J. Ruiz Moreno. 2008. Chemical composition, *in situ* ruminal degradability and post-ruminal disappearance of dry matter and crude protein from the halophytic plants *Kochia scoparia*, *Atriplex dimorphostegia*, *Suaeda arcuata* and *Gamanthus gamacarpus*. Animal Feed Science and Technology 141(3-4)(4/1): 209–19.

Riley, J.J., E.P. Glenn and C.U. Mota. 1994. Small ruminant feeding trials on the arabian peninsula with *Salicornia bigelovii* Torr. Tasks for Vegetation Science, ed. VR Ayoub Squires AT. Vol. 32.

Riley, J.J. and M. Abdal. 1993. Preliminary evaluation of salicornia production and utilization in Kuwait. Tasks for vegetation science., ed. H. AlMasoom Lieth AA. Vol. 28.

Riley, James J., Kevin M. Fitzsimmons and Edward P. Glenn. 1997. Halophyte irrigation: An overlooked strategy for management of membrane filtration concentrate. Desalination 110(3)(9): 197–211.

Rjiba Ktita, S., A. Chermiti and M. Mahouachi. 2010. The use of seaweeds (*Ruppia maritima* and *Chaetomorpha linum*) for lamb fattening during drought periods. Small Ruminant Research 91(1) (6): 116–9.

Sandu, G.R., Z. Aslam, M. Salim, A. Sattar, R.H. Qureshi, N. Ahmad and R.G. Wyn Jones. 1981. The effect of salinity on the yield and composition of *Diplachne fusca* (kallar grass). Plant, Cell & Environment 4(2): 177–81.

Slavich, P.G., K.S. Smith, S.D. Tyerman and G.R. Walker. 1999. Water use of grazed salt bush plantations with saline watertable. Agricultural Water Management 39(2-3)(2/25): 169–85.

Sun, H.X. and D.W. Zhou. 2010. Effect of dietary supplement of seed of a halophyte (*Suaeda glauca*) on feed and water intake, diet digestibility, animal performance and serum biochemistry in lambs. Livestock Science 128 (1-3)(3): 133–9.

Sun, H.X., D.W. Zhou, C.S. Zhao, M.L. Wang, R.Z. Zhong and H.W. Liu. 2011. Evaluation of yield and chemical composition of a halophyte (*Suaeda glauca*) and its feeding value for lambs. Grass & Forage Science 67(2)(06): 153–61.

Swingle, R.S., E.P. Glenn and V. Squires. 1996. Growth performance of lambs fed mixed diets containing halophyte ingredients. Animal Feed Science and Technology 63(1-4)(12/1): 137–48.

Tesfa, Kal'ab N. and Fithawi Mehari. 2011. Comparative feeding value of halophyte as alternative animal feed for small ruminants in Eritrea, pp. 253–260. *In*: Behnassi, M. Shahid, S.A. DSilva, J. (eds.). Sustainable Agricultural Development Chapter 19, 2011. Springer Science + Business Media B.V. 2011.

Toderich, K.N., E.V. Shuyskaya, S. Ismail, L.G. Gismatullina, T. Radjabov, B.B. Bekchanov and D.B. Aralova. 2009. Phytogenic resources of halophytes of Central Asia and their role for rehabilitation of sandy desert degraded rangelands. Land Degradation & Development 20(4): 386–96.

Vasta, V., A. Nudda, A. Cannas, M. Lanza and A. Priolo. 2008. Alternative feed resources and their effects on the quality of meat and milk from small ruminants. Animal Feed Science and Technology 147(1-3)(11/14): 223–46.

Waldron, B.L., J.S. Eun, D.R. ZoBell and K.C. Olson. 2010. Forage kochia (*Kochia prostrate*) for fall and winter grazing. Small Ruminant Research 91(2010): 47–55.

Walker, D.J., S. Lutts, M. Sanchez-Garcia and E. Correal. 2014. *Atriplex halimus* L.: Its biology and uses. Journal of Arid Environments 100(Jan–Feb 2014): 111–21.

Wang, C.J., S.P. Wang and H. Zhou. 2009. Influences of flavomycin, ropadiar, and saponin on nutrient digestibility, rumen fermentation, and methane emission from sheep. Animal Feed Science and Technology 148(2-4)(1/16): 157–66.

Watson, M.C., G.S. Banuelos, J.W. O'Leary and J.J. Riley. 1994. Trace element composition of *Atriplex* grown with saline drainage water. Agriculture, Ecosystems & Environment 48(2): 157–62.

12

Intake and Nutritive Value of Some Salt-Tolerant Fodder Grasses and Shrubs for Livestock: Selected Examples from Across the Globe

Victor R. Squires

SYNOPSIS

A number of salt tolerant grasses and shrubs have been evaluated and field-tested under high salinity regimes in Africa (both North and South), in Asia, especially Pakistan, and in the Middle East. Some work has also been reported from the Americas, Central Asia and from Australia. Several programs and centers have been established to study and evaluate halophytic and salt-tolerant plants as feed for livestock, including non-ruminants. Feeding halophytes is a feasible solution to minimize the problem of feed shortage in arid and semiarid areas. Often small holders use a 'cut and carry' approach to feeding their animals. Cut and carry can mean that newly harvested forages are fed fresh or that they are conserved through oven or air drying before feeding. Either process will result in some changes in nutritive value.

This chapter summarizes the outcome of numerous feeding trials, mainly under dryland conditions. It is complementary to the chapter on the results of trials under irrigation (including seawater irrigation, reported by Riley, this volume).

College of Grassland Science, Gansu agricultural University, Lanzhou.
Email: dryland1812@internode.on.net

Keywords: *Sporobolus, Distichlis, Pennisetum, Sorghum, Chloris, Leptochloa, Spartina, Acacia, Atriplex, Kochia, Prosopis, Cassia, Haloxylon, Medicago, Puccinellia, Panicum, Salicornia, Paspalidium, Paspalum, Suaeda, Leymus, Cynodon, Phoenix, Dichrostachys, Opuntia,* date seed, seaweed, cactus, voluntary feed intake, palatability, salinity, United Arab Emirates, Egypt., Australia, India, Pakistan, North Africa, Mexico, South America, South Africa, antinutritional, drinking water, cannulated sheep, liveweight loss.

1. Introduction

Feed intake may be influenced by a range of physical and chemical properties of the forage, including height, density and architecture of the forage, spatial distribution, sward structure, fiber content, resistance to degradation, digested particle size, energy density, protein content and the presence of toxins or unpalatable secondary compounds. Intake is also influenced by energy demand and therefore physiological state of the animal, as well as climate, disease and most important for livestock grazing halophytes, the availability and quality of the water supply (CSIRO 2007; Ulyatt 1973; Weston 2002). Voluntary feed intake is therefore influenced by many factors that also contribute to nutritive value, meaning that the two components of feeding value are interdependent (Masters, this volume).

2. Salt tolerant grasses and their potential for livestock feeds

2.1 *Sporobolus* sp.

Sporobolus grass (*Sporobolus virginicus*), is well known for its high salt tolerance and has been grown as irrigated forage under high salinity condition in the UAE (Alhadrami et al. 2003, 2010; Al-Dakheel et al. 2006; Al-Shorepy et al. 2010). *Sporobolus virginicus* was thoroughly evaluated and showed consistent yield potential and forage quality under high levels of salinity. Currently, this grass is grown as an irrigated fodder crops on a large scale and mechanically handled and harvested at the UAE University in Al Ain and at the International Center for Biosaline Agriculture (ICBA).

Several experiments were carried out at the UAE University on sheep and goats to investigate the effect of feeding *Sporobolus* and *Distichlis* grasses as well as a mixture of *Atriplex* shrubs (Table 12.1) on growth performance and carcass composition. Five experiments in the UAE aimed to evaluate the performance of sheep and goats fed these plants. Two breeds of sheep (local and Awassi breeds) and one local breed of goat were used. Animals were allocated to several dietary treatment groups to have either *Sporobolus* and *Distichlis* grass hay or a mixture of *Atriplex* shrubs as a source of forage in a replacement series with the conventional forage Rhodes grass *Chloris gayana* commonly used in the region. The data indicated that the productive performance or carcass characteristics of sheep were not influenced by the inclusion of different levels (up to 80%) of *Sporobolus* grass hay in the diets.

Table 12.1. Summary of the reported trials with salt tolerant and halophytic plants. *

Type of feedstuff	Main species fed	Livestock fed	Where?	% in diet	Principal results	Remarks	References
Salt tolerant grasses	*Sporobolus virginicus* / *Distichlis spicata*	Sheep (local and Awassi breeds) Local breed of goat	U.A.E.	Up to 80%	Productive performance or carcass characteristics of sheep/goats were not adversely influenced	Useful role for *Sporobolus* for feeding goats and sheep	Alhadrami et al. 2003, 2010; Al-Dakheel et al. 2006; Al-Shorepy et al. 2010
	Distichlis spicata, Spartina patens, Leptochloa fusca, Sporobolus virginicus		Suez Gulf, Egypt			Sea water irrigation at various dilutions	Roshdy et al. 1997
	Sorghum sudanense, Pennisetum americanum	n/a	South Sinai, Egypt			The focus here was on assessing nutritive value in the laboratory, rather than in feeding trials	Fahmy et al. 20210
	Distichlis spicata	Grazing cattle	Mexico	Up to 100%	Good for beef cattle	Great potential on saltland, especially with improved cultivars	Morales 1980; Llerena 1994; Alhadrami et al. 2010; Abou El Nasr et al. 2006; O'Connell and Young 2002
		Pen-fed goats/ Awassi sheep	U.A.E.		No adverse effect on local livestock		
		Grazing sheep	Australia	Fed *ad libitum*	Horses also do well		
	Distichlis palmeri	Grazing	Mexico, USA	Fed *ad libitum*	Good for beef cattle No adverse effect on local livestock	Great potential on saltland, especially with improved cultivars	Al-Shorepy et al. 2010; Bustan et al. 2005; Pearlstein et al. 2012; Yensen et al. 1985; Yensen 1988

Species	Animal	Country	Feeding	Conclusion	Notes	Reference
Aleuropus lagopoides, Sporobolus tremulus, Paspalum paspalodes, Paspalidium germinatum	n/a	Pakistan	n/a	With careful rationing all test species could be used as supplementary fodder for livestock	The focus was on assessing nutritive value in laboratory, rather than in feeding trials	Khan and Ansari 2008
Panicum turgidum	Calves, goats sheep	Pakistan	Cut and carry *ad libitum*	Calves, goats and sheep responded similarly in successful utilization of this grass in the feed	*Panicum turgidum* can be a major diet component	
Sorghum bicolor	Omani sheep/lambs	Oman	Pen fed, *ad libitum*	Sorghum forage grown under high salinity levels may be used for feeding Omani sheep without adverse effects on health or performance	Irrigated with diluted sea water. Sheep had average daily body weight gains of 68–96 g/day	Al Khalasi et al. 2010
Puccinellia ciliata	Merino Sheep	Australia	Grazed	*Puccinellia* provides a beneficial food option to sheep producers on saltland. Sheep can maintain liveweight and condition score for short periods	A winter active perennial grass that is well suited to saline and waterlogged land	Herrmann and Booth 1997; Edwards et al. 2002
Shrubs						
Pennisetum benthiumo	Buffalo (Nili-Ravi)	Pakistan	Cut and carry	*Pennisetum* is beneficial for buffalo calves	Substitute *Atriplex amnicola* for *Pennisetum* in diet	Bhatti et al. 2009
Atriplex amnicola	Buffalo	Pakistan	Cut and carry	*Atriplex amnicola* can replace *Pennisetum*	Substitute *Atriplex amnicola* for *Pennisetum* in diet	Bhatti et al. 2009

Table 12.1. contd....

Table 12.1. contd.

Type of feedstuff	Main species fed	Livestock fed	Where?	% in diet	Principal results	Remarks	References
	Atriplex halimus	Barki Lambs	Egypt	Pen fed up to 30% of diet	Lambs gained satisfactory growth, when fed *Atriplex* and *Acacia* shrubs supplemented with energy	Salt tolerant feedstuffs can fill 'feed gap'	Shawket et al. 2010
	Atriplex browse (mixed)	Awassi ewes	Jordan	Pen fed	Ewes receiving diets containing saltbush at 50 and 100% lost 0.7 and 4.5 kg in BW, respectively. No significant effect of the proportion of saltbush foliage in diet on milk production	There was no significant effect on birth weight, weaning weight and growth rate of lambs	
	Atriplex nummularia	Sheep, goats, camels	WANA region West Asia Australia South Africa Israel Egypt Chile	Pen fed Pen/field Pen/field Pen Pen Pen Pen	*Atriplex* replaced other roughages and supplements	284 g/day of *A. nummularia* increased the total dry matter intake of ration fed ad lib to Merino sheep by 200 g/day. Halophytes could become important feed resources at moderate inclusion levels if justified agronomically or where harvesting of naturally occurring stands is feasible	USAID 2006; Ben Salem et al. 2010; Weston et al. 1970; Warren et al. 1990; Wilson 1966b; Wilson 1994; Watson et al. 1987; Van der Baan et al. 2004; Du Toit et al. 2004; Abu-Zanat et al. 2006; Atiq-ur-Rehman et al. 1994; Alicata et al. 2002; Du Toit 2004; Abou El Nasr et al. 2006; Degen et al. 2000, 1995; Riley et al. 1994; Madrid et al. 1996; Rankins and Smith 1991

Shrubs					
Atriplex halimus, A. canescens and *A. nummularia*	Sheep/goats	South Africa	Pen fed	The focus here was on assessing nutritive value in the laboratory, rather than in feeding trial but sheep and goats were fed on various rations	The aim of studies was to evaluate the nutritional value of *Atriplex* spp. for small stock production	Van Niekerk et al. 2004a,b; Du Toit et al. 2004
Atriplex barclayana	Sheep/goats	Israel, Mexico, USA	Pen fed	*A. barclayana* had an apparent digestibility of 59% and 56% for dry matter and organic matter, respectively	High inclusion levels the high salt in the leaves and stem lowers its potential as a fodder plant. Gastrointestinal fill limits intake on low-energy diets and sheep cannot compensate for the high ash content by eating more forage	Benjamin et al. 1992; Swingle et al. 1996
Acacia saligna	Sheep/goats	Israel	Pen fed		Acacia saligna could not be used as a sole dietary source for small ruminants because of low intake and negative nitrogen balance. This was due mainly to the high tannin content	Degen et al. 1995, 2000

Table 12.1. contd....

Table 12.1. contd.

Type of feedstuff	Main species fed	Livestock fed	Where?	% in diet	Principal results	Remarks	References
	Salicornia bigelovii	Sheep/goats/ camels Najdi ram lambs	USA, Mexico, Saudi Arabia U.A.E. Egypt Kuwait	Pen fed	The inclusion of up to 30% of the stem and 10% of the spike in a mixed ration was satisfactory		Riley et al. 1994; Riley in this volume; Glenn et al. 1992; Swingle et al. 1996; Ben Salem et al. 2010; Attia et al. 1997; Kraidees et al. 1998
	Kochia spp.	Small ruminants	Pakistan Australia Central Asia	Pen fed	Mixed rations are the best way to use it	Overall, forage *Kochia* has the potential to improve the sustainability of small ruminant production in semiarid regions that frequently experience extended drought and saline conditions	Jami Al Ahmadi and Kafi 2008a; Raisia et al. 1988; Rankins and Smith 1991; Gul and Salehi 2014; Waldron et al. 2010
	Suaeda glauca	Ujmuqin ram lambs	China	Pen fed	Supplementation at 25 or 50% (*Suaeda* hay) significantly increased the digestible energy, DM, OM and ADF	*S. glauca* is a suitable had when added as 25% of the lamb diet	Sun et al. 2011
	Cassia sturtii	Sheep/goats	South Africa	Pen fed	*Cassia* sp. have also been successfully used in mixed rations in Egypt, Israel and South Africa		Van Niekerk et al. 2004a,b; Degen et al. 2010

Unconventional feedstuffs	*Atriplex halimus* and (*Phoenix dactylifera*) ground date pits	Camels, goats, dairy cows	U.A.E. Egypt Oman Algeria		An inclusion rate of 30% date pits gave the best weight gain in sheep fed *Atriplex halimus* hay and a concentrate	Date seeds constitute a good source of cellulose (energy) but low content of N. It is necessary to add protein source	Al-Owaimer et al. 2011
Unconventional feedstuffs	*Opuntia ficus-indica*	Barbarine sheep	Morocco		Combinations of *Opuntia* and *Atriplex* gave satisfactory outcomes		Nefzaoui 2000; Ben Salem et al. 2002
	Ruppia maritima and *Chaetomorpha linum* (seaweeds)	Barbarine sheep	Tunisia	Pen fed	Seaweeds evaluated in this study (i.e., *Ruppia maritima* and *Chaetomorpha linum*) could be used as alternative feed resources for growing lambs during drought periods	The aim was to evaluate the nutritional value for small stock production	Rijiba et al. 2010

* Note: See more detailed description of the work reported here in the main text and also in Riley, this volume and Attia-Ismail, this volume.

2.2 Evaluation of forage quality among salt tolerant grasses in Egypt

Field trials (Roshdy et al. 1997) were conducted at the Saline Agriculture Experimental Station on the Suez Gulf on sandy soil using irrigation with 12.5, 25.0, 37.5 and 50% seawater applied weekly in the hot seasons and bi-weekly in the cooler seasons. The principal purpose was to assess tolerance of these four grass species *Leptochloa fusca* (local), *Spartina patens*, *Sporobolus virginicus* (Smyrna-smooth) and *S. virginicus* (Dixie-coarse) (exotic) to saline water irrigation. *Medicago sativa* (Alfalfa) was used as a control or to contrast with halophyte species. *Leptochloa fusca*, *Spartina patens* and *Sporobolus virginicus* (Smyrna) appeared to be promising halophytic plants for feeding goats and sheep.

In a study reported by Fahmy et al. (2010) three annual salt-tolerant grasses: Sudan grass (*Sorghum Sudanense*), Pearl Millet grass (*Pennisetum americanum*), and a hybrid Sorghum grass (*Sorghum bicolor* x *S. halapense*) were cultivated on salt-affected soil on the South Sinai Research Station, Egypt. The focus here was on assessing nutritive value in the laboratory, rather than in feeding trials (Table 12.1).

2.3 Distichlis spp. a truly salt tolerant grass as a feed source

Distichlis spp. are found in saline areas, brackish marshes, and in salt flats along the coasts of the Atlantic and Pacific Oceans, the Gulf of Mexico and the along the coast of South America. *Distichlis* spp. inhabits upper/high marsh (irregularly flooded) areas, in which the water levels vary between 8 cm above the soil surface and 15 cm below the soil surface. It is one of the most drought-tolerant species and grows on both organic alkaline and in saline soils. *Distichlis* spp. are grazed by both cattle and horses and it has a forage value of fair to good because it remains green when most other grasses are dry during the drought periods and it is resistant to grazing and trampling. It is cropped both when green and in the dry state; however, it is most commonly used the winter for livestock feed as the primary source of hay. The forage (leaf biomass) component has high nutritional value in terms of protein content and low NaCl content.

Most attention has been given to *D. spicata* and it has been successfully used in Mexico, USA, Egypt, the Gulf region and Australia (see Table 12.1). In a study in the UAE (Alhadrami et al. 2010) the inclusion of *Distichlis* grass hay as the only source of forage up to 100% in the diet did not have any adverse effect on growth performance or carcass composition of growing local goat kids or local Awassi sheep. *Distichlis spicata* was thoroughly evaluated in the UAE and showed consistent yield potential and forage quality under high levels of salinity. Currently, this grass is grown under irrigation on a large scale and mechanically handled and harvested at the UAE University and at ICBA.

In Mexico (Morales 1980; Llerena 1994) *Distichlis spicata*, a native grass, has been used successfully as pasture for salinized areas. Beef cattle, predominantly, were raised on *Distichlis* pastures but sheep and horses have been successfully raised over many thousands of hectares, supported by the great expansion of 'salt rangelands' where this salt tolerant grass has demonstrated its relatively high performance as a forage on difficult soils.

In Australia, an improved cultivar of *Distichlis spicata* (NyPar©) has been used extensively in salt affected pasture lands in southern Australia. It is apparent that '*Distichlis*' is palatable, productive and nutritious for sheep and cattle. *Distichlis spicata* is highly drought resistant and tolerates very heavy grazing. This is particularly significant in winter-rainfall in Southern Australia because the feed is available in summer and autumn when no other source of green feed is available without irrigation.

Distichlis spicata can develop tough stems that are difficult for livestock to digest, but if cut regularly the plants can be managed to produce good-quality forage (Al-Shorepy et al. 2010; Bustan et al. 2005). The results of plant tissue analysis shows a range of crude protein concentrations from 5.7% to 17.2% with a digestibility range between 45.6% and 61% and metabolizable energy of between 6.2 Mj/Kg and 7.5. Production with minimal fertiliser application was estimated at between 12 and 27 tonnes per ha green matter with a dry matter content of between 53.6 and 71.3%. Production from one four ha site fertilised with NPK at about 150 kg per ha was estimated at 25 tonnes per ha green matter. Both protein concentration and digestibility decreases with the age of the material analysed. These results do not adequately measure the potential productivity range of the plants in different conditions of plant nutrition.

Feeding to ruminants was undertaken as a series of grazing trials in Australia to enable observations to be made of palatability, health of the plants and resistance of the plants to grazing pressure and drought. An average of 30 sheep were grazed in 1999 over two 30-day periods on a plot of about two ha. The sheep also had access to nearby marine barley grass (*Hordeum marinum*) on a further five ha. NyPar© grass was more palatable than the barley grass. Cattle grazed *Distichlis* in preference to other temperate pasture such as *Phalaris* spp. To make feeding observations on a larger scale, about 20 ha were planted to *Distichlis* and about 300 sheep were grazed on approximately seven ha during the summer drought of 2000/2001 for about three months. Some reports allude to damage to the mouths of grazing livestock so at the conclusion of the grazing period, the sheep were mustered and the mouths were examined. The sheep did not develop mouth lesions. Appropriate management is important to obtain production from NyPar©. While production is not high in conditions of low soil nutrient status, it improves markedly in conditions of good fertility. Livestock prefer leaf material but will graze all of the plant when no other green material is available. All farmers commented favourably on *D. spicata* as summer 'a green bridge' between late spring and winter. The economic importance of halophyte green feed available in the summer to the whole farm enterprise has been estimated by O'Connell and Young (2002) for Western Australian conditions.

The perennial (*Distichlis palmeri*) is endemic to northern Gulf of California tidal marshes flooded with hypersaline (3.8%) seawater. The forage (leaf biomass) component has high nutritional value in terms of protein content and low NaCl content. Biomass yield and forage quality of *D. palmeri* were similar to values for alfalfa (National Research Council 1986). In a study by Pearlstein et al. (2012) it was reported that the total dry matter yields were 9081 kg/ha with a total protein production of 1300 kg/ha. Shoots were low in ash and sodium, and compared favorably to alfalfa forage in protein, digestible carbohydrates and energy contents. Biomass yield and forage quality of *D. palmeri* were similar to values for alfalfa (National Research

Council 1986). *D. palmeri* would have obvious immediate value in those districts where soil conditions are too saline for rice, or where only brackish drainage water is available for irrigation. Other applications for a halophyte grain and forage crop such as *D. palmeri* could be developed for secondary or primary salinized pasturelands in irrigation districts, and coastal deserts around the world. *D. palmeri* appears to be worth developing as a perennial grain (Pearlstein et al. 2012) and forage crop, especially for salinized, flooded soils. Under favorable soil and moisture conditions, studies have shown that both *Distichlis spicata* and *D. palmeri* are suitable for pastures irrigated with saline water (Pearlstein et al. 2012; Alhadrami et al. 2010).

2.4 Evaluation of forage quality among salt tolerant grasses in Pakistan

Considerable variation in biomass accumulation has been reported in salt tolerant forage grasses in controlled laboratory experiments and this variation could be species specific or an adaptive response to habitat conditions (Arzani et al. 2006; Masters et al. 2007; Yayneshet et al. 2008). Species which thrive in saline conditions have a competitive advantage over others in their natural habitats. In this study in Pakistan, four grasses (*Aeluropus lagopoides* and *Sporobolus tremulus*, *Paspalum paspaloides* and *Paspalidium geminatum*) were evaluated for biomass production, mineral composition and forage quality. Forage quality parameters included neutral detergent fiber (NDF), acid detergent fiber (ADF), crude protein (CP), dry matter digestibility (DMD), and metabolizable energy (ME). Some species had higher ADF and crude protein values but DMD were highest in *Paspalum paspaloides* followed by *Sporobolus tremulus*, *Paspalidium geminatum* and *Aeluropus lagopoides*. Estimated metabolizable energy (ME) was highest in *Paspalum paspaloides* with similar values in other test species. Sulfur content was also assessed because high levels (> 0.4%) S could lead to loss of appetite and increased sulphide production by ruminant microorganism (Bird 1972; Kandylis 1984). Sulfur may also interact with Mo in rumen to reduce Cu availability (Suttle 1991) causing anemia, fragile bones and reproductive disorders. *Sporobolus tremulus* had the highest sulphur (1.42%) while the other three species had considerably lower values (< 0.45) which are within acceptable fodder limits for ruminants. Inland grasses (particularly *Paspalum paspaloides*) appeared to be better forage species producing higher biomass, DMD, ME and crude protein and low ADF and S than the coastal ones. However, with careful rationing all test species could be used as supplementary fodder for livestock.

2.5 Panicum spp. for cattle feed in Pakistan

A search in Pakistan among salt tolerant plant species to find suitable replacement fodder for calves has been successful in identifying the potential of a local perennial grass, *Panicum turgidum* as a feed source for cattle (Khan et al. 2009). *Panicum turgidum* is a perennial grass distributed in salt-affected areas and deserts and is a salt excluder with biomass yields of about 60,000 kg/ha/year (fresh weight) when grown in saline soil (EC 10–15 mS cm^{-1}) irrigated with brackish water

(EC 10–12 mS cm^{-1}). As a grass, it has high potential for use as an animal feed. Plants attain a height of about 1 m in 25–30 days in summer (maximum temperatures generally between 30 and 35°C). During the winter months of December to February (minimum 15–18 8°C) it requires 35–40 days for comparable growth.

Panicum turgidum does not contain high salt, excess harmful organic constituents or secondary metabolites in the harvested foliage. The ash is mostly composed of sodium and chloride, but the amount (13%) is not prohibitive for consumption, and may alleviate some of the need for salt licks typically provided to cattle in the area. For comparison, the ash content of several other local obligate halophytes ranges from 34% to 60% (Karimi and Ungar 1986). Oxalates are present in much lower levels (4–14%) than in many other local halophytes (ranging from 14% to 29% in various species), and the water soluble form can be easily metabolized by the animals. In addition to high salts, some halophytic fodders contain undesirable organic compounds. The advantages of *P. turgidum* as a fodder crop was exhibited when it was used as a complete replacement for maize in a cattle feeding trial and resulted in equivalent growth and meat production. There was no difference between the water requirement of the animals fed a diet with no *Panicum* or 100% *Panicum*. Data indicates that *P. turgidum*, despite being grown with brackish water irrigation, has similar effectiveness to conventional feed such as maize for animal production.

Cattle generally will not eat low quality salty fodder if more palatable feed is available (Wilson et al. 1995; Weston 1996; Wilson 1966b) and this is perhaps the reason behind the exclusive use of goats and sheep (considered more salt tolerant) as test animals in most halophyte feeding trials (Masters et al. 2005; Morecombe et al. 1996). *Panicum turgidum* was harvested and fed to the animals on a scheduled and controlled basis. The results from this study indicate that substituting *Panicum* for the maize component completely in the diet results in equivalent cattle growth and meat production, without the need of low salinity soil for maize production, thus freeing up non-saline soils for other essential crops. There was no difference in yield or taste of meat from the animals fed each diet and the venture has the potential to be economically viable. There is progress to find suitable replacements of other components of the diet and a source for protein concentrate which could be produced using the same saline resources.

Panicum turgidum is a perennial and can be continually harvested to about 60,000 kg/ha/year without reseeding, saving considerable time and resources. Eating high salt diet accentuates animal thirst but chemical analysis of the leaf tissue of *Panicum* indicates low (13%) ash content, which is much lower than that for salt accumulators (34–60%). Harmful secondary metabolites such as alkaloids and oxalate (4% in *Panicum* compared to 14–29% in other halophytes) are present in much lower amounts in foliage compared to many halophytes and therefore are little threat to feeding animals. Animals readily eat *P. turgidum* as a main feed component in place of maize, without any significant effect on animal thirst or growth and meat production. It is often difficult to convince farmers to change their preferences for growing feed. In this case care has been taken to involve local farmers in the decision making process and therefore they have readily accepted *Panicum* as a fodder crop, which had not been previously cultivated. Current trials with goats and sheep indicate a similar pattern of successful utilization of this grass in the feed. *Panicum turgidum* appears to be a high-quality fodder grass for the coastal area of Pakistan, and should be suitable for other sub-tropical regions of the world. It can grow from coastal dunes to inland regions. Its salt tolerance may vary from one location to another due to variability in local populations. The use of *Panicum turgidum* as a major component of cattle feed may be expected to generate a profit, especially when expanded to feed more animals. There is also a greater potential economic impact due to the utilization of poor quality water and saline land resources which are generally considered as waste and otherwise would not be utilized.

2.6 Sorghum as a feed source for sheep in Oman

Sorghum bicolor forage grown under high salinity levels may be used for feeding Omani sheep (Al Khalasi et al. 2010). *Sorghum bicolor* variety Super Dan was planted and irrigated with water containing three different concentrations of salt: 3, 6 and 9 dS/m. The sorghum fodder was termed low salinity sorghum (LSS), medium salinity sorghum (MSS) and high salinity sorghum (HSS). The sorghum was manually harvested, dried and chopped before feeding. Thirty-two, 3-month-old Omani male lambs were used. At the end of the trial the animals were slaughtered. The RGH had higher mineral content than sorghum forage grown under various levels of salinity. Animals fed sorghum-based diets did not show any signs of ill health. There were no differences ($P > 0.05$) in digestibility coefficients of acid detergent fiber, neutral detergent fiber, and ether extract between treatments. This study indicated that sorghum forage grown under high salinity levels may be used for feeding Omani sheep without adverse effects on health or performance.

2.7 Puccinellia ciliata a grass for saline lands

Puccinellia (*Puccinellia ciliata*) is a winter active perennial grass that is well suited to saline and waterlogged land, growing well in areas typically dominated with sea barley grass (*Hordeum marinum*) and where the soil has become bare (Herrmann and Booth 1997). Along with *Distichlis* (see above) and marine couch, it is the most salt-tolerant of the commercially available grasses and the only salt-tolerant grass suitable for highly saline scalds. Although a perennial species, it behaves like an annual by drying off back to the base and going dormant when the surface soil dries out in summer. It is best suited to areas with more than 400 mm annual rainfall and where the watertable is not too deep over summer. *Puccinellia* is highly palatable and has a low salt concentration in the leaves. It forms tussocks up to 40 cm high and wide and has long, thin leaves. Its growing points are embedded in the base of the plant, which is compact and resistant to grazing. *Puccinellia* will shoot with the onset of cooler late autumn temperatures and respond to dew, before the opening rains.

The plants grow from mid autumn to spring and mature (hay off) in November/ December in southern Australia, remaining dormant over summer to early autumn. It has its highest grazing value in winter and spring whilst green and before flowering. Nutritive value declines as the plant flowers, matures and senesces, and further declines through summer and autumn even though it is still palatable. It changes from a high quality, highly digestible feed capable of supporting high animal liveweight gains in spring to less than a maintenance ration in late summer/autumn. Mature stands can be grazed after the opening rains in April-May (when they rapidly produce green feed) and/or more commonly as dry feed in late summer-autumn, although at this stage some supplementation will be needed unless weight loss in the animals is acceptable. Leaving the feed standing over summer shades the soil, reducing the concentration of salts at the soil surface through-evaporation.

Puccinellia provides a beneficial food option to sheep producers on saltland, in most instances providing a pasture sward free of awned grass seeds. In the Upper South-East of South Australia *Puccinellia* is usually grazed in late spring as a means of avoiding barley grass (*Hordeum leporinum, H. marinum*) seed problem in livestock or in autumn to fill the usual feed gap and thereby reduce the reliance on supplementary feeding. Anecdotal evidence suggests that *Puccinellia* is excellent stock feed (Herrmann and Booth 1997). Data from a preliminary study of animal production for sheep grazing *Puccinellia* on moderately saline soil in the Mt. Charles area of South Australia (ECe: 10–25 dSm^{-1}) has been reported (Edwards et al. 2002). These results demonstrate that sheep can graze a senescent *Puccinellia*-based pasture for a short period in late summer to maintain liveweight and condition score but will lose liveweight and condition if left grazing this material for too long.

3. Shrubs

There has been an increasing awareness of the value of shrubs in forage production and for rehabilitation of depleted rangelands. Among the wide range of multipurpose fodder trees and shrubs, oldman saltbush (*Atriplex nummularia*) has received increasing interest as livestock forage and valuable revegetation species on marginal saline lands, especially in arid zones of Australia and in the West Asia and North Africa (WANA) region. Adapted to drought and water and soil salinity, oldman saltbush produces important consumable biomass in areas where other crops cannot grow. To cope with these harsh conditions, this species accumulates high levels of salt and oxalates on its leaves rendering them less palatable and decreasing their nutritive value. Despite this, satisfactory performance of small ruminants fed on *A. nummularia* has been reported in numerous research studies (Le Houerou 1992). *Atriplex halimus*, a North American *Atriplex*, is widely used in the western Asia North Africa (WANA) region, and throughout southwest and Central Asia. Naturally enough, livestock producers use it as fodder or in mixed rations. The feed value of other, lesser known, *Atriplex* as well as some species of *Acacia* and *Kochia* have also been assessed.

Feeding halophytes is a feasible solution to minimize the problem of feed shortage in arid and semiarid areas of countries in the WANA region, like Egypt, in the Gulf and along the Arabian Peninsular. Utilization of such halophytic plants supplemented with non-conventional energy supplements could be recommended to enhance feed materials availability all-round year and to improve animal performance as well under arid and saline conditions of Sinai.

3.1 Atriplex spp. as feed for livestock

3.1.1 Atriplex amnicola as a feed supplement for buffalo in Pakistan

Mott grass (*Pennisetum benthiumo*) is an improved cultivar of Dwarf Napier Grass or Elephant Grass that was introduced in Pakistan in 1987 from USA. Mott grass is a common fodder species in Pakistan. A feeding management experiment (Bhatti et al. 2009) was conducted to determine the effect of substituting *Atriplex amnicola* for *Pennisetum benthiumo* and berseem clover (*Trifolium alexandrinum*) on the performance of Nili-Ravi buffalo heifers. Fifteen buffalo heifers of 8 months age and 120 kg average initial body weight were divided into five equal groups and fed on diets in which Atriplex and other non-conventional forages were fed at five levels/ combinations. Daily water intake was significantly higher (*P* < 0.01) in heifers on saltbush substituted diets. Highest daily weight gain was observed on Berseem alone and on saltbush combination substituted diets.

3.1.2 Impact of feeding Atriplex halimus and Acacia saligna with different sources of energy on lambs performance

A study (Shawket et al. 2010) was conducted in Egypt to evaluate *Atriplex* (saltbush) and *Acacia* supplemented with crushed barley grains (CBG) alone or with crushed date seeds (CDS) for fattening growing Barki lambs. This study indicate that lambs could gain satisfactory growth under harsh condition in arid and semi-arid areas of marginal lands, when fed *Atriplex* and *Acacia* shrubs supplemented with CBG and/ or CDS at ratio 1:1 of MER as fodder sources.

3.1.3 Effect of feeding Atriplex browse to lactating ewes

The study by Abu-Zanat and Tabbaa (2006) was conducted to determine the effect of feeding a diet containing saltbush (*Atriplex*) on the milk yield of Awassi ewes and BW gain of their lambs. The dietary treatments had a significant (*P* < 0.01) effect on the overall DMI of ewes. The ewes receiving diets containing the saltbush increased feed intake significantly compared with intake levels of control animals fed with concentrate ration and tibin (shredded barley straws). The treatments had a significant (*P* < 0.01) effect on DMI of ewes during the different physiological stages: late gestation, suckling and after weaning. The dietary treatments had no significant effect either on the final BW or on the overall BW changes of ewes (the difference between initial and final BW). The ewes receiving the diets containing saltbush at 50% and 100% lost 0.7 and 4.5 kg in BW, respectively. There was no significant effect of the proportion of saltbush foliage in diet on milk production. The treatments had no significant effect on birth weight, weaning weight and growth rate of lambs.

Feeding *Atriplex* browse with concentrate ration to lactating Awassi ewes did not present any significant problems on milk yield or growth rate of lambs; the most important sources of income for stockmen. Substituting saltbush browse for tibin in diets of sheep will reduce the cost of roughage component by US$ 0.22 per head.

3.1.4 Atriplex nummularia in sheep and goat feeding

Among the wide range of multipurpose fodder trees and shrubs, oldman saltbush (*Atriplex nummularia*) has received increasing interest as livestock forage. Satisfactory performance of small ruminants fed on *A. nummularia* has been reported in numerous research studies (Norman et al., this volume).

Plant leaves have acceptable dry matter and organic matter digestibilities, as well as fairly low NDF. Intake and NDF digestibility of *A. nummularia* increased when barley or maize was used as an energy source supplementation. However, the tendency of the energy sources to increase NDF digestibility diminished when the supplemental level was raised from 15% to 30% and from 30% to 45%. Results suggested that barley and maize supplemented at a level of 15% gave the highest incremental increase in DM and NDF digestibilities.

Comparison of chemical composition of *Atriplex nummularia* grown under South African conditions with regard to site, species and plant parts was reported by Van der Baan et al. (2004). This study determined seasonal changes in chemical composition of leaves and stems and difference in quality between localities in semi-arid areas. The IVOMD of *A. nummularia* did not show significant differences compared to the other species with respect to seasonal trends.

In another study (USAID 2006) chemical composition of several *Atriplex* spp. (and ecotypes) grown under South African conditions with regard to site, species and plant parts was reported. *Atriplex nummularia* and *A. canescens* were compared. This study determined seasonal changes in chemical composition of leaves and stems of different *Atriplex* spp. and difference in quality between localities in semi-arid areas. CP concentration of *A. nummularia* was significantly higher than the other species in both seasons.

Other studies (Weston et al. 1970; Warren et al. 1990; Wilson 1966a; Watson et al. 1987) have elaborated the chemical composition, *in vitro* digestibility and other measures of nutritive value. Field studies based on actual diet selected by grazing animals that had esophageal fistulas installed, formed the basis of the work of Wilson et al. (1966). Ben Salem et al. (2010) did a thorough review of the literature on fodder potential of oldman saltbush and highlighted the main constraints and opportunities to make better use of this shrub for feeding sheep and goat under different production systems.

In a study reported from Chile by Meneses et al. (2012) *Atriplex nummularia* replaced alfalfa hay (*Medicago sativa*) in a proportion of 0%, 10%, 20%, 30%, and 40% for 60 creole goat kids weighing 13 k. Over 20% of *A. nummularia* hay content in the diet caused an increase in mineral consumption ($P < 0.05$). Total body weight gain decreased ($P < 0.05$) by adding 20% or more *A. nummularia* hay content in the diet. Incorporating 20% of *A. nummularia* into the diet did not produce a negative effect but higher percentages decreased weight gains attributable to the high Na and Cl content.

3.1.5 Suaeda glauca and its feeding value for lambs

In a trial at Changling, China (Sun et al. 2011) 18 Ujmuqin ram lambs were divided into three treatments of feeding *Suaeda glauca* hay. Supplementation at 25 or 50% (*Suaeda* hay) significantly increased the digestible energy, DM, OM and ADF compared with the treatment without *Suaeda* hay. Water intake, urinary excretion, water content of fresh faeces and sodium excretion from urine linearly occurred with increasing *S. glauca* content, but potassium excretion from urine decreased. This study indicates that *S. glauca* is suitable when added as 25% of the lamb diet.

3.1.6 Atriplex and Acacia shrubs as feedstuffs in mixed rations for ruminants

Fodder and forages/browse from salt-tolerant and halophytic species have important roles in many livestock production systems (see below). Incorporation into mixed diets can minimize potential adverse effects of high anti-nutritional components (Attia-Ismail, this volume) high salt, low energy concentrations and would likely yield higher economic returns than would be possible from direct grazing of these resources. Animal feeding studies demonstrate that organic constituents of halophytes are highly digestible (see below) and that diets containing moderate levels of halophytic ingredients are readily consumed by livestock. The problems of feeding halophytic feedstuffs seem to be largely overcome by feeding them as a supplement or as part of mixed rations (Swingle et al. 1995).

The effects of secondary compounds (Attia-Ismail, this volume) are diluted in mixed diets and livestock performance then reflects the true nutritional value of the combined ration. Halophyte forages have produced less positive results when included as the sole dietary component or with just an energy supplement. In an *in vivo* digestibility trial with sheep, *Atriplex barclayana* had an apparent digestibility of 59% and 56% for dry matter and organic matter, respectively (Benjamin et al. 1992). It was concluded that at high inclusion levels the high salt concentrations in the leaves and stem markedly lowers its potential as a fodder plant. The inclusion rate of the halophyte forage was well in excess of those likely to be used in mixed rations for fattening livestock and underestimates the potential value of halophytes at lower inclusion rates, as gastrointestinal fill limits intake on low-energy diets and sheep cannot compensate for the high ash content by eating more forage.

Swingle et al. (1995, 1996) have fed mixed rations containing several *Atriplex* and *Salicornia* components to sheep with favorable outcomes. The growth rates of lambs fed on diets containing halophyte components were assessed in two trials of 84 days duration each, from weaning to slaughter weight. Three halophyte forages, *Atriplex barclayana*, *Suaeda esteroa* and *Salicornia bigelovii* straw, were compared with *Cynodon dactylon* hay at 30% of the diet. Halophyte forages were much higher in mineral content than *Cynodon* hay (24–34% vs. 5%). The trials also compared *Salicornia* seed meal with cottonseed meal at 10% inclusion. All diets were high in concentrate (70%) and contained 12.5–15% protein. Dry matter intake was higher for lambs fed diets containing halophyte forages than for lambs fed on the grass control diet. Because of the increased intake, halophyte-fed lambs were able to gain at the same rate as the control lambs, but, as expected, feed efficiency was lower and water intake was higher. Carcass merit of all lambs was excellent and was not affected by the inclusion of halophyte forages in the diet. In one trial an additional control treatment was included in which *Cynodon* hay was supplemented with NaCl to attempt to isolate the effect of excess salt on feed intake and growth. However, the simulated halophyte diet (*Cynodon* + NaCl) supported lower weight gains and intake rates than the control or natural halophyte diets, showing that the form in which salts are present in halophyte forages is important to their acceptability to animals.

Other feeding trials involved penned animals (Abu-Zanat et al. 2006; Atiq-ur-Rehman et al. 1994; Alicata et al. 2002; Du Toit 2004) and different supplements (Abou El Nasr et al. 2006; Degen et al. 2000, 1995; Riley et al. 1994; Madrid et al. 1996; Rankins and Smith 1991). *Acacia* spp. and *Cassia* sp. have also been successfully used in mixed rations in Egypt, Israel and South Africa. It is concluded from these studies that halophytes could become important feed resources at moderate inclusion levels wherever the production of these plants can be justified agronomically or where harvesting of naturally occurring stands is feasible. Leaves of three *Atriplex* spp. and *Cassia sturtii* grown in two different locations in South Africa were analysed for certain nutritive characteristics. The aim of studies was to evaluate the nutritional value of *Atriplex* spp. for small stock production. Selected plants were harvested and analysed for crude protein, *in vitro* digestibility and leaf to stem ratio. Other studies (Van Niekerk et al. 2004c) compared chemical composition of *Atriplex* spp. grown under South African conditions with regard to site, species and plant parts while Du Toit et al. (2004) assessed the effects of type of carbohydrate supplementation on intake and digestibility of *Atriplex nummularia* cv. De Kock were investigated. Ten rumen cannulated sheep were fed different increments of maize and barley supplements (0%, 15%, 30%, 45%) to a basal diet of *A. nummularia*. Supplementation of *A. nummularia* cv. De Kock with an energy source tended to increase NDF digestibility, decrease rumen pH and with maize as a supplement, increase intake. Leaves of three *Atriplex* spp. and *Cassia sturtii* grown in two different locations in South Africa were analysed for certain nutritive characteristics. Crude protein values ranged from 176 to 234 g/kg DM for the *Atriplex* spp. and for *C. sturtii* from 114 to 147 g/kg DM. Similar work was done by Degen et al. 2010 in Israel in cafeteria trials comparing livestock preference for three salty feedstuffs.

Acacia saligna was examined by Degen et al. (1995, 2000) as potential fodder for sheep and goats raised in arid and semi-arid areas. This leguminous tree remains green all year and can be grown in deserts using only runoff water. Dry matter, organic matter and energy digestibilities were low in both species but were higher for goats than for sheep. Total water intake and output were higher in sheep than in goats. Extractable tannins were virtually absent in faeces in both species, however, output of condensed tannins and protein in the ADF and ADL fractions were substantially higher. This showed the presence of tannin-protein complexes in these fractions which explained the negative digestibilities of ADF and ADL. It was concluded that *Acacia saligna* could not be used as a sole dietary source for small ruminants because of low intake and negative nitrogen balance. This was due mainly to the high tannin content. However, the tree might have a potential as a supplementary fodder due to its high crude protein content.

3.1.7 Metabolism of sheep fed saltbush

Old man saltbush (*Atriplex nummularia*) a native to inland Australia, is planted in saline farming systems worldwide, and is commonly utilized as fodder for livestock (Norman et al., this volume). Saltbush contains very high concentrations of NaCl and KCl, however, the complex interactions between minerals means that sheep grazing saltbush may be susceptible to mineral deficiencies (see Attia-Ismail, this volume, and Squires, this volume for an examination of the water requirements and other physiological consequences of high salt loads).

The effects of type of carbohydrate supplementation on intake and digestibility of *Atriplex nummularia* were investigated in South Africa by Du Toit et al. (2004). Ten rumen cannulated sheep were fed different increments of maize and barley supplements (0%, 15%, 30%, 45%) to a basal diet of *A. nummularia* cv. De Kock. Supplementation of *A. nummularia* with an energy source tended to increase NDF digestibility, decrease rumen pH and, with maize as a supplement, increase intake.

The effect of high NaCl and KCl intakes on the apparent absorption of some macro-minerals (Mg, Ca, P) by sheep fed formulated high-salt diets or old man saltbush was assessed to help identify likely mineral deficiencies. For example several experiments were conducted in Australia and elsewhere to assess the impact. It was concluded by Wilson (1966a, 1995); Du Toit et al. (2004) and others that saltbush, as a sole source of feed, may be unsuitable for sheep with high nutritional demands, and further research is required to fully assess the mineral balance of sheep grazing saltbush pastures but newer cultivars are now being developed that can improve the outcomes (Norman et al., this volume).

3.1.8 Feeding trials with Salicornia bigelovii

A body of published work exists on the potential value of *Salicornia* biomass and by-products, including seed meal in livestock diets. By far, most were about *Salicornia* grown under irrigation with saline water. *Salicornia bigelovii* has demonstrated good agronomic potential in trials on the Arabian Peninsula when irrigated with seawater (Riley et al. 1994; Riley, this volume) and in North America (Glenn et al. 1992). It is viewed as a potential source of oilseed and forage with a biomass production of over 15,000/ha over a growing period of about 10 months. The seed is about 10% of the total biomass and has a high oil and protein content with a low salt content. The vegetative portion, less the seed, has a low protein content (6%) and a high salt content (30%).

Attia et al. 1997 included *Salicornia* seed meal in rations for poultry (broilers). A feeding trial was conducted using broiler chickens to evaluate the nutritional value of SM. SM was incorporated in the diet at levels of 0, 30, 60 and 90 g kg^{-1}. The 60 g kg^{-1} SM diet was fed either with or without cholesterol (5 g kg^{-1}) whereas the 90 g kg^{-1} diet was fed only with cholesterol (5 g kg^{-1}). SM in broiler diets caused a depression ($P < 0.01$) in growth and feed intake proportional to the level of SM in the diet. However, the growth depressing activity of SM was counteracted by the inclusion of cholesterol. These results indicate that SM may be used as an unconventional feed ingredient in broiler diets when supplemented with cholesterol.

Kraidees et al. 1998 studied the influence of inclusion of *Salicornia* biomass in diets for rams on digestion and mineral balance. Sixty-three Najdi ram were used to evaluate the effect of dietary inclusion of *Salicornia bigelovii* by-products on growth performance, carcass characteristics and mineral and water intake. Either the dry stems or spikes of this seawater-irrigated halophyte were incorporated into isonitrogenous diets. Feeding *Salicornia* stems up to 300 g kg^{-1}, or spikes at 100 g kg^{-1}, did not affect DMI, compared to control (0 g kg^{-1} *Salicornia*); however, the inclusion of spikes at levels above 100 g kg^{-1} decreased DMI.

3.1.9 Forage Kochia spp. for fall and winter grazing

Kochia scoparia, a dicotyledonous erect annual herb with high genetic diversity and great potential as fodder, is found scattered in salt and/or drought affected areas. The fast vegetative growth and drought/high temperature tolerance of *Kochia* indicate that this plant has a high potential to be adopted as an important forage and fodder crop, especially in desert areas (Al Ahmadi and Kafi 2008).

Forage Kochia (*Kochia prostrate*), also known as prostrate kochia, or prostrate summer cypress is a long-lived, perennial, semi-evergreen, half-shrub well adapted to the temperate, semiarid and arid regions of central Asia and the western U.S. In these areas it has proven to be a valuable forage plant for sheep, goats, camels, cattle, and horses. Forage kochia is a C_4 plant that is extremely drought and heat tolerant, in part due to a taproot that can extend up to 6.5 m in depth. It is also very salt tolerant and well adapted to some ecosystems dominated by halophytic species. It has been reported (Waldron et al. 2010) to be very productive when grown in soils with salinity electrical conductivity (EC) levels approaching 20 dS/m, and capable of persisting at much higher EC levels. Forage kochia's biomass yield depends upon the subspecies and environment, but reports generally range from 1000 to 1800 kg/ha in environments receiving 100–200 mm annual precipitation. Studies and practical experience have shown that forage kochia is very palatable and nutritious, especially during the late summer through winter period. Its nutritional characteristics include fall and winter crude protein levels above 70 g/kg needed for gestating ruminants. It also has low tannins and oxalates, and has not been reported to be a nitrate accumulator. When fed alone, it has acceptable fiber qualities, but research has shown that it can improve digestion kinetics when in a mixed diet with the low quality grasses as is common during late summer, fall, and winter months. Overall, forage Kochia has the potential to improve the sustainability of small ruminant production in semiarid regions that frequently experience extended drought and saline conditions.

It appears that these halophytic plants may not have enough digestible energy for high producing ruminants. *Kochia* has some toxicity (Rankins and Smith 1991) and mixed rations are the best way to use it. Gul and Salehi (2014) poses the question as to the role of *Kochia* in harsh environments.

4. Unconventional feedstuffs

Mixed rations based on salt tolerant or halophyte feedstuffs can involve less common products.

4.1 Atriplex halimus and ground Date (Phoenix dactylifera) pits as alternative feeds for sheep

Date seeds (*Phoenix dactylifera*) are tough and need to be processed before being fed to animals. They have low protein levels, with about 5–11% DM crude protein. Oil content is in the 4–14% DM range. Date pits contain high and variable quantities of fiber, with a high level of lignification: crude fiber 16–51% DM, NDF 58–90% DM, ADF 41–46% DM and ADL 4–18. Varietal differences are significant. Date pits contain appreciable amounts of K followed by P, Mg and Ca and a low Na content. Of the micro elements Fe, Mn, Zn and Cu are the most important ones.

The nutritive value of date seeds has been quite extensively studied due to their high availability in the countries where date production is important. It should be noted that most of the research concerning the utilization of date pits in ruminants is relatively old, not readily accessible and therefore difficult to assess. The renewed interest in those products is already resulting in new research that suggests that ground date seeds can be used up to 75% in ruminant rations provided that a good protein supplement (such as cottonseed cake) or urea is added. They can also be suitable to balance up a diet where the basic components are too rich in protein, such as young pasture. The crude protein and crude fat of date seeds are low but not negligible, and the seeds need to be processed so that the hard seed coat is no longer an obstacle to their digestion.

The nutritive values of date seeds reported by the literature are variable, with *in vivo* DM digestibility in sheep ranging from 58% to 70%. OM digestibilities values higher than 80% have been reported. *In vitro* DM digestibility using the rumen fluids of goats, sheep and dromedaries were found to be much lower for dromedaries and sheep (30–35%) than for goats (52–60%). Protein digestibility was generally low (< 40%) or not measurable.

The optimal level of date pits in sheep diets is disputed, probably due to the large variety of experimental protocols used by researchers. In a trial Al-Owaimer et al. (2011) found that an inclusion rate of 30% date pits gave the best weight gain in sheep fed *Atriplex halimus* hay and a concentrate. The isonitrogenous replacement of barley grain by ground date pits, included at up 45% in the diet had no effect on total feed intake and *in vivo* diet DM digestibility despite a high increase in the NDF and ADF content of the diet. Fiber digestibility increased with the inclusion of date pits, which suggests that fiber digestibility of date pits was higher than that of barley grain, and that date pits may better prevent acidosis than barley. See above for other studies where date seeds were also included as part of mixed rations (Suliman and Mustafa 2014). The literature suggests that ground date seeds can be included in the ration of local growing goats up to 15% without any detrimental effect on performance. However, the economic benefit could be obtained up to 45% ground date seeds.

4.2 Nutritive value of spineless cactus

Nefzaoui (2000) used 15 Barbarine wethers in Morocco that were randomly allotted to three equal groups, and fed diets based on cactus (*Opuntia ficus-indica* var. *inermis*; 80% of the diet) and *Atriplex nummularia*. Digestibility coefficients of organic matter (OMD) and crude protein (CPD) were relatively high for all diets, averaging 68, 74, and 75%, respectively. In contrast, fiber digestibility was low, probably because of the soluble carbohydrates in cactus, which may have depressed cellulolytic activity in the rumen. In conclusion, sheep may be fed diets based on cactus (rich in energy) and *Atriplex* (rich in nitrogen). Both feeds are available during drought years.

Ben Salem et al. (2002) supplemented spineless cactus (*Opuntia ficus-indica* f. *inermis*) based diets with urea-treated straw or oldman saltbush (*Atriplex nummularia*) and assessed the effects on intake, digestion and sheep growth. Combinations of *Opuntia* and *Atriplex* gave satisfactory outcomes.

4.3 The use of seaweeds for lamb fattening during drought

The objective of this experiment reported by Rijiba et al. (2010) was to evaluate the nutritive value of two seaweeds (*Ruppia maritima* and *Chaetomorpha linum*) and the possibility of including them in a concentrate for partial replacement of common ingredients. Each species was incorporated in a concentrate at 20% level with barley and soybean meal. Nitrogen retention was not affected by diet composition (12 g/day), it representing 47% of N intake. Average daily gain (ADG) was not affected by diet composition ($P > 0.05$) and lambs grew at rates 183, 172 and 138 g with control, *Ruppia* and *Chaetomorpha* diets, respectively. Feed conversion ratio was similar for control and *Ruppia* diets (6.1) and slightly higher for animals fed diet including *Chaetomorpha* (7.3).

It is concluded that seaweeds evaluated in this study (i.e., *Ruppia maritima* and *Chaetomorpha linum*) could be used as alternative feed resources for growing lambs during drought periods.

5. Development of sheep and goat production system based on the use of salt-tolerant plants and marginal resources

While there are many publications that provide qualitative information on performance of salt tolerant pastures and some quantitative information on plant growth and nutritive value, information on the performance of grazing livestock is scarce (O'Connell and Young 2002; Norman et al., this volume). The studies summarized in this chapter showed that incorporating salt tolerant and halophytic feedstuffs such as *Atriplex* spp. shrubs and *Sporobolus* grass hay into mixed diets enhanced growth performance of livestock, especially small ruminants. It can be concluded that when managed properly, *S. virginicus* and *D. spicata*, have the economic and environmental potential to be used in an integrated forage-sheep and goats system, particularly in marginal environments with low quality soil and water resources. However, such approach needs intensive and long-term investigation as those done by Al-Shorepy et al. (2010) in the Arabian Gulf region and Norman et al. (this volume) in Australia. It can be concluded that when managed properly, salt tolerant grasses such as *S. virginicus, D. spicata*, and *Puccinellia* and shrubs such as *Atriplex* and *Salicornia* have the economic and environmental potential to be used in an integrated forage-sheep and goats system particularly in marginal environments with low quality soil and water resources.

Under semi-arid conditions in southern Australia where chenopod (*Atriplex* and *Maireana*) shrubs have widespread natural occurrence (Wilson and Graetz 1980; Squires 1989; Warren and Casson 1992, 1993; Warren et al. 1994, 1996) have assessed the value to free ranging sheep and cattle. Rogers et al. (2005) reviewed the whole question of the potential of saltland pastures and O'Connell and Young (2002) assessed the economic importance of halophyte green feed available in the summer to the whole farm enterprise in southern Australia. Pearce et al. (2010) discuss the role of saltbush-based pasture systems for the production of high quality sheep and goat meat. Their review examines the roles of halophytic forage shrubs such as saltbush (*Atriplex* spp.) in the production of high quality sheep or goat meat. With careful consideration of production targets to minimise liveweight loss, this review has outlined potentially useful ways in a farming system to generate these meat quality benefits. Significant research is needed to understand the grazing conditions. Difficulties arise because forage halophytes are associated with low to moderate energy values, so they are incapable of supporting the levels of liveweight gain required to produce commercially desirable carcass weights, unless the animals are supplemented with either high quality pasture or grains. This review has highlighted the untapped potential for the goat meat industry to derive a meat product from halophytic forage shrubs. Goats may have an increased potential for liveweight gain and higher carcass lean and achieve a similar elevation in muscle vitamin E levels to sheep because of their browsing ability. Masters et al. (2005b) and Masters et al. (2001) reviewed the potential of salt tolerant forage/fodder in saline agro-ecosystems. Gintzburger et al. (1966) and Le Houerou (1992) assessed the role of fodder shrubs, including halophytes in arid and semi arid regions, especially in the WANA region. In Pakistan on highly saline land it was found by Khan and Ansari (2008) that when salt tolerant grass was grown with a salt accumulator (*Suaeda fruticosa*) in adjacent rows and with frequent irrigation, this system may be sustainable in terms of soil salt balance, with little change in soil salinity detected. There is scope for using salt tolerant plants both grasses, trees and shrubs in intercropping systems.

6. Summary and conclusions

Use of halophytes as animal fodder attracted attention of scientists toward the latter half of last century and papers on effect of brackish drinking water and salty feed on animal health and meat quality/quantity started and the problems encountered in such studies were identified (Weston et al. 1970; Weston 1996; Wilson and Kennedy 1996; Riley, this volume). Subsequently, scattered reports appeared in the literature where scientists from Australia, India, Pakistan, Middle East, Africa and North/South America working in the relevant fields have attempted replacing the regular fodder with one or another halophyte. There is now a large body of data from many sources and with several specialist monographs and special issues of various journals that document experiences on many aspects related to using salt tolerant and halophytic feedstuffs in animal nutrition. Despite the difficulties associated with the practice of using such feedstuffs (as outlined in this volume, and elsewhere) there is enough encouragement to do more work.

References and Further Readings

Abdou, A.R., E.Y. Eid, A.M. El-Essawy, A.M. Fayed, H.G. Helal and H.M. El-Shaer. 2011. Effect of feeding different sources of energy on performance of goats fed saltbush in Sinai. Journal of American Science 7(1).

Abou El Nasr, S.A. Al-Shorepy, G.A. Alhadrami and A.J. Al-Dakhell. 2006. Optimizing management practices for maximum production of two salt-tolerant grasses: *Sporobolus virginicus* and *Distichlis spicata*. Proc. 7th Annual UAE University Res. Conf. 2: 44–50.

Abou El Nasr, H.M., H.M. Kandil, E. El Kerdawy, A. Dawlat, H.S. Khamis and H.M. El-Shaer. 1996. Value of processed saltbush and *Acacia* shrubs as feed fodders under arid conditions of Egypt. Small Rumin. Res. 24: 15–20.

Abouheif, M.A., M. Al-Saiady, M. Kraidees, A. Tag Eldin and H. Metwally. 2000. Influence of inclusion of *Salicornia* biomass in diets for rams on digestion and mineral balance. Asian–Aust. J. Anim. Sci. 13: 967–973.

Abu-Zanat, M.M.W. and M.J. Tabbaa. 2006. Effect of feeding *Atriplex* browse to lactating ewes on milk yield and growth rate of their lambs. Small Ruminant Research 64(1-2): 152–161.

Abu-Zanat, M.M., F.M. Al-Hassanat, M. Alawi and G.B. Ruyle. 2003. Mineral assessment in *Atriplex halimus* L. and *Atriplex nummularia* L. in the arid region of Jordan. African Journal of Range & Forage Science 20(3): 247–251.

Al-Ahmadi, M.J. and M. Kafi. 2008. Kochia (*Kochia scoparia*): To Be or Not To Be! pp. 119–162. *In*: M. Kafi and M.A. Khan (eds.). Crop and Forage Production Using saline Waters. Nam S&T Centre, New Delhi.

Al Khalasi, S.S., O. Mahgoub, I.T. Kadim, W. Al-Marzouqi and S. Al-Rawahi. 2010. Health and performance of Omani sheep fed salt-tolerant sorghum (*Sorghum bicolor*) forage or Rhodes grass (*Chloris gayana*) Small Ruminant Research 91(1): 93–102.

Al-Dakheel, A.J., G.S. Al-Hadrami, S.A. Al-Shorepy and G. AbuRumman. 2010. Growth performances and carcass characteristics of indigenous lambs fed halophyte Sporobolus virginicus grass hay. Asian-Australasian Journal of Animal Sciences.

Alhadrami, G.A. 2003. *In situ* dry matter and fiber degradation of salt tolerant Sporobolus grass hay in camels fed yeast culture. J. Camel Practice and Res. 10(2): 139–144.

Alhadrami, G.A., S.A. Al-Shorepy and A.M. Yousef. 2010. Growth performance of indigenous sheep fed *Sporobolus virginicus* grass hay grown in saline desert lands and irrigated with high salt content ground water. Tropical Animal Health and Production 42(8): 1837–1843.

Alhadrami, G.A., A.A. Nigm, A.M. Kholif and O.M. Abdalla. 1997. Effect of roughage to concentrate ratio on performance and carcass characteristics of local lambs in the United Arab Emirates. Arab Gulf J. Sci. Res. 15: 137–148.

Alhadrami, G.A., M.H. Abdel-Gawad and J. Jumma. 1998. *In-situ* SM and fiber degradation in camels and *in-vitro* gas production of two grasses irrigated with sea water. Third Annual Meeting for Animal Production Under Arid Conditions.

Alhadrami, G.A., S.A. Al-Shoropy, M.A. Ayoub and A.M. Yousef. 2004. Growth performance of sheep fed *Sporobolus* grass hay grown in saline desert lands and irrigated with high salt content water. Proc. Seventh Annual UAE University Res. Conf. 2: 27–34.

Alicata, M.L., G. Amato, A. Bonanno, D. Giambalvo and G. Leto. 2002. *In vivo* digestibility and nutritive value of *Atriplex halimus* alone and mixed with wheat straw. J. Agric. Sci. 139: 139–142.

Al-Owaimer, A.N., A.M. El-Waziry, M. Koohmaraie and S.M. Zahran. 2011. The use of ground date pits and *Atriplex halimus* as alternative feeds for sheep. Aust. J. Basic Appl. Sci. 5(5): 1154–1161.

Al-Shorepy, D.S.A. and G.A. Alhadrami. 2008. The effect of dietary inclusion of halophyte *Distichlis* grass hay *Distichlis spicata* (L.) on growth performance and body composition of Emirati goats. Emir. J. Food Agric. 20(2): 18–27.

Al-Shorepy, G.A. Alhadrami and A.I. El Awad. 2010. Development of sheep and goat production system based on the use of salt-tolerant plants and marginal resources in the United Arab Emirates 91(1): 39–46.

Al-Shorepy, S.A., G.A. Alhadrami and A.J. Al-Dakheel. 2003. Growth performances and carcass characteristics of indigenous lambs fed Halophyte *Sporobolus virginicus* grass hay. Asian-Aust. J. Anim. Sci. 23(5): 556–562.

Arieli, A., E. Naim, R.W. Benjamin and D. Pasternak. 1989. The effect of feeding saltbush and sodium chloride on energy metabolism in sheep. Anim. Prod. 49: 451–457.

Arzani, H., M. Basiri, F. Khatibi and G. Ghorbani. 2006. Nutritive value of some Zagros Mountain rangeland species. Small Ruminant Res. 65: 128–135.

Arzani, H., M.R. Sadeghimanesh, H. Azarniv, G.H. Asadian and E. Shahriyari. 2008. Study of phenological stages effect values of twelve species in Hamadan rangelands. Iran. J. Range Desert Res. 16: 86–95.

Ashour, N.I., M.S. Serag, A.K. Abd El-Haleem and B.B. Mekki. 1997. Forage production from three grass species under saline irrigation in Egypt. Journal of Arid Environments 37: 299–307.

Atiq-ur-Rehman, J.B. Mackintosh, J.A. Fortune and B.E. Warren. 1994. Can the voluntary feed intake of wheat straw in sheep be improved by mixing with saltbush (*Atriplex amnicola*)? Proc. Aust. Soc. Anim. Prod. 20: 175–177.

Attia-Imail, S.A. (in press). Plant secondary metabolites of halophytes and salt tolerant plants. pp. 127–142, this volume.

Attia-Ismail, S.A. (in press). Rumen physiology under high salt stress. pp. 348–360, this volume.

Attia, F.M., A.A. Alsobayel, M.S. Kriadees, M.Y. Al-Saiady and M.S. Bayoumi. 1997. Nutrient composition and feeding value of *Salicornia bigelovii* Torr. meal in broiler diets. Anim. Feed Sci. Technol. 65: 257–263.

Ben Salem, H., A. Nefzaoui and L. Ben Salem. 2002. Supplementing spineless cactus (*Opuntia ficus-indica f. inermis*) based diets with urea-treated straw or oldman saltbush (*Atriplex nummularia* L.). Effects on intake, digestion and sheep growth. J. Agric. Sci. Camb. 138: 85–92.

Ben Salem, H., H.C. Norman, A. Nefzaoui, D.E. Mayberry, K.L. Pearce and D.K. Revell. 2010. Potential use of oldman saltbush (*Atriplex nummularia* Lindl.) in sheep and goat feeding. Small Ruminant Research 91: 13–28.

Ben Salem, H., A. Nefzaoui and L. Ben Salem. 2004. Spineless cactus (*Opuntia ficus-indica f. inermis*) and oldman saltbush (*Atriplex nummularia* L.) as alternative supplements for growing Barbarine lambs given straw-based diets. Small Rumin. Res. 51: 65–73.

Ben Salem, H., A. Nefzaoui and L. Ben Salem. 2002. Supplementation of *Acacia cyanophylla* Lindl. Foliage-based diet with barley or shrubs from arid areas (*Opuntia ficus-indica f. inermis* and *Atriplex nummularia* L.) on growth and digestibility in lambs. Anim. Feed Sci. Technol. 96: 15–30.

Benjamin, R.W., E. Oren, E. Katz and K. Becker. 1992. The apparent digestibility of *Atriplex barclayana* and its effect on nitrogen balance in sheep. Anim. Prod. 54: 259–264.

Bhatti, J.A., M. Younas, M. Abdullah, M.E. Babar and H. Nawaz. 2009. Feed intake, weight gain and haematology in Nili-Ravi buffalo heifers fed on Mott grass and berseem fodder substituted with saltbush (*Atriplex amnicola*). Pakistan Vet. J. 29(3): 133–137.

Bird, P.R. 1972. Sulphur metabolism and excretion studies in ruminants X. Sulphide toxicity in sheep. Aust. J. Biol. Sci. 25: 1087–1098.

Cohen, R.D.H., A.D. Iwaasa, M.E. Mann, E. Coxworth and J.A. Kernan. 1989. Studies on the feeding value of *Kochia scoparia* (L.) Schrad. hay for beef cattle. Can. J. Anim. Sci. 69: 735–743.

Davis, A.M. 1981. The oxalate, tannin, crude fiber, and crude protein composition of young plants of some *Atriplex* species. J. Range. Manage. 34: 329–331.

Degen, A.A., S. El-Meccawi and M. Kam. 2010. Cafeteria trials to determine relative preference of six desert trees and shrubs by sheep and goats. Livestock Science 132: 19–25.

Degen, A.A., K. Becker, H.P.S. Makkar and N. Borowy. 1995. *Acacia saligna* as a fodder for desert livestock and the interaction of its tannins with fiber fractions. J. Food Sci. Agric. 68: 65–71.

Degen, A.A.D., R.W. Benjamin, T. Mishorr, M. Kam, K. Becker, H.P.S. Makkar and H.J. Schwartz. 2000. *Acacia saligna* as a supplementary feed for grazing desert sheep and goats. J. Agric. Sci. 135: 77–84.

Du Toit, C.J.L., W.A. Van Niekerk, N.F.G. Rethman and R.J. Coertze. 2004. The effect of type and level of carbohydrate supplementation on intake and digestibility of *Atriplex nummularia* cv. De Kock. South African Journal of Animal Science 34(5): 35.

Edwards, N.J., J.E. Hocking, E. Sanders and D.K. Revell. 2002. Sheep production on puccinellia-based pastures in South Australia. Anim. Prod. Aust. 24: 293.

Fahmy, A.A., K.M. Youssef and H.M. El Shaer. 2010. Intake and nutritive value of some salt-tolerant fodder grasses for sheep under saline conditions of South Sinai, Egypt. Small Ruminant Research 91: 110–115.

Galitzer, S.J. and F.W. Oehme. 1978. *Kochia scoparia* (L.) Schrad. Toxicity in cattle: a literature review. Vet. Hum. Toxicol. 20: 421–423.

Gintzburger, G., M. Bounejmate and A. Nefzaoui (eds.). 1996. Fodder Shrub Development in Arid and Semi-arid Zones. Proceedings of the Workshop on Native and Exotic Fodder Shrubs in Arid and Semi-arid Zones, 27 Oct–2 Nov 1996, Hammamet, Tunisia. ICARDA, Aleppo (Syria) I: 184–186.

Glenn, E., R. Tanner, S. Miyamoto, K. Fitzsimmons and J. Boyer. 1998. Water use, productivity and forage quality of the halophyte *Atriplex nummularia* grown on saline waste water in a desert environment. Journal of Arid Environments 3845–62.

Glenn, E.P., W.E. Coates, J.J. Riley, R.O. Kuel and R.S. Swingle. 1992. *Salicornia bigelovii* Torr.: a seawater-irrigated forage for goats. Anim. Feed Sci. Technol. 40: 21–30.

Grattan, S.R., C.M. Grieve, J.A. Poss, P.H. Robinson, D.L. Suarez and S.E. Benes. 2004. Evaluation of salt-tolerant forages for sequential water reuse systems. III. Potential implications for ruminant mineral nutrition. Agric. Water Manage 70: 137–150.

Gul, B., R. Ansari, I. Aziz and M.A. Khan. 2010. Salt tolerance of *Kochia scoparia*: a new fodder crop for highly saline regions. Pakistan Journal of Botany 42: 2479–2487.

Gul, B.M. and M. Salehi. 2014. Kochia (*Kochia scoparia* (L.) Schrad) unwanted or wanted plant for forage production in harsh environments. pp. 155–174. *In*: Sabkha Ecosystems: Volume IV: Cash Crop Halophyte and Biodiversity Conservation Tasks for Vegetation Science Volume 47.

Herrmann, T.N. and N. Booth. 1997. Puccinellia – perennial sweet grass (Primary Industries, South Australia).

Jami Al Ahmadi, M. and M. Kafi. 2008. Kochia (*Kochia scoparia*): to be or not to be? pp. 119–162. *In*: M. Kafi and M.A. Khan (eds.). Crop and Forage Production Using Saline Water (NAM S&T Centre). Daya Publishing House, Dehli, India.

Kafi, M. and M.A. Khan. 2008. Forage Production Using Saline Waters (eds.): NAM S&T, Delhi.

Kandylis, K. 1984. The role of sulphur in ruminant nutrition. Review. Livest. Prod. Sci. 11: 611–624.

Karimi, H. and I.A. Ungar. 1986. Oxalate and inorganic ion concentrations in *Atriplex triangularis* Willd. Organs in response to salinity, light level and aeration. Bot. Gaz. 147: 65–70.

Khalil, J.K., W.N. Sawaya and S.Z. Hyder. 1986. Nutrient composition of *Atriplex* leaves grown in Saudi Arabia. J. Range Manage. 39: 104–107.

Khan, M.A. and R. Ansari. 2008. Potential use of halophytes with emphasis on fodder production in coastal areas of Pakistan. Biosaline Agriculture and High Salinity Tolerance. Edited by C. Abdelly, M. Ozturk, M. Ashraf and C. Grignon, BirkhuserVerlag, Switzerland, pp. 163–175.

Khan, M.A., R. Ansari, H. Ali, B. Gul and B.L. Nielsen. 2009. Panicum turgidum a potentially sustainable cattle feed alternative to maize for saline areas. Agric. Ecosyst. Environ. 129: 542–546.

Kraidees, M.S., M.A. Abouheif, M.Y. Al-Saiady, A. Tag-Eldin and H. Metwally. 1998. The effect of dietary inclusion of halophyte *Salicornia bigelovii* Torr. on growth performance and carcass characteristics of lambs. Anim. Feed Sci. Technol. 76: 149–159.

Le Houérou, H.N. 1992. The role of salt bushes (*Atriplex* spp.) in arid land rehabilitation in the Mediterranean basin: a review. Agroforestry Systems 18: 107–148.

Llerana, V.F.A. 1994. Massive propagation of halophytes (*Distichlis spicata* and *Tamarix* spp.) on highly saline-alkaline soils in the ex-lake Texcoco, Mexico. pp. 289–292. *In*: V.R. Squires and A.T. Ayoub (eds.). Halophytes as a Resource for Livestock and for Rehabilitation of Degraded Lands. Kluwer Academic Madrid, J., F. Hernandez, M.A. Oulgar and J.M. Cid. 1966. Nutritive value of *Kochia scoparia* L. and ammoniated barley straw for goats. Small Rumin. Res. 19: 213–218.

Masters, D.G., A.J. Rintoul, R.A. Dynes, K.L. Pearce and H.C. Norman. 2005b. Feed intake and production in sheep fed diets high in sodium and potassium. Australian Journal of Agricultural Research 56: 427–434.

Masters, D.G., E.S.E. Benes and H.C. Norman. 2007. Biosaline agriculture for forage and livestock production. Agriculture, Ecosystems & Environment 119(3-4): 234–248.

Masters, D.G., A.J. Rintoul, R.A. Dynes, K.L. Pearce and H.C. Norman. 2005. Feed intake and production in sheep fed diets high in sodium and potassium. Australian Journal of Agricultural Research 56: 427–434.

Masters, D.G., H.C. Norman and E.G. Barrett-Lennard. 2005a. Agricultural systems for saline soil: the potential role of livestock. Asian–Aust. J. Anim. Sci. 18: 296–300.

Masters, D.G., H.C. Norman and R.A. Dynes. 2001. Opportunities and limitations for animal production from saline land. Asian–Aust. J. Anim. Sci. 14: 199–211 (Special issue).

Mayberry, D., D. Masters and P. Vercoe. 2010. Mineral metabolism of sheep fed saltbush or a formulated high-salt diet. Small Ruminant Research 87–92.

Meneses, R., V. Gabriel and H. Flores. 2012. Evaluating the use of *Atriplex nummularia* hay on feed intake, growth, and carcass characteristics of creole kids. Chilean Journal of Agricultural Research 72(1) January–March.

Moinuddin, M., S. Gulzar, I. Aziz, A. Alatar, A.K. Hegazy and M.A. Khan. 2012. Evaluation of forage quality among coastal and inland grasses from Karachi. Pakistan Journal of Botany.

Morales, R.J.A. 1980. Evaluacion del Pasto Salado (*Distichlis spicata*). Como Fuente de forage Patra Ruiminates. Tesis Professional. Uach. Chapingo, Mexico.

Morecombe, P.W., G.E. Young and K.A. Boase. 1996. Grazing a saltbush (*Atriplex-Maireana*) stand by Merino wethers to fill the 'autumn feed gap' experienced in the Western Australian wheat belt. Aust. J. Exp. Agric. 36: 641–647.

National Research Council (NRC). 1986. Nutrient requirements of domestic animals: Nutrient requirements of sheep. National Academy of Sciences, Washington D.C., USA.

Nefzaoui, A. 2000. Nutritive value of spineless cactus (*Opuntia ficus-indica* var. *inermis*) and Atriplex (*Atriplex nummularia*) based diets for sheep. pp. 518–523. *In*: G. Gintzburger, M. Bounejmate and A. Nefzaoui (eds.). Fodder Shrub Development in Arid and Semi-arid Zones. Proceedings of the Workshop on Native and Exotic Fodder Shrubs in Arid and Semi-arid Zones, 27 October–2 November 1996, Hammamet, Tunisia. ICARDA, Aleppo (Syria). Vol. II.

Norman, H.C., E. Hulm and M.G. Wilmot. (in press). Improving the feeding value of old man saltbush for saline production systems in Australia. pp. 79–88, this volume.

O'Connell, M. and J. Young. 2002. The role of saltland pastures in the farming system: a whole-farm bio-economic analysis. National Conference and Workshop on the Productive Use and Rehabilitation of Saline Land (8th: 2002: Fremantle, W.A.).

Öztürk, M. et al. (eds.). 2011. *Sabkha Ecosystems*, Tasks for Vegetation Science vol. 48, 95 Springer Science+Business Media B.V.

Pearce, K.L., D.G. Masters, G.M. Smith, RH. Jacob and D.W. Pethick. 2005. Plasma and tissue a-tocopherol concentrations and meat colour stability in sheep grazing saltbush (*Atriplex* spp.). Aust. J. Agric. Res. 56: 663–672.

Pearce, K.L., H.C. Norman and D.L. Hopkins. 2010. The role of saltbush-based pasture systems for the production of high quality sheep and goat meat. Small Ruminant Research 91: 29–38.

Pearlstein, S.L., R.S. Felger, E.P. Glenn, J. Harrington, K.A. Al-Ghanem and S.G. Nelson. 2012. Nipa (*Distichlis palmeri*): A perennial grain crop for saltwater irrigation. Journal of Arid Environments 82: 60–70.

Pearlstein, S.L., R.S. Felger, E.P. Glenn, J. Harrington, K.A. Al-Ghanem and S.G. Nelson. 2012. Nipa (*Distichlis palmeri*): A perennial grain crop for saltwater irrigation. Journal of Arid Environments 82: 60–70.

Rankins, D.L. and G.S. Smith. 1991. Nutritional and toxicological evaluations of Kochia hay (*Kochia scoporia*) fed to lambs. J. Anim. Sci. 69: 2925–2931.

Revell, D.K., H.C. Norman, P.E. Vercoe, N. Phillips, A. Toovey, S. Bickell, E. Hulm, S. Hughes and J. Emms. 2013. Australian perennial shrub species add value to the feed base of grazing livestock in low- to medium-rainfall zones. Animal Production Science 53(11): 1221–1230.

Riasi, A., M. Danesh Mesgaran, M.D. Stern and M.J. Ruiz Moreno. 2012. Effects of two halophytic plants (*Kochia* and *Atriplex*) on digestibility, fermentation and protein synthesis by ruminal microbes maintained in continuous culture. Asian-Australas J. Anim. Sci. May 2012; 25(5): 642–647.

Riley, J.J., E.P. Glenn and C.U. Mota. 1994. Small ruminant feeding trials on the Arabian peninsula with *Salicornia bigelovii* Torr. pp. 273–276. *In*: V.R. Squires and A.T. Ayoub (eds.). Halophytes as a Resource for Livestock and for Rehabilitation of Degraded Lands. Kluwer Academic Publishers.

Riley, J.J. (in press). Review of Halophyte Feeding Trials with Ruminants, pp. 177–217, this volume.

Rjiba Ktita, S., A. Chermiti and M. Mahouachi. 2010. The use of seaweeds (*Ruppia maritima* and *Chaetomorpha linum*) for lamb fattening during drought periods. Small Ruminant Research 91(1): 116–119.

Rogers, M.E., A.D. Craig, R. Munns, T.D. Colmer, P.H.G. Nichols, C.V. Malcolm, E.G. Barrett-Lennard, A.J. Brown, W.S. Semple, P.M. Evans, K. Cowley, S.J. Hughes, S.J. Snowball and S.J. Bennett, G.C. Sweeney, S. Dear and A. Ewing. 2005. The potential for developing fodder plants for the salt-affected areas of southern and eastern Australia: an overview. Australian Journal of Experimental Agriculture 45(4): 301–329.

Roshdy, A.F., T. El-Sheikh, E.T. Kishek and O.M. Abd El-hafez. 2013. Forage production Potential of Saltbush (*Atriplex halimus* L.) grown under stress conditions at Ras Sudr, South Sinai. Annals of Agric. Sci. 51(1): 1–8.

Shawket, S.A., M.H. Ahmed and M.A. Ibrahim. 2010. Impact of feeding *Atriplex halimus* and *Acacia saligna* with different sources of energy on lambs performance. Egypt. J. of Sheep & Goat Sci. 5(1): 191–208.

Squires, V.R. 1989. Australia: distribution and characteristics of shrublands. pp. 61–92. *In*: C.M. McKell (ed.). Biology and Utilization of Shrubs. Academic Press, New York.

Squires, V.R. (in press). Water requirements of livestock fed on halophytes and salt tolerant forage and fodders. pp. 287–302, this volume.

Suliman A.I.A. and S.M.S. Mustafa. 2014. Effects of ground date seeds as a partial replacer of ground maize on nitrogen metabolism and growth performance of lambs. Egyptian Journal of Sheep & Goat Sciences 9(2): 23–31.

Sun, H.X. and D.W. Zhou. 2010. Effect of dietary supplement of seed of a halophyte (*Suaeda glauca*) on feed and water intake, diet digestibility, animal performance and serum biochemistry in lambs. Livestock Science 128(1-3): 133–139.

Suttle, N.F. 1991. Mineral Nutrition of Livestock. CABI Wallingford, 587 p.

Sweeney, G.C., B.S. Dear and M.A. Ewing. 2005. The potential for developing fodder plants for the salt-affected areas of southern and eastern Australia: an overview. Aust. J. Exp. Agric. 45: 301–329.

Swingle, R.S., E.P. Glenn and J.J. Riley. 1995. Halophytes in mixed feeds for livestock. pp. 97–100. *In*: V.R. Squires and A.T. Ayoub (eds.). Halophytes as a Resource for Livestock and for Rehabilitation of Degraded Lands. Kluwer Academic Publishers. Dordrecht.

Swingle, R.S., E.P. Glenn and V. Squires. 1996. Growth performance of lambs fed mixed diets containing halophyte ingredients. Anim. Feed Sci. Technol. 63: 137–148.

USAID. 2006. Establishment of productive rangeland with high-yielding fodder shrubs, Final Report, 67 p.

Van der Baan, A., W.A. Van Niekerk, N.F.G. Rethman and R.J. Coertze. 2004. The determination of digestibility of *Atriplex nummularia* cv. De Kock (Oldman's Saltbush) using different *in vitro* techniques. S. Afr. J. Anim. Sci. 34(Suppl. 1): 95–97.

Van Niekerk, W.A., C.F. Sparks, N.F.G. Rethman and R.J. Coertze. 2004a. Mineral composition of certain *Atriplex* species and *Cassia sturtii*. South African Journal of Animal Science; 2004 Supplement 34: 105.

Van Niekerk, W.A., C.F. Sparks, N.F.G. Rethman and R.J. Coertze. 2004b. Qualitative characteristics of some *Atriplex* species and *Cassia sturtii* at two sites in South Africa. South African Journal of Animal Science 34(5): 108.

Van Niekerk, W.A., P.J. Vermaak, N.F.G. Rethman and R.J. Coertze. 2004c. Comparison of chemical composition of *Atriplex* spp. grown under South African conditions with regard to site, species and plant parts. South African Journal of Animal Science 34(5): 98.

Waldron, B.L., J.-S. Eun, D.R. ZoBell and K.C. Olson. 2010. Forage Kochia (*Kochia prostrata*) for fall and winter grazing.

Warren, B.E., C.J. Bunny and E.R. Bryant. 1990. A preliminary examination of the nutritive value of four saltbush (*Atriplex*) species. Proc. Aust. Soc. Anim. Prod. 18: 424–427.

Warren, B.E. and T. Casson. 1992. Performance of sheep grazing salt tolerant forages on revegetated saltland. Proc. Aust. Soc. Anim. Prod. 19: 237.

Warren, B.E. and T. Casson. 1993. Saltbush quality and sheep performance. pp. 71–74. *In*: N. Davidson and R. Galloway (eds.). Productive Use of Saline Land, Proceedings No. 42, Australian Centre for International Agricultural Research, Canberra.

Warren, B.E. and T. Casson. 1994. Sheep and saltbush—are they compatible? pp. 125–129. *In*: Third National Conference on the Productive Use and Rehabilitation of Saline Land, Echuca, Australia.

Warren, B.E., T. Casson and D.W. Abbott. 1996. A preliminary study of sheep production from pastures on waterlogged and moderately saline land. Proc. Aust. Soc. Anim. Prod. 21: 177–180.

Warren, B.E., T. Casson and D.H. Ryall. 1994. Production from grazing sheep on revegetated saltland in Western Australia. pp. 263–265. *In*: V.R. Squires and A.T. Ayoub (eds.). Halophytes as a Resource for Livestock and for Rehabilitation of Degraded Lands. Kluwer Academic Publishers, Dordrecht.

Watson, M.C., J.W. O'Leary and E.P. Glenn. 1987. The evaluation of *Atriplex lentiformes* (Torr.) S. Wats. and *Atriplex nummularia* Lindl. as irrigated forage crops. J. Arid. Environ. 13: 293–303.

Weston, R.H. 1996. Factors limiting the intake of feed by sheep. 1. The significance of palatability, the capacity of the alimentary tract to handle digesta, and the supply of glucogenic substrate. Aust. J. Agr. Res. 17: 939–954.

Weston, R.H., J.P. Hogan and J.A. Hemsley. 1970. Some aspects of the digestion of *Atriplex nummularia* (saltbush) by sheep. Proc. Aust. Soc. Anim. Prod. 8: 517–521.

Wilson, A.D. 1994. Halophytic shrubs in semi-arid regions of Australia. pp. 97–100. *In*: V.R. Squires and A.T. Ayoub (eds.). Halophytes as a Resource for Livestock and for Rehabilitation of Degraded Lands. Kluwer Academic Publishers, pp. 101–113.

Wilson, A.D. 1966a. The tolerance of sheep to sodium chloride in food or drinking water. Aust. J. Agric. Res. 17: 503–514.

Wilson, A.D. 1966b. The value of *Atriplex* (Saltbush) and *Kochia* (Bluebush) species as food for sheep. Aust. J. Agric. Res. 17: 147–153.

Wilson, A.D. and R.D. Graetz. 1980. Cattle and sheep production on an *Atriplex vesicaria* (Saltbush) community. Aust. J. Agric. Res. 31: 369–378.

Wilson, J.R. and P.M. Kennedy. 1996. Plant and animal constraints to voluntary feed intake associated with fiber characteristics and particle breakdown and passage in ruminants. Aust. J. Agric. Res. 47: 199–226.

Yayneshet, T., L.O. Eik and S.R. Moe. 2008. Feeding *Acacia Etbaica* and *Dichrostachys cinerea* fruits to smallholder goats in Northern Ethiopia improves their performance during the dry season. Livest. Sci. 119: 31–41.

Yensen, N.P. 1988. A review of *Distichlis* spp. for production and nutritional values in arid lands: Today and Tomorrow. Proc. of an International Res. and Dev. Conf. Tucson, Arizona. Oct 20–25th 1985.

Yensen, N.P., S.B. Yensen and C.W. Weber. 1985. A review of *Distichlis* spp. for production and nutritional values in arid lands today. E.E. Whitehead, C.F. Hutchinson, B.N. Timmermann and R.G. Varady (eds.). Westview Press Boulder, pp. 809–822.

Yousef, K.M., A.A. Fahmy, A.M. El Essawy and H.M. El Shaer. 2009. Nutritional studies on *Pennisetum americanus* and *Kochia indicus* fed to sheep under saline conditions in Sinai, Egypt. American-Eurasian J. Agric. Environ. Sci. 5(1): 63–68.

Halophyte and Salt Tolerant Plants Feeding Potential to Dromedary Camels

Safinaz M. Shawket and *H.M. El Shaer**

SYNOPSIS

Camels are well adapted to arid and semi-arid regions, particularly in desert areas, where other animal species do not thrive and perhaps do not survive. They have the capacity to utilize low quality feed resources such as halophytes and convert them into animal protein and other products. Camels provide a wide range of useful products such as milk, meat, wool and leather. Therefore, it is very important to utilize such natural marginal resources (saline soils, saline water and camels) efficiently for providing secured and enough food (meat, milk, etc.) for humans living in arid and semi-arid regions.

Few studies have been carried out on fattened camels fed on salt tolerant fodders and halophytes. The objective of this chapter is to shed some light on the impact of feeding halophytes and salt tolerant fodder crops on the nutrition and productivity of pregnant and lactating she-camels and calves. It covers many issues in terms of feed intake and feed conversion; body weight and growth performance in addition to nutrients, digestion coefficients, nitrogen utilization and nutritive value. We show clearly that halophytes and salt tolerant fodder crops can be successfully and economically used in feeding camels in arid and semi-arid zones. Utilization of nutrients, body weight gain and feed conversion, in terms

Desert Research Center, Mataria, Cairo, Egypt.
Email: hshaer49@hotmail.com

of kg TDN/kg weight gain and Kg DCP/kg gain, of calves fed halophytes and salt tolerant forages supplemented with available sources of energy such as barley grains or molasses are greatly improved.

Keywords: arid regions, desert, nutrition, feed intake, digestion, nutritive value, fodder crops, saltbush, water intake, milk, supplements, rumen, diet selection, nitrogen utilization, meat.

1. Introduction

The arid regions are classified as the poorest in the world with regard to the availability of renewable fresh water resources. Animal production provides the main source of income for many in the region. Animal production systems in such countries are based to a large extent on natural vegetation as the principal animal feed resource. Large areas of rangeland, support salt and/or drought tolerant plant species, and many are severely degraded, mainly due to human and environmental factors (Gihad and El Shaer 1994). The future prosperity of feed resources in countries located in these areas relies on the economically feasible use of marginal and long-neglected resources such as halophytic plants. Halophytes have the potential as a good animal feed resource (El Shaer 2010). Feeding halophytes particularly to camels is a feasible solution to minimize the problem of feed shortage in arid and semi-arid regions. Halophytes include several nutritious and palatable fodder shrubs and salt tolerant grasses and legumes (El Shaer 1995).

Camels are famed for their peculiar adaptation to the harsh conditions prevailing in arid zones. They possess distinct behavioral, physiological and nutritional adaptive mechanisms that enable them to withstand extreme direct and indirect environmental stresses (Schmidt-Nielsen 1964; Farid et al. 1979, 1997). Camels provide a wide range of useful products such as milk, meat, wool and leather and are basically used for transportation. They graze a broad spectrum of fodder plants, including thorny species, halophytes and aromatic species generally avoided by other domestic herbivores (Schwartz 1992a,b). They have the capacity to utilize low quality feed resources as halophytes; convert them into animal protein and other products (Shawket 1999b) and can maintain themselves on natural ranges based mainly on halophytic plant species (Wilson 1984).

2. She-Camel's performance

2.1 Feed intake, digestion coefficients, nutritive value and body weight change

Camels have the ability to graze a wide range of naturally occurring plants in the arid regions (i.e., thorny species, halophytes and aromatic species generally avoided by other domestic herbivores) (Schwartz 1992a,b).

Camels prefer to consume salty bushes which are rich in moisture and salt. Pirzada et al. (1989) and Williamson and Payne (1990) confirmed that salts present in such plants help to meet the physiological functions of camels. The higher moisture content of salt bushes insures a good portion of camel water requirement in areas where water is the most limiting factor for animals (Gauthier-Pilters and Dagg 1981). Feeding halophytes particularly to camels could be one of suitable approaches in arid regions to minimize the problem of feed shortage. Camels, as well as other livestock, have adapted microbial communities in their rumens that enable them to utilize non-protein nitrogen present in halophytes (Fig. 13.1).

Unfortunately, the available information concerning the effects of feeding halophytic plants on camel performance and productivity outside of Egypt is limited. However, some studies were conducted on *Atriplex* species, mainly *A. halimus* and *A. nummularia* fed to camels. These saltbushes are evergreen shrubs, widely distributed in arid and semi-arid areas. *Atriplex* spp., are generally characterized by high contents of ash, fiber, high crude protein, but are deficient in energy. So, animals fed *Atriplex* spp. should be given an energy supplement, e.g., barley grain, yellow corn grain, date seed (Hassan and Abd El Aziz 1979; Kandil and El Shaer 1988). The influence of energy supplementation levels on the utilization of saltbush (*Atriplex* spp.) by sheep, goats and camels has been evaluated (El Shaer and Kandil 1990; Shawket and Ahmed 2009). The effect on the voluntary intake of *Atriplex nummularia* fed to adult she-camels was studied (Shawket and Ahmed 2001). Yellow corn grain was the

Fig. 13.1. Camels are an important livestock species and can survive and produce in harsh environments.

energy supplement fed at levels of 0, 20, 40 and 60% of maintenance requirements of energy. When she-camels were fed fresh *Atriplex nummularia* solely, dry matter intake (DMI) was 53.7 g/d/kg$^{0.75}$ and it progressively increased as the level of energy supplementation went from 20 to 40 and 60%, of maintenance requirements ($P < 0.01$) (29.1, 55.2 and 85.4, respectively). Body weight changes of camels were significantly ($P < 0.01$) improved by adding yellow corn grain as a source of energy to halophytic diets. The body weight changes improvement increased significantly ($P < 0.01$) by increasing the level of energy supplementation; due to the comparable increases in DMI from saltbush as a result of energy supplementation. The trends of these results are generally in agreement with those reported by many workers on sheep, goats and camels grazing saltbushes with different levels and types of energy supplementation (Farid and Hassan 1976; Hassan and Abd El Aziz 1979; El Shaer and Kandil 1990; Shawket I999a,b). Supplementation with yellow corn grain to she-camels fed fresh saltbush resulted in a significant increase in the digestibility coefficients of DM ($P < 0.01$), OM ($P < 0.01$), crude fiber ($P < 0.01$), CP ($P < 0.05$) and NFE ($P < 0.05$) (Shawket and Ahmed 2001), while EE digestion was not affected by level of energy supplementation. Total digestible nutrients (TDN) per cent of the saltbush significantly ($P < 0.01$) improved as a result of energy supplementation due to the improvement of the nutrients digestibility. The amelioration in digestion coefficients noticed due to yellow corn addition was expected as it introduced a readily available source of carbohydrate for the rumen microflora (Blaxter 1962). The necessity of using energy supplementation source in the diets of animals fed saltbushes was recommended by many workers (McLaren et al. 1965; Hassan and Abd El-Aziz 1979) because such supplementation improves the digestibility and, therefore, the nutritive value of saltbush. Similar findings and recommendations were reported by some investigators on sheep, goats and camels fed saltbushes (Hassan and Abd El-Aziz 1979; Shehata et al. 1987, 1988; El Shaer and Kandil 1990; Shawket 1999a).

The long-term feeding of *Atriplex halimus* to she-camels was evaluated by Shawket and Ahmed (2009). These authors fed berseem[1] hay (as a control group) for 360 days while the second and third groups were offered *Atriplex halimus* for 180 days (short-term) and 360 days (long-term), respectively. All the three groups were supplemented with a mixture of ground barley and yellow corn grain. The daily DMI (g/d/Kgw$^{0.75}$) of long-term *Atriplex* fed group decreased ($P < 0.05$) in comparison with short-term fed group which had similar values to the control group (67.0 vs. 75.6 and 70.7 g/d/Kgw$^{0.75}$, respectively). Nonetheless, the average body weight change (g/d) of camels fed *A. halimus* for long-term was higher ($P < 0.05$) than that of the short-term and control groups. Effect of prolonged feeding of *Atriplex halimus* to camels on digestibility, nutritive value and nitrogen utilization was evaluated by Shawket and Ahmed (2009). DM, OM and NFE digestibilities were not influenced by either changing the type of roughage (berseem hay vs. *A. halimus*) or the total period of feeding *Atriplex* (Shawket and Ahmed 2009). Decreasing CP, CF and EE digestibilities by feeding *Atriplex* instead of berseem hay may be due to the higher salt content of *Atriplex* which is the major negative component in *Atriplex* species (Wilson 1966a). This has led to increased water intake and shortened rumen turnover times with

---

[1] *Trifolium alexandrinum.*

consequential influences on rumen physiology and metabolism (Warner and Casson 1992; Konig 1993; Attia Ismail, this volume). Such results might be also attributed to the secondary chemical compounds in *Atriplex*, which accumulated 6.6% of oxalates and 5.2% DM of hydrolysable tannins which might decrease the production of volatile fatty acids, microbial DNA and RNA in the rumen (Abu-Zanat and Tabbaa 2005; Attia-Ismail, this volume). Camels can tolerate the negative impact of prolonged grazing of *Atriplex* (saltbush) when supplemented with a suitable source of energy (Shawket and Ahmed 2009). Improvement in CP, CF and EE digestibilities when *Atriplex* was fed for long-term might be due to adaptation of rumen micro-organisms to secondary metabolites present in *Atriplex*. Nutritive value (expressed as TDN %) increased ($P < 0.05$) with feeding *Atriplex* for long, rather than short-term, and was comparable to the value of the control group. Generally, feeding *Atriplex* instead of berseem hay tended to increase DCP (%) values, but decreased ($P < 0.05$) with long-term feeding of *Atriplex* to be nearly similar to that of control group. Similar trends were reported on camels fed saltbush and supplemented with a source of energy (Shawket et al. 2010). These findings may be attributed to many factors:

1. Camels appear to need more salt, probably more than other herbivores, which is present in higher proportion in this halophytic plant. This fact, demonstrated previously by Chamberlain (1989), indicating that camels require salt six to eight times more than that required by other livestock; camels without regular access to salty feed require about 140 g of salt per day. These findings explain the higher ($P < 0.05$) intake of DM, OM, TP and CF when camels were fed *Atriplex* in comparison to their mates fed either Berseem hay or straw.
2. In comparison with bovines, camels saliva contain a varying content of high molecular weight mucin-glycoprotein (MGP) that confers protection to the mucosa of the digestive tract from mechanical injuries and fixes the plant tannins preventing their negative effects on protein metabolism in the rumen (Schmidt-Witty et al. 1994).
3. *Atriplex* being a lush green plant was more palatable and preferred by camels in comparison to the dry long clover hay as reported by Abdel-Wahed (2014).

2.2 Water intake

Most arid and semi-arid regions in developing countries have already reached the absolute water scarcity level where the available water per capita is less than 1000 m³/year.[2] Water for livestock is often scarce and highly saline water utilization is a crucial parameter for halophytes and salt tolerant fodders consumption by animals particularly in arid regions (Squires et al., this volume). Trends of water consumption, excretion and retention vary among animals according to several factors; species, seasons of the year, types and form of feed materials, etc. For instance, in she-camels, the amounts of free water intake (ml/d/Kg$^{0.82}$) were not affected by any level of energy supplementation when fed *Atriplex nummularia* (Shawket and Ahmed 2001) whereas, the feed water intake (FWI) and consequently, total water intake (TWI)

---

[2] http://www.gwp.org/en/GWP-Mediterranean/The-Challenge/test/Water-Scarcity/.

were significantly ($P < 0.01$ and $P < 0.05$, respectively) increased as a result of energy supplementation (types and levels). These increments in FWI were 20, 48 and 60% for camels fed saltbush supplemented with corn grains at levels 20, 40 and 60%, of energy maintenance requirements, respectively. The increases in FWI were attributed to the increasing intake of *A. nummularia*, which had high moisture content (74.9%). Similar results are obtained by Hassan and Abd El-Aziz (1979) on sheep and Shawket (1999a) on goats fed saltbushes. Meanwhile, the impact of long and short term feeding of *Atriplex* on water utilization revealed that water intake was significantly affected ($P < 0.01$) by either type of roughage or period of feeding (Shawket and Ahmed 2009). Camels fed *Atriplex* either for short or long-term generally consumed more feed water ($P < 0.01$) by about 14 to 18 times, respectively than camels fed berseem hay diet (6.6, 78.8 and 69.2 ml WI/d/Kgw$^{0.82}$ for control, short and long-term of feeding *Atriplex* groups, respectively). Berseem hay contains less moisture than *Atriplex* (12.7% vs. 72.4 for berseem hay and *Atriplex*, respectively). The higher free drinking water intake by camel groups fed *Atriplex* either for the short or long-term may be due to the higher seasonal contents of Na which ranged between 5.6 and 6.7% in the dry matter of *Atriplex* foliage for spring and fall seasons, respectively (Abu-Zanat et al. 2003). This high level of salt in *Atriplex* browse may force animals to increase their daily water intake (Konig 1993; Abu-Zanat and Tabbaa 2005). Noteworthy, data derived from Shawket and Ahmed (2009) on water intake of she-camels fed *Atriplex*, indicated that prolonged feeding *Atriplex* to camels decreased their water intake compared with those fed *Atriplex* for short-term. This fact confirmed the early conclusion of Farid et al. (1985) and Abou El-Nasr et al. (1988) that camels were apparently better adapted for handling salt loads, being more economic in their use of water intake (Siebert and Macfarlane 1971). In this regard, Shawket et al. (2005) indicated that growing camel calves groups fed *Atriplex nummularia* plus *Acacia saligna* (AN-AS) and *Atriplex nummularia* (AN) rations retained higher ($P < 0.05$) water (ml/Kgw$^{0.75}$) than growing camel groups fed *Acacia saligna* (AS) and treated rice straw (TRS) rations (141.17 and 118.87 vs. 68.33 and 65.45 ml/Kgw$^{0.75}$, respectively). These results revealed that camel's kidneys seem adapted to handling salt load. Moreover, the high and constant water content of salty plants make them a preferred feed source for camels (Wardeh 1990). This high water content might furnish 40–60% of water requirements of animals under desert conditions (Macfarlane 1964).

2.3 Nitrogen utilization

Several halophytic fodder crops, in particular saltbushes, are characterized by moderate crude protein, and low soluble carbohydrates contents (Wilson 1966a, 1977; Hassan and Abd El-Aziz 1979; El Shaer 1981). Such low levels of readily available carbohydrates together with the rapid fermentation of its crude protein in the rumen (Weston et al. 1970; Masters, this volume) may be responsible for the poor utilization of such feed nutrients. In this regard, values of nitrogen intake, nitrogen excretion and nitrogen retention are expected to be affected and vary by the type of fodder crops.

Nitrogen intake (NI), nitrogen excretion and nitrogen retention of she-camels fed *A. nummularia* were significantly ($P < 0.01$) affected by energy supplementation

(Shawket and Ahmed 2009). Nitrogen intake from *A. nummularia* significantly ($P < 0.01$) increased as a result of increasing DMI from this saltbush which is due to energy supplementation. These increments in NI were reflected on increasing fecal and urinary nitrogen excretions. These results indicated that feeding *Atriplex* as a sole diet is insufficient to provide camels with their nitrogen requirements. Therefore, *Atriplex nummularia* (saltbush) could be a good green fodder for feeding camels when supplemented with suitable source of energy at a level not less than 40% of their maintenance energy requirements especially during dry seasons in arid and semi-arid areas (Shawket 1999a; Shawket and Ahmed 2009).

Camels fed *Atriplex* for long-term retained higher ($P < 0.05$) nitrogen values than those fed *Atriplex* for the short period and the control group (berseem hay group) as reported by Shawket and Ahmed (2009). In this respect, digestible protein and metabolizable energy utilization of *A. nummularia* supplemented with a non-conventional concentrate mixture were 5.04 g DCP and 0.16 Mcal ME/kg^{75} which were enough to cover the maintenance and milk production requirements of she-camels during lactation stage as reported by (Shawket and Ibrahim 2013). The present results indicated that camels can be successfully adapted to prolonged feeding of saltbushes with no negative impact on their performance.

2.4 Milk production

Camels are reliable milk producers with a long lactation period and they maintain milk production throughout the long dry spells when milk from cattle and goats is scarce. Estimates of the amount of milk produced by lactating camels provide information for the implementation of optimum management and feeding strategies for camel and their calves. The most important factors affect camel milk production and compositions are feed and water availability. *Atriplex* species could be one of the main forage resources for feeding camels in arid and semi-arid regions. It is important to understand the impact of feeding such feed materials on animal production, especially milk yield. Milk is one of the major sources of income for the pastoral communities in these regions. Therefore, Shawket and Ibrahim (2013) evaluated the effect of long-term feeding of *Atriplex* (saltbush) to sixteen lactating camels on their milk production, chemical composition, physical properties and minerals content. Animals were divided (by weight) into two groups, where they were offered a concentrate diet consisting of 60% crushed yellow corn plus 40% crushed barley grain. For maintenance of their energy requirements *ad lib.* berseem hay was given to the first group (BHG) and *ad lib. Atriplex halimus* was given to a second group (AHG). Results indicated that changing the roughage type of diet by replacing hay with *Atriplex* increased ($P < 0.05$) daily milk yield, total solids, total protein, lactose, Na, K and Ca content while it decreased ($P < 0.05$) titrable acidity (%) and conductivity (ms/cm). Camel's milk fat, ash, Zn, Fe, and Mn content were not significantly affected by changing the roughage type. Calculating the digestible protein and metabolizable energy utilization of the *Atriplex* diet were 5.04 g DCP and 0.16 Mcal ME/kg$^{0.75}$ which were enough to cover the maintenance and milk production requirements. It was concluded that feeding *Atriplex* (saltbush) to lactating camels for a long-term did not negatively affect the milk production or change its chemical composition and physical properties. Therefore, *Atriplex* species could be a cost-effective solution for nourishment of camels during dry seasons particularly in arid regions.

3. Nutritional and growth performance of calves

Data on the impact of feeding halophytes and salt tolerant fodders on the performance of camel calves (females and males) are available. Some information on the nutritional, growth and productive performance of camel calves are presented as an example. Growth performance, efficiency of feed utilizations and carcass characteristics of three groups of fattened male camel calves (average body weight, 178 Kg) fed on saltbush (*Atriplex halimus*) with different energy sources for 180 days was evaluated by Shawket (1999b). The animals were offered: concentrate mixture + berseem hay (A); barley grain + fresh saltbush (B) and barley grain + olive cake + fresh saltbush (C). Feeding of fresh saltbush to male camel calves caused the weight gain of group C to decrease significantly; the average daily gain values were 750, 730.2 and 692.2 g for animals fed diets A, B and C, respectively. Digestion coefficients of dry matter (DM), crude protein (CP) and nitrogen free extract (NFE) improved for diets B and C compared with the control diet (A) as reported by Shawket (1999b). This was a reflection of the different nutritive values of these diets. Daily intakes from DM, total digestible nutrients (TDN) and digestible crude protein (DCP) were not affected by the experimental diets. Feed conversion in terms of kg TDN/kg weight gain and kg DCP/kg gain improved for animals fed saltbush in (B) and (C) groups which led to appreciable reduction in feeding cost for producing one-kg of body weight. The author concluded that fresh saltbush which is naturally available in many arid areas can be successfully and economically used in feeding camel calves provided that available energy sources are added to balance the diets.

Diet selection, feed intake capacity (FIC) and animal performance were evaluated when concentrates (corn grain and commercial concentrates mixture) and roughages (*Atriplex nummularia*, clover hay or rice straw) were fed, *ad lib* free-choice, to growing she-camels in a cafeteria feeding system (Farid et al. 2010). The results indicated that type of roughages and concentrate levels, and their interaction, affected ($P < 0.05$) FIC and diet selection, and consequently live weight gain. Average total and roughage DMI were 78.9 and 16.1, 83.9 and 22.5, 96.4 and 33.4 g DM/day/Kgw$^{0.75}$ for straw, hay and *A. nummularia* groups, respectively. The *Atriplex*-fed growing she-camels recorded the highest ADG, followed by the hay fed ones and the straw fed mates grew the least, 516, 429 and 240 g/d, respectively. The authors pointed out that restricting the level of concentrates offered decreased significantly ($P < 0.05$) the ADG (691, 305 and 189 g/d in camels fed 100, 75 and 50% of *ad lib* concentrate intake) respectively. These results seemed to indicate that growing camels having free choice to select their diets from both concentrates and roughages were capable of regulating their voluntary food intake predominantly through physiological mechanisms to satisfy their energy requirements. This was true for *Atriplex* and hay groups but not for the straw group or when limited concentrates were offered.

Recently, the effect of type of roughage (*Atriplex*, berseem hay or rice straw) and level of concentrate supplements (95% and 50%) on feed digestion and utilization by growing female camels was studied by Abdel-Wahed (2014). Reducing concentrates offered from 95% to 50% tended to decrease ($P < 0.05$) feed intake of concentrates, digestion coefficients of all nutrients, ME intake, DCP intake, nitrogen intake and nitrogen balance as well as average daily gain, while feed intake of roughage, crude fiber percent, rumen un-degradable protein (RUP), TP % and urinary nitrogen were significantly ($P < 0.05$) increased. The growing female camels fed berseem hay were more efficient ($P < 0.05$) digesters than those fed *Atriplex* and straw in dry matter, organic matter, crude fibers, ether extract and nitrogen free extract digestion, whereas digestion coefficient of crude protein was not affected (Abdel-Wahed 2014); the camels fed *Atriplex* had greater ($P < 0.05$) ash digestion than those fed hay or straw. These results are similar to those reported by Kandil (1984) and Kewan (2003) who found that camels fed hay diet were able to digest dry matter, crude protein, crude fiber and nitrogen free extract better than their mates fed the concentrate-straw mixed diet, and found that all hay fed camels were able to digest crude fiber better than those fed the concentrate-straw mixed diet. The same author also reported that limiting concentrate feeding from 95% to 50% *ad lib* decreased dry matter, organic matter, crude protein, crude fibers, ether extract and nitrogen free extract digestions, ash was not affected. Generally, similar trends were found by many workers (El-Banna 1995; Shawket and Ahmed 2001; Mosaad et al. 2003) who reported that increasing the level of energy supplementation improved nutrient digestibility, but decreased the CF digestibility. It is believed that this apparent discrepancy could be due to the fact that the higher percentage level of grain to replace berseem hay led to the decrease in CF % and the increase in concentration of readily available carbohydrates in the experimental camel rumens. Readily available carbohydrates are more utilizable by rumen micro flora than berseem hay crude fiber which has a lesser apparent digestibility than grain. *Atriplex* species do not have well-balanced nutrient composition considering its high crude protein content and a low energy concentration. Moreover, the crude protein in *Atriplex* contains high levels of non-protein nitrogenous compounds such as glycine, betaine and proline, derived from the physiological mechanisms of salt-tolerance (Le Houerou 1992). These compounds are degraded in the rumen and, therefore, utilized by rumen microorganisms to synthesize microbial proteins with the help of highly degradable energy present in *Atriplex*.

The impact of feeding two types of halophytic and salt tolerant plants on the performance of growing male camel calves was evaluated by Shawket et al. (2005). Voluntary feed intake, growth performance, feed conversion, digestion coefficients, nutritive value of fresh *Atriplex nummularia* (AN), fresh *Acacia saligna* (AS) and a mixture of both shrubs *Atriplex* plus *Acacia* (AN-AS) were determined using four equal groups of growing male camels (Shawket et al. 2005). Voluntary DM intake was significantly ($P < 0.05$) different, 103.46; 88.95; 82; 72 and 69.05 g/day/Kgw$^{0.75}$ for camels groups fed on AN-AS, AN, AS and treated rice straw (TRS) rations, respectively. The body weight gains varied significantly ($P < 0.05$) (719, 680, 589 and 525 g/day for AS, AN-AS, TRS and AN, respectively). The authors also reported that the highest ($P < 0.05$) nutritive value was recorded for AN-AS ration followed by AS ration being 77.92, 12.61 and 73.05, 10.02% TDN, DCP, respectively. Feed conversion, in terms of DM and TDN (kg/kg gain) was the best ($P < 0.05$) for TRS and AS camel groups. Nitrogen retention was the highest ($P < 0.05$) for growing calves fed AN-AS ration followed by those fed on *Acacia* ration (1.53 and 1.09 g/kgw$^{0.75}$, respectively). The above results indicated that feeding saltbushes (*Atriplex* species) with low salt plants (like *Acacia* species) is desirable for animals and improves the utilization of the saltbushes (Shawket 1999b). This has led to high nitrogen retention as a percentage of nitrogen intake for camel calves fed AN-AS ration.

The effect of feeding *Atriplex halimus*, supplemented with ground barley grains to pregnant camels on intake, digestibility, milk production of lactating dams and the growth of their calves was studied by Shawket et al. (2010). The body weights of she-camels at calving, after calving, at weaning (40 weeks) and after weaning (44 weeks) and the milk yield of the lactating dams on *Atriplex* diet were significantly ($P \leq 0.05$) higher than the corresponding values of the camels fed berseem hay diet (control group). Moreover, the weaned camel calves fed *Atriplex halimus* diet showed higher ($P \leq 0.05$) DCP intake and nitrogen utilization, but lower ($P \leq 0.05$) TDN intake and nutrient digestibility (OM, CP and CF) than the calves fed berseem hay diet. *Atriplex* diet was richer ($P \leq 0.05$) in protein (DCP) but poorer ($P \leq 0.05$) in energy (TDN) than the berseem hay diet (Shawket et al. 2010). The finishing body weight at 640 days and the feed conversion efficiencies (DMI and TDN) in a period of 360 days of the weaned camel calves on *Atriplex* diet was significantly higher ($P \leq 0.05$) than the calves fed berseem hay diet. The conclusion was that *Atriplex halimus* diet (supplemented with barley) could cover the nutritional needs for growth and milk production in camels.

References and Further Readings

Abdel-Wahed, A.M. 2014. Effect of type of roughage and level of concentrate and supplements on feed digestion and utilization in growing female dromedary camels. J. American Sci. 10(6): 198–206.

Abou El-Nasr, H.M., S.M. Shawket, M.F.A. Farid and H.S. Khamis. 1988. Studies on saline drinking water for camels and sheep: 1- Water intake and excretion. Alex. J. Com. Sci. Dev. Res. 23: 131–139.

Abu-Zanat, M.M.W. and M.J. Tabbaa. 2005. Effect of feeding Atriplex browse to lactating ewes on milk yield and growth rate of their lambs. Small ruminant Research. 1–10.

Abu-Zanat, M.M.W., F.M. Al-Hassanat, M. Alawi and G.B. Ruyle. 2003. Mineral assessment in *Atriplex halimus* L. and *Atriplex nummularia* L. in the arid region of Jordan. Afr. Range Forage Sci. 20: 1–5.

Attia-Ismail, S.A. (in press). Plant secondary metabolites of halophytes and salt tolerant plants. pp. 127–142, this volume.

Blaxter, K.L. 1962. Energy Metabolism In Ruminants. Hutchinson, Scientific and Technical, London.

Chamberlain, A. 1989. Milk production in the tropics. Intermediate Tropical Agriculture Series 202–210.

El Shaer, H.M. 1981. A comparative nutrition study on sheep and goats grazing Southern Sinai desert range with supplements. Ph.D Thesis, Fac. of Agric., Ain Shams Univ., Egypt.

El Shaer, H.M. 1995. Potential use of cultivated range plants as animal feed in the Mediterranean zone of Egypt. Proc. Of the 8th Meeting of the FAO Working Group on Mediterranean Pasture and Fodder Crops. Sylvo Pastoral Systems, Environmental, Agriculture and Economic Sustainability, 29 May–2 June. Avignon, France.

El Shaer, H.M. 2010. Halophytes and salt-tolerant plants as potential forage for ruminants in the Near East region. Small Rumin. Res. 91: 3–12.

El Shaer, H.M. and H.M. Kandil. 1990. Comparative study on the nutritional value of wild and cultivated *Atriplex halimus* by sheep and goat in Sinai. Com. Sci. and Dev. Res. 29: 81–90.

El-Banna, H.M. 1995. Effect of dietary energy, protein and their Interaction on nutrient utilization by sheep, goats and camels. Camel Newsletter No. (11): 16.

Farid, M.F.A. and N.L. Hassan. 1976. The supplemental feeding of growing sheep under simulated high conditions. Alex. Agric. Res. 24: 465.

Farid, M.F.A., A.M. Abdel-Wahed, Safinaz M. Shawket and N.I. Hassan. 2010. Diet selection, feed intake capacity and performance of growing female camels: Effects of type of roughage and level of concentrates offered. J. American Sci. 6: 317–326.

Farid, M.F.A., A.O. Sooud and H.I. Hassan. 1985. Effects of types of diet and protein intake on feed utilization in camels and sheep. Proc. Third AAAP Anim. Sci. Congress, Seoul, Korea, pp. 781–783.

Farid, M.F.A., H.S. Khamis, H.M. Abou El-Nasr, M.H. Ahmed and S.M. Shawket. 1997. Diet selection and food intake capacity of stall-fed sheep, goats and camels in relation to some physical properties of foods and their potential digestion in the rumen. Options Mediterraneennes, Seri A 34: 109–114.

Farid, M.F.A., S.M. Shawket and M.H.A. Abdel-Rahman. 1979. Observations on the nutrition of camels and sheep under strees. *In*: Camels. IFS Symp., Sudan, 125–170.

Gauthier-Pilters, Hilde and A.I. Dagg. 1981. The camel: its evaluation, ecology, behavior and relationship to man. University of Chicago Press, 208 pp.

Gihad, H.M. and H.M. El Shaer. 1994. Nutritive value of halophytes. pp. 281–284. *In*: V.R. Squires and A.T. Ayoub (eds.). Halophytes as a Source for Livestock and for Rehabilitation of Degraded Lands. Klumer, Academic Publishers.

Hassan, N.I. and H.M. Abd El-Aziz. 1979. Effect of barley supplementation on the nutritive value of saltbush (*A. nummularia*). World Rev. of Anim. Prod. 15: 47.

Kandil, H.M. 1984. Studies on camel's nutrition. Ph.D. Thesis, Fac. of Agric., Ain Shams Univ., Egypt.

Kandil, H.M. and H.M. El-Shear. 1988. The utilization of *Atriplex nummularia* by goats and sheep in Sinai. Proc. of the Inter. Symp. on the Constraints and Possibilities of Ruminant Production in Dry Subtropics, Nov. 5–7, 1988 Cairo. Egypt.

Kewan, Kh. Z.M. 2003. Studies on camel nutrition. Ph. D. Thesis, Fac. Agric., Alexandria University.

Konig, K.W.R. 1993. Influence of saltbush (*Atriplex* spp.) as diet component on performance of sheep and goats under semi-arid range conditions. PhD dissertation, Reihe Agrarwissenchaft Institute for Animal Production in the Tropics and Subtropics, Aachen, Germany (ISBN: 3–86111–706–1).

Le Houerou, H.N. 1992. The role of saltbushes (*Atriplex* spp.) in arid land rehabilitation in the Mediterranean Basin. A review. Agrofor. Syst. 18: 107.

Macfarlane, W.V. 1964. Terrestrial animal in dry heat: Ungulates. In Dill Handbook of Physiology, 4: 207–539. American Physiol., Soc. Washington.

McLaren, C.A., C.C. Anderson, L.I. Tsai and K.M. Barth. 1965. Level of readily fermentable carbohydrates and adaptation of lambs in all urea supplemental rations. J. Nutrition 87: 331.

Masters, D.G. (in press). Assessing the feeding value of halophytes, pp. 89–105, this volume.

Mosaad, G.M., A.N. Sayed and D.R. Ibrahim. 2003. Relationship between the dietary energy and the nutrients utilization, blood biochemical changes and follicular dynamics in dromedary she-camel (*Camelus dromedaries*). Assiut Vet. Med. J. 49(98): 46–70.

Pirzada, W.H., R.N. Khan and A. Ghaffar. 1989. Camel-an animal resources of Islamic World. 1- S&T in the Islamic World 7(2): 67–80.

Raza. H.S., M. Riaz and P.N. Raza. 2000. Effect of saltbush (*Atriplex amnicola*) on performance of goats on saline rangelands. J. Anim. Sci. 78(Suppl. 1): 126.

Schmidt-Nielsen, K. 1964. Desert Animals. Oxford University Press. London.

Schmidt-Witty, U., R. Kownatki, M. Lechner-Doll and M.L. Enss. 1994. Binding capacity of camel saliva mucins for tannic acid. J. of Camel Practice and Research, December, 121.

Schwartz, H.J. 1992a. Productive performance and productivity of dromedaries (*Camelus dromedaries*). Anim. Res. Dev. 35: 86–98.

Schwartz, H.J. 1992b. The biology of the camel. In The One Humped Camel in Eastern Africa: A pictorial guide to diseases, health care and management. Verlag Josef Margraf, Germany.

Shawket, S.M. 1999a. Effect of energy level supplementation on the utilization of some pasture plants by goats. J. Agric. Sci. Mansoura Univ. 24: 4565.

Shawket, S.M. 1999b. Fattening of camel calves on saltbush (*Atriplex halimus*) with different energy sources. J. Agric. Sci. Mansoura Univ. 24: 1751.

Shawket, S.M. and A.H. Ibrahim. 2013. Impact of long-term feeding *Atriplex* (saltbush) on camel's milk production efficiency under arid condition. Egyptian Journal of Nutrition and Feeds 16(2) Special Issue: 149–158.

Shawket, S.M. and M.H. Ahmed. 2001. The influence of the level of energy supplementation on the utilization of saltbush (*Atriplex nummularia*) by camels. Egyptian Journal of Nutrition and Feeds 4: 557–565.

Shawket, S.M. and M.H. Ahmed. 2009. Effect of prolonged feeding *Atriplex* (saltbush) to camels on digestibility, nutritive value and nitrogen utilization. Egyptian Journal of Nutrition and Feeds 12: 205–214.

Shawket, S.M., M.K. Youssef and M.H. Ahmed. 2010. Comparative evaluation of Egyptian clover and *Atriplex halimus* diets for growing and milk production in camel. Animal Science Reporter 4(1): 9–21.

Shawket, S.M., Z.K. Kewan, M.A. Nour and A.A.S. Mamdouh. 2005. *Atriplex* and *Acacia* shrubs as feedstuffs for young male camels under Egyptian semi-arid condition. Egyptian J. Nutrition and Feeds 8(1) Special Issue: 225–241.

Shehata, E., A. El-Serafy, A. Heider and F. Swidan. 1988. Nutritive evaluation of *Atriplex nummularia* by sheep and goats in the North West rangeland of Egypt. Proc. 2nd Conf. Agric. Develop. Res., Ain-Shams University, Shoubra El-Khiema, Egypt, pp. 120.

Shehata, E., A. Heider, I. Gomaa and F. Swidan. 1987. Utilization of *Atriplex* shrubs in feeding desert sheep and goats. 2nd Intern. Conf. on Desert Development, Jan. 25–31, Cairo, Egypt.

Siebert, B.D. and W.V. Macfarlane. 1971. Water turnover and renal function of dromedaries in the desert physiological zoology 44(4): 225–240.

Squires, V.R. (in press). Water requirements of livestock fed on halophytes and salt tolerant forage and fodders. pp. 287–302, this volume.

Wardeh, M.F. 1990. The nutrient requirements of the dromedary camels. Third Intr. Symp.: Relationship of feed information Centers (INFIC), Jun. 25–29, 1990, Univ. of Saskatchewan, Saskatoon, Canada. ACSAD/AS/P100/1990.

Warner, B.E. and T. Casson. 1992. Performance of sheep grazing salt tolerant forages on vegetated salt land. Aust. Soc. Anim. Prod. 19: 237–241.

Weston, R.H., T.P. Hogan and J.H. Hemsley. 1970. Some aspects of the digestion of *A. nummularia* by sheep. Proc. Aust. Soc. Anim. Prod. 8: 517.

Williamson, G. and W.I.A. Payne. 1990. An Introduction to Animal Husbandry in the Tropics. Longman, London.

Wilson, A.D. 1966a. The value of *Atriplex* (saltbush) and *Kochia* (blue bush) species as food for sheep. Aust. J. Agric. Res. 17: 147.

Wilson, A.D. 1966b. The tolerance of sheep to sodium chloride in food or drinking water. Aust. J. Agric. Res. 17: 503–514.

Wilson, A.D. 1977. The digestibility and voluntary intake of the leaves of trees and shrubs by sheep and goats. Aust. J. Agric. Res. 28: 501.

Wilson, R.T. 1984. The camel. Harlow, Essex, Longman Group Ltd., UK.

Wilson, R.T. 1992. Factors affecting weight and growth in one-humped camels. Proc. 1th Int. Camel Conf., Dubai, UAE. Pub. R&W Publications Ltd., 309.

PART 4

Physiological Aspects

Part 4 of 6 chapters reviews and synthesizes knowledge on the physiological impacts of heavy salt and toxin loads on digestion, reproduction and health, with special attention to rumen function. Water is both an important nutrient that is vital to body functioning (excretion, temperature regulation, food and energy transport, etc.) and obviously is affected by mineral ash intake that may increase when livestock eat feedstuffs high in sodium (Na) and potassium (K) and have access to water that may contain other electrolytes and minerals such as magnesium (Mg), calcium (Ca), etc. The role of water as a regulator of feed intake, and rumen function is explored in both ruminants and non ruminants. Reproductive physiology, including impact of high salt loads on fertility of both male and female, milk production and skin, hair and wool production in ruminants. Sheep, goats and camels are the main livestock considered here.

Impact of Halophytes and Salt Tolerant Plants on Physiological Performance of Livestock

G. Ashour,[1,]* M.T. Badawy,[2] M.F. El-Bassiony,[2]
A.S. El-Hawy[2] and H.M. El Shaer[2]

SYNOPSIS

The chapter covers different issues related to the effects of feeding halophytes and salt tolerant fodder crops on the physiological aspects of ruminants. It deals with the salt tolerance of different animal species and breeds, sources of salt intake, impact of salinity stress on animal physiological responses, hemo-biochemical parameters such as blood pressure, blood picture, liver and kidney function and hormones, etc. Data related to hemo-biochemical parameters such as blood pressure, picture and metabolites, liver and kidney function and hormones are also presented and discussed.

Generally, it is found that there are large differences in salinity tolerance between animal species or breeds within species. There are specific problems for livestock associated with consumption of salt in feed and water, for most salts, but particularly sodium, potassium and chloride. The animal's body has little or no capacity to store excess electrolytes or to actively excrete through the faeces.

[1] Faculty of Agriculture, Cairo University, Giza, Egypt.
[2] Desert Research Center, Cairo, Egypt.
* Corresponding author: gashour57@yahoo.com

Salinity tolerance of some livestock species (mg total soluble salts/liter water) and tolerance of salty drinking water by different livestock species are summarized. Animals can tolerate much higher salt concentrations if they have become accustomed to a salty water supply, than if forced to drink the more concentrated solutions without a preliminary conditioning period.

Keywords: livestock, salinity tolerance, salinity stress, blood picture, liver function, kidney function, feeding.

1. Introduction

Domestic animals are exposed to saline stress in many instances such as when grazing halophytes or when the saline water from underground wells is the only available drinking water. The ability of ruminants to withstand this type of stress is affected by numerous factors (Squires, this volume). The age and the physiological status are critical factors, where young and productive animals are more sensitive. The presence of other types of stress such as heat and feed shortage will augment the effect of saline load. In other words, salinity tolerance of animals could vary with environmental conditions, type of feed, or physiological stages (Anon 1969).

Sheep and cattle grazing green feed can tolerate higher concentrations of salt in water than the same stock on dry feed. Stocks grazing on saltbush are less tolerant than those grazing other types of pastures. Pregnant, lactating or young animals have a poorer tolerance to saline water than dry mature animals (Inglis 1985). Peirce (1959, 1962, 1966, 1968) and Wilson (1975) also reported that ewes and lambs are less tolerant of saline water than the non-pregnant and older sheep.

Salt tolerance has been interpreted as the absence of a depression in the intake of a certain ration. Under grazing condition, animals may ingest excessive salt through feed, drinking water, and ingestion of soil (Howell 1996).

The ability of an animal to withstand saline load depends on its ability to eliminate excess minerals, especially sodium; which in turn depends on the efficiency of kidney to concentrate the urine. Salt stress was found to affect the efficiency of domestic animals through affecting the steady state of the interior environment. Excessive salt intake (e.g., sodium) is one of the more commonly encountered problems and often causes loss of appetite, reduced milk production and reduced growth (Xu et al. 1994).

2. Salinity tolerance between animal species and breeds

There are large differences in salinity tolerance between animal species and breeds within species (NRC 1980; Inglis 1985; Judson and McFarlane 1998).

The NRC board of mineral requirements (NRC 1980) reported that the maximum tolerable concentrations were 35 and 16–35 g/kg DM for sheep and cattle, respectively. For sheep different tolerances were observed when fed on different rations or levels of intake due to individual differences in their ability to excrete or tolerate sodium chloride (Wilson 1966).

Sheep can tolerate 1.5% of sodium chloride in their drinking water as reported by Wilson (1978); the level of 1.3% is recommended as a safe maximum concentration (Peirce 1959), although 2.0% could be tolerated (Wilson 1966). In this respect camels proved to have the most ability to tolerate salt stress followed by sheep, while cattle, especially lactating ones are less tolerant.

Sheep with previous experience of saline water are more tolerant than young and inexperienced sheep. Inglis (1985) summarized the salinity tolerance of livestock based on previous research studies (Table 14.1).

Therefore, the safe values for salinity of 1.2% on grassland and 0.8% on saltbush were recommended by Wilson (1975), but these values should be further reduced for ewes and lambs, if lambs are born in summer.

Table 14.1. Salinity tolerance of some livestock species (mg total dissolved salts/liter water).

Animal	Maximum concentration* for healthy growth	Maximum concentration to maintain condition	Maximum concentration tolerated
Sheep	6000	13000	Depends on type of feed
Beef cattle	4000	5000	10000
Dairy cattle	3000	4000	6000
Horses	4000	6000	7000

*Parts per million (ppm) of Total dissolved salts (TDS).

3. Sources of salt intake

3.1 Saline drinking water

Many arid and semi-arid areas actually do have sources of water, but the available water is usually brackish (500–5000 ppm salt by weight) or saline (3–5% salt). The water may be present in underground aquifers or as seawater along coastal deserts. Water salinity might be considered an important factor in determining the suitability for a particular source in livestock (Ray 1989; El-Bassiony 2013). Surface water supplies are not regarded as a major source of minerals for the grazing animal. However, evaporation, which is more significant in summer, from streams and lakes, results in a concentration of the dissolved solids.

Groundwater is, generally, more saline than surface water from being in contact with soil and rock for longer periods and has been concentrated by plants through transpiration. Judson and Mcfarlane (1998) indicated that the quality of underground water is variable and significant quantities of Na, Ca and Mg salt which may be consumed by grazing animals. Saline water may often be the only sources of drinking water for animals in the deserts of many arid and semi-arid regions in the developing countries such as Egypt.

The salinity of well water also may vary from 360 to 30082 ppm TDS (Atwa 1979; Aggour 1990). Around 10000 mg/liter of total dissolved salts is the maximum tolerable concentration for mature non-lactating livestock (Judson and McFarlane 1998).

Water with the higher cation concentration could provide about 20–100% of Na and about 5–20% of the daily Ca and Mg needs of the mature animals. Total salinity is higher in summer than in winter.

Brackish water refers to water that is saltier than fresh water, but not as salty as sea water. It may result from mixing of seawater with fresh water, or it may occur naturally. Technically, brackish water contains between 0.5 and 30 grams of salt perlitre (5000 to 30000 ppm TDS). King (1983) explained that the response of livestock to highly saline drinking water is to increase their water intake, but at a certain concentration the appetite becomes depressed and fluid intake is reduced.

The reason is that the higher concentration of salt requires a greater proportion of the water ingested to be used for salt excretion, until not enough water is left for other functions. The salt concentration at which this depression occurs is a measure of arid adaptation in a species (McFarlane 1971; Maloiy 1972) (as shown in Table 14.2). In practice, if animals become accustomed to a salty water supply they can tolerate much higher salt concentrations than if forced to drink the more concentrated solutions without a preliminary conditioning period (French 1956).

Table 14.2. Tolerance of salty drinking water by different livestock species.

Species	% total salts in drinking water
Camel	2.5
Goat	1.5
Sheep	1.3–2.0
Cow	1.0–1.5
Donkey	1.0
Horse	0.9

3.2 Halophytes and salt tolerant forages

A halophyte is a plant that naturally grows where it is affected by salinity in the root area or by salt spray. Animals can ingest salt from water, pastures, soils (Wilson 1978) and other mineral supplements, making it difficult to quantify the salt intake under grazing conditions. Salt intake of grazing animals, as reported by El Shaer (2004) is often influenced by environmental conditions, pasture management, and animal factors (maturity stage, selective grazing, species or breed, etc.).

The sodium content in saltbush is much higher than the level recommended for beef cattle (0.06%, NRC 2000); for goats (0.05%, NRC 1981) and for sheep (0.04–0.10%, NRC 1975). However, these plants can be used as animal feeds (Masters, this volume).

There is a variation in nutritional quality among species of saltbushes. Digestible dry matter ranges from 74.5% to 78.8% and digestible energy from 3.2 Mcal/kg to 3.4 Mcal/kg (Khalil et al. 1986). Rehman (1992) also found that among 8 saltbush species, the dry matter digestibility ranged from 33% to 62%, and nitrogen digestibility varied from 51% to 81%; crude protein content varies from 10% to 20% and ash content varies from 13% to 23% (Masters, this volume).

El Shaer and Gihad (1994) reported that some salt bushes in Egyptian deserts are of nutritional importance due to their moderate contents of crude protein. The lower palatability of such plants by different animal species were found to be due to the presence of secondary plant metabolites such as acid detergent fibers, acid detergent lignin, and hemicellulose (Meyer and Karazof 1991; Kandil and El Shaer 1990; Masters, this volume).

The authors' board of National Research Council (NRC 1980) explained that all mineral elements, whether essential or nonessential, can adversely affect an animal if included in the diet at excessively high levels. Tolerance levels vary from animal to animal. Many factors, such as age and physiological status of the animal (growth, lactation, etc.), nutritional status, levels of various dietary components, duration of exposure and biological availability of the compound, influence the level at which a mineral element causes an adverse effect.

Maximum tolerable salinity level is defined as the dietary level that, when fed for a limited period, will not impair animal performance and should not produce unsafe residues in human food derived from the animal. Greater sensitivity to high mineral levels can be expected in animals that are young, pregnant, lactating, malnourished, or diseased.

There are specific problems for livestock associated with consumption of salt in feed and water (El-Bassiony 2013; El-Hawy 2013) for most salts, but particularly sodium, potassium and chloride (Attia-Ismail, this volume). The animal's body has little or no capacity to store excess electrolytes or to actively excrete through the faeces (Al-Khalasi et al. 2010). Table 14.3 summarizes the recommended tolerable levels suggested by NRC (1980) for sheep and cattle.

Table 14.3. Maximum tolerable level of dietary minerals for cattle and sheep.

Element, ppm	Species	
	Cattle	Sheep
Boron	150	150
Calcium	20000	20000
Cobalt	10	10
Copper	100	25
Iron	1000	500
Lead	30	30
Magnesium	5000	5000
Manganese	1000	1000
Phosphorus	10000	6000
Potassium	30000	30000
Selenium	2	2
Silicon	2000	2000
Sodium Chloride	Lactating 40000 Dry 90000	90000
Sulfur	4000	4000
Zinc	500	500

4. Effect of salinity on hemo-biochemical parameters

The blood system is sensitive to the changes which take place during the stress reactions and it is used as an indicator to monitor health (Zamet et al. 1979), reproduction (Parker and Blowey 1976), nutrition (Lee et al. 1976; Lee and Twardock 1978) and physiological status of the animals (Hussein et al. 1992; El-Bassiony 2013). Blood analysis is considered as a good tool for studying impact of salinity on animal health. In some cases the animals may suffer from subclinical elements toxicity without any distinguished symptoms.

4.1 Hematological profile

4.1.1 Erythrocytes cell count

The average erythrocytes cell count of Barki × Merino rams subjected to low (0.7% NaCl) or high (1.3% NaCl) salt water was slightly greater as compared to that of the control group subjected to fresh water as reported by Hussein (1987). Erythrocytes count of rams was elevated due to drinking (0.7% NaCl) salt water and an opposite trend was observed by Hussein et al. (1990) when salt concentration was 1.3% NaCl.

El-Hassanein and El-Sherif (1996) found that saline-treated sheep showed a higher red blood cell count (RBC) than non-saline treated group. In comparison between camels and sheep drinking saline water, Assad et al. (1997) showed that the RBCs count was consistently but not significantly lower in both animal species that received low or high salt water than in those that received fresh water. The count of RBCs was 8.26×10^6/ml of camels that received fresh water, 7.81×10^6/ml for those that received low salt water (7700 ppm TDS) and 7.50×10^6/ml for those that received high salt water (13500 ppm TDS). They, also, reported that this effect was less in camels than in sheep.

Moreover, a study was conducted by Abdelhameed et al. (2006) on four camel groups (Treatments) fed berseem hay[1] (Control, T1), while the other three groups were fed on fresh *Atriplex nummularia* (T2), fresh *Acacia saligna* (T3) and mixture from *Atriplex* with *Acacia* (T4). The results of this study revealed that camels that were fed fresh halophytic plants (*Atriplex* or *Acacia*) showed highly significant decrease of RBCs whether in summer or in winter seasons than that of the control group. When buffalo calves drank saline water, Ibrahim et al. (1991) noticed that red blood cells count in male buffalo calves was not affected significantly by drinking saline water containing 6000 ppm TDS.

Similar results were recorded in goats that drank saline water (NaCl supplementation in drinking water) in comparison with goats that received either fresh or high salty water (Ibrahim 1995).

4.1.2 Leukocytes cell count (WBCs)

As reported by several studies, salinity in drinking water or in animal feed significantly affects the WBCs count of all animal species as there was a decrease in the total

---

[1] *Trifolium alexandrinum.*

white blood cells count in animals (sheep, goats, camels, and cattle) treated by salty water or fed halophytes (Kawashti et al. 1983; Hussein et al. 1990; Assad et al. 1997; El-Bassiony 2013; El-Hawy 2013). A decrease in total differential leukocytes cell counts was recorded in sheep after 3 months of drinking salt water (Hussein 1987).

Similarly, the total white blood cells count decreased in rams subjected to 0.7 or 1.3% NaCl in drinking water for 4 months as reported by Hussein et al. (1990). Assad et al. (1994) found that sheep and camels receiving saline water containing 7650 and 13535 ppm TDS for 40 days exhibited a considerable elevation in white blood cell counts (9.5×10^3 and 18.6×10^3) as compared to the control group (16.3×10^3 and 31.6×10^3). El-Hassanien and El-Sherif (1996) reported that saline treated growing rams showed a marked reduction in white blood cells, segmented neutrophils, eosinophils, basophils and lymphocytes counts as compared to the values of those that drank fresh water.

The WBCs count of fresh water group was 37.2×10^3/ml, while that of low salt-water group (7700 ppm) was 17.0×10^3/ml and for high salt-water group (13500 ppm) was 16.6×10^3/ml (Assad et al. 1997b).

In contrast, Blood and Rodostitis (1989) reported that sheep and cattle drinking saline water showed a syndrome often described as chronic salt poisoning, which was usually accompanied by diarrhea and gastroenteritis. These symptoms mostly related to leukocytes. In other words, it may be related to the increase of the leukocytes cell count. Ibrahim (1995) reported that in goats, the increase in neutrophils might reflect an incidence of inflammation in response to drinking salty water. This author found non-significant drop in lymphocytes of goats due to drinking saline water containing 9178 or 13760 ppm TDS suggesting a stressful effect of the salt load, consequently increasing the release of ACTH, which resulted in a destruction of lymphocytes according to Makman et al. (1967).

Values of total leukocytes cell count, lymphocytes, monocytes and eosinophils of three treated camel groups (fed on *A. nummularia* T2; *Acacia saligna* T3 and a mixture of both plants T4) were more than that of the control one (fed berseem hay, T1) in summer or winter seasons (Abdelhameed et al. 2006).

On the other hands, Al-Khalasi et al. (2010) reported that animals fed on Rhodes grass (*Chloris virgata*) hay or salt tolerant *Sorghum* irrigated with water containing three different concentrations of salt (3, 6 and 9 ds/m) did not show any significant differences in WBCs counts.

4.1.3 Hemoglobin concentration

Hemoglobin (Hb) in red blood cells is the second most important buffer in blood, after bicarbonate buffer system. The function and importance of Hb in oxygen transport system emphasize the importance of its significance as an internal part of the metabolic system in mammals. In addition, it is possible to rely on Hb concentration for selection within a breed under hot desert climate or for importing high producing breeds to arid and semi-arid areas to increase productivity in these regions (Seerley et al. 1972).

Hb varies within breed, individuals and age as affected by climatic season and environmental temperature. Reece (1991) stated that the average Hb concentration in domestic animals is about 12 g/dl. Hussein (1987) reported that Hb concentration in group of rams receiving a relatively high level of salty water (1.3% NaCl) was significantly lower than that of rams that received a lower level of salt water (0.7% NaCl). Moreover, Hussein et al. (1990) and El-Hassanein and El-Sherif (1996) working on sheep, recorded an elevation in the concentration of Hb due to drinking saline water.

Similarly, Ibrahim (1995) in goats and Badawy (1999) in rams found that Hb decreased due to drinking saline water and this might be due to greater phase of hemodilution especially under hot conditions and increased water intake.

On the other hand, El-Sherif (2003) found that drinking diluted seawater (12494 ppm TDS) for growing sheep resulted in appreciable increase in blood Hb (from 9.95 to 10.77 g/dl). However, Assad et al. (1994, 1997b) found that Hb concentration in groups of sheep and camels subjected to a relatively high level of salt water (13535 ppm TDS) did not exhibit significant differences compared to those subjected to a lower level of 7650 ppm TDS. Ibrahim et al. (1991) reported similar findings in male buffalo calves drinking well water containing 6000 ppm TDS for 60 days.

Four camel groups were used by Abdelhameed et al. (2006); the first served as a control that fed on berseem hay while the other three treated groups (T2, T3 and T4) were fed on fresh *Atriplex nummularia* (T2), fresh *Acacia saligna* (T3) and mixture from *Atriplex* with *Acacia* (T4). They found that Hb concentration wasn't significantly affected by feeding halophytic plants.

4.1.4 Packed cell volume (PCV %)

Packed cell volume (PCV) or hematocrit (Ht): the relative proportion of cells to plasma is considered to be the most useful for helping to distinguish abnormal conditions of the animal's body (Reece 1991). The volume of PCV is directly related to the erythrocyte size count and Hb content. Most species of domestic animals have PCV values from 38 to 45% with a mean of 40% (Reece 2005). Many factors are known to affect hematocrit such as sex, age, species, breed, feeding, environmental temperature, dehydration, pregnancy and lactation (Amakiri and Funsho 1979).

The effect of salt intake on the packed cell volume showed much more alteration as compared to other hematological parameters. In this respect, Kawashti et al. (1983a) recorded an increase in the PCV % of sheep due to drinking salt water at concentration of 1 or 1.5% NaCl. Similar findings were obtained by Omar (1969) in camels and donkeys, Badawy (1999) using mature rams and El-Hassanein and El-Sherif (1996) using growing rams and El-Bassiony (2013) using goats.

Furthermore, Assad et al. (1994) found that sheep and camels receiving saline water containing 7650 and 13535 ppm TDS for 40 days exhibited a considerable elevation in PCV values (40.9% and 35.3% for sheep and camels respectively) as compared to the control animals (38.3% and 30.0%, respectively). Camels and sheep that received diluted seawater containing 7700 and 13000 ppm TDS for consecutive periods of 40 days each showed elevation, but it was non-significant, in PCV as compared to the control animals (fresh water) as reported by Assad et al. (1997b). An opposite trend was observed in sheep as demonstrated by Hussein (1987) and Hussein et al. (1990). They recorded a decrease in PCV % due to drinking saline water. On the other hand, Singh and Taneja (1981) and Hussein et al. (1990) in sheep and Ibrahim et al. (1991) in buffalo calves did not record any appreciable change in PCV % due to drinking saline water.

4.2 Erythrocytes indices

Wintrobe indices include: mean corpuscular volume (MCV), mean corpuscular hemoglobin (MCH) and mean corpuscular hemoglobin concentration (MCHC). Values of MCV, MCH and MCHC of rams decreased due to drinking saline ground water, especially at the high level of 1.3% NaCl as reported by Hussein (1987). The same author found a drop in percentage of change in proportional to zero time in MCV and MCH due to saline water (1.3 NaCl).

Meanwhile, Hussein et al. (1990) showed that in rams receiving 0.7 or 1.3% NaCl in water for 6 months, there was a decrease of MCV and an increase in MCH and MCHC at the 4th month of age. At the 5th month, there was a decrease in MCV, MCH and MCHC in animals receiving 0.7% NaCl and an elevation in those receiving 1.3% NaCl in water. Similarly, El-Hassanein and El-Sherif (1996) working on growing rams that were subjected to prolonged saline water treatment (13100 ppm TDS), recorded a higher rate of elevation in MCH and MCHC, but marked reduction of MCV in comparison to control animals. Moreover, El-Sherif (2003) found that drinking diluted seawater (12494 ppm TDS) for growing sheep resulted in increasing MCHC (from 30.4% to 33.4%).

MCV, MCH and MCHC of sheep and camels that drank saline water were tested (Assad et al. 1997b) and it was observed that their concentrations increased in salt treatment groups over those drinking fresh water. Average values of MCV of camels that drank fresh water (280 ppm TDS), low (7700 ppm TDS) and high (13500 ppm TDS) saline water were 38.0, 40.3 and 47.3 fl, respectively. Values of MCH were 13.4, 14.5 and 16.7 pg, respectively. For MCHC the corresponding values were 35.2, 36.0 and 35.3 g/dl, respectively. Values of MCV were less in camels than in sheep. Average values of MCV of sheep changed between 40.9 to 62.3 fl. The other two parameters, MCH and MCHC, were greater in camels than in sheep.

The results reported by Abdelhameed et al. (2006) indicated that there were increases in MCV and MCH of the treated camels groups that were fed fresh halophytic plants (*Atriplex nummularia*; *Acacia saligna* or a mixture of these plants) compared to those fed on berseem hay (control) in both summer and winter seasons; these values of the treated groups were less than the control in summer and more than the control group in winter which might be due to hyperchromic anemia (i.e., anemia in Hb weight). However, Ibrahim (1995) noticed that in goats, the percentage change in MCV relative to zero time in low (9178 ppm TDS) and high (13760 ppm TDS) saline treated animals showed an increase in both first and second months from the start of the experiment.

4.3 Blood metabolites

Several studies showed that blood metabolites including glucose (Glu), Total proteins (TP), total lipids (TL) and total cholesterol (TC) could be used as good and reliable indicators of the adequacy of feeding management. These metabolites fluctuate, either by increasing or decreasing according to the nutritional status (quantity and quality).

4.3.1 Glucose

Glucose (Glu) concentration decreased in animals fed salt tolerant plants (STP) or drinking saline water (Assad and El-Sherif 2002; Ismail et al. 2003; Kewan 2003; Shaker et al. 2008).

Barki sheep fed STP mixture had a lower ($P < 0.01$) Glu level by about 18% than those fed control ration, 49.9 vs. 39.8 mg/dl (Shaker 2014). Also, feeding STP mixture of sun dried shrubs (*Prosopis Juliflora; Acacia saligna* and *Leucaena leucocephala*) to Shami goats reduced ($P < 0.05$) Glu values by about 24%, 51.1 vs. 41.3 mg/dl (Shaker et al. 2014). Conversely, Al-Khalasi et al. (2010) found that serum Glu levels were higher ($P < 0.01$) in Omani sheep fed high salinity levels especially at the end of the feeding period (11 wks) compared to those fed other diets. They found the Glu levels ranged between 33 and 62 mg/dl in animal fed sorghum fodder grown at different levels of salinity compared to those fed a control diet of Rhodes grass (*Chloris*) hay which had a range of 25 to 50 mg/dl. However, El-Bassiony (2013) found no significant differences in Shami male goats fed STP and drinking SW (49.6 mg/dl) and those fed berseem hay and drinking FW (50.6 mg/dl).

The reduction in Glu values could be related to high content of tannins in STP. Ortiz et al. (1993) reported that tannins could adversely influence digestibility and absorption of nutrients and the activity of digestive enzymes. High salt content in STP is another reason of reducing Glu levels as evidenced in rams receiving SW (Assad et al. 1997b) and in camels fed *Acacia* plus *Atriplex* (Kewan 2003). Furthermore, high salt content in STP may be the major factor which limits intake and reduces digestibility by shortening rumen turnover rates (Warren and Casson 1992).

4.3.2 Total proteins

Blood total protein (TP) and its fractions, albumin (Alb) and globulin (Glb) have a great importance as good indicators of nutritional status. Feeding STP decreased TP in ruminant animals (Ibrahim 2001; El-Bassiony 2013; El-Hawy 2013). Also, feeding fresh acacia decreased TP, Alb and Glb values in growing Barki lambs (Badawy et al. 2002; Shaker et al. 2008) and in Baladi kids (Badawy et al. 2002). Ibrahim (2001) reported that goats fed *Atriplex* or *Acacia* had lower value of TP compared with those fed BH.

Shaker et al. (2008) studied the effect of feeding STP on blood TP of three groups of Barki male lambs. The three groups were fed on berseem hay (BH), G1, fresh mixture of *Atriplex halmus* and *Acacia saligna* (G2) and silage of this mixture (G3). They found that TP significantly decreased in G2 compared with the other two groups with values of 8.6, 6.5 and 7.8 g/dl for G1, G2 and G3, respectively. Whereas Alb insignificantly decreased in G2 compared with the other two groups with values of 3.9, 3.1 and 3.5 g/dl for G1, G2 and G3, respectively. Values of Alb were higher ($P \leq 0.05$) in G2 followed by G1 and G3. In addition, Shaker et al. (2014) found that Shami goats fed salt tolerant shrubs mixture had a slight and insignificantly lower concentration of TP, Alb, Glb and A/G ratio than their counterparts of control group that were fed BH. These decreases ranged between 4.7 to 13.8% in all protein fractions. In this range, the reduction in TP, Alb and Glb were about 5%, 7% and 2%, respectively in Barki sheep fed STP mixture of *Atriplex nummularia, Sorghum bicolor* and pearl millet (*Pennisetum glaucum*) in comparison with control group. However, the differences were not significant (Shaker 2014). In contrast, Fayed et al. (2010) studied the utilization of alfalfa (*Medicago sativa*) and *Atriplex* for feeding sheep under saline conditions of South Sinai. They found that sheep fed 100% *Atriplex* plus barley[2] (R4) or 100% alfalfa (R5) had insignificantly higher values of TP (6.6 and 7.7 g/dl) and Alb (3.2 and 3.3 g/dl), respectively compared to those fed BH plus CFM which had a concentration of 6.4 g/dl for TP and 2.8 g/dl Alb. Whereas, values of Glb (g/dl) were higher in sheep fed R5 and lower (3.1) in those fed R4 compared to those in R6 group (3.6). On the other hand, Al-Khalasi et al. (2010) reported that no significant differences were found in TP values of Omani sheep fed sorghum fodder grown under various levels of salinity compared to those fed control diets.

The combined effect of both feed and water salinity had more deleterious effect on blood TP as reported by El-Bassiony (2013). He found Shami male goats fed STP and drinking salty water (SW) had a lower value of TP than those fed BH and drinking fresh water (FW) by about 20.7%. The water salinity adversely affected TP than that of feed salinity. El-Bassiony (2013) proved that goats fed STP and drinking SW or fed BH and drinking SW had lower TP values (6.6 and 7.4 g/dl) than those fed BH and drinking FW or those fed STP and drinking FW (8.3 and 8.1 g/dl respectively). Furthermore, El-Hawy (2013) found that doe goats fed STP had lower (P < 0.05) TP values than those fed BH (6.0 vs. 6.5 g/dl) and those drinking SW had lower TP than those that drank FW (5.9 vs. 6.6 g/dl). However, the influence of drinking SW on serum profiles of Holstein steers was studied by Robert et al. (1992). He found that TP, Alb, Glb and A/G ratio were not different between control steers which drank water that contained 350 ppm TDS and treated steers which drank water containing 2300 ppm TDS.

With respect to Alb concentrations, Ibrahim (1995) found that serum Alb was lower in goats that received diluted seawater (13760 ppm TDS) compared to the control group that received fresh water and those that received diluted seawater (845 ppm TDS). Also, El-Hawy (2013) found that Alb decreased by feeding STP compared to BH in doe goats (2.6 vs. 3.3 g/dl). Also, drinking SW decreased Alb levels compared to drinking FW (2.8 vs. 3.1 g/dl). Moreover, Shami male goats fed STP or drinking SW had insignificantly lower values of Alb compared to the control group (El-Bassiony 2013).

Regarding Glb concentrations, values of Glb increased in doe goats by feeding STP compared to BH feeding (3.4 vs. 3.2 g/dl) and decreased by drinking SW compared to FW (3.1 vs. 3.5 g/dl), however the differences were not significant (El-Hawy 2013). Conversely, El-Bassiony (2013) found that male goats fed STP or BH, but drinking SW had a decrease in their Glb values compared to those fed STP or BH, but drinking FW, respectively. So, the effect of water salinity was more obvious than that of feed salinity.

---

[2] *Hordeum vulgare.*

Concerning A/G ratio, Badawy et al. (2002) reported that the animals fed *Atriplex nummularia* or *Acacia saligna* showed lower A/G ratio, but the differences were not significant. Ratios of A/G ranged from 0.9 to 1.9 in doe goats (El-Hawy 2013) and from 1.5 to 2.9 in male goats (El-Bassiony 2013), both male and female goats received different combinations of STP, BH, SW and FW.

The reduction of TP and its fractions in animals fed STP may be attributed to the high content of tannins which form complexes with dietary proteins which probably decreased the digestibility of crude protein (Muller et al. 1989; Reede et al. 1990). Also, this reduction may be attributed to the high content of salt in STP. Badawy (1999) reported that drinking SW significantly decreased TP, Alb, and Glb of sheep, suggesting that the decline in Alb may be due to the decrease in active transport of amino acids needed for building Alb rather than Glb in hepatic tissue. It is well known that Alb is synthesized only in the liver. So, the decline in Alb concentrations might reflect reasonable impairment of liver function.

Based on the concentrations of TP, Alb and Glb in previous studies, it is clear that all values are within the normal reference ranges of these metabolites. These values indicating that the experimental animals were in good nutritional status and their liver had no damage since the TP, particularly Alb are mainly synthesized in the liver. The normal ranges in sheep and goats are 5.9–7.8 and 6.1–7.5 g/dl for TP, 2.7–3.7 and 2.3–3.6 g/dl for Alb and 3.2–5.0 and 2.7–4.4 for Glb, respectively (Latimore 2003).

4.3.3 Total lipids

The plasma total lipids (TL) are chiefly triglycerides, phospholipids and cholesterol. Wide variations existed in blood TL levels among different animals.

Feeding salt tolerant plants (STP) to ruminants either significantly or insignificantly decreased TL values were reported by many authors in goats (El-Bassiony 2013; Shaker et al. 2014), in sheep (Abdelhameed et al. 2006; Shaker et al. 2008) and in camels (Abelhameed et al. 2004). Shami goats fed STP had lower (P < 0.01) TL values than their counterparts of control group that were fed BH by about 15%, 300 vs. 270 mg/dl (Shaker et al. 2014). Meanwhile, Barki sheep fed STP mixture decreased (P < 0.01) their TL by 19.3 compared to the control one, 300 vs. 254 mg/dl (Shaker 2014).

Increasing salt load greatly decreased TL concentrations as evidenced by El-Bassiony (2013). He demonstrated that the TL values ranged from 207.8 to 403.1 with an average of 283.7 mg/dl in Shami male goats fed STP and drinking SW. Whereas, those fed BH and drinking FW had a range of 274.7 to 513.6 with an average of 374.4 mg/dl. The salt loaded group had a lower value than that of control group, on average, by about 27%.

Blood TL concentrations depend mainly on the quality and quantity of the consumed ration. Since the STP had a lower fat content, the animals fed on these STP utilized their stored body fat for energy requirements.

4.3.4 Total cholesterol

Feeding STP significantly reduced blood total cholesterol (TC) concentrations in comparison with those fed traditional rations in sheep (Shaker et al. 2008; Fayed 2009; Shaker et al. 2014) and in goats (El-Bassiony 2013; Shaker et al. 2014).

The reduction in TC differed according to species and type of STP in addition to salt load. It was 9.4% in Barki sheep fed STP mixture compared to control ration (93.7 vs. 102.6 mg/dl) as found by Shaker (2014). Also, it was found to be about 18% in Shami goats fed STP mixture compared to the control ones, 84.0 vs. 99.1 mg/dl (Shaker et al. 2014). Moreover, this reduction increased enormously with increasing salt load by feeding STP and drinking saline water as demonstrated by El-Bassiony (2013). He found that Shami male goats fed STP and drinking saline water showed a marked decrease in TC concentrations by about 37.2% compared to those fed Berseem hay and drinking fresh water (82.8 vs. 131.9 mg/dl).

Based on the results of the previous studies, it could be noticed that all values of TC concentrations in animals fed either STP or control rations are within the normal range for sheep—60 to 150 mg/dl, as reported by Reece (2005).

The reduction in blood TC may be caused by the anti-nutritional factors in STP that affected lipids profile indirectly. Tannins play a profound role in lipids digestibility by complexing with fatty acids (Romero et al. 2000). Decreasing TC absorption and increasing fat excretion (Bravo et al. 1993). In addition, saponins content of acacia and different sources of STP may affect TC levels (Mastsura 2001; Salem et al. 2004; El Shaer et al. 2005; Fayed 2009; 2010; Ben Salem et al. 2010). The hypocholesterolaemic action of saponins was also reported by Francis et al. (2002). This effect of saponins could be achieved through inhibiting the TC absorption causing reduction in plasma non-high density lipoprotein cholesterol fraction (Morehouse et al. 1999), and/or delaying the intestinal absorption of dietary fat by inhibiting pancreatic lipase activity (Han et al. 2000). Furthermore, it could affect the digestibility of fats in ruminants which is limited by the lack of emulsifying agents in the rumen (Cheeke 1999).

4.4 Blood electrolytes

Minerals play an important role in the regulation of body fluids, acid-base balance and metabolic processes (Milne 1996). The effect of salt load either in feed and/or in water on blood minerals obviously varied according to several factors.

Blood sodium (Na) increased by feeding STP and/or drinking SW as reported by many workers (Rasool et al. 1996; Badawy et al. 2002; Nasr et al. 2002; El-Bassiony 2013). This increase was about 17% in Barki sheep (Shaker 2014). The increase in blood Na could be attributed mainly to the high content of Na in STP. Mohammed (1996) found that Atriplex contains high content of Na (6.5%) and Cl (7%). Ben-Salem et al. (2010) reported that most of the salt in oldman saltbush (*Atriplex nummularia*)is NaCl and KCl. Also, blood potassium (K) increased by feeding STP (Badawy et al. 2002; El-Bassiony 2013). This increase was about 40% in Barki sheep

(Shaker 2014). Ewes fed fresh form of *Atriplex* and *Acacia* had higher Na and K values than control one (Shaker et al. 2008). It is well known that Na and K levels are high in STP (El Shaer 1981; Donia and Ibrahim 2013). And consequently excess intake of the electrolytes is accompanied by their excess excretion through the kidneys. An excess of K can aggravate a marginal Na deficiency. This can even occur when high forage (pasture, hay or silage) diets are fed. Certain pasture may have up to 18 times more K than Na (Berger 2006).

El-Bassiony (2013) found that Shami male goats fed STP and/or drinking SW in different combinations (three groups) showed higher blood electrolytes than those in control group. In addition, he found that the salt load of the combined effects of consuming STP and drinking SW (G4) were more pronounced and markedly increased blood electrolytes compared to those fed BH and drinking fresh water (G1). These increases were 43.5% (155.3 vs. 108.2 mEq/L) for Na, 19.6% (7.9 vs. 6.6 mEq/L) for K, 77.5% (10.0 vs. 5.6 mg/dl) for Ca, 59.6% (10.3 vs. 6.4 mg/dl) for P, 18.3% (113.1 vs. 95.6 mEq/L) for Cl, 72.1% (3.1 vs. 1.8 mg/dl) for Mg and 15.8% (5.4 vs. 4.7 µg/ml) for Zn. Moreover, he noticed that the magnitude of changes in Ca, Mg, P and Na due to feeding STP and/or drinking SW were more obvious than those in Cl, K and Zn.

In this connection, El-Hawy (2013) found that Shami doe goats fed STP in comparison with those fed BH (control) had higher blood values (mg/dl) either significantly for Mg (2.3 vs. 1.9), Cl (101.9 vs. 92.9), K (9.1 vs. 8.1) or insignificantly for Ca (2.8 vs. 2.7) and Na (162.0 vs. 161.5). But, P level tended to be decreased in STP groups (6.0 vs. 6.1). On the other hand, he found that doe goats that drank SW compared to those that drank FW had higher blood values (mg/dl) either significantly for Mg (2.3 vs. 1.9) and Na (164.3 vs. 159.8) or insignificantly for Cl (98.1 vs. 96.7) and Ca (2.8 vs. 2.7). However, K (8.4 vs. 8.8) and P (5.86 vs. 5.93) slightly and insignificantly decreased in SW groups. Levels of Ca and P seemed to remain almost constant with feeding STP or drinking SW, in spite of the insignificant and slight changes in their levels.

The stabilization, or at least minor changes, of Ca and P under salinity conditions agree with the previous studies. Jaster et al. (1978) reported that Ca and P levels in blood were unchanged and remained relatively constant in cows drinking SW (2500 ppm NaCl). Also, Amer (1990) found, in goats, that Ca and Mg were not affected by drinking SW. Furthermore, El-Hassanein et al. (2002) demonstrated that levels of Ca exhibited non-significant differences in animals fed halophytes in fresh or silage forms and their control one (BH).

Shami goats fed STP (*Sorghum vulgare*) and pearl millet (*Pennisetum glaucum*) showed higher serum values of Na, K, Mo and Zn and lower values of Ca, Mg, Cd, Cr and Pb than those fed wheat straw hay (Donia and Ibrahim 2013). Also, Donia et al. (2014) proved that Shami goats fed STP (the same previous plants) significantly increased blood serum K and decreased Ca, while Na and Mg concentrations increased insignificantly compared to control group that were fed traditional diets (wheat straw). In addition, goats fed STP showed higher values of Mo, Pb and Zn, while they had lower values of Cd and Cr than their counterparts of the control group. However, Shaker et al. (2014) found that Na, K and Ca were comparable with non-significant differences in Shami doe goats fed STP and those fed BH.

Barki sheep fed silage form of STP had lower (P < 0.01) plasma Mn and Se and higher Cu and Zn concentrations compared to those fed BH (Amer et al. 2014). He concluded that feeding silage form of STP was not hazardous and recommended that supplementing trace elements to the rations to improve the antioxidants capacity (deference mechanism) and to enhance growth performance and animal productivity.

The effect of feeding STP on blood electrolytes differed according to the physiological status of the animals and type of plants. Lactating ewes fed saltbush showed a net loss of Ca, but an increase in the blood serum concentration of P, Se and Mo (Alazzeh and Abu-Zanat 2004). Also, Baladi doe goats fed STP showed non-significant differences with those fed control ration in Ca levels during early pregnancy, early lactation and late lactation. But, during late pregnancy the differences were significant. Level of Ca (mg/dl) ranged from 8.3 to 20 during all different physiological stages of pregnancy and lactation being higher (13.7 to 20) in the pearl millet hay group and lower (8.3–12) in the *Kochia indica* hay group compared to the clover hay group (10.3 to 16.7). This may be due to the presence of oxalate in *Kochia* which decreases calcium availability (Hanafy et al. 2007).

Masters et al. (2007) have warned about the risk of Cu deficiencies in sheep fed saltbush due to the high salt content of the forage. Mayberry et al. (2009) reported that sheep grazing saltbush for an extended period of time with supplementation could develop Mg, Ca and P deficiencies. This could cause hypomagnesemictetany (grass tetany) or hypocalcaemic (milk fever), both of them decrease animal production and may result in animal death. High tannins content in STP were found to be disturbing the absorption of minerals through the gastrointestinal tract or increasing the endogenous losses of minerals such as Ca, Mg and P (Mansoori and Acamovic 1997).

5. Kidney functions

The kidneys play an important role in the regulation of water balance, electrolyte balance, acid-base balance and maintenance of osmotic pressure of body fluids and in the removal of metabolic waste products and other toxic substances (Sherwood 1997). Along with the liver, the kidneys are part of the waste processing system of the body. Urea (BUN) and creatinine (CR) are the two main nitrogenous compounds eventually excreted by kidney. Accordingly, any change of their concentration would reflect impaired glomerular filtration and/or insufficiency of renal tubules (Kaneko 2005). Therefore, BUN and CR concentration are the main indicators of kidney functions. The normal BUN for animals were 8–40 mg/dl (Kaneko 2005).

Animals fed STP showed lower values of BUN than those fed control rations (Azamel 1997; Patra et al. 2002; Cook et al. 2008; Pearce et al. 2008; Shaker 2014; Shaker et al. 2014). This decrease was 6.2% in Barki sheep fed STP mixture than those fed control ration (36.8 vs. 42.7 mg/dl). Whereas, in another study this decrease reached about 26% in Shami doe goats fed STP mixture of sun-dried *Prosopis*, *Acacia* and *Leucaena* in comparison with the control group (35.5 vs. 44.8 mg/dl) as found by Shaker et al. (2014). Also, Azamel (1997) found that growing lambs fed *Atriplex nummularia* had significantly lower BUN concentration in comparison with control group. Patra et al. (2002) reported that BUN concentration in Leucaena meal group was lower (P < 0.01) compared to the control group. Furthermore, Cook et al. (2008) reported that adding *Prosopis glandulosa* in goats' ration by 30, 60 and 90% resulted in decreasing BUN concentration as compared to control ones. The decrease was correlated positively with increasing *Prosopis* provision.

Conversely, BUN increased due to STP feeding (Bayoumi et al. 1990; El Shaer et al. 1991; Badawy et al. 2002). The last named author found that animals fed fresh *Atriplex nummularia* had high BUN compared with control group.

On the other hand, no significant difference was found between animals fed STP and those fed control rations. Al-Khalasi et al. (2010) reported that there were no significant differences in BUN of Omani sheep fed salt-tolerant Sorghum forage grown under various levels of salinity or Rhodes grass hay. Also, El-Bassiony (2013) and Ashour et al. (2014) found that feeding STP did not show any adverse effects on BUN concentrations in Shami goats. In addition, feeding mixture of *Sorghum vulgare* and Pearl millet irrigated with saline water as STP did not affect significantly BUN concentrations (Donia et al. 2014).

The decrease in BUN as a result of feeding STP could be attributed to the presence of condensed tannins, which reduced the ruminal proteins degradation (Mashudi et al. 1997).

With respect to blood CR levels, feeding STP significantly decreased CR levels compared to control group (Melladoa et al. 2006; Donia et al. 2014; Shaker et al. 2014). This decrease (P < 0.01) was found to be about 23% in Shami doe goats fed STP mixture compared to those fed control ration (1.01 vs. 1.24 mg/dl, Shaker et al. 2014).

On the other hand, some studies showed a reverse trend of CR changes due to feeding STP. Badawy et al. (2002) and Shaker et al. (2008) found that animals fed fresh *Atriplex* or *Acacia* had higher CR than those fed control ration. This increase (P < 0.01) had reached about 32% in Barki sheep fed STP mixture in comparison to that in control group (1.40 vs. 1.06 mg/dl) as proved by Shaker (2014). This increase in CR concentration may be attributed to the anti-nutritional factors and/or high salt content in STP. Zhu and Filippich (1995) found that CR was significantly increased in sheep dosed intraperitoneally by 0.19 tannic acid/kg body weight. Moreover, high amounts of oxalates in STP caused destruction of renal nephrons as reported by Clark and Clark (1978) in animals fed *Atriplex* species. The common effect of oxalate is to cause kidney damage through blocking of tubules by crystals of calcium oxalate (Cheeke 1995).

Furthermore, high salt content in STP may be another reason for increasing blood CR in ruminants. Many studies confirmed that drinking saline water significantly increased CR levels in sheep, goats and camels (Hussein 1987; Assad et al. 1989; Abdel-Gawad 1993).

Aforementioned studies clearly revealed that changes in both BUN and CR, either increase or decrease, are within the normal reference ranges as 8–40 mg/dl for BUN and 1.2–1.9 mg/dl for CR (Kaneko 2005; Reece 2005). Therefore, feeding ruminant animals STP can be used safely without any serious and adverse complications on their health.

6. Liver functions

Activities of liver enzymes including alanine aminotransferase (ALT), aspartate aminotransferase (AST), alkaline phosphatase (ALP) and gamma glutamyletransferase (GGT) are conventionally used for diagnosing hepatic functions and damage. ALT is particularly useful in measuring hepatic necrosis and increases in serum when cellular degeneration or destruction occurs (Lassard et al. 1986).

Feeding STP significantly increases the activities of both ALT and AST in sheep and goats compared to their counterparts of control groups as confirmed by several studies (Badawy et al. 2002; Shaker et al. 2008, 2014; Ashour et al. 2013; El-Bassiony 2013; Donia et al. 2014).

Badawy et al. (2002) found that feeding lambs on fresh acacia increased the activity of ALT by 14.1% and AST by 13.2%. Shaker et al. (2008) proved that the activities of both ALT and AST were higher (P < 0.01) in lambs fed on fresh or silage form of Atriplex and acacia mixture compared to the control group. The increases of both ALT and AST activities were 17.2% and 10% in Barki sheep fed STP mixture than those fed control rations (Shaker 2014). Whereas, these increases were found to be 16.3% in both enzymes (Shaker et al. 2014). Also, El-Hawy (2013) found that both ALT and AST activities increased from 37.2 and 53.1 U/L in doe goats fed Berseem hay to 46.1 and 72.5 U/L in those fed STP, respectively. Contrarily, Al-Khalasi et al. (2010) reported that there were no significant differences in ALT and AST activities of Omani sheep fed diets containing sorghum fodder that grew under conditions of various levels of salinity.

In this regard, drinking saline water significantly increased ALT and AST activities (Assad et al. 1997; Assad and El-Sherif 2002). In contrast, Badawy (1999) reported that drinking saline water decreased AST and ALT by about 12 and 30% in rams, respectively. On the other hand, Assad et al. (1994) found no significant effect of drinking diluted sea water on serum AST and ALT activities in sheep and camels.

The increases of ALT and/or AST activities may be caused by high oxalates (McIntosh and Potter 1972), tannins (Tripathi et al. 1984), alkaloids (Craig et al. 1991) and salt (Radostits et al. 1994) in such STP. Additionally, high salt content of STP may be another reason for increasing liver enzymes as demonstrated by many studies on different animals, Ibrahim (1995) on goats, Hussein (1995) on sheep, Ibrahim (2001) on camels and Ibrahim et al. (1991) on male buffalo calves.

ALP activity was decreased in animals fed STP in either fresh or silage form compared to the control groups as reported by several authors (Shaker et al. 2008; Shaker 2014; Shaker et al. 2014). This decline was found to be about 11% in Barki sheep fed STP mixture in comparison with those fed control ration (Shaker 2014). The decrease in ALP activity may be attributed to the presence of tannins in such STP, which react with this enzyme and rendering its activity (Horigome et al. 1988; Abdel-Halim 2003). Furthermore, excess of oxalate is another inhibitor of ALP activity (Mc-Comb et al. 1979). Abu-Zanat and Tabbaa (2006) reported that *Atriplex* spp. contains high concentrations of oxalate which can be toxic to livestock. Al-Khalasi et al. (2010) found that the GGT was high at the end of the trial (11 wks) indicating a probable effect on liver in Omani lambs fed *Sorghum* fodder.

From the aforementioned studies, it is obvious and clear that water salinity was more deleterious and stressful than feed salinity on liver functions as confirmed by El-Bassiony (2013) and Ashour et al. (2013). They, working on four groups of goats, G1 fed Berseem hay (BH) and drinking tap water (TW), G2 fed on BH and drinking saline water (SW), G3 fed on STP and drinking TW and G4 fed on STP and drinking SW, found that AST activities were 30.1, 38.9, 36.5 and 44.1 U/L in the four groups, respectively. The corresponding activities for ALT were 14.3, 16.5, 15.6 and 17.4 U/L, respectively.

According to the previous studies, the activity of ALT ranged between 13 and 47 U/L in animals fed STP in comparison with those fed the traditional rations that had a range of 10–37 U/L. Meanwhile, the corresponding activities for AST were 26 and 73 U/L versus 22–53 U/L. In spite of these differences, all the activities values either for treated or control groups were within the normal reference ranges. Latimer et al. (2003) reported that the normal ranges of ALT were 15 to 44 U/L in sheep and 15–52 U/L in goats. The corresponding ranges for AST were 49–123 in sheep and 66–230 in goats. This indicates that salt load either by feeding STP and/or drinking saline water did not seriously affect the liver functions and in turn the animal's health.

7. Hormonal profile

7.1 Thyroid hormones

Values of triiodothyronine (T3) and thyroxin (T4) hormones as good indicators of thyroid gland activity showed a contradictory response to STP feeding. Recently, Shaker et al. (2014) found that Shami does fed STP showed insignificant higher values of T3 and T4 compared to the control group. These increases were 37% and 11.2% in T3 and T4 values, respectively. In contrast, Barki sheep fed STP showed lower (P < 0.01) values of both T3 and T4 levels than those fed control rations by about 31 and 47%, respectively (Shaker 2014). On the other side, no significant changes were reported in sheep (Pessoa et al. 1988), in goats (Rai et al. 1991; Sanani 1997) and in cattle (Kailas 1999) consuming *Leucaena* leaf meal.

High content of salt in STP may be responsible for the lower secretion of T3 and T4 levels due to the reduction in feed intake. This suggestion is supported by the findings of Metwally (2001) who found that camels drinking saline water decreased their T3 and T4 hormones. Contradictions among these different studies may be related to the experimental animals, sex, age, physiological status, kind of STP, water quality, environmental conditions, etc.

7.2 Insulin hormone

Animals fed STP had lower level of the insulin (Ins) hormone in comparison with those fed control ration. Pearce et al. (2008) reported that sheep fed *Atriplex* rations had lower (P < 0.01) plasma Ins. This decrease (P < 0.01) reached to about 36% (12.6 vs. 9.2 µU/ml) in Barki sheep fed STP mixture (Shaker 2014).

Water salinity effect was more stressful than that of feed salinity on Ins levels as proved by El-Bassiony (2013) in Shami male goats. He found that both SW groups either fed BH (G2) or STP (G4) had lower Ins levels than those that drank FW either fed BH (G1, control) or STP (G3). Levels of Ins (μU/ml) decreased significantly in G2 (18.1) and insignificantly in G4 (20.4) by 22% and 12%, respectively compared to that in G1 (23.3). Whereas, this level was higher in G3 (25.7) compared to the control group (G1) by 10%.

The decrease in Ins levels may be due to the high salt intake which affect markedly the feed intake and consequently the energy balance (Masters et al. 2005; Blache et al. 2007; Digby et al. 2010a,b). A decrease either in voluntary feed intake or in fat reserves is usually associated with a decrease in the concentration of metabolic hormones such as Ins and leptin (Chilliard et al. 2005). In addition, a high salt ingestion has been shown by Blache et al. (2007) to affect energy metabolism through changes in Ins concentration of sheep fed high salt (20% NaCl).

7.3 Aldosterone hormone

Blood aldosterone (Ald) level increased significantly in animals fed STP than those fed traditional rations as reported by several workers, Shaker et al. (2014) working on Shami doe goats fed salt-tolerant fodder shrubs.

This increase in Ald level due to feeding STP ranged between 30 and 47%. Whereas, Shaker (2014) found that Barki sheep fed STP increased (P < 0.01) their Ald level by about 30% compared to the control group (800.4 vs. 614.6 pg/ml). The corresponding levels in goats were 782 vs. 533 pg/ml with an increase (P < 0.01) of 47% (Shaker et al. 2014). The increment in Ald level could be attributed mainly to the high content of salt in such STP, particularly K content. Aldosterone regulates the K^+ concentration of the extra cellular fluids (ECF) by its activity in the cortical collecting tubules and in medullary collecting ducts. In this regard, Ald is secreted in response to elevated K^+ concentrations in the ECF. Although Na^+ reabsorption is coupled with K^+ secretion (not a 1:1 exchange), Ald is not involved with the regulation of Na^+ concentration. The ADH thirst mechanism (osmoregulation) regulates the concentration of Na^+ concentration in the ECF (Reece 2005). Therefore, Ald secretion and concentration of Na and K in urine and blood plasma would be changed when the ratio of Na and K intake changed.

On the other hand, several studies showed that Ald decreased with feeding STP or drinking SW. Digby (2007) found that ewes fed high salt diet had a significantly lower Ald concentration than ewes fed the control diet.

El-Bassiony (2013) found that the three groups of male Shami goats that received salt either in water (G2, BH + SW), or in feed (G3, STP + FW) or in both water and feed (G4, STP + SW) had lower (P < 0.05) level of Ald compared to the control group that received no salt (G1, BH + FW). Also, he found that the levels of Ald hormone (pg/ml) in G2 (30.5), G3 (26.4) and G4 (24.4) were greatly decreased than that in G1 (39.1) by about 22.1%, 32.4% and 37.6%, respectively. In addition, he noticed that the effect of feed salinity was more obvious than that of water salinity, where both G3 and G4 had lower Ald level than that of G1 and G3.

Furthermore, El-Hawy (2013) found that levels of blood serum Ald prior to insemination were insignificantly decreased by about 9% in does fed STP than those fed BH (25.4 vs. 27.8 pg/ml). While during pregnancy period, levels of Ald in does fed STP were insignificantly lower by about 14% than those fed BH (27.0 vs. 31.4 pg/ml). Meanwhile, in the last month of pregnancy levels of Ald significantly decreased by about 41.9% in does fed STP than those fed BH (22.5 vs. 38.7 pg/ml).

The animals fed a high-salt diet managed the physiological effects of salt retention for pregnancy and salt excretion for an overload of salt by reducing their plasma Ald concentration by approximately 50% of control values. When pregnant ewes are fed a normal or low salt diet, Ald is required to increase water reabsorption and thus increase extracellular fluid (ECF). When pregnant ewes are fed a high-salt diet, Ald is not required as the high salt from the diet results in increased water intake and thus ECF is increased.

7.4 Cortisol hormone

Cortisol hormone (Cort) is the principal stress hormone associated with the activation of hypothalamic-pituitary-adrenal axis, and it induces changes in immunity, metabolism and reproduction of animals (Parker et al. 2003).

Cortisol levels increased markedly with increasing salinity load, particularly when SW was consumed, as proved by El-Bassiony (2013). They worked on four groups of Shami male goats, the first one (G1) fed on BH and drinking FW, served as a control group. The G2 was fed on BH and drank SW and G3 was fed on STP and drank FW. While the G4 was fed on STP and drank SW. The overall mean of the four groups, respectively, were 13.9, 21.2, 18.2 and 24.5 mg/dl. It could be noticed that animals exposed to the combined stress of salinity (G4) had obviously a higher level than that of control group by 76.4%. This percentage of increase was moderately alleviated when the animals were fed on STP and drank FW (G3), whereas, G3 group had higher Cort level than that of G1 by 31.0%. Also, it is interesting to notice that the effect of drinking SW was more stressful than that of feeding STP as shown by the groups that drank SW (G2 and G4) than those drinking FW (G1 and G3) either fed BH or STP. This is supported by the higher Cort level in G2 (BH + SW) than that of G1 (BH + FW) by 53% and in G4 (STP + SW) than that of G3 (STP + FW). The previous findings are in agreement with those reported by Abdelhameed et al. (2006) who found that Cort levels in camels were higher in the three treated groups that were fed STP compared to those fed BH. Values of Cortin shown in the previous studies are higher than the normal average that has been reported by Kaneko et al. (2008) as 2.24 ± 0.36 in sheep and 2.35 ± 0.29 in goats. It is indicative that, animals suffer severely from salinity stress where the Cort hormone is a good indicator of stressful conditions.

8. Conclusion

In conclusion, feeding halophytes or salt tolerant forages and/or providing SW to drink, slightly affected the physiological performance of the animals compared to feeding the traditional feeds. Most of the physiological traits are within the normal reference ranges. Therefore, STP could be utilized safely without adverse or serious effects on the biological processes, physiological performance and animal health as confirmed by normal liver and kidney functions. In addition, it should be noticed that water salinity was more deleterious and stressful than feed salinity. However, animals that have been accustomed to feeding on STP and/or drinking SW, develop some adaptive mechanisms which enable them to withstand salt loading (Squires, this volume).

References

Abdel-Gawad, M.M.H. 1993. Some nutritional aspects of camels offered saline water. M.Sc. Thesis, Fac. of Agric., Cairo Univ., Egypt.
Abdel-Halim, A.M. 2003. Studies of some anti-nutritional factors affecting utilization by ruminants. Ph.D. Thesis, Fac. Sci., Ain Shams Univ., Egypt.
Abdelhameed, A.E.A., M.S. Shawket and Kh. Z.E. Kewan. 2004. Effect of feeding Atriplex halimus and Acacia saligna on some biochemical parameters of growing camels under desert environmental conditions. Society of Physiological Science and their Application, Second Scientific Conference, July.
Abdelhameed, Afaf A.E., Shawket, M. Safinaz and K.Z. Kewan. 2006. Physiological studies of feeding halophytes on some blood parameters and body fat of camels under desert condition. 4th Sci. Conf., 29-30 July, Giza, Egypt, pp. 157–180.
Abu-Zanat, M.M.W. and M.J. Tabbaa. 2006. Effect of feeding *Atriplex* browse to lactating ewes on milk yield and growth rate of their lambs. Small Rumin. Res. 64: 152–161.
Aggour, T.A. 1990. Responses of geomorphologic and geological feature on ground water in Wadi Araba, Eastern Desert, Egypt. M.Sc. Thesis, Fac. Sci., Ain Shams Univ.
Alazzeh, A.Y. and M.M. Abu-Zanat. 2004. Impact of feeding saltbush (*Atriplex* sp.) on some mineral concentrations in the blood serum of lactating Awassi ewes. Small Ruminant Research 54(1): 81–88.
Al-Khalasi, S.S., M. Osman, T.K. Isam, W. Al-Marzooqi and S.A. Al-Rawahi. 2010. Salt tolerant fodder for Omani sheep: Effect of salt tolerant *Sorghum* on performance, carcass, meat quality and health of Omani sheep. A Monograph on Management of Salt-Affected Soils and Water for Sustainable Agric. Ministry of Agriculture and Fisheries, Oman Sultanates, pp. 67–81.
Amakiri, S.F. and O.N. Funsho. 1979. Studies of rectal temperature, respiratory rates and heat tolerance in cattle in the humid tropics. Anim. Prod. 28: 329–335.
Amer, M.M. 1990. Amelioration of heat stress in lactating goats. M.Sc. Thesis, Fac. Sci., Ain Shams Univ., Egypt.
Anon. 1969. Quality aspects of farm water supplies. Victoria Irrigation Research and Advisory Services Committee (The Department of National Development: Canberra).
Ashour, G., M.T. Badawy, Y.M. Hafez, M.F. El-Bassiony and *Ibrahim*. 2013. Reproductive performance of growing Shami male kids under salinity conditions in South Sinai. The Egyptian Society for Animal Reproduction and Fertility, 23rd Annual Conf. Proc., February 3–7, Cairo/Ain El-Sokhna, Part (2): 195–214.
Assad, F. and M.A. El-Sherif. 2002. Effect of drinking saline water and feed shortage adaptive responses of sheep and camels. Small Ruminant Research 45: 279–290.
Assad, F., M.T. Bayonmi, M. El-Begowryi and S. Deels. 1989. Pathological changes in ewes and their lambs drinking seawater. Egyptian J. Comp. Pathol. Clin. Pathol. 2(2): 61–76.
Assad, F., M.T. Bayoumi and H.M. Abou El-Naser. 1994. Comparative studies on saline drinking water for camel and sheep. Egypt. J. Comp. Pathol. Clin. Pathol. 7(2): 337–351.
Assad, F., M.T. Bayoumi and H.S. Khamis. 1997b. Impact of long-term administration of saline water and protein shortage on the hemograms of camels and sheep. J. of Arid Environments 37: 71–81.
Attia-Ismail, S.A. (in press). Mineral balance in animals as affected by halophyte and salt tolerance plant feeding. pp. 143–160, this volume.
Atwa, S.M.M. 1979. Hydrogeology and hydrogeochemistry of the northwestern Coast of Egypt. Ph. D. Thesis, Fac. of Sci., Alex. Univ., Egypt.
Azamel, A.A. 1997. Physiological responses and daily gain of growing Barki lambs fed on *Atriplex halimus*, a natural desert shrubs. Egypt. J. Appl. Sci. 12(2): 1–9.
Badawy, H.S.M. 1999. Digestive function and heat regulation in Saidi sheep. M.Sc. Thesis, Assiut University, Egypt.
Badawy, M.T., H.A. Gawish and A.A. Younis. 2002. Some physiological responses of growing Barki lambs and Baladi kids fed natural desert shrubs. International Symposium on Optimum Resources Utilization in Salt-Affect Ecosystems in Arid and Semi-arid Regions. 8–11 April, 2002, Cairo, Egypt, pp. 496–503.

Bayoumi, M.T., H.M. El Shaer and Fawzia Assad. 1990. Survival of sheep and goats fed salt marsh plants. J. of Arid Environment 18: 75–78.

Ben-Salem, H., H.C. Norman, A. Nefzaoui, D.E. Mayberry, K.L. Pearce and D.K. Revell. 2010. Potential use of oldman saltbush (*Atriplex nummularia* Lindl.) in sheep and goat feeding. Small Rumin. Res. 91: 13–28.

Berger, L.L. 2006. Salt and Trace Minerals for Livestock, Poultry and Other Animals. Published by the Salt Institute, 700 North Fairfax Street, Suite 600, Alexandria, Virginia 22314-2040.

Blache, D., M.J. Grandison, D.G. Masters, M.A. Blackberry and G.A. Martin. 2007. Relationships between metabolic systems and voluntary feed intake in Merino sheep fed a high salt diet. Aust. J. Agric. Res. 47: 544–550.

Blood, D.C. and O.M. Rodostitis. 1989. Veterinary Medicine. 7th Ed. BaillereTindall Ltd. London. Philadelphia, Toronto, Sydney, Tokyo.

Bravo, L., E. Manas and F.S. Calixto. 1993. Dietary non-extractable condensed tannins as indigestible compounds: Effect of fecal weight and protein and fat excretion. J. Sci. Food Agric. 63: 63–68.

Cheeke, P.R. 1995. Endogenous toxins and mycotoxins in forage grasses and their effects on livestock. J. Anim. Sci. 73(3): 909–918.

Cheeke, P.R. 1999. Actual and potential applications of *Yucca Schidigera* and *Quillaja Saponaria* saponins in human and animal nutrition. Proc. A. Soc. Anim. Sci. 1: 1–10.

Chilliard, Y., C. Delavaud and M. Bonnet. 2005. Leptin expression in ruminants: nutritional and physiological regulation with energy metabolism. Domestic Anim. Endoc. 29: 3–22.

Clark, E.G.C. and M.L. Clark. 1978. Veterinary Toxicology. Baillere Tindall, London.

Cook, R.W., C.B. Scott and F.S. Hartmann. 2008. Short-term Mesquite pod consumption by goats does not induce toxicity. Rangeland Ecology & Management 61(5): 566–570.

Craig, A.M., Pearson, C. Meyer and J.A. Schmitz. 1991. Serum liver enzyme and histopathologic changes in calves with chronic and chronic-delayed *Senecio jacobaea* toxicosis. Am. J. Vet. Res. 52(12): 1969–1978.

Digby, Serina N. 2007. High Dietary Salt During Pregnancy in Ewes Alters the Responses of Offspring to an Oral Salt Challenge. Ph.D. Thesis, Fac. Agric., Adelaide Univ., Australia.

Digby, S.N., D. Blache, D.G. Masters and D.K. Revell. 2010a. Responses to saline drinking water in offspring born to ewes fed high salt during pregnancy. Small Ruminant Research 91: 87–92.

Digby, S.N., D.G. Masters, D. Blache, P.I. Hynd and D.K. Revell. 2010b. Offspring born to ewes fed high salt during pregnancy have altered responses to oral salt loads. Animal 4(1): 81–88.

Donia, G.R. and N.H. Ibrahim. 2013. Assessment of some macro and micro elements and their impact on environmental health in southern Sinai, Egypt. Arab Water Council Journal 4(2): 1–9.

Donia, G.R., N.H. Ibrahim, Y.M. Shaker, F.M. Younis and Hanan, Z. Amer. 2014. Liver and kidney functions and blood minerals of Shami goats fed salt tolerant plants under the arid conditions of southern Sinai, Egypt. Journal of American Science 10(3): 49–59.

El-Bassiony, M.F. 2013. Productive and reproductive responses of growing Shami goat kids to prolonged saline conditions in South Sinai. Ph.D. Thesis, Fac. Agric., Cairo Univ., Egypt.

El-Hassanein, E.E., H.M. El Shaer, A.R. Askar, H.M. El-Sayed and H.S. Soliman. 2002. Physiological and reproductive performance of growing Barki rams fed halophytic plants as basal diet in Sinai. Inter. Symp. on Optimum Resources Utilization in Salt-Affected Ecosystems in Arid and Semi-Arid Regions, 8–11 April, 2002, Cairo, Egypt, pp. 504–514.

El-Hassanein, E.E. and M.A. El-Sherif. 1996. Effect of prolonged drinking saline water on blood picture of growing lambs. Proceeding of the 4th Scientific Congress, Vet. Med. and Human Health, 3–6 April, 1996 Cairo Univ., Egypt. Vet. Med. J. 44(28): 435–441.

El-Hawy, A.S. 2013. Reproductive efficiency of Shami goats in salt affected lands in South Sinai. Ph.D. Thesis, Fac. Agric., Ain Shams Univ., Egypt.

El Shaer, H.M. and E.A. Gihad. 1994. Halophytes as animal feeds in Egyptian desert. pp. 281–284. *In*: V.R. Squires and A.T. Ayoub (eds.). Halophytes as a Resource for Livestock and for Rehabilitation of Degraded Lands. Kluwer Academic, Dordrecht.

El Shaer, H.M. 1981. A comparative nutrition study on sheep and goats grazing Southern Sinai desert range with supplements. Ph.D. Thesis, Faculty of Agriculture, Ain Shams University, Cairo, Egypt.

El Shaer, H.M. 2004. Potentiality of halophytes as animal fodder under arid conditions of Egypt. Rangeland and Pasture Rehabilitation in Mediterranean Areas, Cahiers OPTIONS Mediterraneenes 62: 369–374.

El Shaer, H.M. 2005. Halophytes as cash crops for animal feeds in arid and semi-arid regions. Proc. International conference on Biosaline Agriculture & High Salinity Tolerance, Mugla, Turkey, 9–14 Jan. 2005, pp. 7–15.

El Shaer, H.M., H.M. Kandil and H.S. Khamis. 1991. Saltmarsh plants ensiled with dried broiler litter as a feedstuff for sheep and goats. J. Agric. Sci. Mansoura Univ. 16: 1524–1534.

El-Sherif, M.M.A. 2003. Effect of drinking saline water and falvomycin supplementation on some physiological responses, growth performance and carcass traits of Barki ram lambs. J. Agric. Sci. Mansoura Univ. 28(6): 4429–4447.

Fayed, A.M. 2009. *In vitro* and *In vivo* evaluation of biological treated salt plants. American-Eurasian J. Agric. & Environ. Sci. 6(1): 108–118.

Fayed, Afaf M., Abeer M. El-Essawy, E.Y. Eid, H.G. Helal, Ahlam R. Abdou and H.M. El-Shaer. 2010. Utilization of alfalfa and atriplex for feeding sheep under saline conditions of South Sinai, Egypt. J. Americ. Sci. 6(12): 1447–1461.

Francis, G., Z. Kerem, H.P.S. Makkar and K. Becker. 2002. The biological action of saponins in animal system: a review. British J. Nut. 88: 587–605.

Han, L.K., B.J. Xu, Y. Kimura, Y.N. Zheng and H. Okuda. 2000. Platycodi radix affects lipid metabolism in mice with high fat diet induced obesity. J. Nutrition 130: 2760–2764.

Hanafy, M.A., A.A. Fahmy, M.S. Farghaly, A. Afaf and El Sheref. 2007. Effect of using some fodder plants in diets on goats' performance under desert conditions of Sinai. Egyptian J. Nutr. Feeds 10: 151–163.

Horigome, T., R. Kumar and K. Okamoto. 1988. Effect of condensed tannins prepared from leaves of fodder plants on digestive enzymes *in vitro* and in the intestine of rat. British J. of Nutrition 60: 275.

Howell, J.M. 1996. Toxicities and excessive intakes of minerals. pp. 95–117. *In*: D.G. Masters and C.L. White (eds.). Detection and Treatment of Mineral Nutrition Problems in Grazing Sheep. ACIAR Monograph No. 37 (Watson Ferguson & Co.: Brisbane).

Hussein, F.M., G.A. El-Amawi, I.E. El-Bawab, S.A. Hattab, A.K. Kadoom, F.A. El-Keraby and M.Y. Abboud. 1992. Studies on some biochemical serum constituents during the calving conception interval in Frisian cows. Egyptian Soc. Anim. Reprod. Fert., 4th Annual Cong., Cairo, Egypt, pp. 28–30.

Hussein, N.M. 1987. Haematological studies on sheep drinking salty water. M.Sc. Thesis, Fac. of Agric., Al-Azher Univ., Egypt.

Hussein, N.M., F. Assad, Abdel-Megeed and A.M. Nassar. 1990. Variation in blood cellulose due to drinking salty water in sheep. Egypt J. Comp. Pathol. Clin.Pathol. 3(1): 55–63.

Ibrahim, I.A., O.A. Abd-Alla and A.M. El-Nahia. 1991. Biochemical and Hematological studies in male buffalo—calves drinking well water. Egyptian J. Comp. Pathol. Clin. Pathol. 4(2): 237–244.

Ibrahim, N.H.M. 2001. Studies on some physiological and behavioural aspects in camels. M.Sc. Thesis. Fac. Agric. Minufiya University.

Ibrahim, S.M.N. 1995. Clinopathological studies in goats drinking salty water under desert conditions. M. Sc. Thesis, Fac. of Vet. Med., Cairo Univ., Egypt.

Inglis, S. 1985. Livestock water supplies. Fact sheet 82/77. Department of Agriculture, South Australia.

Ismail, E., M.M. Anwar and S.S. Aboul EL-Ezz. 2003. Change in some blood constituents in sheep fed on rations containing *Acacia saligna* irrigated naturally or on sewage water. Alex. J. Agric. Res. 48(3): 35–40.

Jaster, E.H., J.D. Schuh and T.N. Wegner. 1978. Physiological effect of saline drinking water on high producing dairy cow. J. Dairy Sci. 61: 66–74.

Judson, G.J. and J.D. McFarlane. 1998. Mineral disorders in grazing livestock and the usefulness of soil and plant analysis in the assessment of these disorders. Australian Journal of Experimental Agriculture 38: 707–723.

Kailas, M.M. 1991. Studies on the use of Leucaena leucocephala (Subabul) in the feeding of cattle. Ph.D Thesis, Indian Veterinary Research Institute, Deemed University, Izatnagar, Bareilly (U.P.), India, pp. 78–82.

Kandil, H.M. and H.M. El Shaer. 1990. Comparison between goats and sheep in utilization of high fibrous shrubs with energy feeds. Proc. Int. Goat Prod. Symp. October 22–26. Tallahassee Fl.

Kaneko, J.J. 2005. Clinical Biochemistry of Domestic Animals. Academic Press.

Kaneko, J.J., J.W. Harvey and M.L. Bruss. 2008. Clinical Biochemestry of Domestic Animals. 6th ed. Academic Press, San Diego. 904p.

Kawashti, I.S., M.E. Badawn, S.M. Mageed and M.M. Omer. 1983a. Salt tolerance of desert sheep. 6-Effect of saline water administration on intake-urinary and fecal water losses and body water distribution. Desert Inst. Bull., Egypt. 14: 392–408.

Kewan, Kh. Z.M. 2003. Studies on camel nutrition. Ph.D. Thesis, Fac. Agric., Alexandria University.

Khalil, J.K., W.N. Sawaya and S.Z. Hyder. 1986. Nutrient composition of Atriplex leaves grown in Saudi Arabia. Journal of Range Management 39: 104–107.

King, J.M. 1983. Game domestication for animal production in Kenya: Field studies of the body water turnover of game and livestock. J. Agric. Sci. (Cambridge) 93: 71–79.

Latimer, K.S., E.A. Mahaffey and K.W. Prase. 2003. Duncan and Prasse's Veterinary Laboratory Medicine: Clinical Pathology. 4th ed., Iowa State Press, Iowa.

Lee, A.J. and A.M. Twardock. 1978. Blood metabolic profiles: their use and relation to nutritional of dairy cows. J. Dairy Sci. 61: 1652–1670.

Lee, J.A., J.D. Roussell and J.F. Beaty. 1976. Effect of temperature season on bovine adrenal cortical function, blood cell profile and milk production. J. Dairy Sci. 59: 104–114.

Lessard, P., W.D. Wilson and H.J. Olander. 1986. Clinicopathologic study on horses surviving P. A. (Senecio vulgaris) toxicosis. Aim. J. Vet. Res. 47: 1779–1780.

Macfarlane, W.V. 1971. Isotopes and the ecophysiology of tropical livestock. pp. 805–827. *In*: International Symposium on Use of Isotopes and Radiation in Agriculture and Animal Husbandry Research, New Delhi. Vienna, IAEA.

Makman, M.H., S. Nakagaew and A. White. 1967. Studies of the mode of action of adernal steriods on lymphocytes.Recent Progress Hormone Res. 23.

Maloiy, G.M.O. 1972. Renal salt and water excretion in the camel (*Camelus dromedarius*). Symp. Zool. Soc. Lond. 31: 243–259.

Mansoori, B. and T. Acamovic. 1997. The excretion of minerals from broilers fed tannic acid with and without gelatin. Proc. of Spring Meeting of the World Poultry Sci. 25–26.

Mashudi, Brooke, I.M., C.W. Holmes and G.F. Wilson. 1997. Effect of Mimosa bark extracts containing condensed tannins on rumen metabolism in sheep and milk production by grazing cows. Proceeding o the New Zealand Society of Anim. Prod. 57: 126.

Masters, D.G., S.E. Benes and H.C. Norman. 2007. Biosaline agriculture for forage and livestock production. Aust. Ecosys. Environ. 119: 234–248.

Masters, D.G. (in press). Assessing the feeding value of halophytes. pp. 86–105, this volume.

Masters, D.G., A.J. Rintoul, R.A. Dynes, K.L. Pearce and H.C. Norman. 2005. Feed intake and production in sheep fed diets high in sodium and potassium. Aust. J. Agric. Res. 56(5): 427–434.

Matsuura, M. 2001. Saponin in garlic as modifiers of the risk of cardiovascular disease. J. Nut. 131: 1000–1005.

Mayberry, D.E., D.G. Masters and P.E. Vercoe. 2009. Saltbush (*Atriplex nummularia* L.) reduces efficiency of rumen fermentation in sheep. Options Méditerranéennes Serie A: Séminaries Méditerranéennes 85: 245–249.

McComb, R.B., G. Bower and S. Posen. 1979. Alkaline phosphatase. New York, Plenum Press.

McIntosh, G.H. and B.J. Potter. 1972. The influence of 1.3% saline ingestion upon pregnancy in the ewe and lamb survival. Proc. of the Aust. Physiol. and Pharmacol.Society 3: 61.

Melladoa, M., L. Olivaresa, A. Rodrigueza and J. Melladoa. 2006. Relation among Blood Profiles and Goat Diets on Rangeland. J. of Applied Animal Research, Vol. (30): 1, pp. 93–98.

Metwally, N.H. 2001. Studies on some physiological and behavioral aspects in camels. M.Sc. Thesis, Fac. Agric., Minufiya Univ., Egypt.

Meyer, M.W. and W.H. Karazof. 1991. Deserts. *In*: T. Psalo and C. Rabbins (eds.). Plant Defenses Against Mammalian Herbivory. CRC Press, Boca Raton.

Milne, D.B. 1996. Trace elements. pp. 485–496. *In*: C.A. Burtis and E.R. Ashood (eds.). Tietz Fundamentals of Clinical Chemistry.

Mohamed, M.I. 1996. Studies on Desert roughages on camels and small ruminants nutrition. Ph.D. Thesis, Fac. Agric. Cairo Univ., Egypt.

Morehouse, L.A., F.W. Bangerter, M.P. Deninno, P.B. Inskeep, P.A. McCarthy, J.L. Pettini, Y.E. Savoy, E.D. Sugarman, R.W. Wilkims, T.C. Wilson, H.A. Woody, L.M. Zaccaro and C.E. Chandler. 1999. Comparison of synthetic saponin cholesterol absorption inhibitors in rabbits: evidence for non-stoichiometric, intestinal mechanisms of action. J. Lipid Res. 40: 464–474.

Muller, H.M., E. Leinmuller and U. Rittner. 1989. Effect of tanniferous plant material on protein and carbohydrate degradation in rumen fluid *in vitro*. In Recent Advance of Research. In Antinutritional Factors in Lequme Seeds. (Huisman, J.; Van der Poel, T. F. B. and Liener, I. E. eds.), pp. 156–159, Wageningen.

Nasr, S.M., E.A. Ibrahim, A. Bakeer and M.I. Dessouky. 2002. Clinicopathological and histopathological studies in goats fed on Atriplex halimus raised in the Egyptian desert. International Symposium on Optimum Resources Utilization in Salt—Affect Ecosystems in Arid and Semi-arid Regions. Cairo, 8–11, April, 515–525.

NRC. 1975. Nutrient requirements of domestic animals. No. 5. Nutrient requirements of sheep. National Research Council. Washington D.C.

NRC. 1980. Mineral tolerance of domestic animals. National Academy of Sciences. National Research Council. Washington D.C. USA.

NRC. 1981. Nutrient requirements of domestic animals. No. 15. Nutrient requirements of goats. National Research Council. Washington D.C.

NRC. 2000. Nutrient Requirements of Beef Cattle: Seventh Revised Edition: Update 2000. National Research Council. National Academy Press, Washington D.C.

Omar, M.M. 1969. Some factors affecting body temperature and water economy of camels and donkeys endogenous to the Egyptian Western Desert. M.Sc. Thesis, Faculty of Agricultural, Cairo University, Egypt.

Ortiz, L.T., C. Centeno and J. Tervino. 1993. Tannins in faba bean seeds: effect on the digestion of protein and amino acids in growing chicks. Animal Feed Science and Technology 41: 271–278.

Parker, A.J., G.P. Hamlin, C.J. Coleman and L.A. Fitzpatrick. 2003. Dehydration in stressed ruminants may be the result of a cortisol-induced diuresis. J. Anim. Sci. 81: 512–519.

Parker, B.N.J. and R.W. Blowey. 1976. Investigation on the relationship of selected blood components to nutrition and fertility of the dairy cow under commercial farm conditions. Vet. Record 98: 394–404.

Patra, A.K., K. Sharma, N. Dutta and A.K. Pattanaik. 2002. Effect of partial replacement of dietary protein by a leaf meal mixture containing Leucaena leucocephala, Morus alba and Azadirachta indica on performance of goats. Asian-Aust. J. Anim. Sci. 5(12): 1732–1737.

Pearce, K.L., D.W. Pethick and D.G. Masters. 2008. The effect of ingesting a saltbush and barley ration on the carcass and eating quality of sheep meat. Animal 2(3): 479–490.

Peirce, A.W. 1962. Studies on salt tolerance of sheep. IV. The tolerance of sheep for mixtures of sodium chloride and calcium chloride in the drinking water. Australian Journal of Agricultural Research 13: 479–486.

Peirce, A.W. 1966. Studies on salt tolerance of sheep. VI. The tolerance of wethers in pens for drinking waters of the types obtained from underground sources in Australia. Australian Journal of Agricultural Research 17: 209–218.

Peirce, A.W. 1968. Studies on salt tolerance of sheep. VIII. The tolerance of grazing ewes and their lambs for drinking waters of the types obtained from underground sources in Australia. Australian Journal of Agricultural Research 19: 589–595.

Peirce, A.W.C. 1959. Studies on salt tolerance of sheep. II. The tolerance of sheep for mixture of sodium chloride and magnesium chloride in the drinking water. Aust. J. Agric. Res. 10: 725–735.

Pessoa, J.M., W.M. Cardoso, C.E.S. Velez and N.M. Rodroguez. 1988. Serum levels of thyroxine and triiodothyronine determined by radio-immuno assay in sheep fed Leucaena leucocephala. Aquiro Brasileiro de Medicina Veterinaria Zootech 40: 431–436.

Rai, S.N., H. Kaur, A.S. Harika and T.K. Walli. 1991. Influence of substituting groundnut cake with leucaena leaf meal in concentrate mixture on live weight gain and thyroxine profile in crossbred kids. In: Proceedings of First International Animal Nutrition Workers Conference for Asia and Pacific, 23–28 September 1991, Bangalore, India, p. 108 (Abstract No. 175).

Radostitis, O.M., D.C. Blood and C.C. Gay. 1994. Veterinary Medicine. 8th Ed., BailiereTindall, London, 563–616.

Rasool, E., S. Rafique, I.U. Haq, A.G. Khan and E.F. Thomson. 1996. Impact of fourwing saltbush on feed and water intake and on blood serum profile in sheep. A.J.A.S. 9(2): 123–126.

Ray, D.E. 1989. Interrelation among water quality, climate and diet on feedlot performance of steer calves. J. Anim. Sci. 67: 357–361.

Reece, W.O. 1991. Physiology of Domestic Animals. Lea and Febiger, Philadelphia, USA, pp. 285–316.

Reece, W.O. 2005. Functional Anatomy and Physiology of Domestic Animals. Ames, Iowa Univ. Press, Ithaca, USA, 590 p.

Reed, J.D., H. Soller and A. Woodward. 1990. Fodder tree and straw diets for sheep: intake, growth, digestibility and the effects of phenolics on nitrogen utilization. Anim. Feed Sci. Technol. 30: 39.

Rehman, A. 1992. Factors affecting preference of sheep for saltbush (Atriplex spp.). Masters Thesis, University of Western Australia.

Robert, M., S. Keating, J. Anibal, G. Pordomingo, C. Schneberger, D. Gleen and D. Joe. 1992. Influence of saline water on intake, digesta kinetics and serum profiles of steers. J. Range Manage. 45: 514–518.

Romero, M.J., J. Madrid, F. Hernandez and J.J. Ceron. 2000. Digestibility and voluntary intake of Vine leaves (*Vitisvinifera* L.) by sheep. Small Ruminant Res. 38: 191–195.

Salem, A.F.Z.M., Y. Gohar, M.M. El-Adawy and M.Z.M. Salem. 2004. Growth inhibitory effect of some anti-nutritional factors extracted from *Acacia saligna* leaves on intestinal bacteria activity in sheep. Proc. 12th Scientific Conf., Egypt. Soc. Amin. Prod. 41: 283–300.

Seerley, R.W., C.W. Foley, D.J. Wlliams and S.E. Curtis. 1972. Hemoglobin concentration and thermostability in neonatal piglets. J. Anim. Sci. 34: 82–84.

Shaker, Y.M. 2014. Live Body Weight Changes and Physiological performance of Barki Sheep fed salt tolerant fodder crops under the arid conditions of Southern Sinai, Egypt. J. Am. Sci. 10(2s): 78–88.

Shaker, Y.M., N.H. Ibrahim, F.E. Younis and H.M. El Shaer. 2014. Effect of feeding some salt tolerant fodder shrubs mixture on physiological performance of Shami goats in Southern Sinai, Egypt. J Am Sci. 10(2s): 66–77.

Shaker, Y.M., S.S. Abou El-Ezz and A.L. Hashem. 2008. Physiological performance of Barki male lambs fed halophytes under semi-arid conditions. J. Agric. Sci. Mansoura Univ. 33(9): 6393–6408.

Sherwood, L. 1997. Human Physiology: From Cells to Systems. Wadsworth Publishing Company, London, UK.

Singh, N. and G.C. Toneja. 1981. Variability for salt tolerance in pure-breed and crossbred sheep of Rajasthan desert. Indian J. Anim. Sci. 51(3): 324–327, March, 1981.

Squires, V.R. (in press). Water requirements of livestock fed on halophytes and salt tolerant forage and fodders. pp. 187–302, this volume.

Tripathi, Y.B., O.P. Malhotra and S.N. Tripathi. 1984. Thyroid stimulating action of z-guggulsterone obtained from *Commiphora mukul*. Planta Med. 50: 78–80.

Warren, B.E. and T. Casson. 1992. Performance of sheep grazing salt tolerant forages on revegetated salt land. Aust. Soc. For Animal Production 19: 237–241.

Wilson, A.D. 1966. The intake and excretion of sodium by sheep fed on species of *Atriplex* (saltbush) and *Kochia* (blue bush). Australian Journal of Agricultural Research 17: 155–163.

Wilson, A.D. 1975. Influence of water salinity on sheep performance while grazing on natural grassland and saltbush pastures. Australian Journal of Experimental Agriculture and Animal Husbandry 15: 760.

Wilson, A.D. 1978. Water requirements of sheep. pp. 179. *In*: K.M.W. Howes (eds.). Studies of the Australian Arid Zone. CSIRO.

Xu, Z.Y., S.X. Yang and S. Yang. 1994. Animal Nutrition (Chinese Agricultural Publishing Co., Beijing), pp. 44.

Zamet, C.N., V.F. Colenbrander and R.E. Eib. 1979. Variable associated with peripartum traits in dairy cows. III. Effect of diets and disorders on certain blood traits. Theriogenology 11: 261–272.

Zhu, J. and L.J. Filippich. 1995. Acute intra-abomasal toxicity of tannic acid in sheep. Vet. Hum. Toxicol. 37(1): 50–54.

Water Requirements of Livestock Fed on Halophytes and Salt Tolerant Forage and Fodders

Victor R. Squires

SYNOPSIS

Livestock require more drinking water if they are fed on halophytic feedstuffs or on salt tolerant plants growing on saline soils. To flush the salt through their systems livestock drinking saline water and eating salty food need extra water to flush salts through. This simple fact has implications for both grazing/browsing and for pen fed livestock. There is a hierarchy of tolerance to salinity both between species and between breeds (ecotypes) within species of livestock. Camels are most tolerant and sheep and goats are more tolerant than cattle. The impact of high salinity on rumen physiology, water utilization and turnover rates are reviewed. The implications of ameliorating saline water by mixing with fresh are explained. This chapter reviews research work and experience on behavior, physiology and productivity of livestock on salty diets.

Keywords: saline drinking water, sodium, potassium, sulfates, camels, goat, sheep, cattle, water turnover rate, ameliorating water quality, tolerance to salt, heat dissipation, physiology, rumen microflora, appetite suppression.

Gansu Agricultural University, Lanzhou, China.
Email: dryland1812@internode.on.net

1. Introduction

In subtropical drylands (the main regions where pastoralism is practiced) both forage/ fodder and drinking water can have a high salt content so that livestock can have an excessive intake of sodium, potassium or even magnesium salts. Livestock with a restricted water supply may be able to avoid water deficiency by selectively grazing/ browsing plants high in moisture content. However, in many of the pastoral areas high moisture content is often associated with a high salt content, and selectivity of a low salt diet may differ from that of a high moisture diet. The tolerance of livestock to salt intake is matter of considerable interest in pastoral zones (Edwards et al. 1983). There is a hierarchy of tolerance within livestock species. Camels are most tolerant and cattle and horses least tolerant with sheep and goats in between (Silanikove 1989). Given the capacity of sheep to survive in arid regions, their regulation of salt and water balance may be geared towards coping with salt loads or water shortages, and hence the opportunity to increase the consumption of (or preference for) high-salt diet.

Of course there are also differences within breeds and ecotypes within livestock species. Sheep and camels commonly show greater tolerance than cattle and thus they are grazed in those areas where almost sole reliance is placed on underground (well) waters of high salinity. Tolerance of livestock to saline diets depends on the chemical composition of the salt (in both food and water), the physiological status of animals, and previous experience (Revell et al., this volume). Non-pregnant, non lactating sheep for example, tolerate 1.3–1.5% NaCl in the drinking water without serious depression of food intake. Severe reduction of food intake and possibly death may result from higher concentrations. Where saline food represents a large proportion of the diet, a safe limit for drinking water falls to 0.8%. If $Mg\ SO_4$ is a major constituent of the drinking water, the level of tolerance is much reduced.

2. Role and function of water in the body

Water molecules are by far the most numerous of any in a mammal and represent about 99% of all molecules. Five to ten more moles of water move through the cells than moles of oxygen, and the water flux is about 100X as great as the molecular turnover of all other substances. The rate of energy use in the cell is linked with and probably determines the flux of water in which the other cell processes take place. According to MacFarlane and Howard (1974) about 45% of bodyweight is intracellular water and 5% is extracellular, divided between plasma (5%) and interstitial fluids (20%).

2.1 Water as a coolant

Water possesses certain characteristics which are essential to the maintenance of body temperature. The specific heat of water is considerably higher than that of any other liquid (or solid). Many animals rely on the cooling capacity of water as it gives up latent heat during evaporation, by panting or sweating. As 1 g of water changes from liquid to vapor, whether by panting or sweating, it binds about 2,425 J of heat. In terms of heat exchange this is very efficient use of water when it is realized that to heat 1 g of water from freezing to boiling point requires only 490 J. Because of this great capacity to store heat, any sudden change in body temperature is avoided. Water has greater thermal conductivity than any other liquid and this is important for the dissipation of heat from deeply situated regions of the body. Many ruminants (such as sheep, goats, and cattle) dissipate internal and absorbed heat by evaporation of body water. These physical properties of water which make it ideal as a heat-regulating medium are enhanced by other purely physiological factors. The mobility of blood and the rapidity with which it may be redistributed quickly in the body together with the special physical properties makes it a highly efficient body temperature regulator. The functions of water within the ruminant are basically for intermediary metabolism, or for cooling.

2.2 Water as a nutrient

Water is defined as an essential nutrient because it is required in amounts that exceed the body's ability to produce it. All biochemical reactions occur in water. It fills the spaces in and between cells and helps form structures of large molecules such as protein and glycogen. Water is also required for digestion, absorption, transportation, dissolving nutrients, elimination of waste products and thermoregulation. Water is the main constituent of animals, normally comprising between 50 and 80 per cent of body weight. The relative mass of water decreases with age. A reduction in this body water below a certain critical level is more life threatening than a shortage of any other component in the body, with the exception of oxygen. Requirements for water are related to metabolic needs and are highly variable. They depend to some extent on individual metabolism but also on genetics. Some livestock ecotypes are particularly tolerant (Mittal and Ghosh 1980; Degen 1977; Macfarlane 1965).

2.3 The role of metabolic water

Oxidation of nutrients and tissues leads to the formation of water from the hydrogen (H) present. Wager thus formed may also aid in supplementing water supply. The overall impact though is low, because 1 kg of fat has to be oxidized to produce 1.2 L of water, and 1 kg of protein and carbohydrate only produces 0.5 L. In addition, there is the problem of respiratory loss of water when oxygen is inspired. It has been calculated that in a hot dry climate (ambient temperature 26°C, relative humidity 10%) an animal loses 23.5 g of respiratory water in the process of producing 12.3 of metabolic water. As well as water, metabolic heat is generated (418 KJ). Part of this heat (13.6%) is offset by the heat of vaporization of the expired water. If the remainder (361 KJ) had to be dissipated by sweating, it would cost 249 ml of water. The relationship between metabolic water yield and the water that could be required to dissipate the heat of combustion varies with the organic matter being oxidized. Thus, 1 g of fat yields 1 ml of water, but could require 14 ml for vaporization; 1 g of protein or carbohydrate yields about 0.5 g of water and could require 6.5 ml of sweat.

To maintain the body-water pool within homeostatically acceptable limits, any water lost must be replaced. The proportion or quantity of water utilized in a given unit of time is known as the turnover and varies according to the species, size, and physiological status of the animal, and the environmental conditions. In general, animals adapted to dry environments have lower turnover rates than species in temperate zones, but in the case of many livestock a greater heat load gives rise to substantial water loss by evaporative heat dissipation so that turnover rates may in reality be much higher. Daily turnover can be expressed as millilitres per litre of body water (ml l^{-1} d^{-1}) or as ml per kg of body weight (ml kg^{-1} d^{-1}). The disadvantage of the former is that total body water must be known and assumed to be constant. In practice, no accurate method of establishing body-water content *in vivo* has been found for large animals and validating estimates by the dilution method using total body dissection has revealed variable and substantial over-estimates. In contrast, body weight is an easily determined variable.

Solid foods contribute approximately 20% of total water intake. The remainder of the dietary intake comes from free water and/or other fluids. An additional small amount of water is also made available to the body from metabolism (water of oxidation). The body must retain a minimal amount to maintain a tolerable solute load for the kidneys. The concentration of water in the animal body must be kept as constant as possible for normal tissue function to be maintained.

Food and water intake are closely linked (Anderson 1978; Andersson and Larsson 1961). Enough drinking water is a prerequisite for animals to utilize food efficiently. Goat and sheep breeds which are native to dry regions have developed strategies to cope with scarcity of food and water. However, animals from temperate regions may encounter problems in adjusting to a dry environment. This is obvious during late pregnancy and lactation, when the strain on the fluid regulatory mechanisms becomes more severe (Olsson et al. 1997).

3. Water use and turnover

The rate at which an animal uses water in given environment depends on the genetically-determined drives from the limbic cortex and the hypothalamus. These determine the water intake, while the gut-kidney machinery regulates output.

3.1 Quantitative requirements

As there are many factors that affect water consumption, it is difficult to determine the quantitative requirements of livestock. One method is to correlate the intake of water with the quantity of fodder in a dry weight basis. Various loading are used in this approach to account for physiological changes such as, age, pregnancy and lactation. Young animals require more water than mature animals. The values derived by this method are generally only valid in temperate regions, where complications such as extremely high ambient temperatures and saline food and water do not apply. An alternative is to measure the actual consumption by different types of livestock under various combinations of forage/fodder type, season, and geographic location. Not surprisingly, this amount of documentation has only been collected for a few localities. There is a dearth of published information on the water consumption by livestock under large scale *commercial* management, e.g., on ranches.

Excluding perspiration, the normal turnover of water is approximately 4% of total body weight in adults (Phillis 1976). There is a logarithmically linear relationship between water turnover and body weight amongst mammals. Water turnover has been studied in camels, sheep and some desert animals (MacFarlane et al. 1971). Mammals inhabiting desert regions might be expected to have similar physiology, particularly in rates of water use. Groups of ungulates in the field, however, show a wide range of water consumption. Macfarlane et al. (1971) measured, over periods of weeks, the uninhibited intake of water and food by groups of animals grazing together on the same territory and this was integrated with metabolic water from oxidation of hydrogen. Using this approach they found that buffalo (*Bos bubalus bubalis*) have the highest water turnover of any unstressed animal so far measured—200 ml./kg/24 h. Bostaurus in the same tropical climate were next in rank order, then *Bibos banteng* and the least water-demanding of the cattle group were *Bos indicus* (123 ml./kg/24 h). Cattle (*Bos indicus* and *Bos taurus*) turn over two or three times more water than other ruminants of comparable weight (McDowell 1972). There appear to be inherent physiological differences between major groups of mammals.

The range of turnover rates in four livestock species is shown in Table 15.1. Camels have the lowest turnover, zebu cattle and sheep have comparable rates. Large animals with a lower metabolic rate per unit weight and a low surface-to-volume ratio would be expected to have correspondingly low water turnover rates, and in this respect cattle appear to be the least efficient users of water.

Table 15.1. Water Turnover rates in four livestock species.*

Animal type	Turnover rate (ml kg^{-1}d^{-1})
Dromedary camel	37–74
Bos indicus cattle	70–197
Bos taurus cattle	9–230
Mutton sheep	74–200
Desert sheep	65–130
Goats	90–170

* adapted from Nicholson 1985.

There is a linear relationship between the log of body-water pool and the log of water turnover, the regression being: Water turnover (1 d^t) = antilog [0.836 log body water (1) – 0.619].

Another variable which enters into the patterns of water use is the ambient temperature or any other form of heat load. In the summer deserts of Australia and northern Kenya, the highest water turnover rates occurred in cattle, with sheep at about half their rate, goats somewhat less and camels (*Camelus dromedarius*) least in their water requirements, at half to one-third the rate of cattle (Table 15.1). Siebert and Macfarlane (1971) showed that water turnover, however, is low in camels relative to cattle and varies inversely as the fat content relative to body weight. McDowell 1972, noted that goats native to tropical areas seem to have a lower water turnover rate, 11 per cent lower than sheep at high temperatures. Given the capacity of sheep to survive in arid regions, their regulation of salt and water balance may be geared towards coping with salt loads or water shortages, and hence the opportunity to increase the consumption of (or preference for) high-salt diet.

Experimentally, camels at rest in Sahara (Schmidt-Nielsen et al. 1957) and in central Australia (MacFarlane et al. 1967; Siebert and Macfarlane 1971; Nielsen 1967) maintained adequate functions during 10–15 days without water when daily temperatures reached 39–42°C. Aspects of the physical architecture of the camel that fit it for desert life are the wide range of plants (including halophytes) that it will eat, its heat stage/storage?? capacity, its reflectant summer coat, local storage of energy as fat in its hump, and conservation of nitrogen. But the distribution and turnover of water and functions of kidney and gut in relation to water and electrolyte handling are features critical to survival in water scarce areas.

Table 15.2. Effect of increasing ambient temperature on water requirements of mature cattle.*

Environmental temperature °C	Water intake L/kg DM Intake
10–15	3.6
15–21	4.1
21–27	4.7
Over 27	5.5–6.4

* Cattle over 200 kg liveweight, non-pregnant, non-lactating on fresh (non-saline water)

In temperate environments the same rank order of genera and species was sustained. Because of different degrees of fatness (body solids) it is useful to rank the animals by $mol/l^{0.82}$, which relates to the number of molecules passing through the size-adjusted volumes of body water (Schmidt-Nielsen 1977).

Hydration status, assessed by plasma or serum osmolality is the indicator of choice to assess water requirements. However, the animal's needs vary widely according to environmental conditions, physical activity and individual metabolism. The body can also compensate in the short term for over or under-hydration, so it is difficult to establish experimentally. There is no single level of water intake that would ensure adequate hydration and optimal health in all environmental conditions. Activities such as walking to and from water or grazing/browsing sparse vegetation can generate a need for more water (Squires and Wilson 1971). Water economy and water turnover varies with breed of livestock (Squires 1980).

3.2.1 Water loss

Body water is usually lost by four routes: by evaporation, in the faeces, in the urine, and in the milk. In addition, major haemorrhaging, excessive salivation, and mucous or serous exudation can give rise to additional and serious water losses (Warner and Stacy 1977; Tomas and Potter 1975) as a result of traumatic or pathological conditions, but these are not normal and will not be considered further.

Evaporative water loss rises in response to an increased heat load when the other heat dissipation mechanisms, such as re-radiation, convection, conduction, and vasodilation, are insufficient for thermoregulation (Lowe et al. 2002). The effects of temperature elevation and water deprivation on lamb physiology, welfare, and meat quality were assessed by Lowe et al. 2002).

In hot areas, water loss and thermoregulation are inextricably linked in domestic animals. The efficiency of other features designed to lower heat absorption or increase heat dissipation will affect the extent of evaporative water loss and should be mentioned briefly. They include animal coats, coat colour, and appendages. Animal coats act differently in protection from heat loads: the short, smooth coats of *Bos indicus* (zebu cattle) and goats reflect light and facilitate convective loss; in contrast, the tight-packed wool of the Merino sheep re-radiates in the long wavelengths while the loose wool of the Awassi sheep absorbs more heat while convective heat loss is more effective (Degen 1977).

Heat loss (or lowering of heat absorption) has also been associated with anatomical features such as the narrow dorsal profile of the camel, dewlaps, large ears, and skinfolds of *Bos indicus* or the long limbs of goats, sheep, camels, and pigs in hot regions which reduce absorption of reflected heat from the ground and presumably aid convection. Whilst the efficiency of some of these adaptations has been questioned there is no doubt that the overall effect would be to reduce evaporative water loss, especially since re-radiation and convection will be enhanced by larger surface areas. As a means of dissipating heat, evaporation is efficient: vaporization of 1 g of water releases 2.43 kJ but it is at the risk of accelerating dehydration, as up to 4 or 5 per cent of body weight per day can be lost by evaporation in hot conditions. About 80 per cent of the water loss in tropical ruminants may be from evaporative cooling.

Evaporation therefore becomes an important source of water loss in hot areas especially when an animal is exposed to solar radiation and generating metabolic heat. Evaporative losses can be cutaneous or respiratory and the contribution of each route varies according to species. Whereas in donkeys and camels sweating accounts for almost all evaporative heat loss, panting is the major method of evaporation in dogs and pigs (panting is more efficient for smaller animals).

4. Effect of water quality on livestock performance

Drinking is a vital part of the daily activities of livestock in summer. Most ruminant livestock are inefficient users of water and have high rate of water turnover if allowed unrestricted access to water. It is usual under free range conditions for at least one drink to be taken every day in summer, some livestock, e.g., camels can go without drinking for several days. Some ecotypes of sheep and goats are more water efficient and watering may occur every other day. For a watering regime to be adequate for ruminants eating dry forage the following criteria must be met: the degree of dehydration must not exceed the temporary water holding capacity of the alimentary tract, the animal must have time to drink its fill' and the frequency of watering must be such as to prevent body water loss from reaching the stage of clinical dehydration. Dehydration does not reach a critical level in desert-adapted animals until they have lost 30% of the bodyweight (King 1983). In general ruminants can replace 15–20% of their bodyweight at the first drink and 20–25 within 1–2.5 hrs. The capacity and speed of fluid replacement appears to be higher in more desert-adapted animals. When water availability is restricted, there is a reduction in food intake. The effects of water restriction are felt mainly in the areas of energy production and thermoregulation. There is some evidence that the efficiency of digestion of fiber increases when water is restricted. When the quality of forage is low, freeranging cattle and sheep voluntarily restrict the intake and turnover of water, thereby controlling the nitrogen balance and achieving protein maintenance on diets which would normally be below maintenance requirements. On a low nitrogen diet, a high water intake and a high urine volume can flush urea out of the plasma so that it is not available for recycling to the rumen to stimulate microbial digestion (see Degen and Squires, this volume).

The quality of drinking water can seriously affect livestock productivity, and falls into three categories, total dissolved salts, toxic and contaminating substances, and disease-producing organisms. Different species vary in their tolerance to dissolved salts (Table 15.3) and this is partly related to the ability of the kidney to concentrate urine.

Table 15.3. Tolerance of salty drinking water by different livestock.

Species	Recommended units of total dissolved salts in drinking water (%)
Camel	5.5
Goat	1.5
Sheep	1.3–2.0
Cattle	1.0–1.5
Donkey	1.0
Horse	0.9

Source: Nicholson 1985.

In practice, most species exhibit an adaptation to high salt levels in the water, and the levels given in Table 15.3 are frequently exceeded.

4.1 Influence of salinity on water requirements

Where saline waters are the only source of drinking water, there are additional requirements because extra water is required to flush the salt from the animal's system. Sheep grazing predominantly on salt desert range can ingest up to 200 g of salt/day. They can cope with this load only when adequate water is available to allow the excretory system to its work. In summer sheep under such conditions must drink up to three times as much water as those on non-saline diets.

Infrequent drinking, water restriction, and heat load have an effect on feed intake and halophyte utilization by ruminants (King 1983; Nicholson 1985). The main effects are to reduce appetite and increase feed utilization. Reduction in rumen motility, rumination activity and saliva secretion reduce passage rate, and hence increase the digestibility of structural carbohydrates. However, contrary to a commonly held view, these digestive responses are directly related to the imposed stress, rather than secondarily related to reduction in feed intake. Although water deprivation, water restriction and heat load similarly affect appetite and digestion, the physiological basis of each is different. Water restriction does not disturb plasma tonicity and its effect is mediated by modification of the food-related drinking. Water deprivation effects are mediated through elevation of plasma osmolarity and secondarily, through increase in gut fill and rumen Na concentration. Heat load effect is mediated through elevation of body temperature and secondarily through increase in gut fill. The response to heat stress is more diverse, since heat load affects both the energy and water metabolism, and varies due to differences among ruminants in insulation properties and body size. As water scarcity and high temperatures are characteristic of dry areas, interactions exist between these two factors. Goats, having poor insulation capacity, and sheep, with excellent insulation capacity, represent two physiological models. Breeds of ruminants which are well adapted to arid environments demonstrate a greater capability than non-desert breeds to ameliorate the stressful effects induced by water deprivation and heat load. However, as a general rule, the negative effects of water deprivation and heat stress override the benefits from increased feed utilization.

The greatest amount of physiological strain is placed on mammals during lactation. Efforts of synthesis increase both energy and water consumption rates by 40–60%. For example, lactating camels use 44% more water than non-lactating female camels grazing with them (King 1983).

Water consumption increased nearly two-fold in early pregnancy in ewes fed a high-salt diet (Digby et al. 2008; Blache and Revell, this volume). Concentrations of salt in drinking water, as low as 1.3%, can cause neonatal mortalities in lambs (30 to 50% of lambs born to ewes drinking salty water) (Peirce 1968; McIntosh and Potter 1972; Potter and McIntosh 1974).

4.2 Appetite suppression and its implications

Tolerance of livestock to dissolved chemicals is a matter of considerable interest (Squires and Ayoub 1994) and has been studied in some detail by CSIRO in Australia (Squires 1980, 1995) and in Israel, north Africa (Abu-Zanat et al. 2003; Gihad 1993; Barth and Böer 2002) and the Middle East (Degen 1977; Silanikove 1989; Brosh et al. 1986; SanPietro 1989; Riley et al. 1994; Khan et al. 2011, 2014). Non-field studies have suggested that 0.6% sodium chloride (NaCl) in the drinking water should be considered as the upper limit for sheep entirely dependent on halophytic fodder/forage but evidence is that 0.8% might be tolerated. For sheep on non-saline forage/fodder levels of NaCl up to 1.2% may be tolerated without adverse effects. At higher levels, animals went off their food, scoured, lost weight and became listless (Wilson 1966a,b). Higher concentrations (up to 2%) began to have deleterious effects on lungs, gall bladder and adrenals. Omental and kidney fat showed almost complete absence, as evidenced by post mortem examination. At higher salinity levels wool production and body weight were reduced. In general total salinity is more important than the specific dissolved salts. There are some exceptions and some chemicals are not tolerated as well as the common salts that are found in well water. Magnesium chloride is the most harmful salt and it is recommended that 0.2 to 0.3% should be considered as the upper limit for sheep. High levels of urine osmolarity, loss of kidney function and formation of urinary calculi may occur if livestock are on a high salt regime for an extended period. Cumulative effects are a feature of prolonged exposure to high salinity and thus it is important to account for total salt intake (Wilson 1966a,b).

4.3 Adaptation to saline drinking water

Because of the large external component in the regulation of water intake and loss, behavioural avoidance of extreme climatic conditions plays a large part in water conservation. Internal or physiological processes support behavioural adaptations. The ability to endure harsh environments is made possible by selection of milder conditions (micro habitats) such as shade which lessens the stress on the animal and sometimes removes the need for physiological adaptations. This is particularly critical in arid regions.

Observations on the effect of saline drinking water on various livestock species indicate that animals can adapt after the period during which feed intake is suppressed. Some work suggested that rumen micro-flora may have been adversely affected, but work by CSIRO in Australia where sheep were introduced suddenly to drinking water containing 2% NaCl feed intakes fell sharply for three days, but increased again, and by five to seven days reached a stable level 160 g below the feed intake of a companion group on fresh water (Wilson and Hindley 1968). In a second experiment, rumen liquors for *in vitro* digestion studies were taken from sheep conditioned to saline drinking water (2% NaCl) and from sheep with no experience of saline drinking water. The amount of oven-dried ryegrass that was digested decreased with increasing concentration of the medium, but there was less than the concentration of all ions was increased. It was concluded that the adaptation was physiological rather than attributable to impact on rumen physiology (Wilson and Hindley 1968; Wilson 1966a).

There was also individual variation when sheep were introduced to saline water. Variations in intake tend to be more pronounced when sheep are first given saline but are less pronounced after a long period on the same water. Attempts to separate the effects of taste (palatability) from physiological tolerance by either infusing saline solutions directly into the rumen, or offering it as drinking water, suggest that low food intake by some sheep is attributed to low intake of saline water rather than to low tolerance of NaCl. The combined effects of varying the volume of water ingested and concentration of salts in the water were assessed. The food intakes of the sheep increased as the volumes of either fresh or saline waters were increased from 0.5 to 6 litres per day. The addition of NaCl to the water decreased the food intake, but the decrease was restored by an increase in the volume of water given. The increase was approximately 50% for the 1.5% NaCl, and 100% for the 2% solution. The results showed that the volume of consumption is as important as the degree of salinity. Clearly, much of the variation between animals in tolerance to NaCl can be attributed to the volume of water consumed (Wilson 1966b).

5. Effect of restricted access to water on intake of salty feedstuffs

In subtropical drylands (the main regions where pastoralism is practiced) both forage/fodder and drinking water can have a high salt content so that livestock can have an excessive intake of sodium, potassium or even magnesium salts (Wiley-Taft 1980). The range of turnover rates in four livestock species is shown in Table 15.1. Camels have the lowest turnover, zebu cattle and sheep have comparable rates. Large animals with a lower metabolic rate per unit weight and a low surface-to-volume ratio would be expected to have correspondingly low water turnover rates, and in this respect cattle appear to be the least efficient users of water. Livestock with a restricted water supply may be able to avoid water deficiency by selectively grazing/browsing plants high in moisture content. However, in many of the pastoral areas high moisture content is often associated with a high salt content, and selectivity of a low salt diet may differ from that of a high moisture diet. Animals with low turnover rates, like the camel or the goat, have better chance of survival during periods of water deprivation or drought than animals with a high rate of water turnover such as cattle. Some desert adapted animals such as camels, some sheep and goat (ecotypes) and gazelles, use about the same quantity of water per unit $BW^{0.82}$ when water is not restricted. Under similar conditions bovids use two to three times the water used by wild ungulates.

Diets containing 7.5, 11.25 and 15% NaCl were fed to two different breeds of sheep in southern Australia. The diets were intended to simulate diets containing up to 100% saltbush (*Atriplex* spp.). When access to water was restricted once daily, there was a reduction in feed intake. The reduction was more severe with more severe salty diets. It is apparent from this study that sheep dependent on a saltbush diet need to drink more than once daily if a high level of feed intake is to be maintained. The collective effects of restricted access to saline drinking waters and feedstuffs containing high levels of NaCl are to reduce feed intakes even further. The intake of *Atriplex nummularia* fell to less than half when drinking water was replaced by water containing 0.9 to 1.2% NaCl. The mean Na intake of sheep offered a diet of *Atriplex* and ad *lib* water was 97 g per day and the water required to excrete this was about 6 litres per day (Wilson and Hindley 1968). Rumen capacity is often less than this so twice daily watering is required if intakes are to be maintained (Squires and Wilson 1971). Wool production suffers too if high NaCl loads are imposed. Wool production is adversely affected by salinity in food and drinking water. Wool growth rates decline sharply when salt concentration exceeds 1% (Squires 1980).

When given access to water once every 4 days, black Bedouin goats that were fed dry lucerne hay required, on average, only half the water and 60% of the gross energy they consumed when water was offered to them once daily. Digestibility of the feed in these goats was increased by 4%. The outflow of fluid from the rumen slowed down during the water deprivation period. A flow rate of 1080 $ml/kg^{0.75}$/day was recorded during the 1st day of the period and only 350 $ml/kg^{0.75}$ during the last day. While being offered water once daily the outflow from the rumen amounted to 1243 $ml/kg^{0.75}$/day. Following the subjection of black Bedouin goats to infrequent drinking the mean retention time of the digesta in the rumen was extended by 33% and in the entire gastrointestinal tract by 43%. It is suggested that the slowing down of the ruminal flow that followed the subjection of the goats to infrequent drinking enabled the particulate matter to stay longer in the digestive tract despite their smaller size (Brosh et al. 1986).

6. Ameliorating the effects of salinity

Grazing livestock on halophytic pastures can take in as much as 200 g of salt every day. They cope with this load only when adequate water is available to allow the excretory system to do its work. In summer they must drink up to three times as much water as stock on grasslands and they may need to trek to water points at least twice a day (Squires 1980). In large pastures this salt-induced thirst limits the area that can be grazed (Squires and Wilson 1971).

Knowing that sheep, for example, can drink water containing 1 to 1.3% NaCl without it affecting health or production it may be possible to use the limited supplies of fresh water to greatest benefit by mixing it with saline water. This not a simple matter of proportion, since the consumption of saline water is higher than that of fresh water. If the intake of fresh and saline (1 to 1.3%) water is known then the saving of fresh water by mixing with saline well water may be calculated from the following equation:

$$\text{Increase in number of stock watered (\%)} = \frac{V_f}{V_m} \times \frac{1}{1-(m/b)} \times 100 - 100$$

Where V_f is the volume of fresh water consumed, V_m the volume of mixed water consumed, m the % of salt in the original well water. The increase in the number of livestock that may be watered by mixing the available fresh water (rain water and surface run off or even from a desalination plant) with saline well water is the vital factor. A 25% increase in the numbers of livestock watered would be the minimum to justify the cost, and on this basis, the process would be of general use only where salinity was below 1.5 to 2.0%. For more saline well waters, the applicability would depend on the livestock's characteristics. In areas where the diet contains appreciable amounts of salt (halophyte-dominated areas) the concentration of salt in the drinking water may need to be reduced further than 1.0 to 1.3% salt. Well water of up to 1.4% could be diluted to 0.8% (the 'safe' limit for sheep on saline diets). There are many wells in the halophyte-dominated areas that, because of their high salinity, are of doubtful value but the dilution approach may make it possible to render them fully useable.

For livestock that are pen fed on rations comprised of high levels of halophytic feedstuffs, extra water of low salinity needs to be provided. Lambs in Arizona fed on diets comprised of 30% halophyte drank more water than those on conventional grass hay roughage. Lambs on the halophyte diets (including the artificial halophyte) consumed more water per kg of feed intake than lambs on *Cynodon* hay without added NaCl, but water consumption per kilogram of ash intake was similar among treatments. Water intakes in spring ranged from 6.3–13.21 litres day^{-1} when ambient day temperature reached 25–30°C. Summer consumption could be much higher. The impact of seasonal variation in water intakes was as high as the effect of feeding different halophyte ingredients in the diets (Swingle et al. 1996).

7. Summary and conclusions

Physiology, genetics and behaviour, all interact when livestock are fed salty diets and have access to saline drinking water. Research work from around the world's dryland regions offer some guidance as to levels of tolerance but suggest that the upper limit of tolerance to salt in the diet may be quite restrictive if the production of meat, milk and wool is the goal. Subsistence livestock systems may be able to operate within the constraints imposed. Drought resistant and salt tolerant plants in dryland regions may have limitations as browse or grazing for livestock especially if drinking waters are also saline.

References and Further Readings

Abu-Zanat, M.M., F.M. Al-Hassanat, M. Alawi and G.B. Ruyle. 2003. Mineral assessment in *Atriplex halimus* L. and *Atriplex nummularia* L. in the arid region of Jordan. African Journal of Range & Forage Science 20(3): 247–251.

Andersson, B. 1978. Regulation of water intake. Physiological Reviews 58: 582–603.

Andersson, B. and B. Larsson. 1961. Influence of local temperature changes in the preoptic area and rostral hypothalamus on the regulation of food and water intake. Acta Physiologica Scandinavica 52: 75–89.

Assad, F., M. Bayoumi, M. El-Begowryi and S. Deels. 1989. Pathological changes in ewes and their lambs drinking seawater. Egyptian J. Comp. Pathol. Clin. Pathol. 2(2): 61–76.

Assad, F., M.T. Bayoumi and H.S. Khamis. 1997. Impact of long-term administration of saline water and protein shortage on the haemograms of camels and sheep. J. Arid Environ. 37: 71–81.

Assad, F. and M.A. El-Sherif. 2002. Effect of drinking saline water and feed shortage adaptive responses of sheep and camels. Small Ruminant Research 45: 279–290.

Barth, H.J. and B. Böer (eds.). 2002. Sabkha Ecosystems: Volume I: The Arabian Peninsula and Adjacent Countries, Springer, Dordrecht.

Bass, J.M. 1982. A note on the effects of various diets on the drinking behaviour of wether sheep. Animal Production 35: 293–294.

Blache, D. and D.K. Revell. (in press). Short- and long-term consequences of high salt loads in breeding ruminants. pp. 316–335, this volume.

Brosh, A., I. Choshniak, A. Tadmor and A. Shkolnik. 1986. Infrequent drinking, digestive efficiency and particle size of digesta in black Bedouin goats. The Journal of Agricultural Science 106(3): 575–579.

Brown, G.D. and J.J. Lynch. 1972. Some aspects of the water balance of sheep at pasture when deprived of drinking water. Aust. J. Agric. Res. 23: 669–684.

Calder, F.W., J.W.G. Nicholson and H.M. Cunningham. 1964. Water restriction for sheep on pasture and rate of consumption with other feeds. Canadian Journal of Animal Science 44: 266–271.

Cooper, R.A., S. Evans and J.A. Kirk. 1991. Effects of water additives on water consumption, urine output and urine mineral levels in Angora goats. Animal Production 52: 609.

Cooper, S.J. 1985. Neuropeptides and food and water intake. pp. 17–58. *In*: M. Sandler and T. Silverstone (eds.). Psychopharmacology of Food, Oxford University Press, Oxford.

Degen, A.A., M. Kam, A. Rosenstrauch and I. Plavnik. 1991. Growth rate, total body water volume, dry-matter intake and water consumption of domesticated ostriches (Struthio camelus). Animal Production 52: 225–232.

Degen, A.A. 1977. Fat-tailed Awassi and German mutton Merino sheep under semi-arid conditions: 2. Total body water and water turnover during pregnancy and lactation. The Journal of Agricultural Science 88(03): 699–704.

Degen, A.A and V.R. Squires. (in press). The rumen and its adaptation to salt. pp. 336–347, this volume.

Digby, S., D. Blache, D.G. Masters and D.K. Revell. 2010a. Responses to saline drinking water in offspring born to ewes fed high salt during pregnancy. Small Ruminant Research 91: 87–92.

Edwards, K.S., G.A. Classen and E.H.J. Schroten. 1983. Water quality. pp. 49–58. *In*: The Water Resource in Tropical Africa and its Exploitation. ILCA Research Report No. 6, ILCA. Addis Ababa.

El Shaer, H.M. 2010. Halophytes and salt-tolerant plants as potential forage for ruminants in the Near East region. Small Rumin. Res. 91: 3–12.

English, P.B. 1966. A study of water and electrolyte metabolism in sheep. 1. External balances of water, sodium, potassium and chloride. Research in Veterinary Science 7: 233–257.

Forbes, J.M. 1967. The water intake of ewes. British Journal of Nutrition 22: 33–43.

Gihad, E.A. 1993. Utilization of high salinity tolerant plants and saline water by desert animals. pp. 443–447. *In*: H. Lieth and A.A. Al Masoom (eds.). Towards the Rational Use of High Salinity Tolerant Plants, Vol. 1. Kluwer Academic, Dordrecht.

Gordon, J.G. 1965. The effect of water deprivation upon the rumination behaviour of housed sheep. Journal of Agricultural Science 64: 31–35.

Hamilton, J.A. and M.E.D. Webster. 1987. CSIRO. Food intake, water intake, urine output, growth rate and wool growth of lambs accustomed to high or low intake of sodium chloride. Australian Journal of Agricultural Research 38: 187–194.

Hegarty, R.S., J.V. Nolan and R.A. Leng. 1994. The effects of protozoa and of supplementation with nitrogen and sulfur on digestion and microbial metabolism in the rumen of sheep. Aust. J. Agric. Res. 45: 1215–1227.

Johnson, K.G. 1987. Shading behaviour of sheep: preliminary studies of its relation to thermoregulation feed and water intakes, and metabolic rate. Australian Journal of Agricultural Research 38: 587–596.

Jones, G.B., B.J. Potter and C.S.W. Reid. 1970. The effect of saline water ingestion on water turnover rates and tritiated water space in sheep. Australian Journal of Agricultural Research 21: 927–932.

Khan, M.A., B. Böer, M. Öztürk, T.Z. Al Abdessalaam, M. Clüsener-Godt and B. Gul (eds.). 2014. Sabkha Ecosystems: Volume IV: Cash Crop Halophyte and Biodiversity Conservation, TV:S 47 Springer, Dordrecht.

Khan, M.A., B. Böer, H.J. Barth and G.S. Kust. 2011. Sabkha Ecosystems: Volume II: The Arabian Peninsula and Adjacent Countries, Springer, Dordrecht.

King, J.M. 1983. Livestock water needs in pastoral Africa. ILCA Research Report No.7. Addis Ababa.

Lowe, T.E., N.G. Gregory, A.D. Fisher and S.R. Payne. 2002. CSIRO. The effects of temperature elevation and water deprivation on lamb physiology, welfare, and meat quality. Aust. J. Agric. Res. 53: 707–714.

Lynch, J.J., G.D. Brown, P.F. May and J.B. Donnelly. 1972. The effect of withholding drinking water on wool growth and lamb production of grazing Merino sheep in a temperate climate. Australian Journal of Agricultural Research 23: 659–668.

Macfarlane, W.V. 1976. Water and electrolytes in domestic animals. *In*: J.W. Phillis (ed.). Veterinary Physiology. Wright-Scientechnica, Bristol. 463.

MacFarlane, W.V., B. Howard, H. Haines, P.M. Kennedy and C.M. Sharpe. 1971. Hierarchy of water and energy turnover of desert animals. Nature 234: 525–534.

Macfarlane, W.V. and B.D. Siebert. 1971. Water turnover and renal function of dromedaries in the desert. Physiol. Zool. 44: 225.

Macfarlane, W.V. 1965. Water metabolism of desert ruminants. Studies in Physiology 191–199.

Macfarlane, W.V., B. Howard, H. Haines, P.J. Kennedy and C.M. Sharpe. 1971. Hierarchy of water and energy turnover of desert mammals. Nature 234: 483–484.

MacFarlane, W.V. and B. Howard. 1974. Ruminant water metabolism in arid areas. *In*: A.D. Wilson (ed.). Studies of the Australian Arid Zone. II. Animal Production. CSIRO, Melbourne.

Maltz, E. and A. Shkolnik. 1980. Milk production in the desert: lactation and water economy of the Black Bedouin goat. Physiological Zoology Vol. 53, No. 1, Jan., 1980.

Masters, D.G, Rintoul, J. Allan, Dynes, A. Robyn, Pearce, L. Kelly, Norman and C. Hayley. 2005. Feed intake and production in sheep fed diets high in sodium and potassium. Australian Journal of Agricultural Research 56(5): 427–434.

Maloiy, G.M.O. 1972. Renal salt and water excretion in the camel (*Camelus dromedarius*). Symp. Zool. Soc. Lond. 31: 243–259.

McDowell, R.E. 1972. 'Improvement of Livestock in Warm Climates'. W.H. Freeman Co., San Francisco.

McKinley, M.J., D.A. Denton, D. Gellatly, R.R. Miselis, J.B. Simpson and R.S. Weisinger. 1987. Water drinking caused by intracerebroventricular infusion of hypertonic solutions in cattle. Physiology and Behavior 39: 459–464.

Michell, A.R. and P. Moss. 1988. Salt appetites during pregnancy in sheep. Physiology and Behavior 42: 491–493.

Mittal, J.P. and P.K. Ghosh. 1980. Desert sheep withstand salinity and scarcity of water. Indian Farming 30: 31–33.

More, T., B. Howard and B.D. Siebert. 1983. Effect of level of water intake on water, energy and nitrogen balance and thyroxine secretion in sheep and goat. Australian Journal of Agricultural Research 34: 441–446.

Nicholson, M.J. 1985. The water requirements of livestock in Africa, Outlook on Agriculture 14(4).

Olsson, K., K. Dahlborn, S. Benlamlih and J. Hossaini-Hilali. 1997. Regulation of fluid balance in goats and sheep from dry areas. pp. 159–171. *In*: J.E. Lindberg, H.L. Gonda and I. Ledin (eds.). Recent Advances in small ruminant nutrition. Options Mediterraneenes: Serie A Seminaires Mediterraneenes n. 34.

Olsson, K. and M.J. McKinley. 1980. Central control of water and salt intake in goats and sheep. pp. 161–175. *In*: Y. Ruckehusch and P. Thivend (eds.). Digestive Physiology and Metabolism in Ruminants, MTP Press, Lancaster.

Pierce A.W. 1962. Studies on salt tolerance of sheep. 4. The tolerance of sheep for mixtures of sodium chloride and calcium chloride in the drinking water. Australian Journal of Agricultural Research 13: 479–486.

Robertson, B.M., T. Magner, A. Dougan, M.A. Holmes and R.A. Hunter. 1996. The effect of coal mine pit water on the productivity of cattle. I. Mineral intake, retention and excretion and the water balance in growing steers.

Riley, J.J., E.P. Glenn and C.U. Mota. 1994. Small ruminant feeding trials on the Arabian peninsular with *Salicornia bigelovii* Toit. pp. 273–276. *In*: V.R. Squires and A.T. Ayoub (eds.). Halophytes as a Resource for Livestock and for Rehabilitation of Degraded Lands. Kluwer Academic, Dordrecht.

SanPietro, A. (ed.). 1989. Biosaline Research, International Workshop. Plenum Press, New York.

Schmidt-Nielsen, K. 1959. The physiology of the camel. Scient. Am. 201(6): 140–151.

Schmidt-Nielsen, K. 1997. Animal Physiology: Adaptation and Environment. Camb. Uni. Press.

Schmidt-Nielsen, K., C.R. Taylor and J.L. Raab. 1970. Scaling of the energetic cost of running to body size in mammals. Am. J. Physiol. 219: 1104–110.

Schmidt, P.J., N.T.M. Yeates and D.M. Murray. 1980. The effect of water restriction on some physiological responses of steers during enforced exercise in a warm environment. Australian Journal of Agricultural Research 31: 409–416.

Siebert, B.D. and W.V. Macfarlane. 1969. Body water content and water turnover of tropical Bostaurus, *Bos indicus, Bibos banteng,* and *Bos bubalus bubalis*. Australian Journal of Agricultural Research 20(3): 613–622.

Siebert, B.D. and W.V. Macfarlane. 1971. Water turnover and renal function of dromedaries in the desert. Physiological Zoology 44(4): 225–240.

Siebert, B.D., V.A. Romero, R.A. Hunter, R.G. Megarrity, J.J. Lynch, J.D. Glasgow and M.J. Breen. 1978. Partitioning intake and outflow of nitrogen and water in cattle grazing tropical pastures. Australian Journal of Agricultural Research 29: 631–644.

Silanikove, N. 1989. Interrelationships between water, food and digestible energy intake in desert and temperate goats. Appetite 12(3): 163–170.

Springell, P.H. 1968. Water content and water turnover in beef cattle. Australian Journal of Agricultural Research 19: 129–144.

Squires, V.R. and A.T. Ayoub (eds.). 1994. Halophytes as a Resource for Livestock and for Rehabilitation of Degraded Lands. Kluwer Academic, Dordrecht.

Squires, V.R. 1988. Water and its function, regulation and comparative use by ruminants. pp. 217–226. *In*: D.C. Church (ed.). The Ruminant Animal: Digestive Physiology and Nutrition. Prentice Hall, Englewood Cliffs, New Jersey.

Squires, V.R. 1980. Livestock Management in the Arid Zone. Inkata Press, Melbourne, 281 p.

Squires, V.R. 1989. Australia: Distribution, characteristics, and utilization of shrublands. pp. 61–92. *In*: C.M. McKell (ed.). The Biology and Utilization of Shrubs. Academic Press, San Diego.

Squires, V.R. and A.D. Wilson. 1971. Distance between feed and water supply and its effect on drinking frequency and food and water intake of Border Leicester and Merino sheep. Aust. J. Agric. Res. 22: 283–290.

Squires, V.R. 1993. Australian experiences with high salinity diets for sheep. pp. 449–457. *In*: H. Lieth and A.A. Al Masoom (eds.). Towards the Rational Use of High Salinity Tolerant Plants, vol. 1. Deliberations about High Salinity Tolerant Plants and Ecosystems. Kluwer Academic Publishers, Dordrecht.

Swingle, R.S., E.P. Glenn and V. Squires. 1996. Growth performance of lambs fed mixed diets containing halophyte ingredients. Animal Feed Science Technology 63: 137–148.

Tomas, F.M., G.B. Jones, B.J. Potter and G.L. Langsford. 1973. Influence of saline drinking water on mineral balances in sheep. Aust. J. Agric. Res. 24: 377–386.

Tomas, F.M. and B.J. Potter. 1975. Influence of saline drinking water on the flow and mineral composition of saliva and rumen fluid of sheep. Aust. J. Agric. Res. 26: 585–598.

Warner, A.C.I. and B.D. Stacy. 1968. The fate of water in the rumen. 2. Water balances throughout the feeding cycle in sheep. British Journal of Nutrition 22: 389–410.

Warner, A.C.I. and B.D. Stacy. 1977. Influence of ruminal and plasma osmotic pressure on salivary secretion in sheep. Quarterly Journal of Experimental Physiology 62: 133–142.

Weeth, H.J. and L.H. Haverland. 1961. Tolerance of growing cattle for drinking water containing sodium chloride. J. Anim. Sci. 20: 518–521.

Wiley-Taft, S. 1980. Nutritive value for goats of *Atriplex* species grown with hypersaline water. Master Thesis, Dept. of Animal Science, University of Arizona, Tucson.

Wilson, A.D. 1966a. The intake and excretion of sodium by sheep fed on *Atriplex* (saltbush) and *Kochia* (Bluebush) Aust. J. Agric. Res. 17: 155–163.

Wilson, A.D. 1966b. The tolerance of sheep to sodium chloride in food or drinking water. Aust. J. Agric. Res. 17: 503–514.

Wilson, A.D. 1970. Water economy and food intake of sheep when watered intermittently.

Wilson, A.D. and N.L. Hindley. 1968. The effects of restricted access to water on the intake of salty foods by Merino and Border Leicester sheep. Aust. J. Agric. Res. 19: 597–604.

Wilson, A.D. 1974. CSIRO, Water consumption and water turnover of sheep grazing semiarid pasture communities in New South Wales. Australian Journal of Agricultural Research 25(2): 339–347.

Impact of Halophytes and Salt Tolerant Forages on Animal Reproduction

E.B. Abdalla,[1,]* *A.S. El Hawy,*[2] *M.F. El-Bassiony*[2] and *H.M. El Shaer*[2]

SYNOPSIS

This chapter discusses the impact of feeding halophytes and salt tolerant fodder crops on the reproductive performance of livestock (females and males). The effect on pregnancy rate, lambing percentage, estrous cyclicity, conception rate and sexual hormones profiles of females and reproductive performance of males in terms of the onset of reproductive activity (puberty), semen characteristics and development of reproductive organs are covered.

It is clear that the reproductive performance of livestock is adversely affected by excess salt in drinking water or those fed on halophytic fodders with ewes or does failing to conceive and prevalent mortality of neonatal lambs and kids. However, water salinity was more deleterious than feed salinity with respect to its effect on semen quality and reproductive organ's measurements causing a significant decrease in physical semen quality (volume, concentration, pH and motility).

The combination of salt in animal feed and drinking water is of critical importance. Livestock reproduction is very responsive to changes in animal nutritional supply; there are specific reproductive problems for

---

[1] Animal Production Department, Faculty of Agriculture, Ain Shams University, Cairo, Egypt.
[2] Desert Research Center, Cairo, Egypt.
* Corresponding author: esmat54@hotmail.com

livestock associated with consumption of salt in feed and water. There is some indication that reproducing livestock are more susceptible to high salt intakes than non-reproducing livestock. Moreover, nutrition is considered to be a major factor which controls the output of spermatozoa and accessory fluid production in the male. Little and conflicted information is available concerning the impact of salinity on male fertility. However, there is some information on the effect of halophytes feeding on the various aspects of livestock's reproductive traits of males and females such as estrous and conception rate, sexual hormones profiles, pregnancy, etc.

Keywords: pregnancy, lambing, estrous, conception rate, sexual hormones, puberty, semen, fertility, goats, sheep, water intake.

1. Introduction

Reproduction is very responsive to changes in animal nutritional supply (O'Callaghan and Boland 1999). Animals raised under desert conditions usually face many hazards, e.g., salinity stress from feed and drinking water, fresh water shortage, high temperature and food shortage. The combination of salt in animal feed and drinking water is of critical importance. When the high salt intake comes from feed alone, and there is an unlimited supply of fresh water, the animal can cope by increasing water intake thereby increasing the salt excreting capacity of the kidneys (Squires, this volume). This cannot happen if the salt is present in both feed materials and drinking water. This has significant practical consequences; any level of salt in drinking water will compound the effects of a high dietary salt intake. Such an interaction is likely to be more important during hotter, dry periods of the year (Wilson 1975). Under these challenging conditions, feeding halophytes or/and salt tolerant fodder crops is a feasible solution to minimize the problem of feed shortage especially in arid and semi-arid regions (El-Shaer 2010). The use of halophytic plants (e.g., *Atriplex* spp.; *Acacia* spp.; and *Kochia* spp.; *Nitraria retusa*; *Tamarix* spp., etc.) and salt-tolerant plants (e.g., Sudan grass (*Sorghum halepense*); pearl millet (*Pennisetum glaucum*; *Sorghum* spp); fodder beet (*Beta vulgaris*); *Panicum* spp.; and alfalfa (*Medicago sativa*) represents one of the few options available to re-vegetate salinized lands and re-establish grazing systems (Masters et al. 2007; El Shaer 2010).

There is information on the effect of halophytes and salt tolerant forages feeding on the various aspects of livestock's reproductive traits and responses. This chapter was planned to shed some light on the effect of feeding halophytes and salt tolerant fodder crops on the reproductive performance of livestock (females and males).

2. Effect of salinity on reproductive performance of females

2.1 Estrus behavior and conception rate

In ruminants, the level of feeding, energy intake and protein absorption are important regulators of reproductive function (Stewart and Oldham 1986; Fowden et al. 2006).

There is some indication that reproducing livestock are more susceptible to high salt intakes than non-reproducing livestock (Blache and Revell, this volume).

Concerning the effect of feeding halophytic and salt tolerant forages (HSTF) on different aspects of conception rate and estrus of ruminant, Solomon (2002) reported that the lack of significant differences in estrus behaviour between animals supplemented with *Sesbania seban* and concentrates would suggest that long-term and uninterrupted supplementation of forage trees; *S. sesban* (salt-tolerant plant), did not have any visible negative effects on sexual development and activity of female sheep. However, supplementation of Menz ewes with the shrubs *S. sesban* had a negative influence on their reproduction by compromising manifestation of estrus as reported by Melaku et al. (2004). The effect was related to the anti-nutritional factors in the studied plant. Zarkawi et al. (2003) reported that 50% of the Damascus does fed a diet containing *S. aculeate* hay failed to kid and didn't exhibit estrus after the initial mating. The authors concluded that inclusion of *S. aculeate* hay in the diet either prevented conception or caused early embryonic mortality. Other investigators found that Dwarf does fed Kallar grass (*Leptochloa fusca*) hay had difficulty in delivery and the kids were born weak and did not survive (Khanum et al. 1986). A study was conducted was by Shaker (2009) to determine the effect of feeding *Atriplex nummularia* on the estrous response of Barki ewes to the double injections of prostaglandin $F_{2\alpha}$ (estrumate), and on the plasma progesterone and estradiol 17β profiles. The results presented that feeding *Atriplex nummularia* did not affect the estrus behaviour of Barki ewes adversely despite the energy deficiency in such forage (Kandil and El Shaer 1989). These insignificant differences in estrus behaviour would be a result of the high protein level fed. Conversely, McWilliam et al. (2004) reported a significant improvement of the reproductive performance of Romney ewes with an increased absorption of protein as a result of poplar (*Populus* sp.) supplementation. Weston et al. (1970) reported that *Atriplex nummularia* with energy of 6.1 MJ ME/kg DM, and crude protein of 210 g/kg are sufficient to maintain a certain level of reproduction. Ali et al. (1991) reported that does fed salt tolerant plants had normal conception rate and did not differ significantly with control ones. Shaker (2009) concluded that feeding *Atriplex nummularia* did not have any negative effects on the estrous behavior, estrus cycle, progesterone, and estradiol 17 β profiles. It would be suitable fodder for sheep at early reproductive stages. The same author reported that Barki ewes fed berseem hay (*Trifolium alexandrinum*); the control group, had a insignificant higher percentage of ewes manifesting estrus, in response to the first injection of estrumate, than that of ewes fed *Atriplex* (65 and 60). The estrus duration was 24.5 hrs for control group vs. 25.2 hrs for *Atriplex* group and estrus cycle length was 17.5 day for control group vs. 18.1 day for *Atriplex* group. Recently, Atta et al. (2012) compared feeding pelleted molasses with mash sorghum grain on reproductive performance of Sudan Nilotic does. They found that conception rate was insignificantly higher in groups on pelleted molasses compared to mash sorghum (93.75 vs. 86.67%, respectively), while kidding rate was higher in mash sorghum than pelleted molasses (100 vs. 93.33%, respectively) and pre-weaning kid mortality was lowest in mash sorghum compared to pelleted molasses (6.67 vs. 13.33%) respectively. The impact of salt load on goat reproductive traits was evaluated by El Hawy (2013), where four groups of Shami female goats received berseem hay (Groups 1 and 2) while groups 3 and 4 received alfalfa as a salt tolerant fodder. Groups 1 and 3 drank fresh tap water while those in groups 2 and 4 drank saline ground water (6000 ppm). It was found that conception rate insignificantly increased in alfalfa groups as compared to berseem hay groups. G3 and G4 recorded 90 and 80% while, G2 and G1 recorded 90 and 75% conception rate, respectively. Estradiol-17β (E_2) and progesterone (P_4) profiles during estrus cycle were found to follow the normal pattern reported during follicular and luteal phases in the cycle with insignificant differences between alfalfa groups and berseem hay groups. The author reported that during pregnancy, E_2 levels did not differ significantly among the four experimental groups and remained unchanged till 145 days of gestation, then started to increase sharply till parturition.

2.2 Pregnancy and lambing rates

A number of reports indicate that reproducing ewes are less tolerant of high salt in drinking water than wethers, with consequences including occasional decreases in lambing percentage and increased lamb mortality (Peirce 1968; Potter and McIntosh 1974). The reproductive performance of ewes was adversely affected by excess (1.3% or above) salt in drinking water with ewes failing to conceive and mortality of neonatal lambs was prevalent as indicated by McIntosh and Potter (1972). The influence of salt water ingestion on pregnancy, parturition and subsequent lamb survival in the two age groups of ewes (3 and 7 year old), where each age group drank fresh and saline water (1.3% sodium chloride) was studied by Potter and McIntosh (1974). They found that all ewes were pregnant and lambed but nine out of 18 lambs from the ewes group drinking 1.3% saline died compared with only two deaths from 17 lambs born to control ewes drinking fresh water. Digby (2007) studied the effect of high-salt diet on pregnancy and lambing rate of Merino ewes. Digby reported that there was no significant difference in pregnancy rates across both high-salt and control groups. Digby also found that twenty two lambs out of 38 on high-salt diet had singletons versus twenty four of the 38 in the control group; four out of 38 lambs were twins in the high-salt diet compared to 6 out of the 38 in the control group. Moreover, the impact of feeding either a high salt diet (NaCl 13% of dry matter) or control diet (NaCl 0.5% of dry matter) on reproductive capacity of Merino ewes was evaluated by Digby et al. (2008). They reported that pregnancy rates were not significantly affected by feeding the high-salt diet throughout pregnancy with rates 82 and 73% for control and high salt groups, respectively and there were no significant difference in gestation length (150 days in all groups). See also Blache and Revell, this volume.

Concerning the effect of feeding halophytic and salt tolerant fodder on lambing percentage and litter size at birth of small ruminants, Shetaewi et al. (2001) reported that the kidding percentage and the litter weight at birth did not differ significantly when they used rice straw (G1), *Acacia saligna* (G2) or berseem hay (G3) as a source of roughage in feeding Damascus goat. However, the number of kids weaned per doe kidding and the weight of the kids weaned per doe kidding were significantly higher for G2. They also found that conception rates and gestation length did not differ significantly between treatments. Similarly, reproductive performance of Tunisian

Queue ewes fed on *Acacia cyanophylla* treated with polyethylene glycol and without treatment (control group) were compared (Rekik et al. 2007). No significant differences between the two treatments were detected in terms of ovulation rate, conception rate, lambing rate and percentage of litter size at birth. El Hawy (2013) found that the does exposed to high salt load (drinking saline ground water and offered alfalfa as a salt tolerant fodder recorded high percentage of kidding rate compared to their mates fed berseem hay and drinking saline ground water.

2.3 Fertility

The effect of *Sesbania aculeate* which is grown on salty soil and irrigated by saline water on some reproductive parameters of Syrian Awassi ewes was evaluated by Zarkawi et al. (2005). They concluded that all ewes tested exhibited estrus and mated within 48 h after the removal of sponges, thus mating rate was 100% in both the control and experimental groups. Also, there was no significant (P < 0.05) difference in fertility rate between the control and the experimental groups, with rates being 80 and 72% respectively, indicating that feeding with *S. aculeate* hay had no effect on fertility rate.

3. Reproductive performance of males

3.1 The onset of reproductive activity (puberty)

Puberty is characterized by the beginning of the reproductive activity and it has great importance for the breeding system. However, for reaching its full reproductive potential, the animal needs to reach its sexual maturity, which is a phase when the sexual instinct is shown and the mating capacity and seminal parameters are consistent with the full reproduction (Pacheco et al. 2009).

It is evident that the onset of sexual activity in small ruminants may vary depending on breed, management system and birth date (Madani and Rahal 1988; Abi Saab et al. 1997; Hashem and Hammam 2009).

Gauthier et al. (2001), working with Nigerian Dwarf goats raised under an intensive system, observed that the occurrence of puberty was at about the 20th week of age, whereas Souza et al. (2011) demonstrated in their study that penis detachment occurred at 102.9 ± 15.4 days of age, while the full sequence of sexual behavior (including mating capacity, ejaculation with viable sperm cells), which is understood as reaching puberty was observed at the 20th week (145.2 ± 9.7 days), of age, and sexual maturity at the 38th week of age.

Madani and Rahal (1988) described the influence of the season of birth on the age of puberty in native Libyan goats; the animals born in winter were more precocious than those born in summer (22nd vs. 27th weeks of age, respectively). On the other hand, puberty age of male Shami kids born in spring and autumn (111 ± 3.0 and 112 ± 5.0 days, respectively) was attained significantly earlier than those born in winter (131 ± 4.0 days) as reported by Hashem and Hammam (2009).

3.2 Semen quality characteristics

Semen characteristics of farm animals appear to vary not only among animal species but also among individuals of the same species due to age, sexual preparation, management and frequency of ejaculation. Nutrition is considered to be a major factor which controls the output of spermatozoa and accessory fluid production in the male. Little and conflicted information is available concerning the impact of salinity on male fertility.

In bucks, Ibrahim and Yousri (1992) found that a high concentration of zinc caused a significant decrease in physical semen quality (volume, concentration pH and motility). El-Hassanein (1996) in his study to investigate the accumulative effect of saline drinking water on semen quality of Barki male lambs, found that long term (22 months) drinking of saline water (13.1 g/l TDS) led to a highly significant reduction in sperm cells concentration, percentage of alive sperm-cells and advanced motility of sperms (29.4%, 37.2% and 41.7%) from fresh water-treated group, respectively. These results coincided with a highly significant increase in the percentage of the primary and secondary abnormalities (101.2% and 160.1% from fresh water-treated group values, respectively) and the concentration of fructose (40.6% from fresh water-treated group value). However, drinking saline water had no significant effect on the ejaculate volume.

Exposure to metals has been long associated with low sperm motility, density, increased morphological abnormalities and male infertility as indicated by Mathur et al. (2010). On Barki rams, Zaghloul et al. (2011) found that drinking saline water (10624 ppm TDS) resulted in a significant decrease in ejaculate volume (0.71 ml) as compared to that (0.76 ml) of animals drinking fresh water (300 ppm TDS). They also found that drinking saline water slightly decreased the sperm concentration (5532×10^6/ml), and significantly decreased the percentages of motility and alive sperm-cells (80.0% and 82.9%, respectively) than those of animals drinking fresh water (5609×10^6/ml, 88.1% and 89.1%, respectively).

A study was carried out to evaluate and assess the impact of salinity on reproductive performance of 28 growing Shami male kids (2.5–3.0 months old) from weaning to sexual maturity throughout one year (El-Bassiony 2013). The animals were randomly assigned into four equal groups: the 1st group (G1, as control) was fed on berseem hay (BH) and drank tap water (TW), the 2nd group (G2) was fed on BH and drank saline water (SW), the 3rd group (G3) was fed on salt-tolerant plants (STP, alfalfa) and drank TW and the 4th group (G4) was fed on STP and drank SW. The results pointed out that bucks in G1 and G3 showed the best semen quality, whereas those in G2 and G4 exhibited the worst quality (Table 16.1). These results are confirmed by the observed highest values of ejaculate volume (EV), total sperm per ejaculate (TSE) and total motile sperm (TMS) in G1 bucks in addition to higher values of mass motility (MM), live spermatozoa (LSP) and sperm concentration (SPC) in G3 in comparison with the other two groups (G2 and G4). Whereas, the lowest values of both feed conversion and ASP (abnormal spermatazoa) were obtained by G3 bucks; the highest values of these traits were obtained by G4 bucks. On the other hand, the lowest values of MM, SPC, TSE and TMS were recorded in bucks of G2. Also, the least values of EV and LSP were seen in G4 bucks. From the afore-mentioned results, it seems that both MM

Table 16.1. Semen quality (LSM ± SE) characteristics of Shami male as affected by salinity conditions (El-Bassiony 2013).

Traits	Experimental groups			
	G1	G2	G3	G4
Ejaculate volume (ml)	$1.59^a \pm 0.14$	$1.18^b \pm 0.07$	$1.11^b \pm 0.08$	$1.24^b \pm 0.10$
Sperm motility (%)	$76.7^a \pm 4.71$	$74.4^a \pm 5.56$	$78.9^a \pm 3.51$	$76.7^a \pm 5.77$
Live spermatozoa (%)	$73.6^{ab} \pm 0.96$	$73.7^{ab} \pm 1.53$	$75.4^a \pm 1.21$	$71.8^b \pm 0.72$
Sperm conc. ($\times 10^9$/ml)	$3.07^a \pm 0.07$	$2.99^a \pm 0.07$	$3.12^a \pm 0.09$	$3.02^a \pm 0.14$
Total sperm output ($\times 10^9$)	$4.93^a \pm 0.50$	$3.50^b \pm 0.19$	$3.51^b \pm 0.35$	$3.76^b \pm 0.34$
Total motile sperm ($\times 10^9$)	$3.64^a \pm 0.39$	$2.59^b \pm 0.18$	$2.64^b \pm 0.25$	$2.70^b \pm 0.25$
Fructose conc. (mg/ml)	$1.61^a \pm 0.02$	$1.58^a \pm 0.01$	$1.45^b \pm 0.02$	$1.62^a \pm 0.01$
Sperm abnormalities (%)	$12.9^c \pm 0.42$	$15.8^b \pm 0.81$	$11.7^c \pm 0.36$	$17.8^a \pm 0.59$

G1: fed on berseem hay (BH) and drank tap water (TW), G2: fed on BH and drank saline water (SW), G3: fed on salt-tolerant plant (SP) and drank TW, G4: fed on SP and drank SW.
[a-c] within each row, least square means with different superscripts differ significantly ($P < 0.05$).

and SPC were not affected by salinity conditions. In contrast, EV, TSE and TMS were affected considerably by salinity as shown by the differences ($P < 0.05$) between the G1 bucks and the other three groups which showed insignificant differences among them. With regard to ASP, there was no significant difference between G1 and G3 bucks and both groups differed ($P < 0.05$) than those in G2 and G4 groups.

With respect to the effect of salinity on semen quality and reproductive organs measurements, the present result confirmed that water salinity was more deleterious than feed salinity. The observed results of semen quality are in agreement with those reported by Ibrahim and Yousri (1992) in bucks in which a high concentration of zinc caused a significant decrease in physical semen quality (volume, concentration pH and motility). Also, Mathur et al. (2010) reported that exposure to metals has been long associated with low sperm motility, density, increased morphological abnormalities and male infertility. Zaghloul et al. (2011) found that drinking diluted sea water showed a stressful effect on the animal physiology as it could have a deleterious effect on the accessory sex gland resulting in a lower semen quality. The same trend was previously obtained by El-Hassanein (1996) in Barki rams. On the other hand Yousef et al. (2003) found that drinking saline water containing 800, 3004 or 5284 ppm TDS improved rabbit semen characteristics under desert conditions. The same author reported that the reduced sperm-cells concentration and percentage of motile sperm-cells, in addition to the increased percentage of sperm-cells abnormalities, may reflect an adverse effect of salinity on the spermatocytogenesis process. The long term saline treatment of rams may cause stress on the pituitary gland and lead to an alteration in the gonadotropins releasing cells activity, which controls the androgenic secretion by testes. It has been reported that hypophysectomy results in cessation of spermatogenesis which can be restored by treatment with both FSH and LH or with both FSH and testosterone (Garner and Hafez 1993).

Minerals may operate through hormonal pathways to affect male reproduction, affecting spermatogenesis. Effects may be at different stages of the cell cycle, which in turn may be a cause for sperm abnormalities and hormonal production. In desert rams, Mohamed and Abdellatif (2010) reported that stress directly affects thermoregulation and seminal characteristics which are directly related with physical semen characteristics. The physical changes in semen may be due to hormonal effect of FSH and LH, as a result of GnRH release disturbance due to the effect of salinity on pituitary receptors.

3.3 Effect of salinity on the reproductive organs measurements

3.3.1 Testicular measurements

Size of the testes determines the amount of sperm-producing tissue (Bearden and Fuquay 1997). This, in turn, indicates whether the buck is a good or poor sperm producer. Kridli et al. (2005) in their study to examine scrotal circumference and semen characteristics of Mountain Black goat bucks and its crossbreed with Damascus goat (yearling and matures) prior to the breeding season under arid conditions of Jordan, they found that crossbred bucks (Mountain black × Damascus goats) had scrotal circumference about 25.8 ± 0.8 cm for yearlings and 30.8 ± 1.4 cm for mature with an average of 28.3 ± 1.0 cm.

The testicular activity and in turn semen quality values decreased outside of the breeding season (winter period) but remained acceptable, showing the slight modification in sexual activity for bucks of the Baladi goat breed with seasonal fluctuations as reported by Abi Saab et al. (2008).

Souza et al. (2011) noted in their study that the gradual increase of scrotal circumference occurred in alternating periods of significant growth and stability from the 20th to 44th weeks of age. It has been verified that a high positive correlation between live weight and scrotal circumference (r = 0.94) and also between age and scrotal circumference (r = 0.68) exists. These findings reinforce the idea that, regardless of breed and latitude, the testicular growth is closely correlated to live weight and the age of goats.

A study was conducted to evaluate the effect of feeding salt tolerant fodder plants on the reproductive performance of male goats (El-Bassiony 2013) where 28 growing kids were used in four groups. Animals in Groups 1 and 2 received berseem hay while groups 3 and 4 were offered alfalfa (as a salt tolerant fodder grown in saline lands and irrigated with saline ground water). Groups 1 and 3 were offered fresh tap water while the other two groups (2 and 4) were offered underground saline water (6000 ppm). He reported that the scrotal circumference (SC), testes width (TWD) and testes length (TL) increased gradually with the advancement of age. Bucks in G4 showed the lowest values of all testes measurements compared to the other three groups. This group (G4) differed significantly (P < 0.05) than G1 in all these measurements. The size of the testes determines the amount of sperm-producing tissue (Bearden and Fuquay 1997). This, in turn, indicates whether the buck is a good or poor sperm producer. In this study, gradual increases of scrotal circumference, testes diameter and testes length occurred in alternating periods as age advance showing progressive increases from five to seven months of age. Souza et al. (2011) found that as age increased, there was a high positive correlation between live weight and scrotal circumference (r = 0.94) and also between age and scrotal circumference (r = 0.68). These findings reinforce the idea that, regardless of breed and latitude, the testicular growth is closely correlated to live weight and the age of goats. El-Bassiony (2013) pointed out that there were progressive increases in all testicular measurements from the third month to the fifth of the experimental period (5–7 months of age). These findings are in agreement with those reported by Souza et al. (2011) who noted that the gradual increase of scrotal circumference occurred in alternating periods of significant growth and stability from the 20th to 44th weeks of age.

3.3.2 Anatomical features of the genitalia

The male reproductive system consists of a pair of testes that produce sperm (or spermatozoa), ducts that transport the sperm to the penis and glands that add secretions to the sperm to make semen. The testes consist of a mass of coiled tubes (the seminiferous or sperm producing tubules) in which the sperm are formed by meiosis. Cells lying between the seminiferous tubules (Leydig cells) produce the male sex hormone testosterone. When the sperms are mature they accumulate in the collecting ducts and then pass to the epididymis before moving to the sperm duct or vas deferens. The two sperm ducts join the urethra just below the bladder, which passes through the penis and transports both sperm and urine. Ejaculation discharges the semen from the erect penis. It is brought about by the contraction of the epididymis, vas deferens, prostate gland and urethra.

The accessory genital glands of males in sheep and goat are located along the pelvic portion of the urethra, with their ducts opening and emptying their secretion in to the urethral passage. They include the ampulla, seminal vesicles, prostate gland and the bulbourethral (Cowper's) glands. They contribute greatly to the fluid volume of semen. Their secretions are solution of buffers, nutrients and other substances needed to assure optimum motility and fertility of semen (McDonald 1980; Hafez 2004). These glands produce the greater part of the ejaculate fluid, as a favorable medium for nutrition, and as a buffer against excess acidity of the female genital tract. They elaborate serous and mucous secretion that nourish and activate the spermatozoa, with ducts that empty their secretions into the urethra (Bearden and Fuquay 2000). The enlarged end of the vas deferens near the urethra is the ampulla. Some have suggested that ampulla serve as a short term storage depot for semen. The two ampullas pass under the body of the prostate and open together with the excretory ducts of the seminal vesicles into the urethra with a slit-like orifice on each side of the *colliculus seminalis.*

Seminal vesicles are present in all domestic animal species except the camel, dog and cat. They are large, externally smooth, hollow and a knobby organ. The seminal vesicles are a pair of lobular glands that are easily identified by their knobby appearance (Athure et al. 1996). The paired Cowper's gland is present in all domestic mammalian species except the dog. It consists of right and left club-shaped independent lobes, which lie on the dorsal surface of the caudal part of pelvic urethra and are closely related to the bulb of penis. Hemeida (1985) described the Cowper's gland of Baladi buck as large, dense, spherical organs 1.5 cm in diameter and 2 g in weight.

Khalaf and Merhish (2010) reported that the anatomical parameters of the ampulla, seminal vesicle gland and Cowper's gland in bucks have mean weight 2.18 ± 0.007; 4.18 ± 0.15 and 3.23 ± 0.11 g, respectively. While the corresponding values in rams were 2.24 ± 0.007; 4.42 ± 0.15 and 3.46 ± 0.01 g, respectively.

El-Bassiony (2013) indicated that negative effect of salinity either in water or in forages was reflected in the anatomical measurements of male genitalia (Table 16.2). Bucks in G1, which were fed on berseem hay (BH) and drank tap water (TW), showed the greatest values of genital measurements compared to the other three groups which received salinity in either feed or water of different levels, except only the epididymis weight (EPW); however this difference was not significant. The same author, also, reported that there were no significant differences between G1 and those offered drinking fresh tap water and salt tolerant forage (G3) in most traits, except the ampulla weight (AMW) and seminal vesicles weight (SVW). This result revealed that the effect of feed salinity was less deleterious than that of water salinity. On the other side, bucks in G2 and G4 recorded the smallest measurements, particularly those in G4, compared to bucks in G1 and G3.

Table 16.2. Anatomical feature of the reproductive organs (LSM ± SE) of Shami bucks as affected by salinity conditions (El-Bassiony 2013).

Traits	Experimental groups			
	G1	**G2**	**G3**	**G4**
Testis volume (ml)	$170.8^a \pm 1.61$	$154.2^{ab} \pm 4.81$	$166.7^a \pm 2.71$	$139.2^b \pm 10.67$
Organs weight (g)				
Testis	$87.50^a \pm 2.26$	$74.17^b \pm 3.19$	$81.67^a \pm 1.12$	$75.00^b \pm 1.94$
Epididymis	$16.31^a \pm 0.36$	$13.46^b \pm 0.21$	$17.38^a \pm 1.27$	$13.42^b \pm 0.59$
Ampulla	$1.83^a \pm 0.15$	$1.51^b \pm 0.05$	$1.43^b \pm 0.07$	$1.33^b \pm 0.05$
Seminal vesicles	$7.82^a \pm 0.36$	$6.58^{ab} \pm 0.57$	$5.65^b \pm 0.30$	$6.87^{ab} \pm 0.37$
Cowper's gland	$1.63^a \pm 0.08$	$1.22^c \pm 0.08$	$1.48^{ab} \pm 0.08$	$1.35^{bc} \pm 0.05$

G1: fed on berseem hay (BH) and drank tap water (TW), G2: fed on BH and drank saline water (SW), G3: fed on salt-tolerant plant (SP) and drank TW, G4: fed on SP and drank SW.
[a-c] within each row, least square means with different superscripts differ significantly ($P < 0.05$).

3.3.3 Histological features of the genitalia

Testes are the male gonads, which perform two major functions: production of male gametes (spermatozoa) by seminiferous tubules and production of male sex hormones by interstitial cells of Leydig. Testis of goat buck is an ovoid body weighing 100–150 g (Evans and Maxwell 1987), which varies in size within breeding season. The general structure of buck's testis is that of the mammals.

Each seminiferous tubule is lined by spermatogenic epithelium, surrounding a central lumen and lying on a basement membrane. The epithelium consists of spermatogenic and Sertoli cells. Spermatogenic cells constitute the majority of cells lining the seminiferous tubules. From the periphery to the lumen, they comprise of series of specialized cells arranged in circular layers as; spermatogonia, spermatocytes and spermatids.

Besides the usual connective tissue elements, the stroma of the interstitial tissue contains characteristic cells known as interstitial cells of Leydig. These cells secrete testosterone (androgenic hormone), that is necessary to maintain spermatogenesis in the seminiferous tubules and to regulate reactions of male target tissue throughout the body.

The effect of salinity in drinking ground water and fodder crops (alfalfa) on seminiferous tubule diameters in the testes of Shami bucks have been studied by El-Bassiony (2013). Results showed that there were significant differences (P < 0.05) among all the treatment groups where bucks in the control group, G1 (fed berseem hay and drank fresh water) had the highest values of the average of seminiferous tubule diameters followed by G2 and G3 groups while the lowest value was observed in the G4 group. Moreover, salinity had influenced the percentage of tissue area occupied by seminiferous tubules to the total volume of the testis parenchyma. The present results showed that feeding bucks on STP with drinking TW (G3) increased (P < 0.05) the diameter of the epididymal duct followed by G1 and G2 groups. The lowest diameter of the epididymal duct was observed in the G4 group. The differences among G1, G2 and G4 groups were not significant. Drinking SW alone or combined with feeding SP had adversely affected the muscle layer thickness of the seminal vesicles. There were significant differences among all groups. The highest (P < 0.05) value was observed in the G1 group followed by G2 group and then G3 group. While, the minimum (P < 0.05) value was observed in the G4 groups that fed STP and drank SW.

4. Conclusion

From the presented results, it is concluded that the reproductive performance of livestock is adversely affected, to some extent, by excess salt in drinking water or feeding on halophytes and salt tolerant fodders. Water salinity was more deleterious than feed salinity with respect to the effect of salinity on semen quality and reproductive organs measurements that cause a significant decrease in physical semen quality. However, more studies are needed to evaluate the influence of feeding various halophytes and salt tolerant fodders on reproductive traits and performance of cattle and small ruminants.

References

Abi Saab, S., F.T. Sleiman and K.H. Nassar. 1997. Implication of high and low protein levels on puberty and sexual maturity of growing male kids. Small Rumin. Res. 25: 17–22.

Abi Saab, S., E. Hajj, B. Abi Salloum and E. Rahme. 2008. Seasonal and altitudinal variations on adaptation, growth and testicular activity of Baladi goats with vertical transhumance in Eastern Mediterranean. Lebanese Science Journal 9(1): 99–111.

Ali, M., S.A. Khanum and S.H.M. Naqvi. 1991. Effect of feeding salt tolerant plants on growth and reproduction in goats. pp. 75–88. *In*: Proceeding of the Final Res. Co-ordination Meeting on Improving Sheep and Goat Production with the Aid of Nuclear Techniques (Ed. International Atomic Energy Agency) (IAEA: Vienna, Austria).

Athure, G.H., D.E. Noakes and H. Pearson. 1996. Veterinary Reproduction and Obstetrics. W.B. Saunders Company, USA, pp. 563–564.

Atta, S., C.Y. Zhou, Y. Zhou, M.J. Cao and X.F. Wang. 2012. Distribution and Research Advances of *Citrus tristeza* virus. Journal of Integrative Agriculture 11(3): 346–358.

Bearden, H.J. and J.W. Fuquay. 1997. Applied Animal Reproduction. 4th Ed. Prentice Hall, Upper Saddle River, New Jersey, USA.

Bearden, H.J. and J.W. Fuquay. 2000. Applied Animal Reproduction. 5th ed. (Printes and Hall Eds.), USA, pp. 30–33.

Blache, D. and D.K. Revell. (in press). Short- and long-term consequences of high salt loads in breeding ruminants, pp. 316–335, this volume.

Digby, S.N. 2007. High dietary salt during pregnancy in ewes alters the responses of offspring to an oral salt challenge. Ph.D. Thesis, Fac. Agric., Adelaide Univ., Australia.

Digby, S.N., D.G. Masters, D. Blache, M.A. Blackberry, P.I. Hynd and D.K. Revell. 2008. Reproductive capacity of Merino ewes fed a high-salt diet. Animal 29: 1353–1360.

El Shaer, H.M. 2010. Halophytes and salt tolerant plants as potential forage for ruminants in the Near East region. Small Ruminant Res. 91: 3–12.

El-Bassiony, M.F. 2013. Productive and reproductive responses of growing Shami goat kids to prolonged saline conditions in South Sinai. Ph.D. Thesis, Fac. Agric., Cairo Univ., Egypt.

El-Hassanein, E.E. 1996. Semen quality of rams grown on drinking saline water. 4th Sci. Cong. Proc., Vet. Med. J., Giza 44(2): 495–500.

El-Hawy, A.S. 2013. Reproductive efficiency of Shami goats in salt affected lands in South Sinai. Ph.D. Thesis, Fac. Agric., Ain Shams Univ., Egypt.

Evans, G. and W.M.C. Maxwell. 1987. Handling and examination of semen. pp. 93–106. *In*: Salamon's Artificial Insemination of Sheep and Goats (Butterworths, Sydney, Eds.).

Fowden, A.L., C. Sibley, W. Reik and M. Constancia. 2006. Imprinted genes, placental development and fetal growth. Horm. Res. 65: 50–58.

Garner, D.L. and E.S.E. Hafez. 1993. Spermatozoa and seminal plasma. *In*: E.S.E. Hafez (ed.). Reproduction in Farm Animals, 6th Ed. Lea and Febiger, Philadelphia, USA.

Gauthier, M., J. Pierson and M. Drolet. 2001. Sexual maturation and fertility of male Nigerian Dwarf goat (*Capra hircus*) clones produced by somatic cell nuclear transfer. Cloning and Stem Cells 3(3): 151–163.

Hafez, E.S.E. 2004. Reproduction in Farm Animals. 7th ed. Lea and Febiger, Philadelphia, USA.

Hashem, A.L.S. and A.H. Hammam. 2009. Impact of birth season on puberty in female and male kids of Shami goat in north Sinai, Egypt. J. Agric. Sci., Mansoura Univ. 34(1): 139–150.

Hemeida, N.A. 1985. The reproductive tract of the Caprine Egyptian Baladi bucks. Assuit Vet. Med. J. 14(28): 204–210.

Ibrahim, S.A.M. and R.M. Yousri. 1992. The effect of dietary zinc, season and breed on semen quality and body weight in goats. Inter. J. Anim. Sci. 7: 5–12.

Kandil, H.M. and H.M. El-Shaer. 1989. The utilization of *Atriplex nummularia* by goats and sheep in Sinai. pp. 71–73. *In*: E.S.E. Galal, M.B. Aboul-Ela, M.M. Shafie (eds.). Ruminant production in the dry subtropics: Constraints and potential. Proc. Int. Symp., Cairo, November 1988, EAAP Publications, No 38, Wageningen, Netherlands.

Khalaf, A.S. and S.M. Merhish. 2010. Anatomical Study of the accessory genital glands in males Sheep (*Ovisaris*) and goats (*Caprushircus*). Iraqi J. Vet. Med. 34(2): 1–8.

Khanum, S.A., M. Ali and S.H. MujtabaNaqvi. 1986. Effect of Kallar grass on the health and reproductive of goats. *In*: Paper presented at Seminar on Reproductive Disorders of Dairy Animals. Univ. of Agric. Faisalabad, Pakistan (Cited by Zarkawi et al. 2005).

Kridli, R.T., M.J. Tabbaa, R.M. Sawalha and M.G. Amashe. 2005. Comparative study of scrotal circumference and semen characteristics of Mountain Black goat and its Crossbred with Damascus goat as affected by different factors. Jordan J. Agric. Sci. 1(1): 18–25.

Madani, M.O.K. and M.S. Rahal. 1988. Puberty in Lybian male goats. Anim. Reprod. Sci. 17(3): 207–216.

Masters, D.G., S.E. Benes and H.C. Norman. 2007. Biosaline agriculture for forage and livestock production. Aust. Ecosys. and Environ. 119: 234–248.

Mathur, N., G. Pandey and G.C. Jain. 2010. Male reproductive toxicity of some selected metals: A review. J. Biol. Sci. 10: 396–404.

McDonald, L.E. 1980. Veterinary Endocrinology and Reproduction. 3rd ed. (Lea and Febiger Eds.), USA, pp. 249–250.

McIntosh, G. H. and B.J. Potter. 1972. The influence of 1.3% saline ingestion upon pregnancy in the ewe and lamb survival. Proc. of the Aust. Physiol. and Pharmacol. Soci. 3: 61.

McWilliams, S.R., C. Guglielmo, B. Pierce and M. Klaassen. 2004. Flying, fasting, and feeding in birds during migration: a nutritional and physiological ecology perspective. J. Avian Biol. 35: 377–393.

Melaku, S., K.J. Peters and A. Tegegne. 2004. Supplementation of Menz ewes with dried leaves of *Lablab purpureus* or graded levels of *Leucaena pallida* and *Sesbania sesban* effects on feed intake, live weight gain onoestrus cycle. Anim. Feed Sci. and Technol. 113: 39–51.

Mohamed, S.S. and M.A. Abdelatif. 2010. Effects of Level of Feeding and Season on Thermoregulation and Semen Characteristics in Desert Rams (*Ovis aries*). Global Vet. 4: 207–215.

O'Callaghan, D. and M.P. Boland. 1999. Nutritional effects on ovulation, embryo development and the establishment of pregnancy in ruminant. Anim. Sci. 86: 299–314.

Pacheco, A., A.F.M. Oliveira and C.R. Quirino. 2009. Características seminais de carneiros da raça Santa Inês na prépuberdade, puberdade e na pós-puberdade. Ars Veterinária 25(2): 90–99 (Cited after Souza et al. 2011).

Peirce, A.W. 1968. Studies on salt tolerant of sheep. VII. The tolerance of ewes and their lambs in pens for drinking waters of the types obtained from underground sources in Australia. Aust. J. Agric. Res. 19: 589–595.

Potter, B.J. and G.H. McIntosh. 1974. Effect of salt water ingestion on pregnancy in the ewe and lamb survival. Aust. J. of Agric. Res. 25: 909–917.

Rekik, M., N. Lassoued, H. Ben Salem and I. Tounsi. 2007. Reproductive traits of Tunisian queue fine de i'ouest ewes fed on wheat straw supplemented with concentrate and *Acacia cyanophylla* linda foliage with and without polyethylene glycol (PEG). Livestock Res. for Rural Develop. 19(11).

Shaker, Y.M. 2009. Ovarian response of Barki ewes fed *Atriplex nummularia* to estrus synchronization. Egypt. J. Basic Appl. Physiol. 8(1): 185–203.

Shetaewi, M.M., A.M. Abdel-Samee and E.A. Bakr. 2001. Reproductive performance and milk production of Damascus goats fed Acacia or berseem clover hay in North Sinai, Egypt. Tropical Anim. Health and Produc. 33: 67–79.

Solomon, M.R. 2002. Ocomportamento do consumidor: comprando, possuindo e sendo. Trad. LeneBelonRibeiro. 5ª ed. Porto Alegre: Bookman, 2002.

Souza, L.E.B., J.F. Cruz, M.R.T. Neto, R.C.S. Nunes and M.H.C. Cruz. 2011. Puberty and sexual maturity in Anglo-Nubian male goats raised in semi-intensive system. R. Bras. Zootec. 40(7): 1533–1539.

Squires, V.R. (in press). Water requirements of livestock fed on halophytes and salt tolerant forage and fodders, pp. 287–302, this volume.

Stewart, R. and C.M. Oldham. 1986. Feeding lupins to ewes for four days during the luteal phase can increase ovulation rate. Proceedings Aust. Soc. Anim. Prod. 16: 47–52.

Weston, R.H., J.P. Hogan and J.A. Hemsley. 1970. Some aspects of the digestion of *Atriplex nummularia* (Saltbush) by sheep. Animal Prod. Australia 8: 517–21.

Wilson, A.D. 1975. Influence of water salinity on sheep performance while grazing on natural grassland and saltbush pastures. Aust. J. of Exper. Agric. and Anim. Husbandry 15: 760–765.

Yousef, M.I., M.M. Zeitoun, Z.K. El-Awamry and A.F.M. Ibrahim. 2003. Effect of drinking saline well water on blood testosterone, seminal plasma enzymes and reproductive performance of rabbits. Egyptian J. Androl. Reprod. 17(1): 67–78.

Zaghloul, A.A., Al-Hameed, A. Afaf and K.A. El-Bahrawy. 2011. Effect of drinking saline water and water deprivation on semen quality and some blood parameters of Barki rams. Egypt. J. Basic Appl. Physiol. 10(2): 223–235.

Zarkawi, M., M.R. Al-Masri and K. Khalifa. 2003. An observation on yield and nutritive value of *Sesbania aculeate* and its feeding to Damascus does. Tropical Grasslands 37: 187–192.

Short and Long Term Consequences of High Salt Loads in Breeding Ruminants

Dominique Blache[1] and *Dean K. Revell*[1,2,*]

SYNOPSIS

An adequate intake of macronutrients, including salts such as sodium chloride, is essential for animals to successfully reproduce (Moinier and Drüeke 2008). However, in most mammals, investigations about the impact of macronutrients on reproductive success have focused on the effect of salt deficiency rather that the impact salt load on breeding efficiency. Over the last decades, soil salinity has become an increasing problem in agriculture worldwide (Ghassemi et al. 1995) and a number of studies have investigated the intake of a diet containing high levels of salt on breeding ruminants (Digby et al. 2011). Research on the interactions between high salt load and reproductive capacity has been conducted mainly in sheep because increasing numbers of land holders are grazing small ruminants on halophytic plants such as saltbush to fill a seasonal feed gap (Masters et al. 2006, 2007). In this chapter, we review the impact of high salt load on the reproductive efficiency of ruminants, taking most of our examples from studies conducted using sheep as a model. After a brief review of the physiological consequences of high salt load, we discuss the present knowledge on the short-term impact (within weeks) of salt load on the different phases of the productive cycle in males and females, starting

---

[1] School of Animal Biology, The University of Western Australia, Nedlands WA 6009.
[2] The University of Western Australia, Nedlands WA 6009, Revell Science, Duncraig WA 6023.
* Corresponding author: dean@revellscience.com.au

from puberty to weaning. We then discuss the long-term, lifetime and intergenerational changes triggered by periods of high saltload. We conclude this chapter by discussing the management strategies that are available for the successful feeding of breeding livestock with high salt content plants.

Keywords: saltbush, *Atriplex*, tolerance, adaptation, reproductive success, gestation, voluntary feed intake, rumen function and feed digestibility, reproductive cycle, intergenerational changes, gamete production, puberty, metabolic balance, hormones, water consumption, foetal development, neonatal survival, pre-natal salt exposure.

1. Physiological consequences of high in take of salt

The accepted daily requirement for sodium for sheep is between 0.09% to 0.18% of DM intake (NRC 1985) and the maximum tolerable level of dietary salt has been set at 9.0% (Meyer and Weir 1954; Meyer et al. 1955; NRC 1985) but, as discussed later in this chapter (and elsewhere in this book), sheep can consume diets with a higher salt content. Sheep that eat either plants of high salt content, such as *Atriplex* spp. (saltbush), or experimental diets aimed to mimic the salt level in a diet based on halophytic plants, can have a daily intake of salt ranging between 5 to 25% NaCl (Meyer and Weir 1954; Wilson 1966b; Wilson and Hindley 1968; Hemsley et al. 1975; Masters et al. 2005a; Blache et al. 2007; Digby et al. 2008; Chadwick et al. 2009a, 2009b, 2009c), which is greater than the daily requirement and often in excess of the proposed maximum tolerable level (9%). A high intake of salt (sodium chloride) impacts on the reproductive cycle because of its effects on feed intake and water intake, and on the hormonal control of both energy balance and salt and water balance.

Feed intake begins to decreases when small ruminants consume around 60 g of salt per day, and, as salt concentrations increase further, the feed intake continues to decline (Wilson 1966b; Wilson and Hindley 1968; Masters et al. 2005a; Blache et al. 2007). A salt concentration of 2.0% or greater ingested from drinking salty water alone is associated with a severe reduction in food intake and possibly death (Pierce 1957; Potter 1963; Wilson 1966b; Potter 1968; Wilson and Dudzinski 1973; Hamilton and Webster 1987). Animals with only salty water available cannot increase their intake of (non-saline) water to maintain water balance in their body, which would normally be one of the main physiological responses to deal with ingested salt. This is an important point to consider in practice, as regions where halophytic plants are grazed by livestock may also have saline water as the source of drinking water. The double load of salt in water and feed can have major consequences to production and survival of livestock.

The effects of salt on voluntary feed intake, rumen function and feed digestibility (Masters et al. 2005a; Thomas et al. 2007) can lead to a decrease in energy intake that can indirectly impact reproduction since stages of the reproductive cycle are influenced by energy balance (Blache et al. 2007). If animals are not kept on a high-salt diet for any length of time, then the effects on reproduction via reduced feed will mostly be short-term, but it should be noted that short periods of reduced feed intake can have long-lasting consequences to reproductive performance if it coincides with key developmental stages (Blache and Martin 2009). The mechanisms by which a salt load affects the endocrine control of voluntary feed intake and metabolism are not understood. Insulin, leptin or cortisol do not appear have major roles in the control of feed intake in sheep consuming high levels of salt (20% of dry matter (Blache et al. 2007)). Although salt intake led to a reduction in insulin and glucose beyond the effect that could be attributed to reduced feed intake alone, the drop in insulin would normally be expected to increase rather than decrease appetite and feed intake. However, the decrease of insulin following exposure to a salt load could influence the reproductive system of both males and females since insulin has been shown to stimulate GnRH secretion (Miller et al. 1995).

In addition to this indirect effect of salt load on reproduction via changes in energy intake, the intake of salt can affect reproductive activity through changes in kidney function and the activation of specific systems involved in the control of water salt balance. Ruminants have the potential to tolerate salt because of the adaptive capacity of their kidney function (Potter 1963, 1968). Sodium and chloride ions are excreted at a greater rate following elevated ingestion of salt by (i) an increase in glomerular filtration rate and filtration fraction without any pronounced change in renal plasma flow (Potter 1968), (ii) a reduction in the re-absorption of sodium chloride in individual nephrons of the sheep kidney and (iii) an increase in water retention. The salt and water balance is regulated by the renin-angiotensin system (RAS). In this system, sodium retention is controlled by renin, angiotensin I and II, and aldosterone, while arginine vasopressin (AVP) regulates water re-absorption (see Fig. 17.1 for details). Later in the chapter we will review work that suggests that exposure to salt during pregnancy and early life can affect longer-term activity of the RAS and possibly the capacity of livestock to adapt to salt intake, which can have consequences on survival, growth, and reproduction.

2. High salt load and the reproductive cycle

It has to be noted that adequate intake of salt is essential to successful reproduction. However, in this section we will only discuss the effects of a high-salt diet on the different stages of the reproductive cycle. We will start the reproductive cycle at puberty, when the gametes start to be produced and when the animals start to express sexual behaviour.

2.1 Effect on puberty

No study has been conducted to investigate the impact of a high salt load on puberty in either male or female ruminants. However, since the timing of puberty is strongly related to growth and bodyweight (Foster et al. 1995), it can be suggested that female or male ruminants exposed to high salt intake during their growth phase could experience some delay in puberty due to reduced feed intake and growth rate.

In males, just before puberty, testicular renin, angiotensin converting enzyme (ACE) and angiotensin increase concomitantly with plasma gonadotropins, suggesting

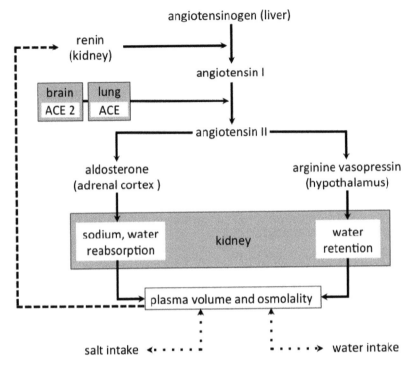

Fig. 17.1. The renin-angiotensin system (RAS) is responsible for the maintenance of water and salt balance. Renin catalyses the conversion of angiotensinogen into angiotensin I which is converted by Angiotensin-converting enzyme (ACE, ACE2) into angiotensin II. Angiotensin II controls the secretion of aldosterone which stimulate sodium retention and arginine vasopressin (AVP) which stimulate water reabsorption by the kidney. Plasma volume and plasma osmolality control salt appetite and drinking behaviour. High-salt intake leads to change in both plasma volume and osmolality which exert a negative feedback on the secretion of renin. Ultimately, high salt consumption induces a decrease in aldosterone concentration that reduces sodium reabsorption and increases sodium excretion. High-salt intake does not induce change in AVP plasma concentration if the intake of fresh water is sufficient to maintain a salt and water balance. Tissues producing each hormone are indicated between brackets.

that these components of the RAS are important for puberty (Hohlbrugger et al. 1982; Parmentier et al. 1983; Speth et al. 1999). That the RAS exists independently in the testis (Okuyama et al. 1988), i.e., not only in the kidney, suggests a potential link between salt and water balance and male reproduction, including delayed puberty or gamete production (see next section). We are not aware of any studies linking high salt load or changes in the RAS on the timing of puberty in females.

2.2 Effects on gamete production

There are limited data available on the effects of high salt on testis function and spermatogenesis. Salt intake could decrease the reproductive capacity of rams because components of the RAS can influence male reproductive capacity including sperm maturation and expulsion, although an effect on fertility is presently unclear

(Leung and Sernia 2003). The effect of salt load on spermatogenesis in ruminants could be different between breeds, according to their adaptation to arid climate as suggested by data obtained in wild rodent species. In male golden spiny mice (*Acomysrussatus*), a desert-adapted species, high dietary salt (3.5% and 5% NaCl for 6–8 weeks) reduced testis mass and spermatogenesis (Wube et al. 2009). By contrast, high dietary salt had no effect on testis function in spiny mouse (*Acomyscahitinus*), a species adapted to a Mediterranean environment (Wube et al. 2009). See also Degen and Kam, this volume.

In females, exposure to salt before mating has been reported to have either negative or no effect on gametes and conception. In a study on a small number of animals, ewes exposed to 1.3% total salts or 0.9% sodium chloride in drinking water before mating, on occasions, failed to conceive (Peirce 1968). No further studies support this result. No effects of salt on ovulation or conception rate were described in a recent study that aimed to provide scientific validation of a popular Bulgarian practice believed to stimulate oestrus by exposing ewes to a sequence of salt-free and high-salt diets for few days (Nedelkov et al. 2012).

2.3 Effects during pregnancy

High salt load can impact the success of gestation because it can affect, in the short-term, the female metabolic status, reproductive endocrinology and control of the salt and water balance, as well as organ development in the foetus.

It has to be noted that the source of salt can influence the impact of high salt consumption on pregnancy. Ingestion of drinking saline water containing 1.3% sodium chloride by pregnant ewes is associated with changes in progesterone (particularly those carrying twins) and electrolyte concentrations of blood plasma, but the biological significance of the association is not clear. The susceptibility of pregnant ewes to salt water ingestion increases with age and/or multiple births (Potter and McIntosh 1974). In contrast to pregnant ewes drinking saline water, pregnant ewes consuming a high-salt diet during gestation can maintain their pregnancy, and the lambs are born within normal measures of weight and crown-rump length (Meyer and Weir 1954; Digby et al. 2008; Chadwick et al. 2009d).

The influence of the source of salt (drinking water versus diet) could be due to a different action on plasma progesterone, a hormone with a critical role during pregnancy. High concentrations of progesterone in early gestation are beneficial to embryo survival (Parr 1992) but a rapid decline in progesterone is essential to the onset of parturition (Weiss 2000). Elevated progesterone levels were observed in twin-bearing ewes that consumed salty water (Potter and McIntosh 1974) while progesterone concentrations throughout pregnancy were not affected in ewes consuming a high-salt feed (Digby et al. 2008).

2.3.1 Metabolic balance

Pregnancy, mainly late pregnancy, is a period of great demands for energy and protein, and any decrease in feed intake could impact on foetal development. Ingestion of high salt decreases both insulin and leptin concentrations through an unknown mechanism independent of changes in feed intake or body weight (Digby et al. 2008). Surprisingly, the decrease in insulin concentrations with high-salt intake during pregnancy was not associated with low glucose levels (Digby et al. 2008). The role and importance of these changes are not understood, but as discussed later in the chapter, careful management of the flock can mitigate the impact of high salt load on intake.

2.3.2 Control of water and salt balance

Pregnant ewes fed high-salt diets are faced with physiological 'conflicts' in RAS activation. Pregnancy is characterised by sodium retention and increased extracellular volume necessary for the maintenance of the mother and growth of the foetus (Davison and Lindheimer 1989). In contrast to responses of the RAS to high salt intake, the RAS during a normal pregnancy (at least in humans) is substantially activated, with both plasma renin activity and plasma angiotensin II concentrations increasing (Wilson et al. 1980). In addition, aldosterone concentrations increase and cause an increase in blood volume resulting in high sodium absorption and water retention, a response in opposition to the described effect of a high intake of salt. Despite the increase in plasma volume, blood pressure falls during normal pregnancy (Wilson et al. 1980); another physiological response opposite to that induced by a high intake of salt. These physiological 'conflicts' result in the activation of the RAS in response to high salt intake during pregnancy. In humans, sodium loading during normal pregnancy results in an increase in body weight, plasma volume, and sodium excretion (Brown et al. 1988). Furthermore, plasma renin activity, as well as plasma and urinary aldosterone are significantly suppressed with salt loading during pregnancy. It has also been shown in rats, that high sodium intake during pregnancy (0.9% or 1.8% in water) results in a reduction of plasma renin activity and aldosterone concentration. However, at 0.9% NaCl, there was no decrease in blood pressure that is essential during pregnancy (Beausejour et al. 2003), whilst at 1.8% NaCl, there was an increase in systolic blood pressure, comparable to the increase in mean arterial pressure in rabbits exposed to dietary sodium loading during pregnancy (Milton et al. 1983). This salt-induced hypertension during pregnancy represents a threat to the health of both mother and foetus.

In ewes fed salt, the regulation of salt and water balance is achieved through a decrease in plasma renin activity which, in turn, lowers the production of angiotensin II and the secretion of aldosterone (Digby et al. 2008; Chadwick et al. 2009d), but does not involve a concomitant effect on AVP. Plasma concentrations of AVP did not increase in these studies, despite high salt ingestion and the water-retaining requirements of pregnancy, possibly because the ewes had free access to fresh water. In such cases, ewes are able to increase their water consumption above normal levels (Hamilton and Webster 1987; Meintjes and Olivier 1992). Water consumption increased nearly two-fold in early pregnancy in ewes fed a high-salt diet (Digby et al. 2008).

The decline in aldosterone concentrations is probably the most important adaptive mechanism that allows ewes to successfully complete pregnancy whilst consuming a high-salt diet. When pregnant ewes were fed a control diet, aldosterone concentrations increased as pregnancy progressed (Digby et al. 2008; Chadwick et al. 2009d). A higher aldosterone concentration increases water reabsorption in the kidneys, and thus extracellular fluid increases as part of a normal pregnancy. When pregnant ewes were fed a high-salt diet, the mechanism of increased aldosterone concentration was apparently not required as the high salt ingestion triggered an increased water intake, which may have sufficiently increased extracellular volume to reach the 'target' level for a pregnant ewe. Therefore pregnant ewes fed high salt are likely able to avoid complications such as hypertension (Rafestin-Oblin et al. 1991) or neonatal mortalities (Potter and McIntosh 1974) by reducing their plasma aldosterone below the concentration in control animals.

2.3.3 Impacts in organ and faetal development during pregnancy

Lambs born from ewes fed a high-salt diet (13% NaCl) seem to have a normal foetal development with a similar crown-to-rump length and birth weight as lambs born of ewes fed standard levels of salt (0.5% NaCl; Digby et al. 2008). Similarly, birth weight of lambs was not affected by grazing ewes on saltbush (Chadwick et al. 2009d) or in lambs exposed to a high-salt diet during the last two month of pregnancy (Mao et al. 2013). However, maternal nutrition during pregnancy can affect kidney development in the foetus that may lead to adverse consequences in the mature kidney, although the adaptive capacity of animals may limit the negative outcomes in the longer term. The ratio of kidney weight/bodyweight was decreased by 21% in foetuses following high-salt diet during pregnancy (Mao et al. 2013), and reduced ratios were also observed in offspring in the early postnatal period and at 5 months of age (decrease of 20 to 25% compared to control; Tay et al. 2007; Mao et al. 2013). The five-month-old offspring of Merino ewes that had been fed a high-salt (10.5% NaCl) diet during pregnancy had altered sodium excretion following a dietary salt challenge, altered glomerular number and size compared to counterparts born from mother fed a control diet (1.5% NaCl; Tay et al. 2012). The high-salt offspring had 20% fewer glomeruli compared to the control offspring, but they also had larger glomerular radii compared to the control offspring. There was no difference in the daily urinary sodium excretion between the two offspring groups, although the high-salt offspring produced urine with a higher concentration of sodium. Together these results suggested that even though maternal high-salt intake during pregnancy can affect fetal nephrogenesis and glomerular number at birth, the kidney efficiency is able to adapt to be able to concentrate and excrete salt.

2.4 Effect on neonatal survival

Concentrations of sodium chloride up to 1.3% in drinking water may cause distress at parturition in ewes, possibly a result of hormonal disturbances such as progesterone concentrations, which in turn results in complicated births (Pierce 1968; Potter and McIntosh 1974). Concentrations of salt in drinking water, as low as 1.3%, can cause neonatal mortalities in lambs (30 to 50% of lambs born to ewes drinking salty water) (Pierce 1968; Potter and McIntosh 1974). By contrast, lambs born from ewes fed a high-salt diet (13% NaCl) had a similar survival rate as lambs born from ewes fed a diet containing typical levels of salt (0.5% NaCl; Digby et al. (2008)). Chadwick et al.

(2009d) did not find any effect on lamb mortality with ewes grazing on saltbush. The reason for the high mortality of lambs from ewes drinking salty water is not known.

The dietary cation-anion difference (DCAD) can have important consequences on reproductive performance, especially around parturition, but in studies that have created a high salt load by the addition of NaCl, DCAD would have remained unchanged, as DCAD is often calculated as $Na^+ + K^+ - Cl^-$. However, in halophytic plants such as saltbush, the DCAD can be high (Ben Salem et al. 2010) and there may be consequences to ewe health and calcium metabolism in the postnatal period.

2.5 Effects of milk production and lamb growth

Ingestion of salt in the diet during lactation had no effect on lamb survival when the amount of salt in the diet was lower than 13.1% (Meyer and Weir 1954). Similarly, the addition of large amounts of sodium chloride to the ewes' rations did not affect body weight gain of the lambs (Meyer and Weir 1954). Milk from ewes exposed to a high-salt diet had a normal fat and protein composition and no increase in Na concentration, but had increased concentrations of potassium, manganese and boron (Digby et al. 2008; Chadwick et al. 2009d). In Damascus goats, a breed very well adapted to arid conditions, exposure to saline water during pregnancy reduced milk yield by 36% without affecting milk composition and the weight gain of the kids, but there was no such effect when feeding on salt-tolerant alfalfa (containing 1.4% Na and 1.7% K; Abdalla et al. 2013).

3. Longer-term consequences on offspring associated with fetal programming

In contrast to the little or no direct effect of salt in the diet on reproduction in ewes or on the physical characteristics of the offspring, high levels of salt intake by the mother during pregnancy can affect the physiology of the offspring by the process of fetal programming. For lambs *in utero* when their pregnant dams are exposed to elevated levels of salt, foetal programming can affect drinking behaviour, endocrine systems controlling both the salt-water balance, and the regulation of metabolic processes. The mechanisms by which high salt loads program the offspring are only starting to be unveiled.

3.1 Foetal programming

Foetal programming is a phenomenon whereby events during gestation can have postnatal consequences that may persist after weaning and, indeed, in some cases for the rest of the offspring's life. Manipulation of maternal dietary intake can lead to behavioural (e.g., diet selection and intake; Curtis et al. 2004) and physiological (e.g., renal function and the renin-angiotensin system; Arguelles et al. 1996) changes in the offspring later in life. From a functional perspective, metabolic programming provides a mechanism to prepare a foetus for the environment in which it will live, including its nutritional environment (e.g., see recent review by Ginane et al. 2015).

The majority of studies on foetal programming have investigated changes in maternal intake during pregnancy, such as undernutrition or overfeeding (i.e., high fat and high protein diets), or exposure to glucocorticoids (see review: Schwartz and Morrison 2005). The mechanism of foetal programming, leading to change in gene expression (an epigenetic effect) is attributed either to a direct effect of the programming agent on the genome, and/or an induced change in the hormonal status by the programming agent (salt, in our case).

The normal growth and development *in utero* is heavily controlled by the endocrine system. Glucocorticoid administration to pregnant ewes for 2 days at the end of the first month of gestation, when the mesonephric kidney is developing, causes a permanent reduction in nephron number and leads to hypertension in the adult offspring (Moritz et al. 2003). It is important to note that there are critical windows of time where events, including changes in diet composition, can have differing effects on long-term health or specific organs and metabolic systems (Symonds et al. 2007).

3.2 Pre-natal salt exposure and behavioural programming— preference for salt

Calves born from cows that received supplementary sodium during pregnancy had a higher appetite for sodium (desire to eat), indicating that appetite regulation of the offspring could be entrained by the sodium intake of the dam during late pregnancy (Mohamed and Phillips 2003). The mechanisms behind the changes in offspring may be due to changes in sensitivity of taste receptors and/or changes in salt and water balance (Contreras and Kosten 1983). In fact, young taste bud cells do not have the same salt response characteristics as mature cells, and a changing neural substrate underlies development of salt taste, both pre- and post-natally (Mistretta and Bradley 1983). In sheep, offspring born to ewes fed high salt during pregnancy did not differ in their preference for dietary salt (Digby et al. 2010b). The absence of any preference for salt in the offspring in this study might have been influenced by the termination of feeding the high-salt diet immediately at birth. In rats, salt preference is increased if the pups are exposed to salt during the postnatal period (Smriga et al. 2002; Curtis et al. 2004). However, the preference for salt in weaned lambs was similarly unaffected in offspring of ewes fed a high-salt diet from day 60 of pregnancy until 21 days of lactation (Chadwick et al. 2009b), suggesting that foetal programming of dietary salt preferences may not occur as strongly in sheep as in rats or cattle. Given the capacity of sheep to survive in arid regions (Zygoyiannis 2006; Webley 2007), their regulation of salt and water balance may be geared towards coping with salt loads or water shortages, and hence the opportunity to increase the consumption of (or preference for) high-salt diet via foetal programming is either limited or not required. Indeed, all of the lambs in the study by Digby et al. (2010b) consumed more than 10 g Na/kg DM, a quantity well above the recommended daily intake of salt of 0.7 g Na/kg DM for sheep (NRC 1985).

3.3 Pre-natal salt exposure and programming of the regulation of salt and water balance

The maternal, foetal and placental units have independent RASs, however the maternal RAS can influence directly and indirectly both foetal and placental RASs. For example, angiotensin II partly regulates the blood flow in the utero placental unit, and thereby indirectly influences homeostasis of foetal volume and oxygenation (Wilkes et al. 1985). The RAS of the mother is essential to the maintenance of the foetal glomerular filtration rate (Lumbers et al. 1996). Both angiotensin converting enzyme inhibitors and angiotensin receptor antagonists administered to the dam cause acute renal failure in the foetus due to the direct effects on the foetal RAS (Lumbers et al. 1996). The foetus is not dependent on the kidney function of the mother for sodium conservation since sodium is readily transported across the placenta. The RAS is active in the foetus and necessary for normal development.

The foetus is very susceptible to changes in maternal fluid balance. For example, the flow rate of foetal urine falls and urine osmolarity rises when the ewe is dehydrated or infused with hypertonic mannitol (Lumbers and Stevens 1983). By contrast, the foetus appears to be protected from both a high maternal salt intake (0.17M NaCl) and moderate salt depletion (Stevens and Lumbers 1986). The efficiency of maternal homeostatic mechanisms is probably responsible for this protection. However, if the salt content of the diet is high enough during late gestation to significantly increase the plasma sodium of the ewe (e.g., a 5 mmol/L rise), the plasma sodium of the foetus will also increase and, to maintain osmotic balance, it will suppress its own plasma renin activity to excrete more salt, as seen in adults.

Rat offspring born from dams exposed to 8% NaCl in their diet through gestation until weaning had increased angiotensin II, higher blood pressure and blood pressure that was less responsive to salt intake (da Silva et al. 2003). Also, rats exposed *in utero* and perinatally to a high salt environment had increased angiotensin II sensitivity (Arguelles et al. 1996). This may have been the result of a feedback mechanism in which angiotensin II receptors are up regulated in the foetus in response to lower activity of the RAS in the mother (Arguelles et al. 1996; Butler et al. 2002). In sheep, two types of foetal programming by prenatal exposure to salt have been suggested (Digby et al. 2010b). The first is programming of the thirst threshold, which is illustrated by the difference in the intake of water over the first four hours after receiving an oral salt dose. Offspring born to ewes fed high salt during pregnancy drank 200 ml less in the first 2 h than controls (ewe not exposed to salt during pregnancy) when exposed to a one-off oral salt dose (Digby et al. 2010a). However, from 2 to 4 h after the oral salt dose, control lambs decreased their water intake by 400 ml whilst the 'salt exposed' lambs decreased water intake by only 100 ml (Digby et al. 2010a). Chadwick et al. (2009d) provided further support for changes in the thirst threshold of offspring born from ewes consuming a high-salt diet during pregnancy. Voluntary consumption of water of 'salt exposed' offspring (at 8 months of age, post-weaning) was up to 35% less than control offspring. If the dams were fed a diet with added NaCl, their offspring consumed 20% less water per unit of feed intake, whereas if the pregnant dams were fed saltbush during pregnancy, their offspring had both lower water intake and lower feed intake, such that the ratio of water:feed consumed was unchanged.

The second type of programming associated with high salt intake during pregnancy is altered regulation of the systems that control salt and water balance. Aldosterone concentration is blunted in response to an oral salt challenge administered after weaning to lambs born from ewes fed high salt during pregnancy. In addition, the basal levels of aldosterone were always higher than control offspring, suggesting the offspring of salt-exposed dams are programmed to retain more salt. The aldosterone response in the study by Digby et al. (2010b) was consistent with the responses in angiotensin II found in rats exposed to perinatal salt overload (da Silva et al. 2003). It seems the sensitivity of the RAS can be programmed in response to low activity of RAS in the mother during pregnancy (Arguelles et al. 1996; Butler et al. 2002). Similarly, offspring born to ewes that consumed a high salt diet (14%) from day 60 of pregnancy until day 21 of lactation, show an attenuated response of renin to a salt dose and tended to retain more salt than normal sheep (Chadwick et al. 2009b). In addition, the plasma concentration of AVP was lower in high-salt offspring, suggesting that the sensitivity of the kidney to changes in osmotic pressure was reduced in offspring born to ewes fed high salt during pregnancy (Digby et al. 2010b). These results combined with those of Desai et al. (2003) suggest that the AVP threshold was altered in offspring in response to changes in water balance or salt ingestion during pregnancy. Furthermore, offspring born to ewes fed high salt during pregnancy presented a modified temporal pattern of hormone secretion when consuming saline water for two days (Digby et al. 2010a). Initially water intake was lower in the offspring born to dams fed high salt during pregnancy, but there was a degree of correction on the second day, such that total water intake did not differ over the 2 days. Salt excretion, urinary output, and aldosterone concentration were similar between both groups on day 1 but, by day 2, urinary output, sodium excretion and aldosterone secretion decreased, in lambs born to ewes fed high salt during pregnancy, whilst they remained unchanged in control lambs (Digby et al. 2010a). These modifications illustrate the programmed adaption to a load of ingested salt.

The possible increase in salt retention of 'salt exposed' offspring is a perplexing possibility. It is generally accepted that extra salt can only be retained in the extracellular compartments of the body in concert with an increase in water retention in order to maintain osmotic pressure of body fluids. But the reduced water intake of 'salt exposed' offspring does not support the concept of increased water retention. One other possibility exists: the retention of 'extra' salt in skin (see commentary by Rabelink and Rotmans 2009). The skin is one of the largest organs of the body, so a small increase in salt retention per unit of tissue can lead to a large increase in overall salt retention. Machnik et al. (2009) showed that sodium can be stored in the connective tissue of the skin and is regulated by the immune system. It is tempting to speculate that events during pregnancy associated with high-salt intake of ewes may affect this phenomenon in the offspring. To date, the only evidence that high-salt feeding during pregnancy may affect skin metabolism of the offspring is a 10% higher wool growth of 'salt exposed' offspring (Chadwick et al. 2009a). Whilst this is a valuable finding for sheep production, it may or may not be related to changes in salt retention in skin.

3.4 Pre-natal salt exposure and programming of metabolism

It has previously been suggested that high salt consumption decreases feed intake. In offspring born to ewes fed high salt during pregnancy, voluntary feed intake was approximately 35% less than controls when consuming 1.3% saline water for 2 days (Digby et al. 2010a). In addition, the secretion of insulin in both groups of offspring suggested that the energy balance of the lambs, defined as the difference between energy expenditure and the sum of energy intake and energy reserves, had drifted towards a negative value because of the decrease in feed intake (Blache et al. 2007). The decrease in insulin secretion was greater in offspring born to ewes fed high salt during pregnancy, possibly due to the lower feed intake. However, previous studies in wethers have revealed that the decrease in insulin following ingestion of high salt diet is independent of a decrease in intake (Blache et al. 2007). Furthermore, plasma concentrations of leptin are lower in offspring born to ewes fed high salt during pregnancy, suggesting not only an increase in energy expenditure and a mobilisation of energy reserves in response to ingestion of saline water, but also an effect of 'salt programming' on leptin. This is very interesting because previous studies have suggested that leptin is not affected by foetal programming from salt ingestion (Blache et al. 2007; Digby et al. 2008, 2010a, 2010b).

3.5 Programming of the RAS

The mechanism by which high-salt exposure during pregnancy affects the RAS in the offspring is not known but the phenomenon of foetal programming has been shown for other physiological systems, whereby a stimulus or insult at critical periods of development affect development of organs and alter their postnatal functions (Godfrey and Barker 2000; Nijland et al. 2008). In sheep, the investigation of the molecular changes in the components of the RAS associated with the foetal programming of the offspring born from ewes fed high salt diet during pregnancy is quite recent.

Lambs born to ewes fed a high-salt diet (10.3% NaCl) during pregnancy had 20%–30% higher DNA methylation in the lung and kidney tissues compared to control lambs born to ewes fed a control diet (1.5% NaCl; Tay 2012). However, the maternal dietary effect (i.e., foetal origin) on DNA methylation in the high-salt offspring was more prominent in the kidney than in the lung tissue (Table 17.1). Alteration of DNA methylation in a targeted organ has been shown in other cases of foetal programming following nutritional or chemical abuse during both the pre- and post-natal life (Van den Veyver 2002; Dolinoy et al. 2006; Bogdarina et al. 2007). A recent study has identified which key components of the RAS are affected by high salt diet during pregnancy (Mao et al. 2013). The RNA expression of all components of the RAS, except renin, were affected by a high-salt diet during the last 2 months of gestation in both the foetuses and the born lambs (Table 17.2; Mao et al. 2013). In our study (Tay et al. 2012), a high-salt diet during the whole pregnancy induced a reduction of renin RNA expression in the kidney of 5 month old offspring suggesting that the length and the intensity of the exposure to high salt diet during gestation could be important in determining the effect on different components of the RAS.

Table 17.1. Global DNA methylation (% of total DNA) of kidney and lung tissues and renin mRNA expression in kidney (expressed as a ratio to cDNA) of 5 month old lambs born to ewes fed a high-salt diet (10.3% NaCl) or a control diet (1.5% NaCl) during pregnancy (Tay et al., unpublished data). Values represent mean ± S.E.M. of 15 lambs per group. Different superscript in the row indicate differences between groups (P < 0.05).

	Diet during pregnancy	
	Control diet	High-salt diet
Global DNA methylation		
Kidney	3.4[a] ± 0.3	4.8[b] ± 0.4
Lung	1.7[a] ± 0.2	2.2[b] ± 0.3
Renin mRNA expression		
Kidney	1.2[a] ± 0.11	0.9[b] ± 0.07

Table 17.2. Changes in the expression of mRNA of key elements of the RAS in lambs born to ewes fed high-salt diet (8% NaCl) compared to that of lambs born to ewes fed normal-salt diet (0.6% NaCl) for 2 months during middle-to-late gestation (70–130 day of gestation). Measurements were taken at 130 day of gestation (foetuses) and 15 and 90 days post natal (5 lambs per group; Mao et al. 2013).

RAS component	Foetuses	Age of the offspring	
		15 days old	90 days old
Renin	=	=	=
Angiotensinogen	↑	↑	↑
ACE	↑	↑	↑
ACE2	↓	↓	↓
AT1	↑	↑	↑
AT2	↑	=	=

The results described above demonstrate that the offspring of ewes fed a high-salt diet during pregnancy have an altered renin-angiotensin system, which may affect the animals' capacity to maintain homeostasis when exposed to a high dietary salt load. The altered responsiveness of the kidney to the post-natal environment, relative to the lung, may reflect the fact that the kidney is the primary organ associated with maintaining salt balance. Thus, the phenomenon of foetal programming is a means to prepare the offspring for the post-natal environment they are likely to encounter after birth (Gama-Sosa et al. 1983; Bateson et al. 2004; Bateson 2007; Unterberger et al. 2009). Consequently, there may be strongly programmed effects on renal function, while changes in other tissues and organs are more likely to change in response to other post-natal triggers. Whether programmed changes continue to be passed on to the next generation has not been investigated.

4. Management options to successfully use halophytic forages for reproductive livestock

Revegetation with saltbush (*Atriplex* species) offers practical options in the management of dryland salinity in Australia (Barrett-Lennard and Norman 2009) and elsewhere. To add economic value to the revegetation, the saltbush needs to be integrated into the production system (Barrett-Lennard and Norman 2009)—for example, it can be used to compensate for the decline in pasture availability during an 'autumn feed gap' in the regions of Western Australia with a Mediterranean-type climate (Morcombe et al. 1993, 1996). In many situations, ewes are pregnant during this period, and so saltbush can be a sustainable and additional feed resource to help meet nutrient requirements of the flock. Pregnant ewes can tolerate 13% NaCl in their feed whilst they have access to fresh drinking water without compromising their reproductive capacity (Digby et al. 2008; Chadwick et al. 2009d).

4.1 Ensure adequate energy intake through complementary feeds

The main nutritional issue that must be addressed when using halophytic forages for breeding animals (and indeed, any class of livestock) is the provision of adequate metabolisable energy. Because a high salt content in the diet limits feed intake (Masters et al. 2005b), it is necessary to ensure that animals with a high energy demand, such as late pregnancy and early lactation (or young, growing animals), are provided with digestible forage or supplementary feed low in salt (Warren et al. 1990; Hopkins and Nicholson 1999; Pearce et al. 2002; Franklin-McEvoy 2007). When sheep fed a high-salt feed are offered a choice of low-salt alternatives, they actively select combinations that improve the overall feeding value of their diet (Thomas et al. 2007). Organic matter intake of the sheep offered a low-salt alternative to complement a high salt feed was about 50% higher and their liveweight gain was at least double that of sheep offered only a high salt feed. This magnitude of response in feed intake has important implications to ensuring adequate reproductive performance of livestock grazing pastures based on halophytic forages.

Where supplementary feed is not available or too expensive to provide in adequate amounts, the use of companion plants that are able to tolerate moderate levels of salinity but that do not accumulate as much salt as, for example, the *Atriplex* shrub species, is an option worth exploring. Other shrub (Revell et al. 2013) or pasture species (Dear et al. 2007) can fit this category, and grazing livestock can be managed to ensure multiple plant species are included in the selected diet to boost liveweight gain (Revell et al. 2013).

Another advantage of providing a range of forage species for grazing livestock is that it helps the animals manage their intake of plant secondary compounds (such as oxalates, nitrates, tannins and saponins), because different plants contain different combinations and concentrations of nutrients and compounds (Provenza et al. 2007). Plant secondary compounds can act to deter herbivory, but the negative effects of secondary compounds on feed intake by livestock can be minimised by avoiding high levels of any one secondary compound through diet diversity (Rogosic et al. 2007) and, indeed, by complementary interactions between different secondary compounds (Lyman et al. 2008).

4.2 Avoid saline water when feeding halophytic forages

For livestock consuming forages with a high salt content, increasing the consumption of fresh water is a critical adaptation process. In sheep, access to fresh water improves the intake of halophytic plants such as *Atriplex* (saltbush) and *Kochia* (bluebush;

Wilson 1966a; Wilson and Hindley 1968). In fact, the daily sodium intake is related to voluntary water consumption with a ratio of sodium chloride intake to total water intake of between 1.8–2.2% (Wilson 1966a). In the field, sheep grazing saltbush can consume up to 200 g of NaCl per day (Wilson 1966a), but this level of salt intake cannot be sustained without access to fresh water. Wilson (1966a) suggests that in these circumstances there should not be more than 0.6% NaCl in the drinking water. The physiological consequences of high salt intake are more severe when the salt is present in drinking water rather than in the feed, to the extent that reproductive performance may be impaired when saline water is provided but not when salty feed is offered.

4.3 Explore the potential for adaptation of animals to high-salt diet through early-life programming

A relatively new area of research has been investigating how high-salt feeding during pregnancy in sheep affects the physiology and performance of the offspring (Chadwick et al. 2009a, 2009c, 2009d; Digby et al. 2010a, 2010b; Tay et al. 2012). There is also biomedical literature on high-salt diets during pregnancy focussing on health implications, such as blood pressure. Combined, there is evidence that the consumption of high levels of salt during pregnancy can alter the preference for salt, water intake, kidney morphology and hormonal regulation of water balance in the weaned offspring. More work is required to fully explore the production consequences of these physiological or metabolic changes, but there is evidence that the offspring born to ewes fed saltbush during pregnancy and early lactation perform better when they graze saltbush-based pasture after weaning (Chadwick et al. 2009a; Norman et al. unpublished data). The potential for providing opportunities for livestock to be better adapted to physiologically deal with halophytic forages by strategically using these forages during pregnancy and lactation, in combination with other forages or feedstuffs, is an attractive proposition that warrants further research under a range of different production scenarios.

5. Conclusion

In this chapter we have discussed the impact of exposure to high salt load on the capacity of sheep to reproduce. We have summarised the impact of high salt load at different stages of the reproductive cycle to illustrate that some effects are dependent of the impact on energy balance and other effects potentially mediated by specific effects of salt (Fig. 17.2). It is important to note that, in contrast to salt in feed, ingestion of salt water has dramatic effects on the reproductive capacity of ewes. Finally, we discussed that the capacity of breeding ewes to tolerate high levels of salt in their diet without compromising reproductive capacity, and how the exposure to salt during pregnancy could give a slight advantage to the offspring's capacity to tolerate salt later in its life. The foetal programming of their RAS, a blunted aldosterone response, changes in the RNA expression of key element of the RAS, and an altered thirst threshold in response to an oral salt challenge may play a role in their adaptation, and possibly that of future generations, to saline environmental conditions.

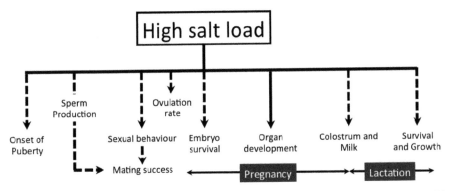

Fig. 17.2. Diagram summarising the effects of high salt load on the reproductive cycle of sheep. Solid lines indicate a direct effect of salt, and dotted lines indicate that the effect is/may also be mediated by the effect of salt intake on metabolism.

References

Abdalla, E.B., H.A. Gawish, A.M. El-Sherbiny, N.H. Ibrahim and A.S. El-Hawy. 2013. Reproductive efficiency of Damascus goats in salt-affected lands in South Sinai, Egypt. Journal of American Science 9: 170–177.

Arguelles, J., P. Lopez-Sela, J.I. Brime, M. Costales and M. Vijande. 1996. Changes of blood pressure responsiveness in rats exposed in utero and perinatally to a high-salt environment. Regulatory Peptides 66: 113–115.

Barrett-Lennard, E.G. and H.C. Norman. 2009. Saltbush for forage production on saltland.

Bateson, P. 2007. Development plasticity and evolutionary biology. The Journal of Nutrition 137: 1060–1062.

Bateson, P., D. Barker, T. Clutton-Brock, D. Deb, B. D'Udine, R.A. Foley, P. Gluckman, K. Godfrey, T. Kirkwood, M.M. Lahr, J. McNamara, N.B. Metcalfe, P. Monaghan, H.G. Spencer and S.E. Sultan. 2004. Developmental plasticity and human health. Nature 430: 419–421.

Beausejour, A., K. Auger, J. St-Louis and M. Brochu. 2003. High-sodium intake prevents pregnancy-induced decrease of blood pressure in the rat. American Journal of Physiology—Heart, Circulation and Physiology 285: H374–H383.

Ben Salem, H., H.C. Norman, A. Nefzaoui, D.E. Mayberry, K.L. Pearce and D.K. Revell. 2010. Potential use of oldman saltbush (*Atriplex nummularia* Lindl.) in sheep and goat feeding. Small Ruminant Research 91: 13–28.

Blache, D. and G.B. Martin. 2009. Focus feeding to improve reproductive performance in male and female sheep and goats—how it works and strategies for using it. Options Mediterraneennes 85: 351–364.

Blache, D., M.J. Grandison, D.G. Masters, R.A. Dynes, M.A. Blackberry and G.B. Martin. 2007. Relationships between metabolic endocrine systems and voluntary feed intake in Merino sheep fed a high salt diet. Australian Journal of Agricultural Research 47: 544–550.

Bogdarina, I., S. Welham, P. King, S. Burns and A. Clark. 2007. Epigenetic modification of the renin-angiotensin system in the fetal programming of hypertension. Circulation Research 100: 520–526.

Brown, M.A., E. Nicholson and E.D.M. Gallery. 1988. Sodium-renin-aldosterone relations in normal and hypertensive pregnancy. British Journal of Obstetrics and Gynaecology 95: 1237–1246.

Butler, D., S. Pak, A. Midgely and B. Nemati. 2002. AT (1) receptor blockade with losartan during gestation in Wistar rats leads to an increase in thirst and sodium appetite in their adult female offspring. Regulatory Peptides 105: 47–57.

Chadwick, M.A., P.E. Vercoe, I.H. Williams and D.K. Revell. 2009a. Programming sheep production on saltbush: adaptations of offspring from ewes that consumed high amounts of salt during pregnancy and early lactation. Animal Production Science 49: 311–317.

Chadwick, M.A., P.E. Vercoe, I.H. Williams and D.K. Revell. 2009b. Dietary exposure of pregnant ewes to salt dictates how their offspring respond to salt. Physiology & Behavior 97: 437–445.

Chadwick, M.A., P.E. Vercoe, I.H. Williams and D.K. Revell. 2009c. Weaner lambs perform better on saltbush if their mothers grazed saltbush while pregnant. Options Mediterraneennes 85: 373–377.

Chadwick, M.A., I.H. Williams, P.E. Vercoe and D.K. Revell. 2009d. Feeding pregnant ewes a high-salt diet or saltbush suppresses their offspring's postnatal renin activity. Animal 3: 972–979.

Contreras, R.J. and T. Kosten. 1983. Prenatal and early postnatal sodium chloride intake modifies the solution preferences of adult rats. Journal of Nutrition 113: 1051–1062.

Curtis, K.S., E.G. Krause, D.L. Wong and R.J. Contreras. 2004. Gestational and early postnatal dietary NaCl levels affect NaCl intake, but not stimulated water intake, by adult rats. American Journal of Physiology 286: R1043–R1050.

da Silva, A., I. de Noronha, I. de Oliveira, D. Malheiros and J. Heimann. 2003. Renin angiotensin system function and blood pressure in adult rats after perinatal salt overload. Nutrition, Metabolism and Cardiovascular Disease 13: 133–139.

Davison, J.M. and M.D. Lindheimer. 1989. Volume homeostasis and osmoregulation in human pregnancy. Baillieres Clinical Endocrinology and Metabolism 3: 451–472.

Dear, B., K. Reed and A. Craig. 2007. Outcomes of the search for new perennial and salt tolerant pasture plants for southern. Australia Australian Journal of Experimental Agriculture 48: 578–588.

Desai, M., C. Guerra, S. Wang and M.G. Ross. 2003. Programming of hypertonicity in neonatal lambs: resetting of the threshold for vasopressin secretion. Endocrinology 144: 4332–4337.

Digby, S., D. Blache, D.G. Masters and D.K. Revell. 2010a. Responses to saline drinking water in offspring born to ewes fed high salt during pregnancy. Small Ruminant Research 91: 87–92.

Digby, S., D.G. Masters, D. Blache, P.I. Hynd and D.K. Revell. 2010b. Offspring born to ewes fed high salt during pregnancy have altered responses to oral salt loads. Animal 4: 81–88.

Digby, S., D. Masters, D. Blache, M. Blackberry, P.I. Hynd and D. Revell. 2008. Reproductive capacity of Merino ewes fed a high-salt diet. Animal 2: 1353–1360.

Digby, S.N., M.A. Chadwick and D. Blache. 2011. Salt intake and reproductive function in sheep. Animal 5: 1207–1216.

Dolinoy, D.C., J.R. Weidman and R.L. Jirtle. 2006. Epigenetic gene regulation: linking early developmental environment to adult disease. Reproductive Toxicology 23: 297–307.

Foster, D.L., D.C. Bucholtz and C.G. Herbosa. 1995. Metabolic signals and the timing of puberty in sheep. pp. 243–257. In: T.M. Plant and P.A. Lee (eds.). The Neurobiology of Puberty. Society for Endocrinology, Bristol.

Franklin-McEvoy, J., W.D. Bellotti and D.K. Revell. 2007. Supplementary feeding with grain improves the performance of sheep grazing saltbush (*Atriplex nummularia*) in autumn. Australian Journal of Experimental Agriculture 47: 912–917.

Gama-Sosa, M., R. Midgett, V. Slagel, S. Githens, K. Kuo, C. Gehrke and M. Ehrlich. 1983. Tissue-specific differences in DNA methylation in various mammals. Biochimica et Biophysica Acta 740: 212–219.

Ghassemi, F., A.J. Jakeman and H.A. Nix. 1995. Salinisation of land and water resources: Human causes, extent, management and case studies. UNSW Press, Sydney, Australia, and CAB International, Wallingford, UK.

Ginane, C., M. Bonnet, R. Baumont and D. Revell. 2015. Feeding behaviour in ruminants: a consequence of interactions between a reward system and the regulation of metabolic homeostasis. Animal Production Science 55: 247–260.

Godfrey, K.M. and D.J. Barker. 2000. Fetal nutrition and adult disease. American Journal of Clinical Nutrition 71: 1344S–1352S.

Hamilton, J.A. and M.E.D. Webster. 1987. Food intake, water intake, urine output, growth rate and wool growth of lambs accustomed to high or low intake of sodium chloride. Australian Journal of Agricultural Research 38: 187–194.

Hemsley, J.A., J.P. Hogan and R.H. Weston. 1975. Effect of high intakes of sodium chloride on the utilization of protein concentrate by sheep. II Digestion and absorption of organic matter and electrolytes. Australian Journal of Agricultural Research 26: 715–727.

Hohlbrugger, G., H. Schweisfurth and H. Dahlheim. 1982. Angiotensin I converting enzyme in rat testis, epididymis and vas deferens under different conditions. Journal of Reproduction and Fertility 65: 97–103.

Hopkins, D.L. and A. Nicholson. 1999. Meat quality of wether lambs grazed on saltbush (*Atriplex nummularia*) plus supplements or lucerne (*Medicago sativa*). Meat Science 51: 91–95.

Leung, P.S. and C. Sernia. 2003. The renin-angiotensin system and male reproduction: new functions for old hormones. Journal of Molecular Endocrinology 30: 263–270.

Lumbers, E. and A. Stevens. 1983. Changes in fetal renal function in response to infusions of a hyperosmotic solution of mannitol to the ewe. Journal of Physiology 343: 439–446.

Lumbers, E., J. Burrell, A. Stevens and C. Bernasconi. 1996. Responses of fetal sheep to reduced maternal renal blood flow and maternal hypertension. American Journal of Physiology—Regulatory, Integrative and Comparative Physiology 271: R1691–1700.

Lyman, T., F. Provenza and J. Villalba. 2008. Sheep foraging behaviour in response to interactions among alkaloids, tannins and saponins. Journal of the Science of Food and Agriculture 88: 824–831.

Machnik, A., W. Neuhofer, J. Jantsch, A. Dahlmann, T. Tammela, K. Machura, J. Park, F. Beck, D. Müller, W. Derer, J. Goss, A. Ziomber, P. Dietsch, H. Wagner, N. van Rooijen, A. Kurtz, K. Hilgers, K. Alitalo, K. Eckardt, F. Luft, D. Kerjaschki and J. Titze. 2009. Macrophages regulate salt-dependent volume and blood pressure by a vascular endothelial growth factor-C-dependent buffering mechanism. Nature Medicine 15: 545–552.

Mao, C., R. Liu, L. Bo, N. Chen, S. Li, S. Xia, J. Chen, D. Li, L. Zhang and Z. Xu. 2013. High-salt diets during pregnancy affected fetal and offspring renal renin–angiotensin system. Journal of Endocrinology 218: 61–73.

Masters, D., N. Edwards, M. Sillence, A. Avery, D. Revell, M. Friend, P. Sanford, G. Saul, C. Beverly and J. Young. 2006. The role of livestock in the management of dryland salinity. Australian Journal of Experimental Agriculture 46: 733–741.

Masters, D.G., S.E. Benes and H.C. Norman. 2007. Biosaline agriculture for forage and livestock production. Agriculture, Ecosystems & Environment 119: 234–248.

Masters, D.G., A.J. Rintoul, R.A. Dynes, K.L. Pearce and H.C. Norman. 2005a. Feed intake and production in sheep fed diets high in sodium and potassium. Australian Journal of Agricultural Research 56: 427–434.

Masters, D.G., A.J. Rintoul, R.A. Dynes, K.L. Pearce and H.C. Norman. 2005b. Feed intake and production in sheep fed diets high in sodium and potassium. Australian Journal of Agricultural Research 56: 427–434.

Meintjes, R.A. and R. Olivier. 1992. The effects of salt loading via two different routes on feed intake, body water turnover rate and electrolyte excretion in sheep. Onderstepoort Journal of Veterinary Research 59: 91–96.

Meyer, J.H. and W.C. Weir. 1954. The tolerance of sheep to high intakes of sodium chloride. Journal of Animal Science 13: 443–449.

Meyer, J.H., W.C. Weir, N.R. Ittner and J.D. Smith. 1955. The influence of high sodium chloride intake by fattening sheep and cattle. Journal of Animal Science 14: 412–418.

Miller, D.W., D. Blache and G.B. Martin. 1995. The role of intracerebral insulin in the effect of nutrition on gonadotrophin secretion in mature male sheep. Journal of Endocrinology 147: 321–329.

Milton, I., M. Lee, J. Calvin and G. Oakes. 1983. Dietary sodium manipulation and vascular responsiveness during pregnancy in the rabbit. American Journal of Obstetrics and Gynecology 146: 930–934.

Mistretta, C. and R. Bradley. 1983. Neural basis of developing salt taste sensation: Response changes in fetal, postnatal and adult sheep. Journal of Comparative Neurology 215: 199–210.

Mohamed, M.O. and C.J.C. Phillips. 2003. The effect of increasing salt intake of pregnant dairy cows on the salt appetite and growth of their calves. Animal Science 77: 181–185.

Moinier, B.M. and T.B. Drüeke. 2008. Aphrodite, sex and salt—from butterfly to man. Nephrology Dialysis Transplantation 23: 2154–2161.

Morcombe, P.W., G.E. Young and K.A. Boase. 1993. Live weight changes and wool growth of sheep grazing saltbush. pp. 79–84. In: N. Davidson and R. Galloway (eds.). Productive Use of Saline Land. Australian Centre for International Agricultural Research, Canberra.

Morcombe, P.W., G.E. Young and K.A. Boase. 1996. Grazing a saltbush (*Atriplex-Maireana*) stand by Merino wethers to fill the 'autumn feed-gap' experienced in the Western Australian wheat belt. Australian Journal of Experimental Agriculture 36: 641–647.

Moritz, K.M., M. Dodic and E.M. Wintour. 2003. Kidney development and the fetal programming of adult disease. BioEssays 25: 212–220.

Nedelkov, K., N. Todorov and N. Vasilev. 2012. The possibility for oestrus synchronization by salt-free-salt diet in some sheep breeds reared in Bulgaria. Bulgarian Journal of Agricultural Science 18: 942–952.

Nijland, M., S. Ford and P. Nathanielsz. 2008. Prenatal origins of adult diseases. Current Opinions in Obstetrics and Gynecology 20: 132–138.

NRC. 1985. Nutrient Requirements of Sheep 6th Edition. National Academy Press, Washington, DC.

Okuyama, A., N. Nonomura, E. Koh, N. Kondoh, M. Takeyama, M. Nakamura, M. Namiki, H. Fujioka, K. Matsumoto and M. Matsuda. 1988. Induction of renin-angiotensin system in human testis *in vivo*. Systems Biology in Reproductive Medicine 21: 29–35.

Parmentier, M., T. Inagami, R. Poschet and J.C. Desclin. 1983. Pituitary dependent renin-like immunoreactivity in rat testis. Endocrinology 112: 1318–1323.

Parr, R.A. 1992. Nutrition-progesterone interactions during early pregnancy in sheep. Reproduction, Fertility, & Development 4: 297–300.

Pearce, K., D.G. Masters, C. Friend, A. Rintoul and D.W.P. 2002. Wool growth and liveweight gain in sheep fed a saltbush and barley ration. Animal Production in Australia 24: 337.

Peirce, A.W. 1968. Studies on salt tolerance of sheep. VIII. The tolerance of grazing ewes and their lambs for drinking water of the types obtained from underground sources in Australia. Australian Journal of Agricultural Research 19: 589–595.

Pierce, A.W. 1957. Studies on salt tolerance of sheep. 1. The tolerance of sheep for sodium chloride in the drinking water. Australian Journal of Agricultural Research 8: 711–722.

Pierce, A.W. 1968. Studies on salt tolerance of sheep. VIII. The tolerance of grazing ewes and their lambs for drinking water of the types obtained from underground sources in Australia. Australian Journal of Agricultural Research 19: 589–595.

Potter, B.J. 1963. The effect of saline water on kidney tubular function and electrolyte excretion in sheep. Australian Journal of Agricultural Research 14: 518–528.

Potter, B.J. 1968. The influence of previous salt ingestion on the renal function of sheep subjected to intravenous hypertonic saline. Journal of Physiology 194: 435–455.

Potter, B.J. and G.H. McIntosh. 1974. Effect of salt water ingestion on pregnancy in the ewe and on lamb survival. Australian Journal of Agricultural Research 25: 909–917.

Provenza, F.D., J. Villabla, J. Haskell, J. MacAdam, T. Griggs and R. Weidmeier. 2007. The value to herbivores of plant physical and chemical diversity in time and space. Crop Science 47: 382–398.

Rabelink, T. and J.I. Rotmans. 2009. Salt is getting under our skin. Nephrology Dialysis Transplantation 24: 3282–3283.

Rafestin-Oblin, M.E., B. Couette, C. Barlet-Bas, L. Cheval, A. Viger and A. Doucet. 1991. Renal action of progesterone and 18-substituted derivatives. American Journal of Physiology—Renal Physiology 260: F828–F832.

Revell, D., H.C. Norman, P. Vercoe, N. Phillips, A. Toovey, S. Bickell, S. Hughes and J. Emms. 2013. Australian perennial shrub species add value to the feedbase of grazing livestock in low-medium rainfall zones. Animal Production Science 53: 1221–1230.

Rogosic, J., R.E. Estell, D. Skobic and S. Stanic. 2007. Influence of secondary compound complementarity and species diversity on consumption of Mediterranean shrubs by sheep. Applied Animal Behaviour Science 107: 58–65.

Schwartz, J. and J.L. Morrison. 2005. Impact and mechanisms of fetal physiological programming.

Smriga, M., M. Kameishi and K. Torii. 2002. Brief exposure to NaCl during early postnatal development enhances adult intake of sweet and salty compounds. Neuroreport 13: 2565–2569.

Speth, R.C., D.L. Daubert and K.L. Grove. 1999. Angiotensin II: a reproductive hormone too? Regulatory Peptides 79: 25–40.

Stevens, A. and E. Lumbers. 1986. Effect on maternal and fetal renal function and plasma renin activity of a high salt intake by the ewe. Journal of Developmental Physiology 8: 267–275.

Symonds, M.E., T. Stephenson, D.S. Gardner and H. Budge. 2007. Long-term effects of nutritional programming of the embryo and fetus: mechanisms and critical windows. Reproduction Fertility and Development 19: 53–63.

Tay, S. 2012. Morphological and molecular changes in the kidney of offspring born to ewes exposed to a high-salt diet during pregnancy. University of Western Australia, Australia.

Tay, S., S. Digby, S. Pain, D. Blache and D.K. Revell. 2007. High-salt intake during pregnancy in ewes alters organ weights and aldosterone concentration in adult offspring. Early Human Development 83: S170–S130.

Tay, S.H., D. Blache, K. Gregg and D.K. Revell. 2012. Consumption of a high-salt diet by ewes during pregnancy alters nephrogenesis in five-month-old offspring. Animal 11: 1803–1810.

Thomas, D.T., A.J. Rintoul and D.G. Masters. 2007. Sheep select combinations of high and low sodium chloride, energy and crude protein feed that improve their diet. Applied Animal Behaviour Science 105: 140–153.

Unterberger, A., M. Szyf, P. Nathanielsz and L.A. Coz. 2009. Organ and gestational age effects of maternal nutrient restriction on global methylation in fetal baboons. Journal of Medical Primatology 38: 219–227.

Van den, Veyver I. 2002. Genetic effects of methylation diets. Annual Review of Nutrition 22: 255–282.

Warren, B., C. Bunny and E. Bryant. 1990. A preliminary examination of the nutritive value of four saltbush (*Atriplex*) species. Proceedings of the Australian Society of Animal Production 18: 424–427.

Webley, L. 2007. Archaeological evidence for pastoralist land-use and settlement in Namaqualand over the last 2000 years. Journal of Arid Environments 70: 629–640.

Weiss, G. 2000. Endocrinology of Parturition. The Journal of Clinical Endocrinology & Metabolism 85: 4421–4425.

Wilkes, B., E. Krim and P. Mento. 1985. Evidence for a functional renin-angiotensin system in full-term fetoplacental unit. American Journal of Physiology–Endocrinology and Metabolism 249: E366–E373.

Wilson, A.D. 1966a. The intake and excretion of sodium by sheep fed on species of *Atriplex* (Saltbush) and *Kochia* (Bluebush). Australian Journal of Agricultural Research 17: 155–163.

Wilson, A.D. 1966b. The tolerance of sheep to sodium chloride in food or drinking water. Australian Journal of Agricultural Research 17: 503–514.

Wilson, A.D. and N.L. Hindley. 1968. Effect of restricted access to water on the intake of salty foods by Merino and Border Leicester sheep. Australian Journal of Agricultural Research 19: 597–604.

Wilson, A.D. and M.L. Dudzinski. 1973. Influence of the concentration and volume of saline water on the food intake of sheep and on their excretion of sodium and water in urine and faeces. Australian Journal of Agricultural Research 17: 245–256.

Wilson, M., A. Morganti, I. Zervoudakis, R. Letcher, B. Romney, P. Von Oeyon, S. Papera, J. Sealey and J. Laragh. 1980. Blood pressure, the renin-aldosterone system and sex steroids throughout normal pregnancy. American Journal of Medicine 68: 97–104.

Wube, T., A. Haim and F. Fares. 2009. Effect of increased dietary salinity on the reproductive status and energy intake of xeric and mesic populations of the spiny mouse, *Acomys*. Physiology and Behavior 96: 122–127.

Zygoyiannis, D. 2006. Sheep production in the world and in Greece. Small Ruminant Research 62: 143–147.

18

The Rumen and its Adaptation to Salt

A.A. Degen[1] and Victor R. Squires[2],*

SYNOPSIS

This chapter considers the functioning of the rumen and its role in digestion and processing of salty feedstuffs. Experience with feeding salt tolerant and halophytic feedstuffs demonstrates that as a result of fermentation, ruminants are able to convert poor quality feed sources such as halophytic feedstuffs, which are unusable or poorly usable by non-ruminants, to high quality meat, milk and fiber. Ruminants usually receive much less of their energy from lipids than do non-ruminants.

Keywords: fermentation, anatomy, rate of passage, rumen microflora, secondary plant metabolites, salt intake, non-ruminants, camels, goats, cattle, horses, non-protein nitrogen, apparent digestibility, particle size, salivation.

1. Introduction

Ruminants exhibit anatomical variations which are the results of evolutionary trends in favour of specific food selectivity and their ability to process low quality feedstuffs that are fibrous and often high in secondary plant metabolites (Attia-Ismail, this volume). A constant supply of substrates is provided by feed consumed by the animal and the large holding capacity of the rumen provides the necessary volume and retention time for complex dietary components (e.g., cellulose and other polysaccharides to

[1] Institutes for Desert Research, Ben-Gurion University of the Negev, Beer Sheva, Israel.
  Email: degen@bgu.ac.il
[2] College of Grassland Science, Gansu Agricultural University, Lanzhou, China.
* Corresponding author: dryland1812@internode.on.net

be degraded and fermented by the rumen microbes). Many salty feedstuffs, even salt tolerant grasses like *Distichlis*, that are fibrous and tough (shrub species like *Atriplex* can be woody) show improved utilization when particle size is smaller (El Shaer, this volume). The rumen's motility patterns allow the processing of low quality plant materials (typical of so many salt tolerant and halophytic feedstuffs).

The difference in the rate of passage of the liquid and the particulate phases leaving the rumen also facilitates efficient fermentation (Mann et al. 1987). While soluble end-products of the fermentation are directly absorbed or rapidly removed with the liquid phase, larger particles are retained until degradation has occurred. Generally, digesta in the dorsal areas of the rumen has a dry matter content of 14–18%, while the ventral areas has a dry matter content of 6–9%. These aqueous conditions provide for optimal microbial interactions and activities of microbial enzymes.

Passage rate influences fermentation patterns as well. The end-products of digestion of complex carbohydrates, lipids and proteins are absorbed as simple compounds, chiefly from the forestomach and small intestine. The physical breakdown of feed is a major function of the ruminant foregut. The operation which is continuous and effective is achieved by two major processes: chewing during eating and rumination and microbial fermentation favoured by maceration and dentrition in the reticulo-rumen. Rumination is defined as the regurgitation, reinsalivation, and remastication and reswallowing of rumen ingesta. During a period of rumination, which may last up to 2 h, this process is repeated approximately once per minute. The length of time a bolus is chewed is relatively independent of the actual amount of coarse material regurgitated and movement of particles out of the reticulo-rumen.

Fodder-fed ruminants spend almost half of their lives chewing during eating and rumination. Without devoting such a large amount of time to chewing, they would be unable to break down fiber into particles small enough to maximize digestion by rumen microbes and to permit passage of indigestible residues. Rumination efficiency measured by units of NDF (cell wall) consumed/per unit of rumination time is influenced by several factors. The level of intake is important; increasing the amount consumed decreases rumination time per unit of cell digested.

Rumen fermentation is currently the only practical means to convert structural carbohydrates to metabolic intermediates which can be used for synthesis to meat, milk, and fiber (wool, mohair, cashmere, etc.). Many factors such as site of digestion, secondary plant compounds (Attia-Ismail, this volume), feed processing and conservation methods (El Shaer, this volume) can greatly affect end product yields from carbohydrate digestion in the ruminant. The physical breakdown of feed is a major function of the ruminant foregut. The rumen's motility patterns allow the processing of low quality plant materials (typical of so many salt tolerant and halophytic feedstuffs) and regulation of particles out of the reticulo-rumen.

Chewing during eating breaks down fodder into particles small enough to be regurgitated; particles 30 cm long are regurgitated by cattle or sheep in significant amounts while 7-cm particles can be returned to the mouth. This is why processing of feedstuffs is so important (El Shaer, this volume). Chewing during rumination is relatively deliberate and at a slower rate than during eating. Chews per minute for sheep and goats are about 70–90. The rate is influenced by the material ingested. Coarser roughage produces slower chewing rates. Particle breakdown occurs during eating as well as during rumination. During eating, from 10–50% of dry matter ingested is broken into particles of < 12 mm depending on the composition and particle size of forages consumed. Many large particles remain after initial eating. The remaining large particles are broken down later.

2. Rumen microflora

Although a myriad of microorganisms are found throughout the digestive tract of the ruminant, it is only with the microbiota in the rumen that a true symbiotic relationship with the host is apparent. These microorganisms, predominately bacteria, protozoa, and anaerobic fungi, depend on the ruminant to provide the physiological conditions necessary for their existence. The bacteria which are found in the rumen number about 10^{10} to 10^{11} cells/g of rumen contents. The majority are obligate anaerobes, but facultative anaerobes may be present in numbers up to about 10^7 to 10^8 cells per/g of rumen contents. Rumen microbes generally contain between 23 and 60% of their dry matter as crude protein. Rumen bacteria as a whole tend to vary only slightly in crude protein content averaging 50% ($\pm$ 5%). Protozoa on the other hand, are much more variable, averaging 40% crude protein with a range of 20 to 60%. The source of N which microbes use for protein synthesis consists of both dietary protein and non-protein nitrogen (NPN) as well as N recycled to the rumen for re-utilization. Cattle, in particular, can grow, reproduce and lactate when the diet contains only NPN. This illustrates the synthetic capacity of ruminal microbes. Microbial crude protein is flushed to the omasum, abomasum and then on to the small intestine for digestion in concert with other residual materials from the rumen. These processes and interactions are particularly relevant to how ruminants process salty food and water (Attia-Ismail, this volume).

3. Intake of halophytes and salt tolerant plants

Herbivores have developed mechanisms which allow them to consume trees and shrubs; however, there are large differences among species in their ability to consume these plants. A good example can be found in the consumption of tannin rich forage. Tannins have wide-ranging effects that reduce consumption and digestibility of forages and can be toxic to herbivores (Clausen et al. 1990; Murdiati et al. 1992). They form precipitates with proteins, resulting in the formation of indigestible tannin-protein complexes (Robbins et al. 1987; Hagerman 1989; Makkar 1993) and can also form complexes with carbohydrates, cellulose, hemicellulose and amino acids reducing their digestibility (Makkar et al. 1987). In addition, tannins react with the protein of the outer cellular layer of the gut, thereby reducing the permeability of nutrients, interact with digestive enzymes rendering them ineffective (Silanikove et al. 1994) and inhibit microorganisms resulting in a reduction in the production of gases and volatile fatty acids (Makkar et al. 1995). Nonetheless, some herbivores can consume substantial amounts of tannins. This is mainly due to proline, an α-amino acid, which has a high affinity for and complexes with tannins, thus neutralizing the negative effect of this secondary compound. Proline is produced in the salivary gland of ruminants and it has been reported that browsers have larger salivary glands than intermediate feeders, which have larger salivary glands than grazers. Consequently, browsers produce more proline and can consume more tannin rich feeds than intermediate feeders and grazers. Both sheep and goats produce very little proline, but goats tolerate tannins better than do sheep as they are able to detoxify tannins or their degraded products to a greater extent (Distel and Provenza 1991). Tannase has been identified in the rumen mucosa of goats (Begovic et al. 1978) and tannin-protein complex degrading bacteria (streptococci, *Enterococcus faecalis*) have been isolated in goat faeces (Goel et al. 2007), which may explain these differences between species.

Herbivores also show differences in their ability to consume oxalates, which can be present in high concentrations in halophytes such as *Atriplex* spp. and *Kochia* spp. Oxalates are needed by plants for osmoregulation of Na, K and Cl ions (Cymbaluk et al. 1986), but can be toxic to livestock (Smith 1992). Oxalate chelates calcium or magnesium to form the almost completely insoluble salt calcium oxalate or magnesium oxalate (Miller and Dearing 2013), which could lead to low serum Ca or Mg levels, hypocalcaemia, osteoporosis, rumen stasis and gastroenteritis (Abu-Zanat et al. 2003; Masters et al. 2009). Furthermore, Ca and Mg oxalate crystals in blood capillaries can cause cellular damage (James and Butcher 1972), and precipitation of these salts in the kidneys can lead to renal failure (Rahman et al. 2013). Consequently, plants containing 10% oxalate or more are considered toxic and should not be consumed (James 1977), while soluble oxalate levels greater than 2% in ruminants and greater than 0.5% in non-ruminants can be considered the upper limits (Rahman et al. 2013). Herbivores do not possess enzymes needed to metabolize oxalate but harbor gut microbiota such as *Oxalobacter formigenes*, a slow-growing gut bacterium, and *Enterococcus faecalis*, which are capable of degrading oxalates into formic acid and carbon dioxide. Ruminants can adapt to higher oxalate intakes by slowly increasing the amount consumed, which will result in an increase in oxalate-degrading bacteria in the rumen (Rahman et al. 2013).

In addition, palatability of halophytes differ among livestock and even differ in seasons within a livestock species. For examples, *Halocnemum strobilaceum* and *Zygophyllum album*, two halophytes in Egypt, are non-palatable for sheep and goats, but can be consumed by camels. In addition, *Tamarix mannifera* is palatable for sheep in the wet season but not so in the dry season (El Shaer 2010).

3.1 Rumen responses

With the intake of high-salt diets, both cattle and sheep increased intra-ruminal salt concentrations and reduced pH, ammonia and volatile fatty acids concentrations. Overall, bacterial numbers were reduced in cattle, but not in sheep. Both cattle and sheep harboured bacterial populations that were able to tolerate 5% salt, with an increase in these populations in sheep when they consumed a high-salt diet. Bacterial diversity was unchanged in cattle, but showed a reduced diversity in sheep (Thomas et al. undated).

Holstein steers fed a low quality mixed hay diet and consuming high saline water had slower particulate passage rates, longer rumen retention times, greater rumen undigested dry matter fill (by about 30%) and had greater rumen fluid volumes than control steers, but their fluid dilution rates did not differ from control steers (Kattnig et al. 1992). However, when mineral salts were infused intraruminally into Holstein steers fed a high-grain basal diet, the dilution rate of rumen fluid increased (Rogers and Davis 1982). Total fluid flow from the rumen also increased with intraruminal salt infusions of sodium bicarbonate and sodium chloride (600 g/day). Rumen fluid osmolalities increased in steers fed a high-roughage diet and infused with salt infusions, but this did not occur when fed a high grain diet. High sodium chloride infusion (600 g/day) either had no effect on ruminal pH (high-grain diet) or caused a decrease (high-roughage diet). In contrast, an increase in ruminal pH was reported when large amounts of sodium chloride (500 g/day) were infused into the rumen of steers fed a high-grain diet (Rogers et al. 1979).

Salt infusions reduced the molar percentage of rumen propionate and increased the molar percentage of acetate in Holstein steers when a high-grain diet was fed (Rogers and Davis 1982). These same treatments were without effect on molar percentages of rumen acids when a high-roughage diet was fed. The change in the molar percentages of acetate and propionate on the high-grain ration was solely due to a reduction in propionate production. The lower production of propionate, from salt infusions, may have occurred because of the washout of readily fermentable materials. This would fit in with the findings of Thomson et al. (1978) that dilution rate was negatively related to the molar proportion of propionic acid and positively related to the molar proportion of acetic acid in rumen fluid in some sheep studies.

The change in molar concentrations of rumen volatile fatty acids could be associated to ruminal microflora changes, which have been observed in some animals fed high-salt diets compared with more conventional diets (Ben Salem et al. 2010). Hungate (1966) reported that the size and number of protozoa were reduced with an increased intake of dietary salt. Selenomonads and bacteriodes were found to be the predominant rumen bacteria; however, with the consumption of mineral salts, selenomonads remained a major component of the microbial population but the number of bacteriodes decreased, being replaced by a large number of chain-forming cocci (Thomson et al. 1978). Succinate is a major fermentation product of bacteriodes and other rumen bacteria and, consequently, the decrease in bacteriodes results in a decrease in the supply of succinate, which could be responsible for the decreased production of propionate.

Rumen micro-flora may be adversely affected by high salt loads, but work by CSIRO in Australia where sheep were introduced suddenly to drinking water containing 2% NaCl showed that feed intakes fell sharply for three days, but increased again, and by five to seven days reached a stable level 160 g below the feed intake of a companion group on fresh water (Wilson and Hindley 1968). In a second experiment, rumen liquors for *in vitro* digestion studies were taken from sheep conditioned to saline drinking water (2% NaCl) and from sheep with no experience of saline drinking water. The amount of oven-dried ryegrass that was digested decreased with increasing concentration of the medium, and was less when the concentration of all ions was increased. It was concluded that the adaptation was physiological rather than attributable to impact on rumen physiology (Wilson and Hindley 1968; Wilson 1966a). Adaptation to high salt loads clearly does occur but aversion and behavioral change may be invoked to reduce intake.

In sheep consuming *Atriplex nummularia*, it was calculated that slightly less than 30% of the volatile fatty acids produced were absorbed from the rumen and that greater than 70% of the dietary N intake left the rumen as non-ammonia N. This is in contrast to the 70% to 80% usually absorbed from the rumen in other feeds (Weston et al. 1970). The authors ruled out residence time as a reason for the impairment of absorption as there was ample time for absorption. They did mention, though, that the 0.3 to 0.9 increase in pH units measured in this study would reduce absorption rate but that "this alone appears unlikely to account for more than a small portion of the discrepancy."

3.2 Protein in animal nutrition

All animals, ruminants and non-ruminants, require amino acids at the tissue level for protein synthesis, with only minor species differences in specific amino acid requirements. But how these amino acids or their precursors are supplied in the diet can vary substantially between ruminants and non-ruminants. One major advantage of ruminants is that the rumen microflora can convert some NPN and/or poor quality proteins to good quality microbial protein. Livestock fed on high salt diets, especially halophytic species with high levels of ash NPN and secondary plant metabolites are faced with more challenges (Attia-Ismail, this volume; Masters, this volume; Nizar et al., this volume). The same rumen fermentation can also cause less efficient use of high quality proteins and energy utilization is lowered by methane production. Methane producing rumen microbes thrive on highly fibrous feeds (e.g., mature pasture and hays). These low digestibility diets ferment to a near-neutral pH producing large amounts of hydrogen gas which the methane-producers require (Mitsumori and Sun 2008).

4. Forage and water intake by livestock consuming halophytes and salt tolerant plants

Halophytes and salt tolerant plants have a greater effect on water and voluntary dry matter intakes than on digestion (Masters et al. 2005, 2007). Both sheep and cattle drank an additional 4 litres of water per day, for every 100 g of salt ingested (Thomas et al. undated). Holstein steers infused intraruminally with 600 g NaCl per day drank approximately two times the volume of water of control animals, independent of diet (Rogers and Davis 1982). However, in a study of yearling Holstein steers offered low quality mixed hay and saline water containing 2,300 ppm total dissolved solids (TDS), the high-saline water did not affect feed or water intake, although there was a tendency for greater consumption of both feed and water in these steers (Kattnig et al. 1992).

In general, halophytes, salt tolerant plants and salts reduce dietary intake. Reasons for the depression in voluntary feed intake remain unclear. Increased rumen osmolality, which occurs with salt and halophyte consumption, has been found to decrease voluntary food intake (Durand and Kawashime 1980). Rogers et al. (1979) measured decreased cellulose degradation due to the increase in rumen osmotic pressure and suggested that this may be the reason for the reduced voluntary feed intake. In addition, it has been suggested that the hypothalamus may receive signals from osmoreceptors found in the rumen or from the hormonal sensation of high osmolality and depress voluntary feed intake. In addition, high concentrations of Na^+ and K^- ions in extracellular fluid of the rumen mucosa, and the involvement of arginine vasopressin may be involved in reducing intake (Masters et al. 2005).

Thomas et al. (undated) reported that voluntary feed intake decreased substantially and progressively in both cattle and sheep when dietary NaCl concentrations exceeded 10%. This decline was more noticeable if the diet was highly digestible, and represents a salt intake threshold of between 120 g and 200 g NaCl/day for young wethers, and about 870 g NaCl/day for Angus steers. Furthermore, Masters et al. (2005) found that sheep would stop consuming salty forage after they had ingested approximately 200 g of salt in a day. These authors calculated that a 50 kg wether consuming oldman saltbush, which has an ash concentration of about 250 g/kg of edible material, can consume only about 800 g of DM. This is well below maintenance requirements. Furthermore, there is added energy expended to absorb Na by the digestive tract and to excrete it by the kidneys. Apparently, this additional energy expenditure for the processing of the elevated salt intake can reduce the metabolizable energy intake by up to 10%. In addition, because of the high ash content of saltbush, metabolizable energy is low and, also, much energy is lost through faeces and methane gas (Arieli et al. 1989; Masters et al. 2005, 2009; Ben Salem et al. 2010).

4.1 Digestibility of halophytes and salt tolerant plants

The proportion of feedstuff which disappears during the passage through the alimentary tract, or apparent digestibility, is often measured and used as an indicator of nutritive value. Apparent digestibility while meaningful, is an oversimplified expression of nutritive value because the factors, both related to and independent of apparent digestibility, also contribute to the overall nutritive value of high-salt feedstuff. As Masters (this volume) shows, the apparent digestibility of salt tolerant and halophytic feedstuffs is often overestimated. Apparent digestibility values likewise do not reflect the nature of the absorbed end-products of digestion or the amount of energy lost as a result of digestive processes, both of which are influenced by sites in the gastro-intestinal tract in which digestion occurs. Retention time of feed residues in the digestive tract has profound effects on both the total tract digestibility and site of digestion. Factors such as level of feed intake, ingredients and nutrient composition of the diet and methods of feed processing also affect rate of passage, apparent digestibility and site of digestion. Apparently, salt consumption affects digestibility in sheep more so than it does in cattle. Organic matter digestibility was depressed by up to 5% when sheep were fed more than 15% NaCl, although this did not occur in cattle (Thomas et al. undated). In addition, *in situ* dry matter disappearance from the rumen of Holstein steers infused with mineral salts (Rogers and Davis 1982) or consuming high saline water of 2300 ppm TDS did not differ from that of steers consuming normal drinking water (Kattnig et al. 1992; Squires, this volume).

Weston et al. (1970) reported that, in sheep, organic matter digestibility of *Atriplex nummularia* containing 22.7% ash content was 60.8%, of which only 33.2% was digested in the rumen while 67.8% was digested in the intestines. Digestion of cell wall constituents followed a similar pattern: whole tract digestibility was 60.4% of which 43% was digested in the rumen and 57% in the intestines. However, generally 55 to 67% of organic matter and 75% to 100% of cell wall constituents are digested in the rumen. The authors offered two options for the low rumen digestibility: (1) reduced particle retention time in the rumen; and (2) impaired metabolism of rumen microbes that digest fibres. As retention time was not affected by the consumption of *Atriplex nummularia*, they concluded that it was due to the latter option. The osmotic pressure of rumen liquor in these sheep was comparatively high at 331 m-osmole/kg, that is, well above the usual rumen osmolality of about 280 mOsm/kg. Favourable osmotic pressure for rumen ciliates is 260 mOsm/kg. Digestion in the rumen is impaired by high osmotic pressure; in particular cellulose degradation (Durand and Kawashime 1980) and that may partly explain the low rumen digestibility.

However, a different pattern from that measured by Weston et al. (1970) was reported by Riasi et al. (2008) and Danesh Mesgarana and Stern (2005), who determined ruminal and post-ruminal digestion of dry matter and protein in four other halophytes, *Atriplex dimorphostegia*, *Kochia scoparia*, *Suaeda arcuata* and *Gamanthus gamacarpus*. They found that, in general, greater than 60% of the digestion took place in the rumen of sheep, in contrast to what was found for *Atriplex nummularia* (Table 18.1).

Table 18.1. Ruminal and post-ruminal digestion of different components of five halophyte species (g/kg DM).

Halophyte species	Measurement	Ruminal	Post-Ruminal	Total	Reference[a]
Atriplex dimorphostegia	Dry matter	472	102	574	1
	Crude protein	557	229	821	1
		370	250	620	2
Atriplex nummularia	Organic matter	202	406	608	3
	CWC[b]	260	344	604	3
Kochia scoparia	Dry matter	444	124	568	1
	Crude protein	517	276	793	1
Suaeda arcuata	Dry matter	553	73	626	1
	Crude protein	557	223	780	1
Gamanthus gamacarpus	Dry matter	663	44	707	1
	Crude protein	667	234	901	1

[a] (1) Riasi et al. (2008), (2) Danesh Mesgarana and Stern (2005), (3) Weston et al. (1970).
[b] Cell wall constituents.

Crude protein content of halophytes is generally relatively high (Ben Salem et al. 2010). However, much of the total N intake is due to non-protein compounds, including nitrates glycinebetaine and proline; the latter two aid the plant in coping with stresses of low water and high salt (Le Houérou 1992). The non-protein compounds can be converted into microbial protein in the rumen, but this requires metabolizable energy in order to do so. However, metabolizable energy is low in halophytes, and, because of the lack of energy, many of the compounds are converted to ammonia, which is absorbed by the ruminant. The ammonia is then converted to urea, which can be excreted in the urine or recycled to the rumen. Consequently, the crude protein value of halophytes is lower than in other feeds of higher metabolizable energy yield (Weston et al. 1970; Ben Salem et al. 2010; Masters, this volume).

5. Nutrient needs of ruminants versus monogastric species

The major differences in nutrient needs of ruminants and non-ruminants are readily apparent (even between ruminants and equines where the duodenum, jejunum and ileum are the main sites for fermentation) (Hess 2010). Ruminants can derive much of the energy from fibrous compounds such as cellulose whereas non-ruminants can utilize only limited amounts of fibrous material (see Degen and Kam, this volume, for a case study on fat sand rats that are wholly dependent on *Atriplex* for their water and energy). One usually thinks of non-ruminants as being unable to utilize cellulose however that is not true of all non-ruminants. Some non-ruminants such as rabbits (through copraphagia), kangaroos and the hippopotami have considerable pregastric fermentation while others such as the horse and rabbit have considerable hindgut fermentation. Degen and Kam (this volume) explain more about the differences between ruminants and non-ruminants and the reader is referred there for this information. Secondly, ruminants do not have a dietary amino acid requirement, only a nitrogen (N) or crude protein requirement. As Masters (this volume) explains, many halophytic feedstuffs are high in ash content and there is often a high proportion of non-protein nitrogen that can cause apparent digestibility to be over estimated. The presence of secondary plant metabolites (Attia-Ismail, this volume) is also a complicating factor. Ruminants do not have a dietary requirement for vitamin B complex vitamins. All three of these differences are due to the microorganisms in the rumen that ferment cellulose and other complex carbohydrates for energy (see above), synthesize microbial protein from a wide variety of N sources including non-protein (NPN) and synthesize B vitamins (Schingoethe et al. 1988).

6. Conclusion

As a result of fermentation, ruminants are able to convert poor quality feed sources such as halophytic feedstuffs, which are unusable or poorly usable by non-ruminants, to high quality, meat, milk and fiber which can be used to support peoples in harsh climatic zones, especially in arid and semi arid regions.

Ruminants usually receive much less of their energy from lipids than do non-ruminants. Among carbohydrates there are also differences in forms and amounts of various carbohydrates which can be used by different species. Camels in particular (Safinaz and El Shaer, this volume) are able to cope with rough, fibrous and dry feedstuffs much better than cattle or sheep and they can do this with less drinking water (Squires this volume).

References and Further Readings

Abu-Zanat, M.M., F.M. Al-Hassanat, M. Alawi and G.B. Ruyle. 2003. Oxalate and tannins assessment in *Atriplex halimus* L. and *A. nummularia* L. Journal of Range Management 56: 370–374.
Arieli, A., E. Naim, R.W. Benjamin and D. Pasternak. 1989. The effect of feeding saltbush and sodium chloride on energy metabolism in sheep. Animal Production 49: 451–457.
Attia-Ismail, S.A. (in press). Plant secondary metabolites of halophytes and salt tolerant plants. pp. 127–142, this volume.
Attia-Ismail, S.A. (in press). Rumen physiology under high salt stress. pp. 348–360, this volume.
Bejovic, S., E. Dusic, A. Sacirbegovic and A. Tatro. 1978. Examination of variations of tannase activity in ruminal content and mucosa of goats on oak leaf diet and during intraruminal administration of 3 to 10% tannic acid. Veterinaria, Sarajevo 27: 445–457.
Ben Salem, H., H.C. Norman, A. Nefzaoui, D.E. Mayberry, K.L. Pearce and D.K. Revell. 2010. Potential use of oldman saltbush (*Atriplex nummularia* Lindl.) in sheep and goat feeding. Small Ruminant Research 91: 13–28.
Calvin, H.W., R.D. Digesti and J.A. Louvier. 1978. Effect of succulent and nonsucculent diets on rumen motility and pressure before, during and after eating. Journal of Dairy Science 61: 1414–1421.
Clausen, T.P., F.D. Provenza, E.A. Burritt, P.B. Reichardt and J.P. Bryant. 1990. Ecological implications of condensed tannin structure: a case study. Journal of Chemical Ecology 16: 2381–2392.
Cymbaluk, N.F., J.D. Miller and D.A. Christensen. 1986. Oxalate concentration in feeds and its metabolism by ponies. Canadian Journal of Animal Science 66: 1107–1116.
Danesh Mesgarana, M. and M.D. Stern. 2005. Ruminal and post-ruminal protein disappearance of various feeds originating from Iranian plant varieties determined by the *in situ* mobile bag technique and alternative methods. Animal Feed Science and Technology 118: 31–46.
Degen, A.A. and M. Kam. (in press). Energy and nitrogen requirements of the fat sand rat (*Psammomys obesus*) when consuming a single halophytic chenopod. pp. 373–387, this volume.
Distel, R.A. and F.D. Provenza. 1991. Experience early in life affects voluntary intake of blackbrush by goats. J. of Chemical Ecology 17: 431–450.
Dumancic, D. and H.N. Le Houérou. 1980. *Acacia cyanophylla* Lindl. as supplementary feed for small stock in Libya. pp. 321–325. *In*: H.N. Le Houérou (ed.). Browse in Africa, the Current State of Knowledge. UNESCO, Paris.
Durand, M. and R. Kawashime. 1980. Influence of minerals in rumen microbial digestion. pp. 375–408. *In*: Y. Ruckebusch and P. Thivend (eds.). Digestive Physiology and Metabolism in Ruminants. AVI, Westport, Connecticut.
Dynes, R.A. and A.C. Schlink. 2002. Livestock potential of Australian species of Acacia. Conservation Science, Western Australia 4: 117–124.
El Shaer, H.M. (in press). Halophytes and salt tolerant forages processing as animal feeds at farm level: Basic Guidelines. pp. 388–405, this volume.
El Shaer, H.M. 2010. Halophytes and salt-tolerant plants as potential forage for ruminants in the Near East region. Small Ruminant Research 91(1): 3–12.
Goel, G., A.K. Puniya and K. Singh. 2007. Phenotypic characterization of tannin-complex degrading bacteria from faeces of goat. Small Ruminant Research 69: 217–220.
Hagerman, A.E. 1989. Chemistry of tannin-protein complexation. pp. 323–333. *In*: R.W. Hemingway and J.J. Karchesy (eds.). Chemistry and Significance of Condensed Tannins. Plenum Press, New York.
Hess, T. 2010. Nutrient metabolism of non ruminants in rangeland ecosystems. pp. 48–83. *In*: V.R. Squires (ed.). Range and Animal Sciences and Resources Management Vol. II, Encyclopedia of Life Support Systems, EOLSS Publishers/UNESCO, UK.
Hungate, R.E. 1960. The rumen and its microbes. Academic Press, New York.
James, L.F. 1977. Oxalate poisoning in livestock. pp. 135–145. *In*: R.F. Keeler, K.R. Van Kampen and L.F. James (eds.). Effects of Poisonous Plants on Livestock. Academic Press, New York, N.Y. USA.
James, L.F. and J.E. Butcher. 1972. Halogeton poisoning of sheep: effect of high level oxalate intake. Journal of Animal Science 35: 1233–1238.
Kattnig, R.M., A.J. Pordomingo, A.G. Schneberger, G.C. Duff and J.D. Wallace. 1992. Influence of saline water on intake, digesta kinetics, and serum profiles of steers. Journal of Range Management 45(6): 514–518.
Le Houérou, H.N. 1980. Planting and management methods for browse trees and shrubs. pp. 351–359. *In*: H.N. Le Houérou (ed.). Browse in Africa, the Current State of Knowledge. UNESCO, Paris.

Le Houérou, H.N. 1992. Forage halophytes and salt-tolerant fodder crops in the Mediterranean basin. pp. 123–137. *In*: V.R. Squires and A.T. Ayoub (eds.). Halophytes as a Resource for Livestock and for Rehabilitation of Degraded Lands. Kluwer Academic Publishers, Dordrecht.

Makkar, H.P.S. 1993. Antinutritional factors in foods for livestock. pp. 69–85. *In*: M. Gill, E. Owen, G.E. Pollott and T.L.J. Lawrence (eds.). Animal Production in Developing Countries. Occasional Publication No. 16—British Society of Animal Production.

Makkar, H.P.S., B. Singh and R.K. Dawra. 1987. Tannin-nutrient interactions. International Journal Animal Science 2: 127–140.

Makkar, H.P.S., M. Blümmel and K. Becker. 1995. Formation of complexes between polyvinyl pyrrolidone and polyethylene glycol with tannins and their implications in gas production and true digestibility in *in vitro* techniques. British Journal of Nutrition 73: 897–913.

Mann, D.L., L. Goode and K.R. Pond. 1987. Voluntary intake, gain, digestibility, rate of passage and gastrointestinal tract fill in tropical and temperate breeds of sheep. Journal of Animal Science 64: 880–886.

Masters, D.G. (in press). Assessing the feeding value of halophytes. pp. 89–105, this volume.

Masters, D.G., S.E. Benes and H.C. Norman. 2007. Biosaline agriculture for forage and livestock production. Agriculture Ecosystems and Environment 119: 234–248.

Masters, D.G., H.C. Norman and E.G. Barrett-Lennard. 2005. Agricultural systems for saline soil: the potential role of livestock. Asian-Australian Journal of Animal Science 18: 296–300.

Masters, D., M. Tiong, H. Norman and P. Vercoe. 2009. The mineral content of river saltbush (*Atriplex amnicola*) changes when sodium chloride in the irrigation solution is increased. *In*: T.G. Papachristou, Z.M. Parissi, H. Ben Salem and P. Morand-Fehr (eds.). Nutritional and Foraging Ecology of Sheep and Goats. Zaragoza: CIHEAM, FAO, NAGREF. Options Méditerranéennes, Série A: Mediterranean Seminars 85: 153–164.

Miller, A. and D. Dearing. 2013. The metabolic and ecological interactions of oxalate-degrading bacteria in the mammalian gut. Pathogens 2: 636–652.

Murdiati, T.B., C.S. McSweeney and J.B. Lowry. 1992. Metabolism in sheep of gallic acid, tannic acid and hydrolyzable tannin from *Terminalia oblongata*. Australian Journal of Agricultural Research 43: 1307–1319.

Mitsumori, M. and W.B. Sun. 2008. Microbial fermentation for mitigating methane emissions from the rumen. Asian-Aust. J. Anim. Sci. 21(1): 144–154.

Nizar, M., G. Hajer and H. Kamel. (in press). Potential use of halophytes and salt tolerant plants in ruminant feeding: A Tunisian case study. pp. 37–59, this volume.

Rahman, R.B. Abdullah and W.E. Wan Khadijah. 2013. A review of oxalate poisoning in domestic animals: tolerance and performance aspects. Journal of Animal Physiology and Animal Nutrition 97: 605–614.

Riasi, A., M. Danesh Mesgaran, M.D. Stern and M.J. Ruiz Moreno. 2008. Chemical composition, *in situ* ruminal degradability and post-ruminal disappearance of dry matter and crude protein from the halophytic plants *Kochia scoparia*, *Atriplex dimorphostegia*, *Suaeda arcuata* and *Gamanthus gamacarpus*. Animal Feed Science and Technology 141: 209–219.

Robbins, C.T., T.A. Hanley, A.E. Hagerman, O. Hjeljord, D.L. Baker, C.C. Schwartz and W.W. Mautz. 1987. Role of tannins in defending plants against ruminants: reduction in protein availability. Ecology 68: 98–107.

Rogers, J.A. and C.L. Davis. 1982. Effects of intraruminal infusions of mineral salts on volatile fatty acid production in steers fed high-grain and high-roughage diets. Journal of Dairy Science 65: 953–962.

Rogers, J.A., B.C. Marks, C.L. Davis and J.H. Clark. 1979. Alteration of rumen fermentation in steers by increasing rumen fluid dilution rate with mineral salts. Journal of Dairy Science 62: 1599–1605.

Safinaz, M.S. and H.M. El Shaer. (in press). Halophyte and salt tolerant plants feeding potential to dromedary camels. pp. 247–260, this volume.

Schingoethe, D.J., F.M. Byers and G.T. Schelling. 1988. Nutrient needs during critical periods of the life cycle. *In*: Church D.C. (ed.). The Ruminant Animal. Prentice Hall.

Silanikove, N., Z. Nitsan and A. Perevolotsky. 1994. Effect of a daily supplementation of polyethylene glycol on intake and digestion of tannin-containing leaves (*Ceratonia siliqua*) by goats. Journal of Agricultural and Food Chemistry 42: 2844–2847.

Shetaewi, M.M., A.M. Abdel-Samee and E.A. Bakr. 2001. Reproductive performance and milk production of Damascus goats fed *Acacia* shrubs or berseem clover hay in North Sinai, Egypt. Tropical Animal Health and Production 33: 67–79.

Smith, G.S. 1992. Toxification and detoxification of plant compounds by ruminants: an overview. J. Range Manage. 45: 25–30.

Squires, V.R. (in press). Water requirements of livestock fed on halophytes and salt tolerant forage and fodders. pp. 287–302, this volume.

Thomas, D., D. Blache, D. Revell, H. Norman, P. Vercoe, Z. Durmic, S. Digby, D. Mayberry, M. Chadwick, M. Sillence and D. Masters. (undated). The impact of high dietary salt and its implications for the management of livestock grazing saline land. http://archive.agric.wa.gov.au/objtwr/imported_assets/content/aap/thomas_salinity_v2.pdf.

Thomson, D.J., D.E. Beever, M.J. Latham, M.E. Sharpe and R.A. Terry. 1978. The effect of inclusion of mineral salts in the diet on dilution rate, the pattern of rumen fermentation and the composition of the rumen microflora. Journal of Agricultural Science, Cambridge 91: 1–7.

Valderrabano, J., F. Munoz and I. Delgado. 1996. Browsing ability and utilization by sheep and goats of *Atriplex halimus* L. shrubs. Small Ruminant Research 19: 131–136.

Van Niekerk, W.A., C.F. Sparks, N.F.G. Rethman and R.J. Coertze. 2004. Qualitative characteristics of some *Atriplex* species and *Cassia sturtii* at two sites in South Africa. SA J. Anim. Sci. 34(Supplement 1): 108–110.

Ventura, M.R., J.I.R. Castanon, M.C. Pieltain and M.P. Flores. 2004. Nutritive value of forage shrubs: *Bituminaria bituminosa, Rumex lunaria, Acacia salicina, Cassia sturtii* and *Adenocorpus foliosus*. Small Ruminant Research 52: 13–18.

Weston, R.H., J.P. Hogan and J.A. Hemsley. 1970. Some aspects of the digestibility of *Atriplex nummularia* (saltbush) by sheep. Proceedings of the Australian Society of Animal Production II: 517–521.

Wilson, A.D. 1966a. The intake and excretion of sodium by sheep fed on species of *Atriplex* (Saltbush) and *Kochia* (Bluebush). Australian Journal of Agricultural Research 17: 155–163.

Wilson, A.D. and N.L. Hindley. 1968. Effect of restricted access to water on the intake of salty foods by Merino and Border Leicester sheep. Australian Journal of Agricultural Research 19: 597–604.

Rumen Physiology Under High Salt Stress

Salah A. Attia-Ismail

SYNOPSIS

When animals are fed halophytes or offered salty water, the rumen fermentative profile will change. The changes that occur involve primarily the microorganisms leading to changes in the products of fermentation. This may have implications on rumen function and therefore the animal's performance. This chapter will look at these aspects in the presence of high salt intakes.

Keywords: VFA, water intake, osmolarity, osmolality, NaCl, ash content, fermentation, microbial, mineral, pH, secondary plant metabolites, *Atriplex nummularia, Kochia scoparia* and *Atriplex dimorphostegia,* fiber, buffering, toxic compounds.

1. Introduction

Dietary salt (NaCl) was used as a mineral supplement for decades. Recommendations for minerals have been set to maximize animal growth rate, milk yield and reproduction. High levels of salt in the feed or water will increase the sodium and chloride and maybe other mineral concentrations and this has certain implications on animal performance. Levels and types of minerals in water/feeds vary greatly, depending on several factors. Salt-tolerant plants may contain over 20% salts in their edible parts (Wilson 1975) and runs up to more than 40% of plant materials (Attia-Ismail 2008a). The high content of ash is a typical characteristic of halophytes. Mineral contents differ in halophytes from those of regular forages.

Desert Research Center, Cairo, Egypt.
Email: saai54@hotmail.com

It is expected that changes in rumen function induced by feeding salty plants and/or drinking salt-water could occur. The mineral compositions of the high ash contents of halophytes have been controversial as to whether they support animal's requirements of certain minerals or they might exceed the requirements to the extent that they may represent a poisoning threat to animals (Attia-Ismail, this volume). The plant secondary metabolites present in halophyte plants may be subjected to chemical transformation not only by rumen microbes but also by enzymatic changes in gut mucosa. Some authors (Roige and Tapia 1996) classified toxic compounds into compounds that may be detoxified in the rumen, toxic compounds produced in the rumen and compounds which affect rumen metabolism. In contrast to the belief that high salt load is a constraint to the animals, it was found that as liquid dilution rate increases, the growth of ruminal microbes increases and the amount of substrate needed for microbial maintenance decreases (Isaacson et al. 1975), thus increasing efficiency of fermentation. This merit was utilized for animals suffering heat stress. Significant increases in production by heat-stressed dairy cows fed higher-than-recommended concentrations of dietary sodium (Na) and potassium (K) have been reported (Schneider et al. 1984; Mallonee et al. 1985; West et al. 1987).

The rumen is a dynamic, continuous fermentation vat. It provides a suitable environment for a variety of species of anaerobic bacteria, protozoa, and fungi (Degen and Squires, this volume). The rumen environment varies among individual animals. Rumen microorganisms have a complex series of interactions with the feeds supplied. In adult ruminants all of the soluble nutrients in feeds are fermented by the rumen microbial population. A typical rumen environment has mineral concentrations as Na, 1.5–3.0; K, 0.6–2.3; Ca, 0.1–0.2; Mg, 0.1; P, 0.3; PO4-3, 1.9; HCO3-, 1.3 mg/ml and S, Cl, etc. are always present. The normal Osmolality of the rumen is < 400 mOsmol/Kg. The volatile fatty acids (VFA) concentrations are within the range of Formic, < 0.02; acetic, 66; propionic, 23; butyric, 20 µmol/ml.

The objective of feeding ruminants nutritionally balanced diets is to provide a rumen environment that maximizes microbial production and growth. In order to optimize animal performance, compromises in feeding the microbes and the ruminant animal may occur. The microbial community of the rumen consists of bacteria, protozoa, and fungi. The majority of the concentration is as bacteria, where their numbers range from 10^{10} to 10^{11} cells/gram of rumen contents. Bacteria can be grouped according to several factors (e.g., shape, size, structure, and substrate). The most important with regard to animal nutrition is the grouping according to the substrates for fermentation. The bacteria, herein, are divided into eight distinct groups; Cellulolytic Species, Hemicellulolytic Species, Pectinolytic Species, Sugar-utilizing Species, Amylolytic Species, Proteolytic Species, Ureolytic Species, Lipid-utilizing Species, Methane-producing Species, Acid-utilizing Species and Ammonia-producing Species (Church 1998).

When animals are fed halophytes or offered salty water, the rumen fermentative profile will change. The changes that occur involve primarily the microorganisms leading to changes in the products of fermentation. This may have implications on rumen function and therefore the animal performance. This chapter will take a look at these aspects.

2. Effect of salt load on rumen pH

Rumen pH value is the net result of several factors interacting within the rumen environment (i.e., the fermentation processes). The normal rumen pH is in the range 5.5–6.8, depending on the type of diet. The production of VFA is the main factor in generating rumen pH. Rumen pH is one of the most variable factors that can influence the microbial population and the levels of VFA produced. VFA are produced as a consequence of ruminal microbial degradation of ingested carbohydrates and proteins. VFA and lactic acid can build up in the rumen and reduce ruminal pH. These VFA are absorbed directly into the blood via the rumen epithelium. Under high concentrate feeding regimes, pH is lowered, total VFA levels arise and rumen motility is diminished. Maintaining rumen pH is very important for the persistence and stability of the gut microbes. The fiber digesters are most active at pH range of 6.2 to 6.8. Cellulolytic bacteria and methanogenic bacteria are reduced when the pH begins to fall below 6.0 while the starch digesters prefer a lower pH of 5.2 to 6.0. Certain species of protozoa can be greatly depressed when pH drops to lower than 5.5. The reduction of the pH in the artificial rumen fluid from 6–78 to 4–79 decreased the sodium, chloride and magnesium transport (Gaebel et al. 1987).

High salt rations decreased the total VFA production and the values were similar to those obtained when animals are fed *Atriplex nummularia* (Masters et al. 2009). Salinity in drinking water (Silvia et al. 2008; Squires, this volume) at TDS levels of: 1,000; 5,000 and 10,000 mg/l; rumen pH remained unchanged (6.37, 6.37 and 6.36 respectively). Added salt linearly increased percentages of butyrate and isobutyrate leading to an increased energy charge of ruminal VFA and reduced pH (Care et al. 1984).

The ruminal pH values were significantly affected by increasing salinity in drinking water. Both salinity levels (8.15 and 12.33 g TDS l[-1]) showed lower ruminal pH than that of fresh water (0.42 g TDS l[-1]) (Attia-Ismail et al. 2008b). When animals were offered treatments containing different amounts of total dissolved salts: 1,000; 5,000 and 10,000 mg/L in the drinking water, the rumen pH remained unchanged (Valtorta and Gallardo 2012) and so were the total VFA concentrations.

3. Acid base balance and osmotic pressure in the rumen

The first to propose that there is an interrelationship between minerals and acid-base status was Shohl (1939). The difference between the amounts of absorbable anions and cations in a diet determines the acid base status of the animal (Pehrson et al. 1999). Therefore, the net influx of any mineral cation or anion into the body of animal changes its acid base status. The acid-base equilibrium requires excretion of excess dietary cations and anions. The balance of Na^+, K^+ and Cl^- plays a major role in determining acid-base balance in biological fluids. The importance of these minerals in ruminant metabolism lies in their indirect participation in osmotic balance and acid base balance (Afzal et al. 2004).

Volatile fatty acids concentrations were directly proportional to osmotic pressure when there was no addition of minerals, because they count for the major part of its variation. However, major minerals contribute more than short chain fatty acids to rumen osmolality (Bennink et al. 1978). Usually, the osmotic pressure in the rumen is rather constant (250–280 mOsmol/kg), but it can increase up to 350–450 mOsmol/kg after feeding certain diets (e.g., lucerne pellets, Bennink et al. 1978). The fasting levels of the rumen osmolality and specific conductivity, the dry weight content and the concentrations of sodium and potassium appeared to be characteristic of the individual animal. When *Kochia scoparia* and *Atriplex dimorphostegia* provided substrates to the microbes in a continuous culture fermentor (Riasi et al. 2012) there was no significant difference in molar production of total VFA; it being 0.037 and 0.026 mol/d, respectively.

In vivo and *in vitro* studies have shown that sodium, chloride and magnesium can be absorbed from the rumen of sheep and cows (Care et al. 1984). It was concluded that the increment in potassium concentrations caused a rise in rumen osmolarity to about 400 mOsmol/kg (Warner and Stacy 1965; Barrio et al. 1991) observed a closer relationship between osmolarity of rumen fluid and potassium concentration than between osmolarity and sodium concentration in sheep. When the animals were fed, the rumen contents rapidly became hypertonic, largely due to the ingestion of potassium salts and the production of VFA (Warner and Stacy 1965).

Several chemical and physical changes take place in a rumen that is subjected to an acid-base imbalance. Added salt decreased arterial blood and ruminal potassium increased the ruminal concentration of sodium (La Manna et al. 1999). Both active transcellular and diffusive paracellular movement of ions can be affected by elevated acidity and osmolarity of the solution bathing the luminal surface. In addition to the changes in microbial population, a rapid fall in the pH of the ruminal content occurs. It was found that the ingestion of a dry ration containing NaCl did not greatly affect the concentration of sodium in the rumen contents (Warner and Stacy 1965). Also, osmotic pressure is believed to influence voluntary dry matter intake (Welch 1982) when it is higher than 400 mOsmol kg H_2O.

If the osmotic pressure is high, the rumen contents draw water from the body, upsetting fluid balance. However, the increase in rumen osmotic pressure is followed by an increase in blood osmotic pressure (Dobson et al. 1976). Camels fed on rations containing *Atriplex nummularia* showed a higher blood Na values (448.3 and 454.3 mg/100 ml), than those of other control groups. Kewan (2003) and Khattab (2007) had the same results when *Atriplex halimus* was fed to sheep. La Manna et al. (1999) found that higher salt concentration, though not altering the dietary cation anion balance of the ration, decreased blood base excess and thereby might increase likelihood of acidosis. Stacy and Warner (1966) concluded that after feeding, there was clear-cut evidence of increased entry of body water into the rumen, the dilution rate increasing from 007–0 11 hr^{-1} to 0 19–0-28 hr^{-1}. Mayberry et al. (2009) found that feeding *Atriplex nummularia* to sheep reduced rumen efficiency and that this reduction in efficiency was not due to high salt concentrations. However, the type of ration affects the changes in osmotic pressure, microbial population and activity (Wilson and Dudzinski 1973).

The high mineral treatment (1.93% potassium, .68% sodium and 1.85% chloride) (Schneider et al. 1988) diets fed to cannulated cows resulted in 17% more water intake suggesting faster liquid dilution rates, but turnover rates of solids were not affected. Their results were interpreted to indicate that high level inclusion of sodium and potassium chloride salts altered digestive, acid-base and mineral status of heat-stressed lactating cows.

4. Buffering capacity of halophytes

Mineral salts (Sodium bicarbonate, Magnesium oxide) could be used to buffer rumen fluid when pH is reduced which cause acidosis in animals. Buffers are compounds which in aqueous solution help resist changes in rumen pH when certain diets are fed (e.g., high grain) that may cause specific acidosis problems. They neutralize the acids produced when high concentrate rations are fed. The buffering capacity of rumen fluid is variable and is generally assumed to depend primarily on sodium bicarbonate and it is very much limited. Buffering capacity of forages can be defined as the degree to which forage material resists changes in pH. The presence of salt in halophytes may help solve a problem like this. Legumes tend to have a higher buffering capacity than grass. Protein, total ash and cation (Na, K) content of forages are good indicators of total buffering capacity. Buffering capacity increases as these measures increase. The feeding of mineral buffers, to lambs and steers during abrupt ration changes (Ralston and Patton 1976) is effective in preventing a rapid drop in rumen pH and can help to maintain the pH above the critical value. Salts in halophytes are present in dissolved forms and can be transmitted through different tissues of the plant. As mentioned in chapter of this book, Na and Cl were the most abundant ions in the plants followed by K and Ca and the other elements depending on soil salt profile and composition of salts in irrigation water. These dissolved salts may act as buffers in the rumen. The effect of different buffers (bentonite, MgO and KCl) on rumen kinetics (Table 19.1) is of sheep fed on halophyte plants (Attia-Ismail 2009). Magnesium oxide raised pH value significantly while bentonite and KCl had almost identical values. Bentonite showed the least value of Ec and TDS (6.74 mmoles/cm and 3.15 mg/l for Ec and TDS) respectively. Elevated concentrations of bentonite did not affect either Ec or TDS significantly.

Table 19.1. *In vitro* effect of buffer type on pH, Ec, TDS and TVFA's of *Atriplex halimus*.

	Bentonite	MgO	KCl
pH	6.56b	7.44a	6.57b
Ec, mmoles/cm	6.74b	7.11a	6.91c
TDS, mg/l	3.15c	3.37a	3.24b
TVFA's	9.61b	9.98b	10.71a

Adapted from Attia-Ismail et al. (2009).
a, b, c Different superscripts at the same row differ significantly ($P < 0.01$).

5. Rumen volume

Rumen volume increased when osmotic pressure was increased through increasing doses of sodium chloride (Lopez et al. 1994). Yet, these authors concluded that the osmotic pressure effect is mainly exerted on the rate and extent at which water enters or leaves the rumen pool rather than on the absolute size of the rumen pool itself. It is, therefore, that the wash out of the digesta from the rumen increases. Harrison et al. (1975) suggested that such an increase in the flow of digesta to the duodenum would increase the efficiency of microbial synthesis in the rumen, and so increase the flow of protein and starch to the duodenum. The provision of drinking water containing 1.3% NaCl to sheep has shown greater voluntary intakes of fluid and consequently in greater flows of fluid through the rumen (Potter et al. 1972). Animals consuming high saline water showed greater rumen fluid volumes than control (Kattnig et al. 1992). There were no significant effects of VFA infusion rate on ruminal volume (López et al. 2003).

6. Methane production under salt stress

The rumen consists of complex, anaerobic microbial populations among which are the methanogenes which constitute 10^8–10^9/ml. Ruminants lose about 2–15% of their ingested energy solely as methane (Moss et al. 2000). Methane comprises between 20 and 30% of total gases produced in the rumen.

The factors which influence the production of methane in the rumen are pH, VFA, diet, feeding frequency, animal species and environmental stresses. The increased passage rate in the rumen due to increased osmotic pressure may reduce methane emission (Benchaar et al. 2001).

Organic acid salts, most commonly fumarate and malate, when offered to ruminants have been shown to produce a decrease in methane emissions. Addition of sodium fumarate *in vitro* decreased methane production (Lopez et al. 1999). Sodium malate which is converted via fumarate in the rumen was found to stimulate propionate production and to inhibit methanogenesis *in vitro* (Martin and Streeter 1995). Encapsulated fumaric acid (EFA), when offered to lambs as 10% of the diet, reduced methane emissions by 75% and increased feed conversion efficiency by 20% (Wallace et al. 2006). While McCourt et al. (2008) found that encapsulated fumaric acid cannot be used as a mitigation strategy to reduce methane production from grazing dairy cows.

Yousef Elahi (2013) found that the maximum cumulative gas volume at 96 hours and b fraction were related to *Aeluropus littoralis* while (Bakhashwain et al. 2010) found that jojoba (*Simmondsia chinensis*) leaves produced less gas volume than other halophytes.

7. Rate of passage and dilution rate

When mineral salts were infused intraruminally into animals the dilution rate of rumen fluid increased (Rogers and Davis 1982). Also, animals consuming high saline water showed lower particulate passage rates and longer rumen retention times, but their fluid dilution rates did not differ from control steers (Kattnig et al. 1992). Croom et al. (1982) have shown that salts of saliva (mainly $NaHCO_3$) and mineral salts increased ruminal liquid dilution rates.

López et al. (2003) concluded that there was a biphasic response of liquid outflow to changes in the total VFA concentration in the rumen as both variables increased together up to a total VFA concentration of 80.1 mM, whereas, beyond that concentration, liquid outflow remained stable at an average rate of 407 mL/h. However, in order for bacteria to maintain their numbers in the rumen, it is necessary that their reproduction time be shorter than the turnover rate of the rumen digesta.

8. Manipulation of ruminal fermentation

Addition of feed additives is a common practise to alter rumen fermentation with an attempt to improve animal productivity. Organic acids are a class of feed additives used to enhance ruminal fermentation. Irrespective of their mode of action, organic acids increase propionate lower lactic acid and methane by rumen microbes. Inhibition of methane would reduce energy yield and, therefore, decrease net microbial growth in the rumen. However, inhibition of methane has been shown to interfere with rumen function by depressing the net energy value for ground maize (Cole and McCroskey 1975).

9. Water intake as affected by salt load

Observations on the effect of saline drinking water on various livestock species are discussed (Squires, this volume; Degen and Squires, this volume). However, water intake was not significantly increased by added salt; water intakes averaged 14 and 30% more with the .25 and .5% salt levels than without added salt (La Manna et al. 1999). On the other hand, water consumption and urine output in sheep remained elevated for 11.5–15 h after ruminal salt loading (Carter and Grovu, undated).

10. Summary and conclusions

The rumen is a complex organ that enables many livestock to utilize sparse, often fibrous or woody vegetation that is commonly saturated with antinutritional compounds derived from secondary plant metabolites and characterized by high ash content. Many livestock the world over, and people who raise them, depend on the rumen's adaptability to harsh environmental conditions, seasonal shortage of nutritious feedstuffs and saline drinking water. Some ruminants, e.g., the camel and the goat (but also some locally-adapted breeds of sheep) are able to cope with the harsh conditions that prevail in arid and semi arid regions and allow the conversion of lower quality plant materials into products (meat, milk, wool) that make it possible for millions of people to inhabit the drylands that cover about 40% of the world's land surface.

References and Further Readings

Afzal, D., M. Nisa, M.A. Khan and M. Sarwar. 2004. A review on acid base status in dairy cows: implications of dietary cation-anion balance. Pakistan Vet. J. 24(4).
Attia-Ismail, S.A. (in press). Plant secondary metabolites of halophytes and salt tolerant plants. pp. 127–142, this volume.
Attia-Ismail, S.A. (in press). Mineral balance in animals as affected by halophyte and salt tolerance plant feeding. pp. 143–160, this volume.
Attia-Ismail, S.A., H.M. Elsayed, A.R. Asker and E.A. Zaki. 2009. Effect of different buffers on rumen kinetics of sheep fed halophyte plants. Journal of Environmental Science 19(1): 89–106.

Attia-Ismail, S.A. 2008. Role of minerals in halophyte feeding to ruminants. *In*: M.N.V. Prasad (ed.). Trace Elements: Nutritional Benefits, Environmental Contamination, and Health Implications.

Attia-Ismail, S.A., (Mrs.) Ahlam R. Abdo and A.R.T. Asker. 2008. Effect of salinity level in drinking water on feed intake, nutrient utilization, water intake and turnover and rumen function in sheep and goats. Egyptian J. of Sheep and Goat Sciences (Special Issue, 2nd Inter. Sci. Conf. on SR Production, 2008) 3(1): 77–92.

Bakhashwain, A.A., S.M.A. Sallam and A.M. Allam. 2010. Nutritive value assessment of some Saudi Arabian foliages by gas production technique *in vitro*. Met., Env. & Arid Land Agric. Sci. 21(1): 65–80.

Barrio, J.P., S.T. Bapat and J.M. Forbes. 1991. The effect of drinking water on food intake responses to manipulations of rumen osmolality in sheep. Proc. Nutr. Soc. 50: 98A.

Benchaar, C., C. Pomer and J. Chiquette. 2001. Evaluation of dietary strategies to reduce methane production in ruminants: a modeling approach. Can. J. Anim. Sci. 81: 563–574.

Bennink, M.R., T.R. Tyler, G.M. Ward and D.E. Johnson. 1978. Ionic milieu of bovine and ovine rumen as affected by diet. J. Dairy Sci. 61: 315–323.

Care, A.D., R.C. Brown, A.R. Farrar and D.W. Pickard. 1984. Magnesium absorption from the digestive tract of sheep. Quarterly Journal of Experimental Physiology 69: 577–587.

Carter, A.R.R. and W.L. Grovum (undated). Effect of ruminal NaCl loading on water intake, urine output and reticular motility in sheep. Proc. Aust. Soc. Anim. Prod. Vol. 19.

Church, D.C. (ed.). 1988. The Ruminant Animal: Digestive Physiology and Nutrition. Englewood Cliffs, N.J. Prentice Hall.

Cole, N.A. and J.E. McCroskey. 1975. Effects of hemiacetal of chloral and starch on performance of beef steers. J. Anim. Sci. 41: 1735.

Croom, W.I., Jr., R.W. Harvey, A.C. Linnerud and M. Froetschel. 1982. High levels of NaCl in beef cattle diets. Can. J. Anim. Sci. 62: 217.

Degen, A.A. and V.R. Squires. (in press). The rumen and its adaptation to salt tolerance. pp. 336–347, this volume.

Dehority, B.A. 2005. Effect of pH on viability of Entodinium caudatum, Entodinium exiguum, Epidinium caudatum, and Ophryoscolex purkynjei *in vitro*. Journal of Eukaryotic Microbiology 52: 339–342.

Dobson, A., A.F. Sellers and V.H. Gatewood. 1976. Absorption and exchange of water across rumen epithelium. Am. J. Physio. 231: 1588–1594.

Dobson, A., A.F. Sellers and V.H. Gatewood. 1976. Absorption and exchange of water across rumen epithelium. American Journal of Physiology 231(Issue 5): 1588–1594.

Franzolin, R., F.P. Rosales and W.V.B. Soares. 2010. Effects of dietary energy and nitrogen supplements on rumen fermentation and protozoa population in buffalo and zebu cattle. Revista Brasileira de Zootecnia 39: 549–555.

Gaebel, G., H. Martens, M. Suendermann and P. Galfit. 1987. The effect of diet, intraruminal ph and osmolarity on sodium, chloride and magnesium absorption from the temporarily isolated and washed reticulo-rumen of sheep. Quarterly Journal of Experimental Physiology 72: 501–511.

Gelfert, C.-C., S. Bender and R. Staufenbiel. 2003. Influence of acidogenic salts and sodium bicarbonate as feeding components on urine composition. Abstracts—Oral presentations at 11th ICPD. Acta vet. scand. Suppl. 98: 188.

Harrison, D.G., D.E. Beever, D.J. Thomson and D.F. Osbourn. 1975. Manipulation of rumen fermentation in sheep by increasing the rate of flow of water from the rumen. The Journal of Agricultural Science 85: 93–101.

Hungate, R.E. 1960. The rumen and its microbes. Academic Press, New York.

Isaacson, H.R., F.C. Hinds, M.P. Bryant and F.N. Owens. 1975. Efficiency of energy utilization by mixed rumen bacteria in continuous culture. J. Dairy Sci. 58: 1645.

Kattnig, R.M., A.J. Pordomingo, A.G. Schneberger, G.C. Duff and J.D. Wallace. 1992. Influence of saline water on intake, digesta kinetics, and serum profiles of steers. Journal of Range Management 45(6): 514–518.

Kewan, K.Z. 2003. Studies on camel nutrition. Ph.D. Thesis, Faculty of Agriculture, Alexandria.

Khattab, I.M.A. 2007. Studies on halophytic forages as sheep fodder under arid and semi arid conditions in Egypt. Ph.D. Thesis. Alexandria University.

La Manna, A.F., F.N. Owens, S. Janloo, T. Shenkoru and S.D. Welty. 1999. Impact of dietary salt concentration on water intake and physiological measurements of feedlot cattle. Animal Science Research Report, pp. 159–164.

Lopez, S., C. Valdes, C.J. Newbold et al. 1999. Influence of sodium fumarate on rumen fermentation *in vitro*. Br. J. Nutr. 81: 59–64.

López, S., F.D. Hovell and N.A. MacLeod. 1994. Osmotic pressure, water kinetics and volatile fatty acid absorption in the rumen of sheep sustained by intragastric infusions. Br. J. Nutr. 71: 153–168.

López, S., F.D.D. Hovell, J. Dijkstra and J. France. 2003. Effects of volatile fatty acid supply on their absorption and on water kinetics in the rumen of sheep sustained by intragastric infusions. J. Anim. Sci. 81: 2609–2616.

Mallonee, P.G., D.K. Beede, R.J. Collier and C.J. Wilcox. 1985. Production and physiological responses of dairy cows to varying dietary potassium during heat stress. J. Dairy Sci. 68: 1479.

Martin, S.A. and M.N. Streeter. 1995. Effect of malate on *in vitro* mixed ruminal microorganism fermentation. J. Anim. Sci. 73: 2141–2145.

Masters, D.G., D.E. Mayberry and P.E. Vercoe. 2009. Saltbush (*Atriplex nummularia* L.) reduces efficiency of rumen fermentation in sheep. Options Méditerranéennes, A/no. 85. Nutritional and foraging ecology of sheep and goats.

Mayberry, D.E., D.G. Masters and P.E. Vercoe. 2009. Saltbush (*Atriplex nummularia*) reduces efficiency of rumen fermentation in sheep. Options Méditerranéennes 85: 245–250.

McCourt, A.R.Y.T., S. Mayne and R.J. Wallace. 2008. Effect of dietary inclusion of encapsulated fumaric acid on methane production from grazing dairy cows. Proc. Br. Soc. Anim. Sci. British Society of Animal Science, Scarborough, UK.

Moss, A.R., J.P. Jouany and J. Newbold. 2000. Methane production by ruminants: its contribution to global warming. Ann. Zootechnol. 49: 231–253.

Pehrson, B., C. Svensson, I. Gruvaeus and M. Rirkki. 1999. The influence of acidic diets on the acid-base balance of dry cows and the effect of fertilization on mineral content of grass. J. Dairy Sci. 82: 1310–1317.

Peirce, A.W. 1963. Studies on salt tolerance of sheep. V. The tolerance of sheep for mixtures of sodium chloride, sodium carbonate, and sodium bicarbonate in the drinking water.

Potter, B.J., D.J. Walker and W.W. Forrest. 1972. Changes in intraruminal function of sheep when drinking saline water. Br. J. Nutr. 27: 75–83.

Quinn, L.Y., W. Burroughs and W.C. Christiansen. 1962. Appl. Microbiol. 10: 583.

Ralston, A.T. and W.R. Patton. 1976. Controlled ruminant response to abrupt ration changes. pp. 140–147. *In*: M.S. Weinberg and A.L. Sheffner (eds.). Buffers in Ruminant Physiology and Metabolism. Church Dwight Co., Inc., New York.

Riasi, A., M. Danesh Mesgaran, M.D. Stern and M.J. Ruiz Moreno. 2012. Effects of two halophytic plants (*Kochia* and *Atriplex*) on digestibility, fermentation and protein synthesis by ruminal microbes maintained in continuous culture. Asian-Aust. J. Anim. Sci. 25(5): 642–647.

Rogers, J.A. and C.L. Davis. 1982. Effects of intraruminal infusions of mineral salts on volatile fatty acid production in steers fed high-grain and high-roughage diets. Journal of Dairy Science 65: 953–962.

Roige, M.B. and M.O. Tapia. 1996. The effect of plant and fungal toxins on rumen metabolism. Archivos-de-Medicina-Veterinaria 28(1): 5–16, 108 ref.

Schneider, P.L., D.K. Beede and C.J. Wilcox. 1988. Effects of supplemental potassium and sodium chloride salts on ruminal turnover rates, acid-base and mineral status of lactating dairy cows during heat stress. J. Anim. Sci. 66: 126–135.

Schneider, P.L., D.K. Beede, C.J. Wilcox and R.J. Collier. 1984. Influence of dietary sodium and potassium bicarbonate and total potassium on heat-stressed lactating dairy cows. J. Dairy Sci. 67: 2546.

Shohl, A.T. 1939. Mineral metabolism. Reinhold Publishing Corp., New York.

Silvia, E.V., R.G. Miriam, A.S. Oscar, R.R. Germán, A. Cristina, L. Perla, E.G. Mónica and J.T. Esteban. 2008. Water salinity effects on performance and rumen parameters of lactating grazing Holstein cows. Int. J. Biometeorol. 52: 239–247.

Squires, V.R. (in press). Water requirements of livestock fed on halophytes and salt tolerant forage and fodders. pp. 287–302, this volume.

Stacy, B.D. and A.C.I. Warner. 1966. Balances of water and sodium in the rumen during feeding: osmotic stimulation of sodium absorption in the sheep. J. Exp. Phyviol. 51: 79–93.

Valtorta and Gallardo. 2012. http://creativecommons.org/licenses/by/3.0.

Wallace, R.J., T.A. Wood, A. Rowe, J. Price, D.R. Yanez, S.P. Williams and C.J. Newbold. 2006. International Congress Series 1293: 148–151.

Warner, A.C.I. and B.D. Stacy. 1965. Solutes in the rumen of the sheep. Quart. J. Exp. Physiol. 50: 169–184.

Welch, J.G. 1982. Rumination, particle size and passage from the rumen. J. Anim. Sci. 54: 885–894.

West, J.W., C.E. Coppock, K.Z. Milam, D.H. Nave, J.M. LaBore and L.D. Rowe, Jr. 1987. Potassium carbonate as a potassium source and dietary buffer for lactating cows during heat stress. J. Dairy Sci. 70: 309.

Weston, R.H., J.P. Hogan and J.A. Hemsley. 1970. Some aspects of the digestibility of *Atriplex nummularia* (saltbush) by sheep. Proceedings of the Australian Society of Animal Production II: 517–521.

Wilson, A.D. 1975. Influence of water salinity on sheep performance while grazing on natural grassland and saltbush pastures. Australian Journal of Experimental Agriculture & Animal Husbandry 15: 760–765.

Wilson, A.D. and M.L. Dudzinski. 1973. Influence of the concentration and volume of saline drinking water on the food intake of sheep, and their excretion of sodium and water in urine and faeces. Aust. J. Agr. Res. 24245–256.

Yousef Elahi, M. 2013. Determination of nutritive value of five species of halophyte plants used by camel in East South Iran. International Research Journal of Applied and Basic Science 4(9): 2721–2725.

PART 5

Focus on Non-ruminants and Uniting Perspectives

Part 5 of 4 chapters is about non-ruminants and considers the special dietary requirements of poultry and rabbits as well as the differences between ruminant and non-ruminant fermentation processes. By comparison some aspects of the special adaptation by a desert dwelling rodent that is almost wholly dependent on a halophyte for its survival is highlighted here. A special case study is presented on the processing equipment used in making salt tolerant and halophytic feedstuffs more acceptable to livestock and which also retain feed value and reduce wastage. We conclude with a brief section on *uniting perspectives* and some suggestions about how to cope with the remaining problems and speculate about the prospects for overcoming the challenges.

Utilization of Halophytes and Salt Tolerant Feedstuffs for Poultry and Rabbits

R.K. Khidr,* K. Abd El-Galil and K.R.S. Emam

SYNOPSIS

Egypt, like other developing countries faces a large deficit in poultry feed resources. Feed cost represents more than 75% of the total production cost exacerbated by the high cost of importing conventional feedstuffs (corn and soya bean, etc.). In addition, there is a competition between human and animal with respect to the diminishing traditional resources (Soliman et al. 2007). Feed supply is a serious constraint for livestock production in North Africa and the Middle East. A major gap exists between the requirements and supplies of fodders for feeding poultry in Egypt. In order to alleviate this shortage, it is essential to utilize some non-conventional feed resources. In this chapter we explore the possibility of using such feedstuffs in non-ruminant nutrition under Egyptian desert conditions.

Keywords: corn, soya bean, non ruminants, proximate analysis, turkeys, *Atriplex, Acacia*.

1. Introduction

The future development of the non ruminants (birds or rabbits) in Egypt, particularly under the desert and newly reclaimed areas, depends mainly on the availability of feedstuffs used in feeding (Abd El-Galil and Khidr 2000, 2001). For any years, some

Desert Research Centre, Cairo, Egypt.
* Corresponding author: Raafatkhidr@yahoo.com

halophytes and salt tolerant feedstuffs have been used in feeding large animals under Egyptian desert conditions. These feedstuffs underwent some chemical, mechanical and biological treatments and proved useful in improving productivity and reproductive traits of animals under desert conditions (Kandil and El Shaer 1988, 1990; El Shaer et al. 1990; Abou El-Nasr et al. 1998).

Data on using halophytes and salt tolerant feedstuffs in non-ruminant nutrition, especially for poultry, is not well documented but such non-conventional feedstuffs could be considered as a promising solution for non-ruminant feeding under the desert conditions. In this chapter we explore the possibility of using such feedstuffs in non-ruminant nutrition under Egyptian desert conditions (see also Degen and Kam, this volume)

2. The proximate chemical analysis of halophytes and salt tolerant plants

The proximate chemical analysis of halophytes and salt tolerant feedstuffs (Tables 20.1 and 20.2) showed a reasonable content of nutrients.

For example, crude protein (CP%) in crushed *Atriplex* leaves had a high content in the range 20.13% (Abd El-Galil et al. 2014a), through 19.95% (Amin 1999) to 18.89% (Abd El-Galil and Khidr 2000).

Acacia leaf meal (ALM) ranged from 16.2 CP % (Abdel Samee et al. 1994) up to 18.9 CP % by Abd El-Galil and Khidr (2000) while a median value (17.3%) was recorded by El-Gandy (1999).

Crude fiber recorded a higher value in *Acacia* leaf meal which might be favorable in formulation of diets for growing rabbits as they tolerate a higher level of crude fiber compared to poultry species (Abd El-Galil and Khidr 2000). Like most halophytes, ash content in ALM showed a pronounced higher value Na % (Table 20.3) (Amin 1999; Abd El-Galil and Khidr 2001; Abd El-Galil et al. 2014b), while Ca % was similar to the values recorded in other studies. Also trace elements values shown in Table 20.2 were low.

In a recent study, Abd El-Galil et al. 2014a estimated the amino acid of ALM compared to those reported by Amin (1999). They concluded that Methionine and Histidine were the first and second limiting amino acid respectively, while Lysine was the third limiting one. This means that ALM is not a reliable source of the essential amino acid methionine while, it has reasonable values of Leucine, Valine and Arginine, respectively (Table 20.3).

However, chemical composition and palatability of *Atriplex nummularia* vary greatly from one area to another due to the the location factor (El-Bassosy 1983). Also, these fluctuations might be controlled by the phenology of plants and by environmental factors (drought, cold, and heat waves). For example Pasternak et al. (1986) explained that ash content of *Atriplex* can be as low as 200 g/kg for shrubs irrigated with sweet water and as high as 300 to 400 g/kg under saline conditions.

The gross energy content of the *Atriplex nummularia* leaves meal (ALM) contained 3277 kcal/kg DM and apparent metabolizable energy (AME) was 1863 kcal/kg DM. ALM containing a low gross energy and apparent metabolizable energy compared to yellow corn (Abd El-Galil et al. 2014a).

Table 20.1. Proximate chemical analysis of *Atriplex* leaf meal in different studies.

Items	Amin (1999)	Abd El-Galil and Khidr (2001)	Abd El-Galil et al. (2014a and b)
CP %	19.95	18.89	20.13
CF %	2.43	6.26	4.01
EE %	3.06	3.11	3.29
NFE %	40.91	41.33	41.53
Ash %	22.05	20.19	18.39
Calcium (Ca) %	1.40	1.99	1.50
Phosphorus (P) %	0.40	0.24	0.35
Magnesium (Mg) %	---	0.94	0.89
Potassium (K) %	3.20	4.82	3.51
Sodium (Na) %	5.90	8.96	2.96
Iron (Fe) ppm	---	21.0	20.00
Copper (Cu) ppm	78.0	28.0	26.57
Zinc (Z) ppm	46.0	---	51.89

Table 20.2. Proximate chemical analysis of *Acacia* leaf meal in different studies.

Items	Abdel Samee et al. (1994)	El-Gendy (1999)	Abd El-Galil and Khidr (2000)
CP %	16.20	17.28	16.58
CF %	21.30	21.60	17.82
EE %	5.50	5.38	3.56
NFE %	42.40	42.62	48.47
Ash %	9.70	9.12	7.51
Calcium (Ca) %	0.42	---	1.64
Phosphorus (P) %	0.27	---	0.11
Magnesium (Mg) %	0.06	---	0.27
Potassium (K) %	---	---	0.89
Sodium (Na) %	0.24	---	0.30
Iron (Fe) ppm	---	---	0.17
Copper (Cu) ppm	---	---	12.0
Zinc (Z) ppm	20.20	---	32.0
Manganese (Mg) ppm	32.40	---	76.0

3. Growth performance and feed utilization

a) Rabbits

Evidence on the value of incorporating *Atriplex* and *Acacia* into rations for poultry and rabbits as reflected in value of body weight, daily gain, feed consumption and feed conversion ratio is contradictory. In this respect Abdel Samee et al. (1992) found that body weight gain and feed utilization did not change by feeding 60% concentrate plus *Acacia* leaf meal *ad lib*. El-Eraky and Mohamed (1996) found that the body weight and daily gain of rabbits fed diet containing 15% *Acacia* leaves was significantly

Table 20.3. Amino acid composition of *Atriplex nummularia* meal in different studies.

Amino acid	Abd El-Galil et al. (2014a)	Amin (1999)
	mg/kg	
Essential amino acid		
Methionine	0.09	1.01
Lysine	2.84	3.20
Arginine	3.54	2.01
Phenylalanine	3.53	3.00
Leucine	5.35	4.42
Histidine	2.78	1.40
Isoleucine	3.32	2.79
Valine	4.15	3.20
Threonine	3.06	2.16
Non essential amino acid		
Aspartic acid	7.11	4.42
Cystine	0.45	0.14
Alanine	4.26	3.30
Glutamic acid	7.72	6.80
Glycine,	4.40	3.09
Serine	3.23	2.32
Tyrosine	1.25	2.01
Proline	3.45	3.44

higher than those fed the control diet or containing 30% *Acacia* leaves. In contrast, El-Gendy (1999) increased this level up to 30% in growing rabbit (from 5–13 weeks of age) without detrimental effects on body weight and feed utilization. However, Abd El-Galil and Khidr (2000) found that feeding rabbits on 20% *Acacia* leaf meal (Table 20.4) resulted in 5.4% higher body weight than that of the control.

In the same way 25% *Atriplex* leaf meal in growing rabbit's diets resulted in (Table 20.5) 5.2% higher in body weight than the control group (Abd El-Galil and Khidr 2001). However, Abdel Samee et al. (1994) recommended a limit of 25% of *Atriplex* as green forage to avoid detrimental effects on growth and physiological responses.

Table 20.4. Growth performance of NZ White rabbits as affected by feeding diets containing different levels of *Acacia* leaf meal.

Items	Levels of Acacia leaf meal					Sig.
	0 (Control)	10%	20%	30%	40%	
Final body wt., g	$1900^b \pm 60$	$1937^b \pm 67$	$2002^a \pm 70$	$1914^b \pm 75$	$1835^b \pm 78$	*
Daily gain, g	$25.41^b \pm 3.1$	$26.43^a \pm 3.3$	$27.45^a \pm 4.1$	$26.00^a \pm 4.3$	$24.29^b \pm 4.9$	*
Feed intake, g/d	$84.44^b \pm 1.8$	$90.12^b \pm 2.0$	$99.50^a \pm 2.5$	$104.28^a \pm 3.1$	$107.45^a \pm 4.1$	*
Feed conversion	$3.32^b \pm 0.15$	$3.41^b \pm 0.27$	$3.62^{ab} \pm 0.29$	$4.01^a \pm 0.35$	$4.42^a \pm 0.39$	*
Mortality rate %	6.67^c	6.67^c	6.67^c	13.00^b	20.00^a	**

[a,b,c] means within a row with different superscripts are significantly different (P < 0.05), Sig. = Significance * = (P < 0.05) ** = (P < 0.01).

Table 20.5. Growth performance of NZ White rabbits as affected by feeding diets containing different levels of *Atriplex* leaf meal.

Items	0 (Control)	10%	15%	20%	25%	30%	Sig.
	levels of *Atriplex* leaves meal						
Final body wt., g	1925[b] ± 65	1961[b] ± 63	1982[b] ± 73	2014[ab] ± 71	2062[a] ± 65	2013[ab] ± 70	**
Daily gain, g	25.82[b] ± 3.5	26.55[b] ± 3.41	27.08[ab] ± 4.4	27.72[a] ± 4.6	28.41[a] ± 5.20	27.6[ab] ± 5.6	*
Feed intake, g/d	86.50[b] ± 1.62	87.88[b] ± 1.9	93.46[b] ± 2.3	97.02[b] ± 2.91	105.67[ab] ± 3.51	113.16[a] ± 4.1	*
Feed conversion	3.35[b] ± 0.13	3.31[b] ± 0.20	3.45[ab] ± 0.26	3.50[a] ± 0.33	3.72[a] ± 0.40	4.11[a] ± 45	*
Mortality rate %	5.00[c]	5.00[c]	5.00[c]	10.00[b]	10.00[b]	15.00[a]	**

[a,b,c] means within a row with different superscripts are significantly different (P < 0.05), Sig. = Significance.
* = (P < 0.05) ** = (P < 0.01).

It is worth noting that increasing *Acacia* and *Atriplex* leaf meal in growing rabbits' diets resulted in an increase in total and daily feed consumption (El-Eraky and Mohamed 1996; Abd El-Galil and Khidr 2000, 2001). The inclusion of *Acacia* and *Atriplex* leaf meal at high percentages led to high mortality rates in rabbits (Tables 20.4 and 20.5).

3.1 Carcass traits

The different percentages of *Acacia* and *Atriplex* leaves meal in rabbit diets revealed no significant (P > 0.05) effects on all chemical compositions as well as slaughter parameters. It is worth noting that the results of moisture, CP, EE and EV of meat were decreased as the levels of *Atriplex* leaf meal increased in the diet. It means that using *Acacia* leaf meal and ALM did not adversely affect dressing percentage, edible giblets, or alimentary tract percentage (El-Gendy 1999). However, this result was previously confirmed by Abdel Samee et al. (1992), El-Eraky and Mohamed (1996), Abd El-Galil and Khidr (2000, 2001).

Table 20.6. Guidelines for feeding of halophytic feedstuffs to rabbits.

Reference	Type	*Acacia* leaf meal	*Atriplex* leaf meal	Type of feedstuff
Abdel Samee et al. (1992)	Growing rabbits	40	---	Wilting green leaves
Abdel Samee et al. (1994)	Growing rabbits	---	25	As green forage
El-Eraky and Mohamed (1996)	Growing rabbits	15	---	Pellet form
El-Gendy (1999)	Growing rabbits	30	---	Pellet form
Abd El-Galil and Khidr (2000)	Growing rabbits	20	---	Pellet form
Abd El-Galil and Khidr (2001)	Growing rabbits	---	25	Pellet form

4. Poultry

Both *Acacia* and *Atriplex* leaf meals have been fed to chicken and to turkeys.

Amin (1999) fed ALM at two levels (6% and 12%) to growing and layer/ing turkeys respectively. The inclusion of *Atriplex* leaves in layer turkey up to 12% significantly increased feed intake compared to that of the control treatment. The feed conversion values were low when high levels of *Acacia* and *Atriplex* leaf meal were used. This might be due to the increase in feed intake and reduction in body weight gain Abd El-Galil et al. (2014b) found that feeding layer hens on 8% ALM improved (P > 0.05) live body weight.

On the other hand, increasing *Atriplex* leaf meal levels in the experimental diets up to 8 and 12% was followed by a decrease in feed intake of layer hens than those of the control treatment (Abd El-Galil et al. 2014b). However, the feed conversion ratio (g feed/g egg mass) was lower at higher percentages of *Atriplex* leaf meal (12%) in diet (Table 20.7). The improvement in feed conversion ratio of 8% ALM level may be due to its highest egg mass and decreased feed intake as compared to 12% level.

Table 20.7. Effect of feeding different levels of *Atriplex nummularia* leaf meal on productive performance (X̄ ± SE) of Sina laying hens.

Items	Level of *Atriplex nummularia* leaf meal				Sig.
	Control	4%	8%	12%	
Body wt. changes (g)	227.18 ± 25.71	228.99 ± 28.89	233.55 ± 29.16	221.80 ± 32.01	ns
Egg weight (g)	44.88[b] ± 0.20	45.51[ab] ± 0.28	45.49[a] ± 0.26	45.57[a] ± 0.35	*
Feed intake (g/hen/day)	98.68 ± 2.15	101.43 ± 2.47	97.88 ± 2.53	95.40 ± 2.74	ns
Egg number (egg/hen/day)	0.635[ab] ± 0.95	0.658[a] ± 1.05	0.662[a] ± 1.07	0.567[b] ± 1.15	*
Egg mass (g)	28.50[ab] ± 0.35	29.96[ab] ± 0.95	30.11[a] ± 1.05	25.87[b] ± 1.13	*
Feed conversion (g feed/egg mass)	3.46[ab] ± 0.11	3.39[ab] ± 0.15	3.25[b] ± 0.13	3.69[a] ± 0.17	*

[a,b]: Means within the same row showing different letters are significantly different. Sig. = Significant, * = (P < 0.05), ns = not significant.

4.1 Digestibility and nutritive values

Digestion coefficients of nutrient contents as well as the nutritive values of the experimental diets as affected by different levels of *Acacia* and *Atriplex* leaves meal are shown in Tables 20.8 and 20.9. There is a significant decrease in most components at higher levels of inclusion of *Acacia* and *Atriplex* leaf meal (Abdel Samee et al. 1992; Abdel Galil and Khidr 2000, 2001; Abd El-Galil et al. 2014b). The decrease in digestion coefficients of nutrient content at higher levels of *Acacia* and *Atriplex* coincided with the decrease in body weight and daily gain when fed at the same level of inclusion. Nutritive values (as indicated by DCP, TDN and DE) decreased due to the lowering of digestion coefficients in nutrient content.

The probable interpretation for the decreasing digestion coefficients of some nutrients is due to presence of some anti-nutritional factors in halophytes (such as tannin, oxalate and nitrates). These substances form insoluble complexes with essential minerals, proteins and carbohydrates, lowering the nutritive value of product (Ferket and Middelton 1999; Redd 1990; Masters, this volume; Attia-Ismail, this volume).

4.2 Egg Production (EP) and Egg Mass (EM)

Egg production and egg mass of laying turkey fed diets with 10 or 15% *Atriplex* leaf meal were lower than the control treatment (Amin 1999). Abd El-Galil et al. (2014b) found that the egg number at a level 12% *Atriplex* leaf meal in diets decreased by 10.6% of that of the control treatment, while the increase was 4.2 or 3.7% with *Atriplex* leaf meal level 8 or 4%, respectively, compared to the control group. It is clear that increasing *Atriplex* leaf meal level to 12% tends to worsen egg production (Table 20.7). This may be attributed to the decrease in feed intake and effect on hormones. A decrease was observed in Estradiol and Progesterone hormones with increasing percentage of *Atriplex* leaf meal in the diets. Egg mass was maximized at 8% *Atriplex* leaf meal diet, while hens fed 12% *Atriplex* leaf meal recorded the lowest (Abd El-Galil et al. 2014b). It is clear that inclusion of 12% ALM in the diet decreased egg mass by 9.2% compared to that of the control treatment. It is worth noting that substitution of ALM at 4, 8 and 12% led to an increase ($P < 0.05$) in egg weight compared to the control treatment. The control diet and the 8% ALM level recorded the lowest egg weight compared to the other ALM levels in experimental diets (Table 20.8).

Table 20.8. Effect of feeding different levels of *Atriplex nummularia* leaves meal on egg quality ($\overline{X} \pm SE$) of Sina laying hens.

Items	Level of *Atriplex nummularia* leaves meal				Sig.
	Control	4%	8%	12%	
Egg weight (g)	45.27 ± 0.66	45.96 ± 0.71	45.98 ± 0.60	46.02 ± 0.79	ns
Albumen (%)	51.54 ± 0.60	51.35 ± 0.47	51.25 ± 0.57	51.11 ± 0.60	ns
Yolk (%)	33.70 ± 0.61	33.79 ± 0.30	33.82 ± 0.32	33.83 ± 0.60	ns
Shell (%)	14.76 ± 0.26	14.86 ± 0.50	14.93 ± 0.57	15.07 ± 0.59	ns
Shape index	76.61 ± 0.51	76.55 ± 0.82	76.60 ± 0.36	76.95 ± 1.26	ns
Yolk index	39.18 ± 0.54	40.72 ± 0.71	40.22 ± 0.41	40.58 ± 0.10	ns
Shell thickness (mm)	$0.401^b \pm 0.04$	$0.421^b \pm 0.05$	$0.453^{ab} \pm 0.03$	$0.482^a \pm 0.08$	*
Haugh unit	91.15 ± 0.40	93.36 ± 0.55	90.58 ± 0.49	93.27 ± 0.62	ns

ns = not significant. a, b: Means within the same row showing different letters are significantly different. * = ($P < 0.05$).

4.3 Egg quality traits

Egg weight (g), albumen, yolk, eggshell %, shape index, yolk index and Haugh unit recorded a non significant ($P > 0.05$) difference. On the other hand, shell thickness (mm) increased ($P < 0.05$) by increasing ALM (Abd El-Galil et al. 2014b). Yolk, eggshell %, shape index, yolk index percentage and shell thickness increased as the level of

ALM increased in the diet from 8% to 12% respectively, while albumen percentage decreased by increasing ALM (Abd El-Galil et al. 2014) (Table 20.8).

4.4 Electrolytes balance and biochemical parameters

Improvement of productive performance in the hens fed 4 or 8% ALM under heat stress conditions (desert conditions) may be attributed to the electrolytes system of hens that is capable of maintaining normal homeostasis, and/or maintaining blood acid-base balance which get disturbed during heat stress conditions (Tanveer 2004).

Abd El-Galil et al. (2014b) found that hens fed 12% ALM recorded the lowest (P < 0.05) values of red blood cells (RBC), hemoglobin (Hb) and mean corpuscular hemoglobin (MCH) as compared to control, 4 and 8% ALM, respectively. However, this decrease may be attributed to the decrease of erythropoitine hormone which damages kidney tissue. This hormone stimulates Marrow bone to produce RBC's (Guyton and Hall 2006).

On the other hand, the decrease in mean corpuscular volume (MCV) of hens fed 4 and 8% ALM occurred in spite of increase in RBC's and decrease in hematocrite (Ht) %. This might indicate the responses efficiency of hens fed 4 and 8% ALM by increasing surface ratio compared with volume unit leading to rapid diffusion of oxygen (Alessandro et al. 2011). The effect of using *Atriplex nummularia* leaf meal on electrolyte balance, Ht, MCV and MCH increased (P < 0.05) in the diet containing 12% ALM compared to that of other treatments. However, Ht was increased (P < 0.05) in the hens of 8 and 4% ALM as compared to control diet. There were no significant differences among the hens fed 8% ALM and control diet on MCV and MCH. Furthermore, Abd El-Galil et al. (2014a) found that there were no significant differences between treatments in sodium (Na) and magnesium (Mg) concentrations. Potassium (K) concentration was significantly lower in the diets containing 4 and 8% ALM by 13.5 and 20.3%, respectively, compared with control one.

Minerals occur in body fluids and tissues as electrolytes, concerned with the maintenance of osmotic pressure, acid-base balance, membrane permeability and tissue irritability (Underwood and Suttle 1999). Potassium concentration decreased (P < 0.05) in the hens fed 8% ALM by 14.5% as compared to those fed on 12% ALM. This decrease in K concentration may occur to protect the body against hyperkalemia and, therefore, protect the body against muscle irritability. So, increased levels of Na/K ratio and Na/(Na/K) index may be due to increase in the rate of glomerular filtration in the kidney for such electrolytes or increase in absorption of such minerals in the gastrointestinal tract and hence a direct increase in their concentration in the blood (Hamdi et al. 1982; Balnave et al. 1989; Ahmed and Abdel-Rahman 2004; Morsy et al. 2012; El-Hawy 2013).

Serum calcium (Ca) concentration was higher (P < 0.05) in the diets containing 8% *Atriplex nummularia* leaves meal by 23.2, 31.1 and 26.5% compared with control, 4 and 12% ALM treatments, respectively. On the other hand, phosphorus (P) concentration showed reverse trend, whereas hens fed 8% ALM showed lower (P < 0.05) concentration of P by 21.8% as compared to control diet. The decreased phosphorus level may be attributed to their reciprocal reverse relationship as the increased blood calcium level resulted in increased parathyroid hormone secretion which inhibits the renal tubules re-absorption of phosphorus (Morsy et al. 2012; Tyler 1979). Chloride (Cl) concentration was lower (P < 0.05) in the hens fed 8% ALM by 14.1% as compared to control diet, while there were no significant differences between the diets containing 12, 4 and 0% ALM. This decrease may be attributed to the fact that Cl is the primary anion, balancing sodium, potassium and other cations (NRC 2005).

The effect of using *Atriplex nummularia* leaf meal as a source of alternative feed resources on Sina laying hens on a diet containing, 12% ALM showed decreased (P < 0.05) serum total protein (30.5, 30.8 and 28.4%) and globulin (45.7, 53.9 and 51.9%) as compared to control 4 and 8% ALM, respectively. However, A/G ratio increased (P < 0.05) in diet of 12% ALM than that of other treatments. El-Hawy (2013) suggested that feeding salty plants such as *Atriplex nummularia* and/or saline water might reduce hepatic synthesis of RNA, which in turn depressed the incorporation of amino acids for protein synthesis, while, Ahmed (1996) and Morsy et al. (2012) postulated that this decrease in TP was due to the insignificant decrease in feed intake or the increase in water intake and consequently dilution of the blood components.

Cholesterol concentration significantly decreased in the diet containing 8 and 12% ALM by 17.9 and 19.3%, respectively as compared to control and it decreased by 11.8 and 13.3%, respectively as compared to 4% ALM. Also, triglycerides decreased (P < 0.05) in the hens fed 8 and 12% ALM by 11.2 and 11.4%, respectively as compared to the control.

Alanine transaminase (ALT) increased (P < 0.05) in the hens fed 12% ALM by 78.1% than that of the diet containing 8% ALM, and it insignificantly increased in the diet containing 12% ALM by 60.3 and 59.0% as compared to control and 4% ALM respectively. Meanwhile, aspartic transaminase (AST) increased (P < 0.05) in the hens fed 12% ALM by 53.6, 64.5 and 48.7% than that of control, 4 and 8% ALM, respectively. This increase being significant only with feeding 12% ALM may be due to the direct effect of tannins on the liver function. The liver and kidney suffer serious damage from feeding tannins. Tannins cause liver polyribosome disaggregation, inhibition of microsomal enzymes, inhibition of protein and nucleic acid synthesis, fibrosis, coagulation and necrosis in the liver cells (Singleton 1981). Also, the increase of ALT or AST concentration might be caused by high tannins, oxalates and alkaloids in salty plants (Craig et al. 1991). Level of 12% ALM showed decreased (P < 0.05) total antioxidant capacity (TAC) by 17.2 and 14.2% as compared to control and 4% ALM respectively. However, no significant differences were detected between 0, 4 and 8% ALM groups in TAC. The lowest TAC decreased antioxidant status in Sina hens fed 12% ALM. Therefore a low antioxidant status has been regarded as one of the major factors negatively affecting bird's performance (Zhao et al. 2011). Albumin concentration showed a non-significant difference between treatments. However, hens fed salt plant had the least concentrations of albumin. It is known that change in albumin levels reflect the change in liver function because the liver is the site of albumin synthesis (Ahlam et al. 2011).

4.5 Hormonal changes

Aldosterone hormone decreased ($P < 0.05$) in the hens fed 12% ALM by 39.5% than that of control group, and it insignificantly decreased in the hens fed 8 and 4% ALM by 22.3 and 17.2% respectively as compared to control group. The hens fed a high ALM (12%) managed the physiology of salt retention and salt excretion for the salt overload by reducing their plasma aldosterone concentration by approximately 50% of control values. Aldosterone is responsible for 50–70% of total mineral corticoids activity as well as regulation and adjustment of water and electrolytes balance among the body compartments (Amal 2003; El-Hawy 2013). Triiodothyronine (T_3), Estradiol and Progesterone hormones recorded insignificant decreases ($P > 0.05$) with increasing ALM levels in the diet. T_3 level decreased in the hens fed 12% ALM by 15.3, 2.7 and 24.8% as compared with control 4 and 8% ALM groups, respectively. Additionally there is little information on the effects of salt plants feeding on the estradiol and progesterone hormones in poultry (Table 20.9). The insignificant decrease might be attributed to the high content of salt in *Atriplex*.

Table 20.9. Effect of feeding different levels of *Atriplex nummularia* leaf meal on the hormonal changes of Sina laying hens.

Traits	Levels of *Atriplex nummularia* leaf meal			
	Control	**4%**	**8%**	**12%**
Aldosterone (pg/ml)	15.03[a] ± 1.98	12.44[ab] ± 2.54	11.67[ab] ± 1.32	9.09[b] ± 1.67
T_3 (ng/ml)	2.93 ± 0.19	2.55 ± 0.19	3.30 ± 0.35	2.48 ± 0.35
Estradiol (pg/ml)	77.32 ± 7.17	77.26 ± 8.21	81.77 ± 5.97	66.33 ± 10.91
Progesterone (ng/ml)	1.30 ± 0.37	0.91 ± 0.31	1.24 ± 0.68	1.01 ± 0.25

T_3 = Triiodothyronine hormone.
[a,b]. Means that with different superscript in the different columns there are significant differences ($P < 0.05$).

5. Optimum recommended levels of using halophytes and salt tolerant plants in poultry feeding

Values of recommended levels of *Acacia* and *Atriplex* leaf meal percentages in poultry feeding could be listed in Table 20.10. The recommended levels for different factors might be related to species, percentage of stems to leaves, physiological reaction of bird, plant processing methods and any other factor related to the experiment.

Table 20.10. Recommended levels of *Acacia* and *Atriplex* leaf meal percentages in poultry feeding.

Reference	Poultry type	*Acacia* leaf meal	*Atriplex* leaf meal	Type of feedstuff
Amin (1999)	Growing turkey	---	6	Mash form
Amin (1999)	Layer turkey	---	12	Mash form
Abd El-Galil et al. (2014b)	Layer hens	---	8	Pellet form

In conclusion, although *Acacia* and *Atriplex* leaf meals are some of the untraditional feedstuffs that could be used in poultry diets, attention must be paid to their chemical composition, particularly their content of the anti-nutritional factors. Such components have pronounced negative effects on performance. These changes vary according to the age and phenological stage of the plant and the other seasonal fluctuations. Some treatments of these halophytes and salt tolerant plants before using it in poultry feeding might be helpful to alleviate their detrimental effects as well as increase the possibility of using them safely in the different poultry species.

References and Further Readings

Abd El-Galil, K. and R.E. Khidr. 2000. Utilization of *Acacia saligna* in feeding growing rabbits under the desert and newly reclaimed areas. Egypt. Poult. Sci. J. 20: 497–515.

Abd El-Galil, K., A.S. Morsy, K.R.S. Emam and Amal M. Hassan. 2014b. Physiological and productive performance of Sina laying hens fed *Atriplex nummularia* Leaves Meal under Arid Conditions of South Sinai. J. Am. Sci. 10(5): 161–170 (ISSN: 1545–1003).

Abd El-Galil, K. and R.E. Khidr. 2001. Utilization of *Atriplex nummularia* in feeding growing rabbits under the desert and newly reclaimed areas. Egypt. Poult. Sci. J. 21: 53–71.

Abd El-Galil, K., R.E. Khidr, S.E.M. El-Sheikh, A.A. Salama, Henda A. Mahmoud, Mona M. Hassan, A.A. Abd El-Dayem and Fayza M. Salem. 2014a. Utilization of atriplex leaves meal as a non-traditional feedstuff by local laying hens under desert conditions. Egypt. Poult. Sci. 34(II): 363–380.

Abdel Samee, A.M., K.M. El-Gendy and H. Ibrahim. 1992. Growth performance and some related physiological changes in rabbits as affected by feeding acacia under subtropical conditions. Egyptian J. of Rabbit Sci. 2: 13–22.

Abdel Samee, A.M., K.M. El-Gendy and H. Ibrahim. 1994. Rabbit growth and reproductive performance as influenced by feeding desert forage (*Acacia saligna* and *Atriplex numularia*) at North Sinai. Egyptian J. of Rabbit Sci. 4: 25–36.

Abou El-Nasr, H.M., Dawlat, A. El-Kerdawy and H.S. Khamis. 1998. Vetch (*Vicia monantha*) straw as a basal diet with *Acacia* or *Atriplex* shrubs for sheep under arid conditions of Egypt. First Inter. Con. on Anim. Prod. & Health in Semi Arid Areas, Suez Canal Univ., Fac. of Environmental Agricultural Sci., El-Arish, North Sinai, 1–3 Sept.

Ahlam, R.A., E.Y. Eid, Abeer M. El-Essawy, Afaf M. Fayed, H.G. Helal and H.M. El-Shaer. 2011. Effect of feeding different sources of energy on performance of goats fed saltbush in Sinai. Journal of American Science 7(1): 1040–1050.

Ahmed, M.H. 1996. Effect of saline drinking water on productive performance of rabbits. M.Sc. Thesis, Fac. of Agric., Zagazig Univ., Egypt.

Ahmed, M.M. and M.A. Abdel-Rahman. 2004. Effect of drinking natural saline groundwater on growth performance, behavior and some blood parameters in rabbits reared in new reclaimed lands of arid areas in Assiut Governorate. Assiut Univ. Bull. Environ. Res. 7(2): 125–135.

Alessandro, Z., S. Salvatore, M. Vanessa, C. Stefania, R. Ambra and P. Giuseppe. 2011. Hematological profile of Messinese goat kids and their dams during the first month post-partum. Animal Sci. Papers and Reports 29: 223–230.

Amal, M. Hassan. 2003. Some physiological and productive effects of promoting growth in rabbits drinking natural saline water. Ph.D. Thesis, Fac. Agric. Cairo, Univ.

Amin, E. 1999. Effects of crossbreeding and feeding different levels of *Atriplex* meal on production characters of local and commercial varieties of turkey. M.Sc. Thesis, Fac. of Agric., Alex. Univ., Alex., Egypt.

Attia-Ismail, S.A. (in press). Mineral balance in animals as affected by halophyte and salt tolerance plant feeding. pp. 143–160, this volume.

Balnave, D., I. Yoselewitz and R.J. Dixon. 1989. Physiological changes associated with the production of defective egg-shells by hens receiving sodium chloride in the drinking water. British J. Nutr. 61(1): 35–43.

Degen, A.A. and M. Kam. (in press). Energy and Nitrogen Requirements of the Fat Sand Rat (*Psammomys obesus*) When Consuming a Single Halophytic Chenopod. pp. 373–387, this volume.

El- Bassosy, A.A. 1983. A study of the nutritive value of some range plants from Saloom to Mersa Mattrouh. Ph.D. thesis, Fac. of Agric., Ain Shams Univ., Egypt.

El-Eraky, W.M. and W.M. Mohamed. 1996. Growth performance, carcass traits and some related physiological changes of growing rabbits fed on *Acacia* and water Hyacinth. Egyptian J. of Rabbit Sci. 6: 87–98.

El-Gendy, K.M. 1999. Effect of dietary inclusion of *Acacia* leaves meal (*Acacia saligna*) on digestibility, growth performance and blood constituents of growing rabbits. Egypt. J. of Rabbit Sci. 9: 271–283.

El-Hawy, A.S. 2013. Reproductive efficiency of Shami goats in salt affected lands in South Sinai. Ph.D. Thesis, Fac. Agric., Ain Shams Univ.

El Shaer, H.M., S.S. Tag El-Din and H.M. Kandil. 1984. Nutritive evaluation of some introduced pasture plants in North Sinai. Ist Egyptian British Conference on Animal and Poultry Production, September, 11–13, Zagazig, Egypt 135–145.

El Shaer, H.M., O.A. Salem, H.S. Khamis, A.S. Shalaby and M.F.A. Farid. 1990. Nutritional comparison studies on goats and sheep fed broiler litter ensiled with desert shrubs in Sinai. Proc. of the Inter. Goat Production Symp., Oct. 22–26, 1990. Tallahassee, Fl, U.S.A., pp. 70–74.

Ferket, P.R. and T. Middelton. 1999. Anti nutrients in feedstuffs. Poultry International 38: 46–55.

Guyton, A.C. and G.E. Hall. 2006. Text Book of Medical Physiology. 6th Edn., London.

Hamdi, H., Y. Abdel-Rahman, A. Malek, I. El-Bagoumy, Z. Ibrahimw and F. Makasem. 1982. Fundamentals of human physiology. Kidney and body fluids. Vol. 7. Atlas Press, Cairo.

Kandil, H.M. and H.M. El- Shaer. 1988. The utilization of *Atriplex nummularia* by goats and sheep in Sinai. Proc. of the Inter. Symp. on the Constraints and Possibilities of Ruminant Production in the Dry subtropics, Nov. 5–7, Cairo, Egypt, pp. 71–73.

Kandil, H.M. and H.M. El-Shaer. 1990. Comparison between goats and sheep in utilization of high fibrous shrubs with energy feeds. Proc. Inter. Goat Prod. Symp., Oct. 22–26, Tallahassee, FL, USA, pp. 75–79.

Khidr, R.E. and K. Abd El-Galil. 2002. Utilization of *Atriplex* and *Acacia* leaf meal in poultry feeding: A Case Study in Egypt, International Symposium on Optimum Resources Utilization in Salt-Affected Ecosystems in Arid and Semi-Arid Regions, 8–11 April, 2002, Cairo, Egypt.

Masters, D. (in press). Assessing the feeding value of halophytes. pp. 86–105, this volume.

Morsy, A.S., M. Hassan. Mona and M. Hassan. Amal. 2012. Effect of natural saline drinking water on productive and physiological performance of laying hens under heat stress conditions. Egypt Poult. Sci. Vol. (32) (III): 561–578.

National Research Council. 2005. Mineral tolerance of animals. National Academies Press, Washington.

Pasternak, D.A., J.A. Nerd, H. Aronson, R. Kolz and R.W. Venkert. 1986. Fodder production with saline water. Report from the years 1984–1985, Institute for Applied Research, Ben- Gurion Uni. of the Negev, Beer Sheva, Israel.

Reed, J.D., H. Soller and A. Woodward. 1990. Fodder tree and straw diets for sheep: intake, growth, digestibility and the effects of phenolics on nitrogen utilization. Anim. Feed Sci. Tech. 30: 39–50.

Singleton, V.I. 1981. Naturally occurring food toxicants phenolics substances of plant origin common in foods. Advances in Food Research 27: 149–242.

Soliman, A.Z.M, R.E. Khidr, M.A.F. El-Manylawi and S.E.M. El-Sheikh. 2007. Studies on date stone meal as an untraditional feedstuff in doe rabbit diets. Egyp. J. of Rabbit Sci. 17: 103–118.

Tanveer, Ahmed. 2004. Effect of different dietary electrolyte balance on performance and blood parameters of broilers reared in heat stress environments. Ph.D. Thesis, Fac. of Institute of Animal Nutrition and Feed Technology, University of Agriculture, Faisalabad.

Tyler, D.D. 1979. Water and mineral metabolism. In: Physiological Chemistry. 17th Ed., Lang Medical Publication, California, USA.

Zhao, X., Z.B. Yang, Yang, W.R. Y. Wang, S.Z. Jiang and G.G. Zhang. 2011. Effects of ginger root (Zingiber officinale) on laying performance and antioxidant status of laying hens and on dietary oxidation stability. Poult. Sci. 90: 1720–1727.

Energy and Nitrogen Requirements of the Fat Sand Rat (*Psammomys obesus*) When Consuming a Single Halophytic Chenopod

*A. Allan Degen** and *Michael Kam*

SYNOPSIS

The fat sand rat (*Psammomys obesus*; Gerbillinae), a diurnal gerbillid, is herbivorous and is able to thrive while consuming only one halophytic chenopod, in particular the saltbush *Atriplex halimus* (Chenopodiacae), a plant relatively low in energy yield and high in ash and oxalate contents. Before consuming *A. halimus* leaves, fat sand rats scrape off the outer layer with their teeth, which removes much of the electrolytes. To overcome the high oxalate content, fat sand rats harbor the bacterium *Oxalobacter* spp. in their intestines, which is capable of degrading oxalates.

When they were fed only on *A. halimus*, the basal metabolic rate (BMR) of the fat sand rats was 168 kJ kg$^{-0.75}$d^{-1}, 57 to 60% of that expected for an eutherian mammal of its body mass, while the average daily metabolic rate (ADMR) was 499 kJ kg$^{-0.75}$d^{-1}, 88% of that expected for an eutherian mammal of its body mass. Field metabolic rate (FMR) ranged between 565 kJ kg$^{-0.75}$d^{-1} in summer and 680 kJ kg$^{-0.75}$d^{-1} in winter, and in summer was 83% of that expected for a desert eutherian mammal of its body mass.

Institutes for Desert Research, Ben-Gurion University of the Negev, Beer Sheva, Israel.
* Corresponding author: degen@bgu.ac.il

Atriplex halimus is high in crude protein and the fat sand rats were easily able to balance their N intake. Metabolic fecal nitrogen was 70.5 mg $kg^{-0.75}d^{-1}$ and endogenous urinary nitrogen was about 171.9 mg $kg^{-0.75} d^{-1}$ and, therefore, minimal N requirements equalled 242.3 mg $kg^{-0.75}d^{-1}$, which is 98% of the amount predicted for a placental animal of its body mass.

The efficiency of utilization of energy of *A. halimus* for maintenance was only 0.32 and for growth only 0.30 and, therefore, their respective heat increments of feeding were 0.68 and 0.70. This low utilization of feed plus its low energy yield and high water content forced the fat sand rats to consume large quantities of fresh and dry matter for maintenance. In spite of these negative aspects, there are several advantages to consuming only *A. halimus*, namely: (1) it provides a more stable diet throughout the year than do seeds; (2) there is little competition for this food resource from other rodents; and (3) fat sand rat burrows are at the base of the plants and, therefore, minimal energy is expended for foraging.

Keywords: oxalate, nitrogen, energy, herbivory, fiber, salt, electrolyte balance, metabolic rate, water, digestibility, foraging.

1. Introduction

Herbivores, animals that can utilize the fibrous portion of plants, are represented in 11 of the 20 mammalian orders, five of which contain only herbivores (Stevens 1989). Fibrous portions such as cellulose and hemicellulose are not digested by enzymes secreted by mammals but are broken down by microbial fermentation involving both bacteria and protozoans. Fermentation of these polysaccharides occurs principally in a single segment of the digestive tract. In foregut fermenters such as ruminants, camels and kangaroos, fermentation takes place in a modified portion of the stomach prior to the "gastric part of the stomach" and small intestine. In hindgut fermenters, fermentation takes place in the posterior part of the gut. In this group, the larger herbivores such as elephants and horses use the colon as the main fermentation chamber, while the smaller herbivores such as rodents mainly use the cecum (Hume 1989). The capybara is the largest cecum fermenter (Borges et al. 1996). Furthermore, foregut fermenters, principally ruminants, predominate in the 10 to 600 kg group size, while hindgut fermenters predominate among small and very large herbivores (Demment and van Soest 1985). Foregut fermenters are considered to be more efficient utilizers of cell wall constituents than hindgut fermenters (van Soest 1982).

Herbivory is characteristic of a large number of medium and large mammals but, less so, in small mammals. This is so since smaller mammals require more energy in proportion to their body mass than do larger mammals (Kleiber 1975); however, gut capacity is related linearly to body mass. Consequently, smaller mammals require more energy per unit gut capacity and often are forced to be more selective in their dietary intake than are larger mammals (Demment and van Soest 1985; Foley and Cork 1992). This was reported in three species of gerbils in which the smallest gerbil selected the most energy-rich diet and the largest gerbil the least energy-rich diet

(Kam et al. 1997). Even among ruminants this is evident where the choice in the quality of diet can be very varied among herbivores. Most of the smaller ruminants, such as the dik-dik and suni, consume diets of high quality which are easily digestible and rich in accessible plant cell contents (Hofmann 1989).

In general, small herbivorous mammals, such as voles (Batzli 1985; Haken and Batzli 1996) and degus (Bozinovic 1995), consume a variety of plant species and, when available, select plants of low fiber content. When consuming high fiber vegetation, rodents compensate for the poor diet by increasing the size of the gastrointestinal tract (Owl and Batzli 1998) and, thus, volume of digesta (Bozinovic et al. 1997), increasing dietary intake (Justice and Smith 1992), by increasing nutrient uptake by the small intestine (Hammond and Wunder 1991; Karasov and Diamond 1988), and by decreasing basic metabolic rate (Veloso and Bozinovic 1993). At times, a nutritional bottleneck can occur because of high fiber diets and, in spite of compensatory responses, herbivores are unable to meet energy requirements. In fact, some small herbivores often experience severe shifts in food quality, in particular fiber content, and concomitant changes in body mass (Karasov 1989).

1.1 Fat sand rats (Psammomys obesus)

In contrast to most herbivorous small mammals, free-living fat sand rats (*Psammomys obesus*) can thrive while consuming only one chenopod species (Degen 1988; Kam and Degen 1989). Chenopods are low in organic matter and energy yield and high in inorganic matter (mainly sodium, potassium and chloride), and, because of these negative characteristics, are often avoided by herbivores. Like other small herbivores consuming a high fiber diet, fat sand rats have a low metabolic rate (Degen et al. 1988). However, because they consume a single item diet, changes in gastrointestinal size, digesta volume and dry matter intake with changes in dietary quality (fiber content) are less applicable to them.

Fat sand rats (*Psammomys obesus*; Gerbillinae) are widely distributed in the Saharo-Arabian deserts where they inhabit *wadis* (ephemeral riverbeds) and sodic areas that support halophytic vegetation (Mendelsshon and Yom-Tov 1987; Nowak and Paradiso 1983). They are unusual among the Gerbillinae in that they are diurnal and wholly herbivorous (Daly and Daly 1973), while other gerbillid species are nocturnal and primarily granivorous (Bar et al. 1984). As adults, fat sand rats are solitary and each inhabits a complex burrow system with several openings (Orr 1972). They are active above ground all year (Ilan and Yom-Tov 1990). In Israel, fat sand rats inhabit arid areas of the Negev and Judean deserts and the Arava (part of the rift valley).

They feed mostly on halophytic vegetation, many of which belong to Chenopodiacae, and usually inhabit areas dominated by *Atriplex halimus* (Daly and Daly 1973). Under controlled laboratory studies, fat sand rats were able to breed and raise 8 pups to weaning when consuming only *Atriplex halimus* (Kam and Degen 1993, 1994). *Atriplex halimus* is a chenopod high in salt content (Degen 1988; Kam and Degen 1988) and also contains high concentrations of oxalate (> 1% of dry matter). Oxalate can be toxic because it chelates calcium to form the almost completely insoluble salt calcium oxalate. To detoxify the high oxalate concentrations, fat sand rats harbor the bacterium *Oxalobacter* spp. in their intestines, which is capable of degrading oxalates (Palgi et al. 2005; Palgi et al. 2008). To reduce their electrolyte intake, fat sand rats scrape off the cuticular and epicuticular layers of leaves with their teeth before consuming them. They scrape off negligible amounts of leaf from moist plants and substantial amounts from dry plants (Degen 1988). A similar behavior has been reported for the New World heteromyid *Dipodomys microps* when it consumes the saltbush *Atriplex confertifolia* (Kenagy 1972, 1973) and the South American octodontid *Tympanoctomys barrerae* when it consumes *Atriplex* spp. (Mares et al. 1997; Ojeda et al. 1999).

Fat sand rats do not drink water. Scraping the leaves, besides reducing electrolyte load, also reduces their water needs. Nonetheless, lactating fat sand rats are highly responsive to dietary preformed water when only consuming *A. halimus*. Growth rate of suckling pups decreased with a decrease in dietary water content (Kam and Degen 1993, 1994).

Interestingly, when fed a high energy concentrate diet, fat sand rats exhibit pathological metabolic changes resembling those of type 2 diabetes (Schmidt-Nielsen et al. 1964; Hackel et al. 1967; Adler et al. 1990) and, today, they are used widely in diabetes research (Scherzer et al. 2011). In addition, fat sand rats have been identified as one of the main reservoirs for the parasite *Leishmania major*, which causes the disease cutaneous leishmaniasis. The vector for the parasite is the female phlebotomine sand fly, *Phlebotomus papatasi* (Gandacu et al. 2014). In an effort to eliminate the disease, many fat sand rats have been killed by authorities over the years.

2. Consumption of chenopods and energy metabolism

The ability of fat sand rats to consume and thrive only on *Atriplex halimus* is due, at least in part, to their behavioural adaptation of scraping the leaves before consumption. The electrolytes are most concentrated on the leaf surface. In this manner, the fat sand rats have some control over the electrolyte intake, being able to scrape off more leaf when water content of the plant is low and less leaf when the water content is high. They do not scrape all the leaves they consume and use different patterns in scraping the leaves. Sometimes they scrape part of the leaf, consume the scraped part, scrape off a little more, consume that part, and so on. Other times, they scrape large segments of the leaf at one time and then consume it. Leaf scrapings are relatively large and could be easily identified and collected (Degen 1988).

Only 0.8% of the leaf, on a dry matter basis, was scraped off when the fat sand rats were offered *Atriplex halimus* of 84% water content, but 6.4% and 14.3% were scraped off the leaves that contained 78% and 69% water content, respectively (Table 21.1).

Ash content of the scrapings, per unit of DM, was approximately 66% higher than that in non-scraped leaves and Na^+, K^+ and Cl^- were 79%, 129% and 85% higher, respectively. Maximum amounts of electrolytes were removed in summer, when water content of the leaves was relatively low. On a dry matter basis, the leaves consumed were 11% lower in ash content, and 13%, 21% and 14% lower in Na^+, K^+ and Cl^-, respectively. On an organic matter basis, ash was reduced by 14% and Na^+, K^+ and Cl^- by 17%, 25% and 18%, respectively (Table 21.2). As leaf scrapings of

Table 21.1. Body mass of *Psammomys obesus* and their dry matter intake (DMI) and scrapings of *Atriplex halimus* leaves of different water content offered in summer, autumn and winter. Values are means ± SD (adapted from Degen 1988).

	Summer	Autumn	Spring
Body mass (g)	170.1 ± 20.8	135.4 ± 7.9	131.6 ± 12.3
Water content of *A. halimus* (%)	69.4 ± 0.7	77.6 ± 1.0	83.6 ± 0.03
ml water/g DM	2.27 ± 0.02	3.46 ± 0.04	5.10 ± 0.00
DMI (g/d)	16.6 ± 1.2	12.1 ± 0.7	12.0 ± 2.1
Scrapings (g DM/d)	2.76 ± 0.85	0.83 ± 0.17	0.10 ± 0.12
% scrapings[a]	14.4 ± 4.7	6.4 ± 1.7	0.8 ± 1.2

[a]% scrapings = [scrapings/(scrapings + DMI)] X 100.

Table 21.2. Ash, Na^+, K^+ and Cl^- intakes and removal by *Psammomys obesus* consuming *Atriplex halimus* leaves of different water content offered in summer, autumn and winter. Values are means ± SD (adapted from Degen 1988).

	Summer	Autumn	Spring
Ash intake (g/d)	4.57 ± 0.33	3.32 ± 0.27	3.60 ± 0.63
Removed (g/d)	1.24 ± 0.38	0.41 ± 0.11	0.05 ± 0.06
% removed[a]	10.88 ± 3.56	4.19 ± 1.19	0.55 ± 0.86
% removed[b]	14.71 ± 4.81	6.33 ± 1.68	0.77 ± 1.20
Na^+ intake (g/d)	0.95 ± 0.08	0.75 ± 0.06	0.77 ± 0.13
Removed (g/d)	0.31 ± 0.10	0.09 ± 0.02	0.01 ± 0.01
% removed[a]	13.00 ± 4.26	5.37 ± 1.43	0.65 ± 1.00
% removed[b]	17.41 ± 5.71	7.21 ± 1.92	0.88 ± 1.35
K^+ intake (g/d)	0.49 ± 0.05	0.43 ± 0.11	0.44 ± 0.08
Removed (g/d)	0.24 ± 0.08	0.07 ± 0.01	0.01 ± 0.01
% removed[a]	21.43 ± 7.16	8.84 ± 2.35	1.07 ± 1.64
% removed[b]	25.01 ± 8.36	10.68 ± 2.84	1.30 ± 1.99
Cl^- intake (g/d)	1.20 ± 0.10	0.89 ± 0.54	1.01 ± 0.18
Removed (g/d)	0.44 ± 0.14	0.13 ± 0.03	0.02 ± 0.02
% removed[a]	14.09 ± 4.62	5.81 ± 1.56	0.71 ± 1.13
% removed[b]	17.83 ± 5.85	7.65 ± 2.05	0.93 ± 1.48

[a] % removed on a dry matter basis.
[b] % removed on an organic matter basis.

Atriplex halimus are lower in organic matter than non-scraped leaves (51 vs. 77% DM), they are also lower in gross energy content (9.0 vs. 13.9 kJ/g DM) and, consequently, consumed leaves are higher in gross energy content than non-scraped leaves by 3.14% when fed leaves are of 77.6% water content (Degen et al. 1988). Air temperature can also affect the amount of leaf scraped off. When maintained at air temperatures between 15°C and 34°C, fat sand rats at 34°C scraped the most leaf material, 14.6% vs. 4.9% at 15°C, and had the highest urine concentration, 4476 vs. 2711 mOsm/kg (Kam and Degen 1992).

2.1 Dry matter digestibility

Because of their dietary habits, it was hypothesized that fat sand rats would be able to digest chenopods and their fibers efficiently and that fiber digestion would contribute substantially to their energy budget. Apparent dry matter digestibility of *Atriplex halimus* averaged 66.1% and apparent digestible and metabolizable energies averaged 64.1% and 62.3%, respectively. Detailed studies were also done on another halophytic chenopod, *Anabasis articulata*. Apparent dry matter digestibility of this plant averaged 62.5% and apparent digestible and metabolizable energies were 56.1% and 55.1%, respectively (Table 21.3).

Table 21.3. Dry matter intake and digestibilities of the halophytes *Atriplex halimus* and *Anabasis articulata* consumed by *Psammomys obesus*. Values are means ± SD, n = 6 for each halophyte (adapted from Degen et al. 1988, 2000).

	Atriplex halimus	*Anabasis articulata*
Body mass (g)	135.4 ± 7.9	131.0 ± 12.0
Dry matter intake (DMI: g/d)	12.1 ± 0.71	10.0 ± 3.6
Faeces (g/d)	4.08 ± 0.37	3.71 ± 1.24
Urine (ml/d)	18.9 ± 0.96	8.9 ± 2.0
DM digestibility (%)	66.1 ± 4.7	62.5 ± 2.1
Gross energy (kJ/g)	14.2 ± 0.06	16.5 ± 0.12
Gross energy intake (GEI: kJ/d)	172.1 ± 8.1	164.6 ± 58.1
Fecal Energy (kJ/d)	61.7 ± 9.9	70.9 ± 22.6
Digestible energy (% GE)	64.1 ± 5.7	56.1 ± 2.9
Digestible energy intake (kJ/d)	110.3 ± 11.4	93.7 ± 35.8
Urine energy (kJ/d)	3.1 ± 0.2	1.3 ± 0.30
Metabolizable energy (% GE)	62.3 ± 5.7	55.1 ± 3.5
Metabolizable energy intake (kJ/d)	107.2 ± 11.4	92.4 ± 35.8

Note: Digestible and metabolizable energies are apparent.

Apparent digestibilities of total fibers (NDF), hemicellulose, and cellulose were 51.6%, 66.4% and 44%, respectively. Fiber digestion was able to fulfill close to all of the BMR requirements of the fat sand rats, one of the highest amounts reported thus far for placental mammals, and above 30% of ADMR requirements.

Table 21.4. Fiber content of *Anabasis articulata* and its digestibility when consumed by *Psammomys obesus*. Values are means ± SD, n = 6 (adapted from Degen et al. 2000).

	Anabasis articulata	Digestibility
Neutral detergent fiber (%)	32.7 ± 3.6	51.8 ± 4.4
Acid detergent fiber (%)	15.2 ± 2.3	35.8 ± 7.5
Hemicellulose (%)	17.6 ± 1.4	66.0 ± 2.6
Lignin (%)	3.8 ± 0.5	−12.9 ± 11.9
Cellulose (%)	11.3 ± 2.5	46.5 ± 9.9

2.2 Basal metabolic rate and average daily metabolic rate

Like other small herbivores consuming a high fiber diet, fat sand rats have a low metabolic rate. BMR, calculated from O_2 consumption, was 281 kJ $kg^{-1}d^{-1}$ or 168 kJ $kg^{-0.75}d^{-1}$, which is 57% of that expected for an eutherian mammal of similar body mass (Degen et al. 1988). The lower than expected BMR was consistent with findings of low BMR for desert rodents (McNab 1979).

Average daily metabolic rate (ADMR) when consuming either *Atriplex halimus* or *Anabasis articulata* was determined by regressing change in body mass (m_b) or change in energy retention (ER) on metabolic energy intake (MEI). ADMR was taken at the point of zero m_b or zero ER change. The regression equation took the form (Degen 1997; Degen et al. 1998):

$$m_b \text{ change (\%/d)} = a + b \text{ MEI (kJ kg}^{-0.75}\text{d}^{-1})$$

The efficiency of utilization of energy for maintenance (k_m) was calculated as BMR/ADMR and the heat increment of feeding for maintenance (HIF_m) was taken as $(1-k_m)$. Heat production (HP) was calculated as (Kam and Degen 1997):

$$HP \text{ (kJ kg}^{-0.75}\text{ d}^{-1}) = MEI \text{ (kJ kg}^{-0.75}\text{ d}^{-1}) - ER \text{ (kJ kg}^{-0.75}\text{d}^{-1})$$

In the regression equation of:

$$ER \text{ change (kJ kg}^{-0.75}\text{d}^{-1}) = a + b \text{ MEI (kJ kg}^{-0.75}\text{d}^{-1})$$

The point of zero ER change equals ADMR and the slope of the line (b) equals k_m when the animals are in negative energy balance (Emmans 1994). When the animals are in positive energy balance, the slope of the line equals the efficiency of utilization of energy for growth (k_f) and ($1-k_f$) equals the heat increment of feeding for growth (HIF_f).

The regression equation of percentage m_b change on MEI for fat sand rats when consuming *Atriplex halimus* was as follows:

$$m_b \text{ change (\%/d)} = -4.353 + 0.0087 \text{ MEI (kJ kg}^{-0.75}\text{d}^{-1})$$

The regression equation for fat sand rats losing body energy when consuming *Atriplex halimus* was as follows:

$$ER \text{ change (kJ kg}^{-0.75}\text{ d}^{-1}) = -161.1 + 0.324 \text{ MEI (kJ kg}^{-0.75}\text{ d}^{-1});$$

and when gaining body energy was:

$$ER \text{ change (kJ kg}^{-0.75}\text{ d}^{-1}) = -147.2 + 0.295 \text{ MEI (kJ kg}^{-0.75}\text{ d}^{-1}).$$

Consequently, ADMR was approximately 499 (kJ $kg^{-0.75}$ d^{-1}), k_m was 0.32, HIF_m was 0.68, k_f was 0.30 and HIF_f was 0.70 (Degen et al. 1988).

The regression equation of percentage body mass (m_b) change on MEI for fat sand rats when consuming *Anabasis articulata* was as follows:

$$m_b \text{ change (\%/d)} = -5.07 + 0.0077 \text{ MEI (kJ kg}^{-0.75}\text{ d}^{-1})$$

The regression equation for fat sand rats losing body energy when consuming *Anabasis articulata* was as follows:

$$\text{ER change (kJ kg}^{-0.75}\,\text{d}^{-1}) = -192.6 + 0.292\ \text{MEI (kJ kg}^{-0.75}\,\text{d}^{-1})$$

From these regressions equations, ADMR was approximately 658.4 kJ kg$^{-0.75}$ d^{-1}, the efficiency of utilization of energy for maintenance, k_m, was 0.29 and HIF$_m$ was 0.71 (Degen et al. 1991).

Because of the low BMR of fat sand rats, it was expected that ADMR would also be relatively low. However, according to the allometric equation generated for rodents, ADMR was about 88% to 110% of that predicted for an animal of its body mass. The relatively high ADMR was a consequence of the high HIF of *Atriplex halimus* and *Anabasis articulata* and low k_m and k_f of these plants.

2.3 Field metabolic rate

Doubly labelled water was used to estimate seasonal water flux, energy expenditure and intake of *Atriplex halimus* of free-living fat sand rats in the Negev Desert (Degen et al. 1991). Rates of energy expenditure in adults were greater in winter than in summer. Mass specific energy expenditure was similar in juveniles and adults and ranged between 0.88 and 1.13 kJ per g daily (Table 21.5), which was 0.90 to 0.99 of that predicted for a desert eutherian. Measurements were taken of juveniles in the winter only.

It was calculated that fat sand rats in the Negev Desert must consume *Atriplex halimus* at a rate of 24.7 to 43.4% of their body mass per day in fresh matter or 9.3 to 11.8% of their body mass per day in dry matter to obtain enough energy to fulfill their requirements. In a study in Tunisia, Ben-Chaouacha et al. (1983) estimated that fat sand rats consumed 59% of their body mass per day in fresh matter, which was slightly higher than the intakes in the Negev Desert. However, a wide range of intakes has been reported for free-living fat sand rats in different studies; 32 to 100% of their body mass daily in fresh matter (Daly and Daly 1973; Happold 1984; Petter 1961).

Table 21.5. Water influx, CO_2 production and daily energy expenditure (FMR) in free living juvenile and adult fat sand rats (*Psammomys obesus*) in winter and summer. (Adapted from Degen et al. 1991).

	Winter		Summer
	Juvenile	**Adult**	**Adult**
Body mass (g)	81.2 ± 2.3	175.7 ± 14.8	165.6 ± 6.1
Water influx (ml/d)	28.4 ± 7.5	41.6 ± 4.3	30.2 ± 4.1
ml g^{-1} d^{-1}	0.351 ± 0.080	0.248 ± 0.008	0.182 ± 0.022
CO_2 production (ml g^{-1} d^{-1})	2.18 ± 0.02	2.03 ± 0.12	1.72 ± 0.39
FMR (kJ/d)	91.8 ± 0.84	184.5 ± 10.9	146.3 ± 13.2
(kJ g^{-1} d^{-1})	1.13 ± 0.02	1.05 ± 0.06	0.88 ± 0.25
Dry matter intake (g^{-1} d^{-1})	9.58	19.26	15.27
(g g^{-1} d^{-1})	0.118	0.110	0.093
Fresh matter intake (g^{-1} d^{-1})	35.01	54.89	40.73
(g g^{-1} d^{-1})	0.434	0.320	0.247

Mass-specific influx of water of fat sand rats in the Negev Desert was higher in winter than in summer (Table 21.5, Degen et al. 1991). It was suggested that the difference in seasonal water influx was a result of differences in preformed water of *A. halimus* consumed. A similar seasonal shift in diet was reported in several Negev Desert rodents (Bar et al. 1984). In addition, water influx reflecting dietary preformed water also was reported in pocket gophers (*Thomomys bottae*, Gettinger 1984) and in antelope ground squirrels (*Ammospermophilus leucurus*, Karasov 1983), both species active during the day. However, all of these animals have a varied diet, feeding on seeds and dry vegetation that have a water content of ca. 8–10% of fresh matter, and on moist vegetation and animal matter that have a content of ca. 70–80% of fresh matter.

The foods eaten by most rodents exhibit seasonal availability, thus influencing maintenance homeostasis of the consumers. This is not the case with fat sand rats that consume only *A. halimus* year-round. *A. halimus* shows a large within-day variation of water content, and, although the plant is usually moister in winter than in summer, there is large overlap in water content between seasons (Degen et al. 1990). Intake of preformed water thus depends on the time of day that the *A. halimus* is consumed and in this way fat sand rats regulate their water influx to some extent. For example, in summer, foraging begins earlier in the day than in winter, a time when water content of the leaves is at a maximum. Fat sand rats are in their burrows at midday. In contrast, in winter, foraging begins later in the day than in summer and fat sand rats are active at midday (Ilan and Yom-Tov 1990).

Calculated water content of the *A. halimus* eaten by fat sand rats in the two seasons was similar, 64.9% in winter and 62.5% in summer. Thus, the increase in water influx in winter was not a consequence of the difference in preformed water in the diet, as reported for other animals, but rather mainly as a consequence of higher energy expenditure of fat sand rats in winter. Adult fat sand rats expended 18% more energy mass-specifically in winter than in summer (1.05 kJ g⁻¹d⁻¹ and 0.089 kJ g⁻¹d⁻¹, respectively); consequently, the greater intake of *A. halimus* was a prime reason for greater water influx.

Juvenile fat sand rats consumed *A. halimus* that possessed a slightly higher water content than that eaten by adults, and juveniles had a higher mass-specific influx of water than adults. Ben-Chaouacha-Chekir et al. (1983) reported a water influx of 0.383–0.406 ml g⁻¹ day⁻¹ in Tunisian fat sand rats weighing 65–87 g and consuming chenopods. These fat sand rats were similar in body mass and had a similar water influx to juveniles in the Negev Desert, which suggests Ben-Chaouacha et al. (1983) also measured juveniles. This also supports the conclusion that juveniles select leaves of higher water content than those selected by adults.

Generally, water influx of fat sand rats was similar to that predicted for herbivores of equal body mass, but tended to be higher than that predicted for eutherian mammals in general, rodents, and desert eutherians. This indicates that water influx often is a reflection of diet selection and that this measurement should be used with caution in assessing the adaptability of an animal to its environment. Furthermore, daily expenditure of energy tended to be lower than that predicted for herbivores and desert eutherians. This was rather surprising because *A. halimus* has a high fiber content and low efficiency of use of its energy for maintenance and growth (Degen et al. 1988; Kam and Degen 1989) suggesting a relatively high energy expenditure for animals consuming this shrub. In contrast, seeds have a low fiber content, high digestibility, and high efficiency of use of energy, and granivorous rodents usually have lower metabolic rates than herbivorous rodents (McNab 1986). The predictive equations of energy expenditure for desert eutherians included several granivorous rodents, yet energy expenditure of fat sand rats was similar or tended to be lower than that predicted for these two groups. Fat sand rats exhibited an energy expenditure lower than that predicted for herbivores indicating that these rodents are well adapted for consuming *A. halimus* as a staple in their diet. In addition, because water influx of fat sand rats tends to be higher than that predicted for herbivores, but their energy expenditure lower than that predicted, suggests that the *A. halimus* consumed by fat sand rats had a higher water content than vegetation consumed by other herbivores.

3. Electrolyte balances and nitrogen requirements

Electrolyte input (either Na^+, K^+ or Cl^-) was calculated as the concentration (g/g DM) of that electrolyte in unscraped *A. halimus* times [scraped *A. halimus* intake (g DM) + leaf scrapings (g DM)] minus the concentration of the electrolyte (g/g DM) in leaf scrapings times leaf scrapings (g). Output of an electrolyte was calculated as the concentration of that electrolyte in faeces (g/g DM) times fecal output (g DM) plus concentration of the electrolyte in urine (g/ml) times urine output (ml). Approximately 81 to 84% of the electrolyte output were eliminated via the urine (Table 21.6).

Table 21.6. The intake of the electrolytes Na^+, K^+ and Cl^- and of nitrogen by a 135 g fat sand rat (*Psammomys obesus*) consuming 12.2 g dry matter of *Atriplex halimus*. Values are means $\pm$ SD, n = 6.

	Na^+	K^+	Cl^-	Nitrogen
Intake (mg/d)	753.5 ± 58.3	426.0 ± 109.2	960.0 ± 93.2	299.3 ± 21.1
Fecal output (mg/g DM)	29.4 ± 11.8	18.4 ±7.4	36.5 ± 13.6	19.7 ± 2.7
(mg/d)	121.8 ± 56.1	76.3 ±33.9	151.4 ± 63.3	79.9 ± 9.5
Urine output (mg/ml)	27.8 ± 3.6	18.6 ± 7.4	43.7 ± 7.7	6.2 ± 1.0
(mg/d)	528.0 ± 87.7	351.7 ± 103.9	823.3 ± 143.1	113.9 ± 15.6
Balance	+103.7 ± 76.7	−1.9 ± 98.4	−14.9 ± 125.6	+105.5 ± 23.6

Because of the high N content of *Atriplex halimus*, the fat sand rats were able to fulfill their N requirements easily (Table 21.6). Minimal N requirements were calculated from the total of metabolic fecal nitrogen (MFN) and endogenous urinary nitrogen (EUN). MFN and EUN were estimated from the point of zero N intake of the linear regressions of fecal-N or urine-N on N intake (Robbins 1983).

The best description of the relationship between fecal-N output and N intake was:

$$\ln \text{fecal-N (mg kg}^{-0.75}\text{d}^{-1}) = 0.0012 \text{ N intake (mg kg}^{-0.75}\text{d}^{-1}) + 4.255$$

and for the relationship between urine-N output and N intake was:

$$\ln \text{urine-N (mg kg}^{-0.75}\text{d}^{-1}) = 0.0008 \text{ N intake (mg kg}^{-0.75}\text{d}^{-1}) + 5.147$$

From the regressions, MFN equalled 70.5 mg kg$^{-0.75}$d^{-1}, EUN equalled 171.9 mg kg$^{-0.75}$d^{-1} and minimal N requirements equalled 242.3 mg kg$^{-0.75}$d^{-1}.

MFN could be dependent on dry matter intake, whereas EUN would depend on basal metabolic rate. Predicted daily MFN for rodents is approximately 2.5 g N/kg DMI and daily EUN for placental mammals is 105 mg N kg$^{-0.75}$d^{-1} (Robbins 1983). Therefore, it would be predicted that a 128.4 g rodent (mean body mass for this study) at maintenance, i.e., consuming 12.2 g DM, would produce an MFN of 30.5 mg d^{-1} or 142.2 mg N kg$^{-0.75}$ d^{-1} and an EUN of 105.0 mg N kg$^{-0.75}$ d^{-1} and would require a minimal N of 247.2 mg N kg$^{-0.75}$ d^{-1}. The calculated requirements for the fat sand rat were 98% of this prediction.

4. General discussion

Herbivory has been studied extensively in ruminants, less so in small mammals. Nonetheless, strict herbivory is a successful strategy used by some small mammals such as voles (Batzli 1985; Hammond and Wunder 1991). In addition, the highest dry matter digestibilities of fibrous diets have been reported in small herbivores. Voles are usually able to achieve dry matter digestibilities of over 70%, whereas among domestic ungulates they range between 50 and 60% (Lee and Houston 1993a,b). When consuming a fibrous diet, the high digestibilities of voles are accomplished mainly through coprophagy and an increase in gut capacity. Few studies are available on digestibility trials comparing desert and non-desert small mammals. In one such study, the desert fat jird (*Meriones crassus*) and the mesic Levant vole (*Microtus guentheri*) had similar digestibilities when offered a high quality forage, but, when offered a poor quality forage, the fat jird had a higher digestibility than the Levant vole (Yahav and Choshniak 1989).

When consuming the chenopod *Anabasis articulata*, a 125 g fat sand rat required 14.9 g dry matter for maintenance which yielded 658.4 kJ kg$^{-0.75}$ d^{-1}. The fat sand rats were able to digest 51.6% of the total fiber (taken as NDF content) in which hemicellulose digestion was 66.4% and cellulose was 44.0%. Assuming 17.6 kJ g^{-1} fiber digested (Southgate and Durnin 1970), then fibers provided 211.7 kJ kg$^{-0.75}$d^{-1}, which was about 32% of the maintenance needs of the fat sand rats. Most of the energy from fiber digestion was derived from hemicellulose.

The higher digestion of hemicellulose than cellulose found in the fat sand rats has been reported for other small non-ruminant herbivorous mammals (Dawson 1989). In general, non-ruminants digest relatively more hemicellulose than cellulose whereas ruminants digest about equal amounts of both polysaccharides (van Soest 1982). Hemicellulose may be hydrolyzed and solubolized in the acids of the stomach. This solubolized hemicellulose, which now is no longer part of the NDF, is transported in the fluid phase of the digesta to the cecum where it undergoes rapid fermentation. Digestion of hemicellulose in this fashion increases the energy derived from herbage without the need of a large fermentation chamber and it is believed that this form of hemicellulose digestion may be widespread in small herbivores. In contrast, it appears that cellulose digestion occurs solely via microbial fermentation (Parra 1978) and that the provision of an adequate fermentation chamber is not a viable option for small mammalian species. Moreover, gerbils grind fibers finely with their teeth. This is followed by fiber separation in the digestive tract where coarse and thick fibers are quickly evacuated whereas the small and thin ones are fermented in the cecum (Naumova et al. 1998).

Fiber digestion can provide a large portion of energy requirements in a number of herbivores. This is especially true for ruminants which are foregut fermenters. In these animals, products of fiber fermentation can provide 70% to 80% of their basal metabolic rate (BMR) requirements (Kempton et al. 1976). In hindgut fermenters, large differences have been reported. For example, energy contributed by fiber digestion can provide from 5% to 7% of the BMR in the dog and rat (Yang et al. 1970; Stevens et al. 1980), 30% in the pony (Glinsky et al. 1976), 33% in the Cape porcupine (Knight-Eloff and Knight 1988) and 87% in the rock hyrax (Rubsamen et al. 1982).

The body energy loss or heat production (HP) of a 125 g fat sand rat at zero energy intake that is close to BMR, was 192.6 kJ $kg^{-0.75}d^{-1}$. This is about 20% higher than BMR measured for fat sand rats consuming *Atriplex halimus* (Degen et al. 1988). If we take 192.6 kJ $kg^{-0.75}d^{-1}$ as being close to BMR, then fiber digestion provided approximately 110% of its BMR.

The efficiency of utilization of energy for maintenance of the chenopods was found to be 0.29 to 0.32. This is relatively low compared to other forages. However, fat sand rats are able to consume large amounts of food, both on a fresh and dry matter basis to meet energy requirements. The fat sand rats were able to digest the fibers efficiently and BMR requirements were wholly covered by the energy derived from fiber digestion. The low efficiency of utilization could be partly due to: (1) fat sand rats have to digest large quantities per unit of energy intake and this results in a high increment of feeding; and (2) the nature of the diet being high in fibers.

Herbivorous rodents generally consume a variety of plants and, when available, select ones with low fiber content. However, unlike most herbivorous rodents, fat sand rats consume only one species of chenopods on which they can thrive, reproduce and females can raise young. The low energy content of the chenopods and the low efficiency of utilization would appear to make this diet undesirable. However, the ability of fat sand rats to consume a single chenopod does have a number of advantages. These are: (1) it is usually readily available throughout the year and provides a more stable diet than do seeds; (2) there are no other gerbils or rodents inhabiting deserts that feed heavily on chenopods, and therefore there is no competition for this food resource; and (3) burrows of fat sand rats are at the base of the chenopod and, therefore, they expend minimal energy in foraging.

This dietary habit of the fat sand rat is also practiced by two other unrelated herbivorous rodent species, showing convergent evolution among three species. The North American heteromyid *Dipodomys merriami* (Kenagy 1972, 1973) and the South American octodontid *Tympanoctomys barrerae* (Mares et al. 1997; Ojeda et al. 1999) can also thrive when consuming only one species of chenopods and they also remove the outer layer of *Atriplex* leaves to reduce electrolyte intake. Thus, this unique behavior of feeding on one species of chenopods has proven successful in deserts in several continents.

References

Adler, J.H., R. Kalman, G. Lazarovici, H. Bar-On and E. Ziv. 1990. Achieving predictable model of Type 2 diabetes in sand rats. pp. 212–214. *In*: E. Shafrir (ed.). Frontiers in diabetes research II. Lessons from animal diabetes III. Smith-Gordon, London.

Bar, Y., Z. Abramsky and Y. Gutterman. 1984. Diets of gerbilline rodents in the Israeli desert. Journal of Arid Environments 7: 371–376.

Batzli, G.O. 1985. Nutrition. pp. 799–806. *In*: R.H. Tamarin (ed.). Biology of New World *Microtus*. Special Publication, The American Society of Mammalogists.

Ben-Chaoucha-Chekir, R., F. Lachiver and T. Cheniti. 1983. Donnees preliminaires sur le taux de renouvellement d'eau chez un gerbillide desertique, *Psammomys obesus*, etudie son environment naturel en Tunisie. Mammalia 47: 543–546.

Borges, P.A., M.G. Dominguez-Bello and E.A. Herrera. 1996. Digestive Physiology of wild capybara. Journal of Comparative Physiology B, Biochemical Systematic and Environmental Physiology 166: 55–60.

Bozinovic, F. 1995. Nutritional energetics and digestive responses of an herbivorous rodent (*Octodon degus*) to different levels of duetary fiber. Journal of Mammalogy 76: 627–637.

Bozinovic, F., F.F. Novoa and P. Sabat. 1997. Feeding and digesting fiber and tannins by an herbivorous rodent *Octodon degus* (Rodentia: Caviomorpha). Comparative Biochemistry and Physiology 118A: 625–630.

Daly, M. and S. Daly. 1973. On the feeding ecology of *Psammomys obesus* (Rodentia: Gerbillinae) in the Wadi Saoura, Algeria. Mammalia 37: 545–561.

Dawson, T.J. 1989. Food utilization in relation to gut structure and function in wild and domestic birds and mammals. Acta Veterinaria Scandinavica, Supplementum 86: 20–27.

Degen, A.A. 1988. Ash and electrolyte intakes of the fat sand rat, *Psammomys obesus*, consuming saltbush, *Atriplex halimus*, containing different water content. Physiological Zoology 61: 137–141.

Degen, A.A. 1997. Ecophysiology of small desert mammals. Springer-Verlag, Berlin Heidelberg, pp. 296.

Degen, A.A., A. Hazan, M. Kam and K.A. Nagy. 1991. Seasonal water influx and energy expenditure of free-living fat sand rats. Journal of Mammalogy 72: 652–657.

Degen, A.A., M. Kam, I.S. Khokhlova, B.R. Krasnov and T.G. Barraclough. 1998. Average daily metabolic rate of rodents: habitat and dietary comparisons. Functional Ecology 12: 63–73.

Degen, A.A., M. Kam and D. Jurgrau. 1988. Energy requirements of fat sand rats (*Psammomys obesus*) and their efficiency of utilization of the saltbush *Atriplex halimus* for maintenance. Journal of Zoology (London) 215: 413–452.

Degen, A.A., M. Kam, I.S. Khokhlova and Y. Zeevi. 2000. Fiber digestion and energy utilization of fat sand rats (*Psammomys obesus*) consuming the chenopod *Anabasis articulata*. Physiological and Biochemical Zoology 73: 574–580.

Degen, A.A., B. Pinshow and M. Ilan. 1990. Seasonal water flux and urine and plasma osmotic concentrations in free-living fat sand rats feeding solely on saltbush. Journal of Arid Environments 18: 59–66.

Demment, M.W. and P.J. van Soest. 1985. A nutritional explanation for body size patterns of ruminant and nonruminant herbivores. American Naturalist 125: 641–672.

Emmans, G.C. 1994. Effective energy: a concept of energy utilization applied across species. British Journal of Nutrition 71: 801–821.

Gandacu, D., Y. Glazer, E. Anis, I. Karakis, B. Warshavsky, P. Slater and I. Grotto. 2014. Resurgence of Cutaneous Leishmaniasis in Israel, 2001–2012. Emerging Infectious Diseases [Internet]. http://dx.doi.org/10.3201/eid2010.140182.

Gettinger, R. 1984. Energy and water metabolism of free-ranging pocket gophers, *Thomomys bottae*. Ecology 65: 740–751.

Foley, W.J. and S.J. Cork. 1992. Use of fibrous diets by small herbivores: how far can the rules be "bent"? Trends in Ecology and Evolution 7: 159–162.

Glinsky, M.J., R.M. Smith, H.R. Spires and C.L. Davis. 1976. Measurement of volatile fatty acid production rates in the cecum of the pony. Journal of Animal Science 42: 1465–1470.

Hackel, D.B., E. Mikat, H.E. Lebovitz, K. Schmidt-Nielsen, E.S. Horton and T.D. Kinney. 1967. The sand rat (*Psammomys obesus*) as an experimental animal in studies of diabetes mellitus. Diabetologia 3: 130–134.

Haken, A.E. and G.O. Batzli. 1996. Effects of availability of food and interspecific competition on diets of prairie voles (*Microtus ochrogaster*). Journal of Mammalogy 77: 315–324.

Hammond, K.A. and B.A. Wunder. 1991. The role of diet quality and energy need in the nutrional ecology of a small herbivore, *Microtus ochrogaster*. Physiological Zoology 64: 541–567.

Happold, D.C.D. 1984. Small Mammals. pp. 251–275. *In*: J.L. Cloudsley-Thompson (ed.). Sahara Desert, Permagon Press Limited, London.

Hofmann, R.R. 1989. Evolutionary steps of ecophysiological adaptation and diversification of ruminants: a comparative view of their digestive system. Oecologia 78: 443–457.

Hume, I.D. 1989. Optimal digestive strategies in mammalian herbivores. Physiological Zoology 62: 1145–1163.

Ilan, M. and Y. Yom-Tov. 1990. Diet activity pattern of a diurnal desert rodent, *Psammomys obesus*. Journal of Mammalogy 71: 66–69.

Justice, K.E. and F.A. Smith. 1992. A model of dietary fiber utilization by small mammalian herbivores, with empirical results for Neotoma. American Naturalist 139: 398–416.

Kam, M. and A.A. Degen. 1988. Water, electrolyte and nitrogen balances of fat sand rats (*Psammomys obesus*) consuming the saltbush *Atriplex halimus*. Journal of Zoology, London 215: 453–462.

Kam, M. and A.A. Degen. 1989. Efficiency of use of saltbush (*Atriplex halimus*) for growth by fat sand rats (*Psammomys obesus*). Journal of Mammalogy 70: 485–493.

Kam, M. and A.A. Degen. 1992. Effect of air temperature on energy and water balance of *Psammomys obesus*. Journal of Mammalogy 73: 207–214.

Kam, M. and A.A. Degen. 1993. Energetics of lactation and growth in the fat sand rat, *Psammomys obesus*: new perspectives of resource partitioning and the effect of litter size. Journal of Theoretical Biology 162: 353–369.

Kam, M. and A.A. Degen. 1994. Body mass at birth and growth rate of fat sand rat (*Psammomys obesus*) pups: effect of litter size and water content of *Atriplex halimus* consumed by pregnant and lactating females. Functional Ecology 8: 351–357.

Kam, M. and A.A. Degen. 1997. Energy requirements and the efficiency of utilization of metabolizable energy in free-living animals: evaluation of existing theories and generation of a new model. Journal of Theoretical Biology 184: 101–104.

Kam, M., I.S. Khokhlova and A.A. Degen. 1997. Granivory and plant selection by desert gerbils of different body size. Ecology 78: 2218–2229.

Karasov, W.H. 1983. Water flux and water requirements in free-living antelope ground squirrels *Ammospermophilus leucurus*. Physiological Zoology 56: 94–105.

Karasov, W.H. 1989. Nutritional bottleneck is a herbivore, the desert wood rat (*Neotoma lepida*). Physiological Zoology 62: 1351–1382.

Karasov, W.H. and J.M. Diamond. 1988. Interplay between physiology and ecology in digestion. BioScience 38: 602–611.

Kempton, T.J., R.M. Murray and R.A. Leng. 1976. Methane production and digestibility measurements in the grey kangaroo and sheep. Australian Journal of Biological Sciences 29: 209–214.

Kenagy, G.J. 1972. Saltbush leaves: excision of hypersaline tissue by a kangaroo rat. Science 178: 1094–1096.

Kenagy, G.J. 1973. Adaptations for leaf eating in the Great Basin kangaroo rat, *Dipodomys microps*. Oecologia 12: 383–412.

Kleiber, M. 1975. The fire of life: an introduction to animal energetics. Kreiger, Huntingdon, New York, pp. 453.

Knight-Eloff, A.K. and M.H. Knight. 1988. Volatile fatty acid production in the Cape porcupine. South African Journal of Wildlife Research 18: 157–159.

Lee, W.B. and D.C. Houston. 1993a. The role of coprophagy in digestion in voles (Microtus agrestis and Clethrionomys glareolus). Functional Ecology 7: 427–432.

Lee, W.B. and D.C. Houston. 1993b. The effect of diet quality on gut anatomy in British voles (Microtinae). Journal of Comparative Physiology B 163: 337–339.

Mares, M.A., R.A. Ojeda, C.E. Borghi, S.M. Giannoni, G.B. Diaz and J.K. Braun. 1997. How desert rodents overcome halophytic plant defenses. BioScience 47: 699–704.

McNab, B.K. 1979. Climatic adaptation in the energetics of heteromyid rodents. Comparative Biochemistry and Physiology 62: 813–820.

McNab, B.K. 1986. The influence of food habits on the energetics of eutherian mammals. Ecological Monographs 65: 1–19.

Mendelsshon, H. and Y. Tom-Tov. 1987. Plants and animals of the land of Israel. Volume 7, Mammals, pp. 112–114. Ministry of Defense Publishing House, Tel Aviv, Israel (in Hebrew).

Naumova, E.I., G.K. Zharkova and N.G. Nesterova. 1998. New approaches to investigate feeding and digestive systems in herbivorous mammals. Zoologicheskii Zhurnal 77: 20–29 (in Russian with English summary).

Nowak, R.M. and J.L. Paradiso. 1983. Walker's Mammals of the World. Fourth edition. The Johns Hopkins University Press.

Ojeda, R.A., C.E. Borghi, G.B. Diaz, S.M. Giannoni, M.A. Mares and J.K. Braun. 1999. Evolutionary convergence of the highly adapted desert rodent *Tympanoctomys barrerae* (Octodontidae). Journal of Arid Environments 41: 443–452.

Orr, T. 1972. The underground burrow of the fat sand rat. Teva Va'aretz (magazine of the Israel Nature Protection Society) 6: 280–284 (in Hebrew).

Owl, M.Y. and G.O. Batzli. 1998. The integrated processing response of voles to fibre content of natural diets. Functional Ecology 12: 4–13.

Palgi, N., H. Taleisnik and B. Pinshow. 2008. Elimination of oxalate by fat sand rats (*Psammomys obesus*): Wild and laboratory-bred animals compared. Comparative Biochemistry and Physiology, Part A 149: 197–202.

Palgi, N., I. Vatnick and B. Pinshow. 2005. Oxalate, calcium and ash intake and excretion balances in fat sand rats (*Psammomys obesus*) feeding on two different diets. Biochemistry and Physiology, Part A 141: 48–53.

Parra, R. 1978. Comparison of foregut and hindgut fermentation in herbivores. pp. 123–130. *In*: G.G. Montgomery (ed.). The Ecology of Arboreal Folivores. Smithonian Institute Press, Washington.

Petter, F. 1961. Repartition geographique et ecologie des rongeurs desertiques (du Sahara occidental a l'Iran oriental). Mammalia 25: 1–222.

Robbins, C.T. 1983. Wildlife feeding and nutrition. Academic Press, New York.

Rubsamen, K., I.D. Hume and W. von Engelhardt. 1982. Physiology of the rock hyrax. Comparative Biochemistry and Physiology 72A: 271–277.

Scherzer, P., S. Katalan, G. Got, G. Pizov, I. Londono, A. Gal-Moscovici, M.M. Popovtzer, E. Ziv and M. Bendayan. 2011. *Psammomys obesus*, a particularly important animal model for the study of the human diabetic nephropathy. Anatomy and Cell Biology 44: 176–185.

Schmidt-Nielsen, K., H.B. Haines and D.B. Hackel. 1964. Diabetes mellitus in the sand rat induced by standard laboratory diets. Science 143: 689–690.

Southgate, D.A.T. and J.V.G.A. Durnin. 1970. Calorie conversion factors. An experimental assessment of the factors used in the calculation of the energy value of human diets. British Journal of Nutrition 24: 517–535.

Stevens, C.E. 1989. Evolution of vertebrate herbivores. Acta Veterinaria Scandinavica, Supplementum 86: 9–19.

Stevens, C.E., R.A. Argenzio and E.T. Clemens. 1980. Microbial digestion: rumen versus large intestine. pp. 685–706. *In*: Y. Ruckebusch and P. Thivend (eds.). Digestive Physiology and Metabolism in Ruminants. MTP Press, Lancaster, U.K.

van Soest, P.J. 1982. Nutritional ecology of the ruminant. O & B Books, Inc., Corvallis, Oregon.

Veloso, C. and F. Bozinovic. 1993. Dietary and digestive constraints on basal metabolism in a small herbivorous rodent. Ecology 74: 2003–2010.

Yahav, S. and I. Choshniak. 1989. Energy metabolism and nitrogen balance in the fat jird (*Meriones crassus*) and the Levant voles (*Microtus guentheri*). Journal of Arid Environments 16: 315–332.

Yang, M.G., K. Manoharan and O. Mickelsen. 1970. Nutritional contribution of volatile fatty acids from the cecum of rats. Journal of Nutrition 100: 545–550.

Halophytes and Salt Tolerant Forage Processing as Animal Feeds at Farm Level: Basic Guidelines

H.M. El Shaer

SYNOPSIS

This chapter aims to introduce some basic guidelines on processing salt tolerant forages and halophytes to produce high quality livestock feeds, particularly at the farm level. Farm-scale treatment methods are those that can be used on the individual farm. They are simple and equipment costs can usually be paid by the individual farmer. The shortage of feed resources is considered a main obstacle for livestock development since farm animals don't get the necessary food requirements to have real productive efficiency. The daily ration of nutrients that an animal receives from a feed may vary from time to time due to a number of reasons. The sources of variation will probably cause variation in the day-to-day level of nutrition received by an individual animal. The presence of certain nutrients, at minimum levels, is mandatory and regulatory officials will be concerned if those required levels are not met. Certain ingredients may be highly toxic. So, it is important to focus on a simplified model of the feed procedures and processing to supply animal feed from traditional or non-traditional feed materials, e.g., halophytes and salt tolerant forages (HSTF), whether natural or cultivated on marginal saline resources.

Desert Research Center, Cairo, Egypt.
Email: hshaer49@hotmail.com

Keywords: feedstuffs, silage, hay, supplements, processing equipment, feedblocks, lick, pellets.

1. Introduction

Natural resources in the Arab region are facing severe degradation. This includes the degradation of land, and plant and animal water resource. Degradation has led to the reduction of biodiversity, and the spread of desertification in the Arab region has reached to about 68.4% and could spread across 20% more in the dry and arid regions (AOAD 2012). Degradation has led to the extinction of vegetation in large areas which in turn leads to the reduction of wood production, energy, animal feed, poverty and raises the pressure on natural resources. In addition, the phenomenon of salinity is affecting large areas of land in Arab countries, particularly the irrigated lands in arid and semi-arid zones. According to the Annual Statistics of AOAD (2012), the studies showed that about 50% of the irrigated land in the Arab world is affected by the spread of salinization.

Natural pastures are the main feed resources in the Arab world. Natural pastures in the Arab region are characterized by arid and dry climate, low rainfall which fluctuates geographically and seasonally and additionally to the prevalence of dry and undeveloped soil. This situation leads to poor, simple and low productivity vegetation. Despite this prevailing situation, natural pastures provide livestock with approximately 39% of food needs with variables between 20 and 100% depending on the properties of the region and its rainfall (statistics from AOAD 2006). Thus, it is contributing effectively in supplying the Arab citizens with meat, milk and fiber, as this region has large numbers of livestock that was estimated by AOAD in 2012 as follows: the total numbers (1000 heads), were: 54164.5, 4304.51, 177036.81, 88981.7 and 16112.24 heads for cattle, buffaloes, sheep, goats and camels, respectively. The total production (1000 MT) was estimated to be 8605.84, 4900.51, 3705.33, 26147.54, 1804.07 and 4185.24 MT, for total meat, red meat, white meat, milk, eggs and fish, respectively. Animal feed is one of the most important factors in animal production investments because a reduction in the cost of animal feed leads to increasing the profit of animal production investors. Livestock development in the Arab world is imperative to cover the current lack of animal products.

2. Utilization of halophytes and salt tolerant forages in animal feeding

Simply stated, animal feed is food given to animal in the course of animal husbandry. There are two basic types of feed, fodder and forage; the word "feed" more often refers to fodder. Traditional sources of animal feed include, generally, different types of roughages such as straw, hay, forages, agricultural by-products and several types of concentrates such as grains, protein concentrates, energy concentrates, industrial— byproducts (food processing industries such as milling and brewing), etc. Such feed materials could be offered to animals as a sole feed stuff or as a compound feed depending on many factors related to both animals and types of feed. Compound feed is a fodder that is blended from various raw materials and additives. These blends are formulated according to the specific requirements of the target animal, e.g., pellets (Fig. 22.1).

Fig. 22.1. Pellets can be of various sizes and composition for small ruminants, poultry or rabbits.

It is worth mentioning that natural pasture and fibrous crop are the main animal feed resources for the smallholder for their survival, growth, reproduction and production. Since quality and quantity of such feed materials vary with season, animals dependent on it are subjected to nutritional stress in the dry season when feed resources are senesced and in short supply leading to decreased animal productivity. On the other hand, several halophytes and salt tolerant forages are characterized by great biomass yield but are less palatable for many livestock species. Therefore, it would be better to process such plants into valuable animal feeds in order to:

- improve the food value and consequently feed utilization by animals.
- store the superabundance of fresh HSTF production to be used during feed shortage seasons or it could be sold to other farmers.

The most common and dominant HSTF species are:

1. Annual forage species such as: *Sorghum* spp., pearl millet (*Pennisetum glaucum*), rye grass (*Lolium perenne*), Sudan grass (*Sorghum halepense*), cow pea (*Vigna unguiculata*), guar (*Cyamopsis tetragonoloba*), berseem (*Trifolium alexandrinum*), fodder beet (*Beta vulgaris*) and wheat, barley and triticale for straws and grains.
2. Perennial forage species such as: alfalfa (*Medicago sativa*), *Medicago arborium*, *Panicum* sp., *Kochia* sp., etc.
3. Fodder shrubs: *Atriplex* sp., *Sesbania*, *Leucaena*, *Tamarix* sp., *Salsola* sp., *Nitraria retusa*, *Zygophyllum album*, *Acacia* sp., *Juncus* sp., *H. strobilaceum*.

3. Manufacturing (processing) halophytes and salt tolerant fodders (HSTF) as animal feed

3.1 Benefits of processing halophytes and salt tolerant fodders

It is known that several halophytes and slightly salt tolerant fodders contain some physical and chemical materials that limit and constrain its palatability and utilization by animals (Attia-Ismail, this volume; Masters, this volume). The main constraints are:

- High ash content (minerals).
- High fiber content (in particular lignin and hemi-cellulose).
- Low protein and energy contents.
- High presence of secondary components produced by plants, such as Tannins, Saponins, Alkaloids, Nitrates, etc. which have a direct impact on the processes of digestion in animals.
- Physical materials such as thorns.

Therefore, it is recommended that the nutritive value of HSTF be determined by chemical analysis before giving to animals to determine the feed quality for better utilization (Norman, Hulm and Wilmot, this volume). These feed materials could be given to animals as the sole fresh diet or mixed with different fresh forage in proportions or could be processed using a specific processing treatment according to HSTF which will be explained later in this chapter. In other words, cultivated HSTF (annual, perennial, fodder shrubs) with good palatability for most animal species and with high nutritional value is given to animals directly when available and can also be saved in the form of silage or hay to be used during other periods of the year or can be mixed with other feed ingredients.

Therefore, the processing of HSTF feed materials, whether cultivated, or naturally grown, in different shapes such as silage, feed cubes, hay, etc. have many benefits namely:

- Utilization of natural unpalatable halophytes or those less palatable with large biomass.
- Improvement of the nutritional value and palatability of forage plants with low nutritional value and palatability. Providing balanced nutritional feed all year round with economical costs.
- Decrease feeding costs by at least 35% and thereby increasing the profit of livestock owners.

3.2 Methods of manufacturing halophytes and salt tolerant forages at farm scale

The manufacture of HSTF into different types of animal feed (such as hay, silage, fodder cubes, etc.) can be achieved through several stages according to the type of feed product. Methods of treating HSTF may be classified broadly into physical, chemical and biological categories. Farm-scale treatment methods are those that can be used on the individual farm. They are simple and equipment costs can usually be paid by the individual farmer. The methods vary in their effectiveness and the reasons for this will be highlighted in the following paragraphs.

All manufacturing operations begin with cutting and shredding plants as a basic step, then grinding into smaller particles 3–5 cm in length (Fig. 22.2).

These chopped materials will be ready to be used for the proper processing forms as silage, feedblocks, hay, etc. For example, in the case of silage production the partial drying and partial wilting has to be done first. Then the remaining steps in manufacturing silage will be undertaken. In the case of grinding plants into other forms such as hay or treatment with urea or feed cubes, the final air drying process must take place (20% is the maximum moisture in chopped dried plants). Then it is crushed as in the case of processing fodder blocks or left to dry in case of hay production until processing into other ways, such as the treatment with urea, molasses, etc. However, the fresh HSTF can be simply processed through: (1) Chemical treatments, (2) physical and mechanical treatments and (3) biological treatments. The following fresh HSTF processing will be briefly and simply, explained herein.

Fig. 22.2. Chopping fodder into lengths about 3–5 cm is first step for making many conserved feedstuffs, especially silage.

4. Treatment options

4.1 Chemical treatments

Chemical methods currently being used include alkali treatment. This method may be classified as wet methods using NaOH or dry methods. The dry methods at farm scale treatment could be used as daily treatment using NaOH, $Ca(OH)_2$, NH_3, or bulk treatment:

1. followed by stacking (NaOH),
2. followed by ensiling (NaOH, $Ca(OH)_2$),
3. followed by stacking under plastic sheet (NH_3).

Wet methods involve soaking straw in 10 liters of chemical solution per kg of dry straw. In dry methods straw is sprayed with 0.1 to 3 l of chemical solution per kg of dry straw or exposed to ammonia vapor. Wet methods employ a higher ratio of chemical to chopped dried plants than do the dry methods.

However, chemical treatments are not very acceptable to farmers due to high operational costs, particularly of chemicals and unavailability of these chemicals at farm levels.

4.2 Physical and mechanical treatments

Harvesting operation can be done manually by workers or by using the serration-mower-scissors such as pruning scissors-cutlery for cutting the top of naturally growing plants, especially their terminal thin parts. The cultivated plants should be harvested at approximately 15 cm above the ground using one of the above mentioned tools or agricultural machinery. Harvested parts are assembled for use in chopping operations that is done manually by using sharp knives or by machines. The fresh chopped materials should have identical lengths (3–5 cm tall) and become ready for drying and processing of different types of feed that will be mentioned later. Generally, the process of chopping or shredding and grinding helps in increasing the palatability of HSTF as it reduces the concentrations of fibrous materials and some plant secondary components such as oxalate, tannins, etc. See Attia-Ismail in this volume for discussion of the effects of these antinutritional components.

4.2.1 Drying process

It is worthy to mention that some fresh feed materials can be fed in the shredded form while other shredded ones could be dried aerobically either in the sun or in the shade, preferably in the shade. Drying process often improves the nutritional value of salt forage plants such as fodder beet leaves. The degree of drying depends on the manufactured feed product. For example, in the case of silage processing the drying is partial (partially wilt with humidity ranging between 60–70%), while in case of other processing such as hay production, ammonia and urea treatments and feedblocks, drying should be complete (humidity must not exceed 10%).

Air drying (dehydration) treatment for hay production: The primary objective of drying (dehydration) is to reduce the moisture content of green plants to (15–20%) to stop the activity of plant enzymes, bacteria and fungi on the plant nutrients so it can be stored without rotting or lessen its nutritional value. Hay is the material produced from drying green fodder; it is one of the simplest practical methods to store forages in dry hot areas. The timing of cutting cultivated green forge to make hay is very important. Early cutting of green fodder produces a crop with low dry matter as a result of high humidity, while delay in cutting plants increases defoliation due to dryness and the high proportion of fiber and lignin resulting in second-rate hay. So, the best time to cut green forage for hay is when it reaches the stage of growth that gives the largest quotient of the total indigestible food and not the biggest crop of dry matter (DM).

Methods of hay making: There are several techniques to make hay, namely: (1) Natural drying (dehydration) and (2) Artificial or the so called industrial drying. Artificial Drying means that the forage plants are exposed to a stream of air until it they dry. It includes: (A) Drying in the field and barn, and (B) Heat drying. The artificial drying process has advantages and disadvantages. Artificial drying advantages are: saving of time and hay keeps most green feed ingredients. On the other hand, the disadvantages of artificial drying are high expenses at farm levels and loss of some nutritional value as a result of high temperatures. Natural drying (dehydration) process is a common, simple and cheap method used by farmers' on a large scale. Therefore, natural drying (dehydration) process will be stressed in this chapter.

Natural drying (dehydration) process: There are different methods of natural drying in the field which are explained as follows:

A. **On Ground:** Spread the cut parts of HSTF in the field in thin layers for 4–5 days to wilt and lose portions of humidity. Stir twice a day and move it to a spacious place in the early morning with some dew to keep and preserve the leaves. The dried materials are collected in small piles with diameter 1.5–2 meters long and 1 meter height and left for 2–3 day. Turn over the pile till its base becomes dry. After two days, drying process is done by gathering the small piles into large square piles with dimensions (8 x 8 m) and 3 m height, placed on a layer of maize or cotton stalks or other field waste about 30 cm thick to save the hay from ground moisture or any contaminants. To facilitate ventilation inside the pile make vertical holes by placing bundles of wood with about 60 cm diameter with 2–3 m distance between them. These are removed after constructing the pile leaving gaps which help in ventilation. It is better to cover the pile with a layer of wood or straw to protect it from the climatic elements.

B. **On Tripods:** Cut parts of plants are spread in thin layers on the floor for 4–5 days until they are tender and have lost part of their moisture and then placed on wooden tripods for two weeks until they are dry. They are collected and pressed into bales that can be stored until summer. This method is better than ground drying because of the following: (1) suitability for rainy and winter seasons, (2) fast drying as a result of the air passing through the green plants and (3) the production of good quality hay.

However, several practices can cause loss of the nutritional value during dehydration process, mainly:

- *Mechanical loss:* Leaves dry up earlier than stems and fall off. Since leaves are richer in contents of energy, protein, vitamins and minerals, the dried plants must be transferred carefully in the early morning with some dew which reduces the mechanical loss.
- *Sun exposure:* Excessive exposure of forages to sun's rays leads to losses of natural green color and the loss of carotene.

High quality of hay production can be identified by the following signs:

(1) Green color, (2) Presence of a large proportion of leaves, (3) Good aroma and flavor, (4) Palatable and acceptable to animals and (5) Free of mould, clay, gravel and others foreign materials.

4.2.2 Grinding or mashing process

The grinding or mashing of dried plants is done by a Hammer mill (Grinder) for manufacturing feed cubes only, since the lengths of plants must not exceed 3 cm. With most of dried forages diets, grinding causes an increase in intake and weight gain. These effects are greater for fodders with lower digestibility. The improvement in net energy value, because of grinding treatment, may be more marked with poor quality fodder crops or straws.

4.2.3 Mixing process

Chopped or ground feed materials are mixed to form a homogeneous mixture of feed materials which will be used in the production of different types of feeds. Mixing process could be conducted manually by hands or using a special mixing machine.

4.2.4 Feed blocks processing

It means mixing some ground dried feed ingredients then pressing them into blocks (see Fig. 22.3) for improving palpability and the nutritional value of halophytes, salt tolerant fodders and agricultural residues as animal feed. Such feeds can be stored for a long time.

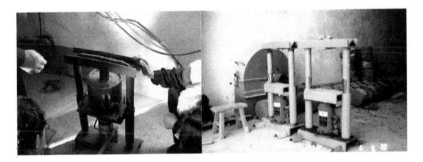

Fig. 22.3. Farm-scale manufacture of feed blocks uses simple equipment.

Types of feed blocks: Feed block can be made in two forms:

1. *Concentrates*: blocks of concentrates that include some agro-industrial by-products such as crushed date stones, olives by-products, wheat bran, etc. with additions of any other materials like molasses, urea, mineral salts and vitamins.

2. *Roughage or enriched roughage with concentrates*: Blocks that are made from mixing any chopped dried forage and agriculture residues materials with wheat bran and/or with addition of some minerals and other nutrients. Bonding or cementing materials (see below) should be added to any sort of feed blocks.

Method of manufacturing feed blocks:

1. *Cutting*: using slicers as already mentioned above.
2. *Drying*: drying the cut forage materials that will be manufactured. Air drying is preferred to solar drying as moisture content do not exceed 20%.
3. *Chopping*: dried forage ingredients must be chopped or ground with lengths from 1 to 3 cm 3. Mixing: mixing ingredients very well to homogenize feed materials and can be done manually or by a mixer (urea should be dissolved in a small amount of water before it is added. Molasses are added last by spraying onto the ingredients and mixing them well), add the bonding materials direct before pressing and mix well. Moisture content of the mixture must be tested as it should be around 65%.
4. *Pressing*: mixed ingredients are pressed by using simple manual Pressing Units (Fig. 22.3).
5. *Drying*: leave the blocks in the shade for a week in summer or two weeks in winter. Turn the fresh blocks over after 24 hours of pressing. The blocks are left on the ground (on concrete or rice-straws) for air dehydration.
6. When completely dry, it should be stored in dry area under the shade for later usage.

Bonding material: There are many bonding materials that could be added to produce feed blocks. For every 100 kg of feed ingredients mixture:

1. 5 kg cement or 5 kg quicklime (concentrate blocks) or 10 kg wheat flour or rice bran (integrated feed blocks).
2. 4 kg of clay can be added to all sorts of blocks as it has other benefits such as absorbing water, reducing diarrhea and intestinal disorders and improving digestion and processing efficiency.

Feed blocks features: (1) Ease of transportation and storage (2) Simplicity and ease of blocks process (no need to complex processing or techniques) (3) Feeding animals on blocks will force animals to consume the whole diet without any feed materials selection (4) Possibility of adding urea, molasses and any other good feed additives. (5) Possibility of being used to feed animals on natural rangelands as feed supplement.

4.3 Biological treatments

The possibility of biological treatments (methods) of HSTF materials has a great appeal as an alternative to the use of expensive (in terms of money and energy) chemicals. Several biological methods have been developed such as producing single-cell protein using straws as an energy source, either directly, by cultivating cellulolytic organisms on it, or indirectly by hydrolyzing its polysaccharides chemically or enzymatically and using the resulting monosaccharides to feed yeasts. Growing yeasts on chemically degraded straws and then feeding the degraded straw along with the yeast to cattle is still under research. However, Fodder conservation through other simple methods such as urea and molasses solution treatment and silage are broadly recommended among farmers due to simplicity and economic costs; therefore these treatments will be stressed in the following paragraphs.

4.3.1 Urea and molasses solution treatment

This treatment is considered the most common of techniques that is used widely in feeding ruminant farm animals because it is a simple technique that any farmer can do without the need for additional manufacturing units. Treatment of poor nitrogen and energy salt tolerant and halophytes fodders with urea and molasses solution would enhance its nutritive value and acceptability by animals as a result of doubling the proportion of protein and breaking the ligno-cellulosic bonds.

1. ***Urea and molasses solution treatment could be conducted as follows:*** (1) Stack the chopped dry plant mixture in layers making a pile; sprinkling on each layer the solution of prepared urea (each 100 kg of the substance needs 4 kg urea melted in 50 liters of water in which 5 kg of molasses is dissolved).
2. Cover this pile with plastic to prevent leakage of ammonia which will be the outcome of urea breakdown and leave the pile covered completely for 2–3 weeks.
3. Then remove the cover from the treated chopped forage plants gradually, to be offered to animals gradually. Notice that the treated pile can be made on the surface of the ground directly or in a trench (1 m x 1 m x 1 m) or silos (building walls as barriers 2 m length and 1 m high).

4.3.2 Liquid nutrient treatment

Liquid nutrients process is used to improve the palatability and feeding value of feed for salt tolerant forages and halophytes fodders by the enrichment with missing nutritional elements, such as energy, non-protein nitrogen and mineral salts. It is a liquid such as molasses and is rich in energy and mineral elements. It is added to chop dried halophytes and salt tolerant fodders that are poor in nitrogen and energy and to other roughages such as agricultural by-products. The liquid nutrient is consisted of molasses (91%), as a source of energy; urea (3.5%), as a source of non-protein nitrogen; mixed minereal elements (0.5%), manganese, cobalt, copper, magnesium, zinc, iodine, potassium and phosphorus; diluted with the urea in 5% water. The liquid mixture is prepared in a special mixing unit.

4.3.2.1 The benefits of Liquid nutrient treatment for the breeders

How to use Liquid nutrient treatment in feeding animals to enable use of large quantities of unpalatable halophytes in animal nutrition after being processed.

*   Ensure the coverage of a part of animal nutritional needs of energy and necessary mineral elements with the appropriate proportions.

- Increase milk production in dairy animals, and growth rates in fattening animals.
- Increase diseases resistance that arises from lack of mineral elements.
- Sprinkling the mixture on low-nitrogen and energy halophytes hay encourages animals to consume larger quantities which reduce feed costs especially in low productivity animals. It is preferable that these animals consume greater amounts of fillers such as or hay or corn stalks.
- Reduce nutrition costs as 0.75 kg of feed can be reduced per day when adding the liquid feeder without any harmful impact on milk or meat production. It is known that each 0.5 liters of feed liquid is worth approximately 0.5 kg of concentrated feed as it contains indigestible energy.

1. Sprinkle the nutritious liquid on low nitrogen and energy chopped dried halophyte and other salt tolerant fodders as well as corn stalks that are cut with a tube that has got holes from the bottom to ensure good distribution on hay and encourages animal to consume hay in front of it.
2. The recommended amounts in animal feeding are: (1) around 0.5–1 kg for large animals; twice daily (morning and evening) and (2) approximately 0.75–1 kg for sheep and goats for every five heads twice daily morning and evening.

Precautions to be observed when using Liquid nutrient treatment:

1. It must be given gradually by spraying small amounts on coarse fodder such as hay halophytes then increasing up to the recommended amount.
2. Must be stored in clean and dry top covered pots to prevent insects and flies from falling in.
3. Animals should not be fed on nutritious liquid directly but gradually and not added to drinking water.

Do not use liquid nutrient treatment in case of baby calves; it can be used after a month of weaning.

4.4 Ensiling (anaerobic fermentation) process

The process of anaerobic fermentation for silage production is a common treatment nowadays among farmers. The process of keeping the green forage materials is called ensiling. The place to store the silage is called a silo. Silage making is feasible under conditions when green fodder is abundantly available over a short period of time in a season.

Many forages species and agricultural wastes are used in manufacturing silage either alone or in combinations; in a specific feed mixture according to the purpose of the feed to be fed to different animal species. The chopped fresh forages are left to lose a large part of humidity (up to 45%) to be ready for mixing with other feed materials then compressed in the silo (Fig. 22.4). The chopped forage is semi dried; it is called Haylage form as it is a compromise between hay and silage, and its nutritional value is also between hay and silage. In the case of fodder where moisture cannot be easily reduced, it must be mixed with some dried materials like straw and agricultural wastes to adjust the moisture content of the mixer.

Fig. 22.4. Farm scale silage making is improved if there is a grass chopper and if silage is made in a pit.

4.4.1 Advantages of silage

(1) Save the green fodder to be offered to animals when needed urgently especially when there is a lack of forage resources in summer. (2) Improve the palatability and nutritional value of some plants that are low in palatability and feeding value. (3) Can be used as a supplement as a part of the feed or is used as the main feed with some other food additives according to the animal productive purpose. (4) It can be done from any forage materials, crop residues and agro-industrial by-products. (5) Silage needs a smaller storage space compared to hay. (6) It can be done in conditions that are not suitable for hay process such as during rainy periods. (7) Keep its substance in juicy fresh form and this is important in summer as it has a moisturizing effect on animals thus alleviating the harmful impact of high temperature. (8) Low loss of feed consumption, as the animal eats the entire plant including stems. (9) Allow to evacuate the farmland quickly for early planting of another crop.

4.4.2 Types of Silos

1. **Pit**: A large round hole dug in the ground; it is narrow at the bottom and wide at the surface, its floor and walls are of cement. Its bottom is higher than ground water level which is at an adequate distance.
2. **Trench**: It is a large rectangle dug into the ground with a tight bottom and wide surface, walls and floor are from cement. The floor slopes downward along the trench to allow the drainage of fluid from the pressure of green feed block. The bottom of the trench must higher than the ground water level.
3. **Bunker**: It looks like a ditch but is built above the ground surface and both narrow sides are open. Its sides are made of wood and must slide from the top to

the outside. The sides and a large part of the silo may be built above the ground and the rest could be under the ground. Generally its bottom is higher than the ground level.

A tractor is used to compress the green chopped forages in layers of 20–30 cm thickness, during the filling of the silo whether the pit, the trench or the bunker. When the silo is filled to the top it is covered well with plastic covers then heavy objects (brick, stone, frames of trolleys, etc.) are placed on top of it.

4. ***Tower:*** It is a vertical cylindrical building; it is built either with concrete or heavy metal panels. It has a set of metal doors on one side of the silo (60 x 60 cm). The doors are apart from each other with a distance of 180 cm. These doors are closed during the silo filling with green fodder and are opened when taking silage from it. A crane machine is used to carry the green fodder when filling the silo tower. The green fodder does not need to be squeezed and pushed in this kind of silo as natural gravity is sufficient for this purpose. The tower of the cylinder bottom has a drain hole to get rid of fluid resulting from feed pressure.

5. ***Other types of simple silo:*** Silage can be made by choosing a high place of the ground (floor), then putting a layer of wood or other plant waste. Silage is made above this layer in the ordinary way. Sometimes, barrels or plastic bags are used in making a small amount of silage for small farms.

4.4.3 Characteristics of high quality silage

Silage quality can be judged through the natural characteristics which must be:

(1) Acceptable acidic smell. (2) Yellow color without black or brown colors. (3) Not moldy. (4) Symmetric and regular moisture within 65%. (5) Palatable to animals.

Silage quality can be, also, judged through chemical characteristics as good silage where:

- Acidity degree approximately pH = 4.
- The Concentration of lactic acid must be about 9% of dry matter and acetic acid must be 2% of dry matter and butyric acid must be 0.2% or less of the dry material.
- Ammonia nitrogen concentration must be at least 10% of nitrogen.

4.4.4 How to make silage

The basis of this process is the fermentation of soluble carbohydrates (of sugars and carbohydrates) in plants by anaerobic bacteria to lactic acid, which reduce the pH number to 4 and the output lactic acid work as a preservative that prevents the growth of bacteria and fungi. Silage can be made as follows:

1. Reduce the moisture of cut plants to about 65–70% (not less than 60%, but no more than 75%) by partial drying (partial wilt) in the sun for 3 to 4 hours in the field in a moderate climate, or by adding hay or rice straw, or chopped maize

stalks, (these substances reduce the nutritional value of silage). If the green forage is drier than needed, water should be added to raise the content of moisture up to 70%.

2. Green plants must be chopped into small parts (3–5 cm) to facilitate the pressure to ease expulsion of air.

3. *Fill the silo*: close the whole silo or cover ground-pile with plastic, then fill the hole with 4–5 layers of chopped material. Each layer must be equivalent in depth, and then press well. When may use filled barrels of water or sand for pressing or it could be done by the feet of menor by a tractor to get rid of as much of air as possible. Finally cover it with plastic materials.

4. The silo must be filled in clear weather and quickly in as little time as possible to avoid the corruption that may occur before the end of the packing and closing the silo.

5. In the case of silage made from moderate protein content plants (such as leguminous plants), add starchy cereal (maize) or sugar (Molasses) with ratio 1–3% (10–30 kg/tons of green fodder) where molasses is mixed with the same amount of water then distributed on feed layers for maximum benefit. One of the demerits of adding molasses is adding unwanted moisture to feed.

6. When making silage from low protein plants content (such as Gramineae plants) it is preferable to add a protein source which analysis easily for balancing the composition of feed and raises its protein equivalent. Limestone powder may also be added with ratio 0.5–1% (5–10 kg/ton green corn) to increase the calcium content in silage.

7. To avoid air presence and corruption caused by it, it is important to distribute the cut green feed equally inside the silo. This equality can be achieved by maintaining a plane level for the surface mass or for it to be slightly higher in the middle. Pressing the mass must be done well. Tractors can be used in the horizontal silos to pass over the plant and get rid of maximum amount of air (good constant pressure is required when pressing material to be manufactured in the form of silage).

8. The top of the silo must be closed tightly to keep the air out of the silage by using a plastic cover in the silo (pit or trench or bunker). If a plastic cover is not available then place a thick layer of straw and mud on top of the silo and add some weights above. This must be done quickly so as not to dry out the surface of the block; the following must be observed:

   • Squeeze the top well especially beside the walls.
   • Cover the top layer with two layers of straws or plastic cover.
   • Cover with plastic which must be with dimensions suitable to the silo.
   • Continuous pressing during the coverage especially beside the walls.
   • Cover the plastic cover with a layer of dirt 20–25 cm thick.
   • It is preferable to put weights on the edges such as bricks or stones.
   • Preferably to be compressed at a later stage by a tractor.

9. After 6–8 weeks at least (depending on the season) open the silo after removing dust and plastic. The required quantities are taken daily; then close or cover the silo with plastic only. Note when opening the silo make a small opening that

allows taking out silage and does not allow air to enter in large quantities. After taking the required quantity the opening should be closed immediately and covered with soil again until the next use. This is repeated as long as you use the hole. When taking the silage out of the hole cut from the top down, not the other way around.

Remarks to be considered when feeding dairy cattle on silage: (1) It is recommended that silage be introduced gradually into the diet. (2) It is recommended to be given after milking (6–8 hours before the next milking). (3) Not storing the silage in the stable where livestock are housed. (4) Taking into account the stable ventilation continuously. (5) Removing the milk directly after the milking to avoid picking up taints from silage.

Feeding silage to livestock: Using silage in ruminants feeding should be done gradually, approximately as follows:

- Sheep and goats: 1 kg/day
- Local cattle: 9 kg/day
- Foreign cattle: 12 kg/day
- Buffalo: 15 kg/day.

Fig. 22.5. Well made silage is a valuable feedstuff for sheep, goats and cattle.

4.4.5 Fodder beet (fresh or silage)

Fodder beet (Fig. 22.6) can tolerate high salinity; it is a winter forage crop that is grown for the purpose of using its leaves and root (tubers) in animal feed. Crop of tubers per feddan[1] ranges from 30–100 tons and of leaves from 5–10 tons.

Because fodder beet spoils quickly and is not able to be stored for a long time, it is left in the ground without harvesting and taken daily as needed. This is a great loss

---

[1] When Egypt adopted the metric system, the *feddan* was the only old unit that remained legal. Currently taken as 0.42 hectares.

Fig. 22.6. Fodder beet is a valuable fodder source that can be used for direct feeding or conserved as silage.

as the land cannot be cultivated with another crop. Thus the best way of storing is the mentioned way for a month and make the remaining into silage.

4.4.5.1 Benefits of growing fodder beet

Fodder beet gives a plentiful yield of up to 100 tons. Its maturity is in June when alternative fodder is lacking after winter and before the maturity of summer fodder crops. Feeding dairy cattle on fodder beet will increase milk due its high moisture and sugar contents. It has a palatable taste because of it is sweetness.

4.4.5.2 The way of feeding fodder beet

Fodder beet could be fed to animals directly or processed.

First: feed directly: (1) Fodder beet is harvested as required, taking the quantity that is sufficient for the animals for about a month. It is stored in a well-ventilated place away from the sun. It must be put on layer hay without over-crowding (to prevent rotting). (2) It is given to animals in its raw state and without cutting to help the animals getting used to it quickly. In case of large size tubers or roots, it must be cut into two or three parts. (3) Feeding operation must be done after cleaning the mud and it is preferable to feed leaves and roots together after cutting the leaves and exposing them to the sun for at least two days to avoid bad influence of oxalates.

Second: making fodder beet into silage: (1) Separate tops from roots (tubers). (2) Leave beet tubers for a week to two weeks in a well-ventilated place away from the sun to reduce moisture. (3) Cover the silage processing place with a layer of rice straw or any other type of dry waste. (4) Cut beet root into medium pieces. (5) Mix the beet pieces with wheat straw or beans straw or corn stalks to increase the proportion of dry material to absorb the juice from beets. (6) It must be well pressed with the insulation layer of plastic. Then follow the same steps in manufacturing normal silage (see above).

4.4.5.3 Remarks for feeding fodder beet

Note that: (1) Feeding on beet can start after 5 weeks of silage processing. (2) Fodder beet is considered by the dairy farmers as a high-quality animal nourishment (cow-buffalo); a head may be given up to 30 kg per day, while fattening cattle may be given up to 5 kg per 100 kg weight. (3) When feeding on fodder beets, a feed rich in protein should be added to fulfill animal protein needs (especially animals with high milk production) as well as the addition of coarse material (wheat straw, rice straw). Calcium carbonate or calcium phosphate should be also added for animals that eat the beet due to deficiency in calcium. Also sodium bicarbonate should be added to fodder beet feeding to equalize the acidity excess. (4) Do not chop beet tubers into small pieces as the animals swallow them without chewing thus they will not be digested. (5) Prevent giving large quantities of beet along with poor protein feed materials because it reduces the percentage of fat in milk as well making its smell not acceptable.

4.4.5.4 Other advantages of silage making

In transitional zones at high altitude where some forages such as maize crop cannot reach maturity because of cold season local forages can be harvested, chopped and preserved as silage. It is successful and has prevented wastage through free choice feeding as fresh fodder and served as an alternative to hay making which was not efficient due to cold season and limited sun exposure.

1. *Making silage from fodder and sugar beet pulp*: Fresh sugar beet pulp contains high moisture 85–90% and due to high sugar contents can be successfully preserved as silage after mixing with dry roughages. Farmers are making beet pulp silage for feeding to milking buffaloes and cows and also to feedlot calves. It is available over a short period of 50–60 days (May-June) and preserving it as silage is the best option because drying of wet beet pulp is very difficult, if not impossible, in the open. Beet pulp silage was found to be a highly economical feed that increases milk production and supports high growth rate in calves and saves expensive concentrate feeding. Some famers also add urea to it at the time of ensiling. The practice of silage making from beet pulp has saved wastage of this valuable feed which occurs when a truck load of beet pulp is dumped at the farm and animals are allowed to consume, through free choice feeding over weeks of lying in open.

2. *Bale Silage as a possible solution for urban/peri-urban commercial dairy farms*: Commercial urban and peri-urban dairy farmers are mostly landless and buy fodder from local markets on daily basis. Farmers near cities having cultivable land can prepare bale silage of fodder for selling to urban and peri-urban dairy farms. This is easy to transport and store and will help these dairy farms in combating production losses during feed scarcity periods.

5. Conclusions and recommendations

Some halophytes and salt tolerant forages species could be used directly in animal feeding due to its good quality and palatability. On the other hands, several halophytes particularly the natural grown ones and salt tolerant fodder shrubs can be fed indirectly after being processed to enhance their feeding values. Such feed materials could be processed into different feed forms such as hay, silage, haylage, Feedblocks, treated hay with urea and molasses. The processed feeds have high nutritional value and palatability that can be offered to all animals at different physiological stages such as pregnancy, milking and fattening, etc. Therefore, the process of manufacturing halophytes and salt tolerant forages has several advantages such as:

1. Better utilization of halophytes and salt tolerant forages with high vegetative production and that which is unpalatable in feeding animals.
2. Better utilization of unpalatable or less palatable halophytes and salt tolerant forages in feeding animals through improving their feeding values.
3. Using agricultural residues in feed mixes with halophytes and salt tolerant feeds.
4. Increase the sources of animal food, and supply animals with good quality and consistent source of feed throughout the year.
5. Reduce the cost of animal feed which helps to improve the profitability of the farmers and improve their income.
6. Improve the productive efficiency of animals.

References

AOAD. 2012. Arab Organization for Agriculture Development (AOAD), Arab Agriculture Statistics Yearbook- Volume 33, Part IV: livestock, Poultry and fish production www.aoad.org/AASYXX.htm.
Attia-Ismail, Salah A. (in press). Plant secondary metabolites of halophytes and salt tolerant plants. pp. v127–142, this volume.
Masters, D.M. (in press). Assessing the feeding value of halophytes. pp. 89–105, this volume.
Norman, H.C., E. Hulm and M.G. Wilmot. (in press). Improving the feeding value of old man saltbush for saline production systems in Australia. pp. 79–88, this volume.

Unifying Perspectives: Halophytes and Salt-tolerant Feedstuffs and their Role in Livestock Production Systems

Victor R. Squires[1] *and H.M. El Shaer*[2]

SYNOPSIS

This chapter brings together some of the main issues related to how we can better utilize this diverse group of plants, that encompasses forbs, grasses, shrubs and trees in a world that is increasingly under pressure to feed the burgeoning human populations and their associated livestock. Drylands occupy about 40% of the world's land surface area. These lands are home to hundreds of millions of people and the livestock on which they depend. Livestock in these drylands is exposed to saline stress in many instances such as when grazing halophytes and/or salt-tolerant fodder/forage plants or when the saline water from underground wells is the only available drinking water. The ability of livestock to withstand this type of stress is affected by numerous factors: genetic, physiological, nutritional and climatological. All these aspects receive attention in this book. The age and the physiological status are critical factors, where young and productive animals are more sensitive. The presence of other types of stress such as heat and feed shortage exacerbate the effect of saline load. In other words, salinity tolerance of animals could vary with environmental conditions, type of feed or physiological state. We explore the issues around biosaline

---

[1] Gansu Agricultural University, Lanzhou, China.
[2] Desert Research Center, Cairo, Egypt.

agriculture that focus on production of salty feedstuffs from abandoned croplands and other 'wastelands, including sabkha and coastal deserts.

Keywords: drylands, salt load, water, ruminants, poultry, non ruminants, sabkha, feedlots, hi-tech, forage harvesting, nutrition, biosaline agriculture, California, Australia, Pakistan, India, Mexico, Central Asia, Mongolia, NW China, Rann of Kutch, Indus, Tigris, Euphrates, Nile Delta, Imperial Valley, Egypt, Gulf region and Arabian peninsula.

1. Introduction

Worldwide, but especially in the drylands that cover about 40% of the earth's land surface, natural resources have been diminishing because of increased demands for food and fiber. Much of this production in drylands comes from subsistence pastoralism (Grice and Hodgkinson 2002). This sort of pressure is on-going and in some places is accelerating, and results from the ever-increasing human population. Inevitably, marginal and long-neglected natural resources have to be re-assessed in preparation for utilization. Not less than 400 Mha and perhaps as much 900 Mha of land in arid and semi arid regions may be salt-affected from natural and anthropogenic causes. Definitive data on the annual worldwide loss of farmland due to salinization and related causes is lacking. However, salinity is unquestionably the most important problem of irrigated agriculture and one-fifth of the world's cropland (about 47 Mha) is salt-affected. Reliable estimates suggest that in the Mediterranean region alone, some 15 Mha have become seriously affected by anthropogenic influences. In India over 6 Mha, of a national total of only 40 Mha of irrigated farmland, have been made useless by salinity and water logging, while over 40000 ha are abandoned each year. In Pakistan, estimates show that about 0.45 Mha of land in Pakistan (as one example) is being rendered useless by salinity every year. Even in USA and Australia there are large areas that lie waste as a result of salinity (some of it non-irrigated or 'dryland salinity'). In USA, well over 500,000 ha of Californian soil particularly in the Coachella, Imperial, Sacramento and San Joaquin valleys have been strongly affected and rendered useless by salinity. Overall, the USA reportedly loses 80–120,000 ha of cropland annually due to salt build up in soil or irrigation water. It is estimated that 7–10% of the world land area is salt affected (Dudal and Purnell 1986). Halophytes survive salt concentrations around 200 mM NaCl or more in order to reproduce in such environments and they constitute about 1% of the world's flora (Flowers and Colmer 2008). Halophytes grow in many arid and semi-arid regions around the world and are distributed from coastal areas to mountains and deserts (see PART 1, this volume).

The shortage of animal feeds is the main constraint to increase indigenous animal production in many of the world's dryland regions. Feed shortage characterizes arid and semi-arid regions which is considered the main constraint to improvement of livestock productivity. Most of the countries in these regions import large amounts of feed materials to cover the nutritional gap for animal and poultry which puts a heavy burden onto the local governments and the farmers. Consequently it decreases the net profits on animal investments due to the high costs of such feed materials. Therefore, intensive efforts have been directed to find alternative feed resources from halophytes.

Halophytes and salt tolerant plants can play a significant role in the well being of different people. The way in which this group of plants are assessed will very much depend on which production system dominates (Squires and El Shaer, this volume). As already mentioned hundreds of millions live at subsistence level keeping highly mobile livestock to "chase forage and water", others are in semi-commercial operations that aim to use forage derived from saline lands (either naturally occurring or under cultivation). Animal husbandry, as the main income resource for pastoralists, is based mostly on the natural vegetation for rearing sheep, goats, camels and other herbivores. Any evaluation must depend on viewing performance in the context (biological/economic) in which it occurs. However, unpalatable halophytes are widely distributed throughout the world. The challenge is to make these plants more acceptable to livestock and effort has been made to do this through a diverse array of approaches. Plant breeding, better agronomic techniques, physical and chemical treatments that render plant material more palatable and nutritious have been evaluated. Halophytic plants such as *Atriplex* spp., *Nitraria*, *Haloxylon*, *Salsola* spp., etc. are generally considered extremely valuable as a fodder reserve during drought.

Attention has been also focused on the use of salt tolerant plants, including obligate halophytes, as candidate species for rehabilitation of saline lands, including abandoned irrigation areas (see below). The role of biomass produced on such rehabilitated lands was also investigated and new ways to use such biomass, especially in mixed rations involving more conventional fodders (Squires and Ayoub 1995). As the loss of arable land to secondary salinity (Zerai 2007) became more widespread and as the agronomy of salt tolerant food crops was advanced, attention turned to utilizing the crop by-products, including oil seed cakes and meals. Sea water irrigation of halophytes gained traction as the search for commercially viable biosaline systems progressed. This development generated a larger quantity of potentially usable feedstuffs both as biomass and as by-product such as seed cake (Glenn et al. 1995). Halophytes have the unique ability to revegetate salt-affected land and offer a low-cost approach to reclaiming and rehabilitating saline habitats previously regarded as useless. Abandoned irrigation areas where salinity has rendered them unsuitable for conventional crops can be used for the purpose of growing fodder for livestock.

Finding suitable forage/fodder (and even grain crops such as *Distichlis palmeri*) which does not encroach upon the land under conventional crops may be useful for cattle raising and meeting the requirement of meat, poultry and dairy products is a challenge and specialized research centers were established, principally in the Gulf region, the WANA region, Pakistan, Central Asia and USA (Arizona) and Australia. Concurrently there was interest in using halophytes for remediation of damaged lands, including mine sites. The potential of salt tolerant plants, including extreme halophytes, for carbon sequestration also received attention. Some of the biomass derived from C sequestration plantations began to enter the supply chain for feedlots where guaranteed supplies of high quantities of suitable feedstuffs are required. In addition to salt tolerant and halophytic plants grown under intensive cultivation there are extensive areas of naturally-occurring shrublands, woodlands and grasslands that are used as fodder reserves or protein supplements, either grazed/bowsed by livestock or in cut-and-carry systems.

The volume of data generated and the interest in filling the 'feed gap' in many animal production systems, especially in arid and semi arid regions, has provided impetus to convene International, regional and national symposia and spawned special issues of journals (e.g., *Small Ruminant Nutrition*) and various Conference Proceedings. The plethora of information generated by nutritionists, animal physiologists, veterinarians, agronomists and livestock specialists is scattered throughout reports, journal articles and some specialist monographs but lacks integration and synthesis into a coordinated body of knowledge. In this book we attempt a synthesis that considers the role and potential of salt tolerant and halophytic feedstuffs and their impacts on nutrition, physiology and reproduction of livestock, including ruminants and non ruminants such as poultry and rabbits.

2. Problems and prospects for utilizing salt tolerant and halophytic feedstuffs

At the local scale the options available are largely dictated by the type of livestock and the animal husbandry system being followed but for halophytic feedstuffs to be commercially viable large quantities and a reliable supply need to be available (Squires and El Shaer, this volume). For full scale commercial utilization, e.g., in feedlots, the feedstuffs need to be produced under irrigation (see below). As indicated in the previous chapters, most feedstuffs require processing of some kind to make them easier to handle, more palatable to the livestock and more nutritious and to minimize the anti nutritional components (Attia-Ismail, this volume; El Shaer, this volume). The option of using salt tolerant fodder plants on saline soils (often abandoned cropland) using saline water opens the possibility for commercialization of the production of such fodders.

2.1 Why irrigate with saline waters

There are 295 Mha of coastal desert in the world of which an estimated 50 Mha (17%) would be suitable for irrigated agriculture in terms of soil type, slope class and lack of competing uses if saline water could be used as a water source. This amount of land would increase the total irrigated land in desert regions of the world by (80%) or even more if the large areas of abandoned irrigation in inland regions like Central Asia are added. Coastal regions of particularly appropriate are the deltas of the great desert rivers, Nile, Indus, Tigris Euphrates. These deltas have rich alluvial soils, large portions of which have become barren and salinized due to the interception of river flows from upstream and irrigation of conventional crops. Saline water halophyte and salt tolerant fodder plants offer a method to bring these lands into agricultural production (O'Leary and Glenn 1994). In the vast inland areas of Central Asia (including NW China) there are millions of ha of abandoned irrigated fields with infrastructure (canals, pumping stations, etc.) in place. Sandy desert seacoasts along the coast of California, including both USA and Mexico, the Red sea, Arabian (Persian) Gulf and the Indian Ocean offer additional larger areas for seawater irrigation (El Shaer and Attia-Ismail, this volume; Squires and El Shaer, this volume). Other potential seawater areas are the Ranns of Kutch in India and Pakistan and the sabkha[1] flats of the Arabian peninsula and the sand coasts of southern and western Australia. Some of these areas are near to large cities Los Angeles, Karachi, Cairo, Bombay, Baghdad, Adelaide and Perth. Saline (including seawater irrigation) based production of animal feeds from these areas could relieve the pressure on scarce water resources and reduce pressure on overgrazed rangelands. A new paradigm for developing agriculture along coastal deserts is to use saline water directly from the sea or from saline aquifers to irrigate halophyte crops. Impetus for this has waxed and waned over the years (Stenhouse and Kijne 2011; Riley, this volume).

2.2 Research outcomes on making saline diets more useful

In the meantime research has gathered momentum and much has been learned about how livestock react to saline diets, as this book demonstrates. The role of water (both access and water quality) in the utilization of forage/fodder derived from salt tolerant and halophyte plants is also important (Squires and El Shaer, this volume). Other research has focussed on improving the nutritional value (Norman et al., this volume; Attia-Ismail, this volume) including measures to overcome the antinutritional components (Attia-Ismail, this volume). Post harvest treatments to improve palatability and digestibility have been the focus of some work and guidelines have been prepared (El Shaer, this volume). The incorporation of saline dietary components into the rations for non-ruminants such as poultry and rabbits is a relatively new development (Khidr et al., this volume). Detailed research on animal physiology has revealed much (Ashour et al., this volume; Attia-Ismail, this volume; Degen and Squires, this volume) including the impact on reproductive success (Abdalla et al., this volume) including the intergenerational consequences (Blache and Revell, this volume).

3. Whither animal production systems based on salt tolerant and halophytic feedstuffs?

The production of fodder, forage and vegetation in general is diminishing worldwide at a very fast rate. One of the causes, that may have catastrophic consequences, is desertification (Heshmati and Squires 2013) and an increase in the area of saline land—the two being part of the one wider problem of accelerated land degradation. Land degradation is threatening rangeland, rain-fed cropland as well as canal irrigated land in arid areas on a global scale. Estimates show that about 0.45 Mha of land in Pakistan (as one example) is being rendered useless by salinity every year as human population doubles every 40 years. UN estimates are that Pakistan will have a population in 2050 of 309 million. The consequence of this trend worldwide in terms of both economy and environmental quality are dire and far reaching, especially for countries that are heavily dependent on agriculture and animal husbandry.

---

[1] Sabkha: Equates to a **playa.**

Nutrition represents one of the most serious limitations to livestock production in developing countries, especially in the tropics. Feed resources are inadequate in both quality and quantity, particularly during the dry seasons. Biotechnological options are available for improving rumen fermentation and enhancing the nutritive value and utilisation of agro-industrial by-products and other forages (Kundu and Kumar 1987).

3.1 Increasing digestibility of low-quality forages

Fibrous feeds, including crop residues, of low digestibility constitute the major proportion of feeds available to most ruminants under smallholder situations in developing countries. The associated low productivity can be overcome to some extent by several means, among which are: balancing of nutrients for the growth of rumen microflora thereby facilitating efficient fermentative digestion and providing small quantities of by-pass nutrients to balance the products of fermentative digestion, enhancing digestibility of fibrous feeds through treatment with alkali or by manipulating the balance of organisms in the rumen and genetic manipulation of rumen micro-organisms, currently acknowledged as potentially the most powerful tool for enhancing the rate and extent of digestion of low quality feeds. Rumen micro-organisms can also be manipulated by adding antibiotics as feed additives, fats to eliminate or reduce rumen ciliate protozoa (defaunation), protein degradation protectors, methane inhibitors, buffer substances, bacteria or rumen content and/or branched chain volatile fatty acids.

Low-quality forages are a major component of ruminant diets in the tropics. Thus, much progress can be made by improving the forage component of the ration. The characteristic feature of tropical forages is their slow rate of microbial breakdown in the rumen with the result that much of the nutrients of the feed are voided in the faeces. The slow rate of breakdown also results in reduced outflow rate of feed residues from the rumen which consequently depresses feed intake. At present, the main treatment methods for forages such as cereal straws are either mechanical (e.g., grinding), physical (e.g., temperature and pressure treatment) or a range of chemical treatments of which sodium hydroxide or ammonia are among the more successful.

The lignification of the cell walls prevents degradation by cellulase or hemicellulase enzymes. Fortunately, it is possible to use lignase enzyme produced by the soft-rot fungus (*Phanerochaete chrysosporium*) which causes a high degree of depolymerisation of lignin. The enzyme acts like a peroxidase and causes cleavage of carbon-carbon bonds. At present the levels of the lignase enzyme produced by the basidiomycete fungi are insufficient for the treatment of straw on a commercial scale. However, it is conceivable that the use of recombinant DNA engineering techniques will allow the modification of the lignase genes and associate proteins to increase their efficiency and stability.

As indicated above, using various biotechnological methods (El Shaer, this volume) it is now possible to convert biomass from salt tolerant and halophytic plants into value added products thus making biomass production from saline soils (or from saline water irrigation) an economic proposition. The biological approach for economic utilization of saline wastelands (coastal deserts, abandoned irrigation areas, etc.) has become a reality as many national agencies and international organizations are keenly pursuing it because of its sustainable and environment-friendly nature. It is significant that high-tech and large-scale approaches have been applied in developed regions such the Imperial Valley in USA (mainly to service demand from feed lots (see earlier) but it is not confined there. The biosaline approach that uses adapted plants not only improves the general ecology of area, including biodiversity conservation and dust storm prevention but in return improves land users revenue stream.

The first question to be answered is whether it is a dream or an economic and ecological imperative. As indicated above (Section 2.1) there is potential to expand the technologies related to saline water irrigation to millions of ha of land. Some successful large scale efforts have been reported from the Gulf region and from northern Mexico (Khan et al. 2014) and the Biosaline research center in Dubai (UAE), Kuwait and Saudi Arabia as well as the Sinai region of Egypt. There is considerable potential to use salt tolerant and halophyte plants in efforts to reclaim and rehabilitate irrigation areas in the Nile Delta where rising sea levels, marine incursions and rapidly rising soil salinity makes it impossible to successfully grow conventional crops (Darwish et al. 2013). Complementary research is under way in Central Asia (Toderich et al. 2013) and in Pakistan (Khan et al. 2014) where large areas of formerly productive land lies idle because of salinity in the soil or water (or both).

So we might conclude from this that it is not entirely a dream. It is technically feasible, economically viable and ecologically sound. Coupled with the pressure of population (both humans and their livestock) especially in arid and semi arid regions, the high costs of feed imports and recent advances in the level of understanding of how and when and in which form to present highly saline feedstuffs it seems inevitable that any previously neglected plant and soil resources will be brought into play. At the small scale (household level) techniques for harvesting, processing and conserving fodder have been developed over decades (El Shaer, this volume) and at the other end of the spectrum, large-scale mechanized harvest of salt tolerant and halophytic plants for processing as feedstuffs for feedlot animals is becoming more common (Fig. 23.1).

4. Conclusions

The area of secondarily salinized lands is increasing at a faster rate over time. Many irrigation districts around the world are shrinking as a result of secondarily salinized soils. This is resulting in crop yield losses. Irrigation practices with low drainage are intensifying this problem. We must conclude that it is technically feasible to utilize these degraded soil/water systems by growing salt tolerant and halophytic plants but its economic viability remains to be established. The biosaline approach most relevant to this book are those related to selection of salt tolerant and halophytic forage/fodder plants of commercial value from existing flora of conventional and non-conventional halophytic plants for use in saline soils or that utilize saline irrigation water (Yamaguchi and Blumwald 2005). Aronson (1989) identified over 1560 species in 550 genera and 117 families that show potential. However, his list only included plants that had potential as food, forage, fuelwood, or soil stabilization crops. Less than 100 of these are in common use as forage/fodder. We cannot stress enough that understanding

Fig. 23.1. Highly mechanized forage harvesting, handling and processing have enabled feeding of mixed rations to livestock in feedlots.

the production system into which biosaline agriculture must fit is the key to success (see discussion in Norman et al., this volume; Squires, this volume). Expert opinion (Stenhouse and Kijne 2011) suggests that saline irrigated agriculture is most likely to succeed in the West Asia/North Africa region as a complement to small-scale mixed livestock and cropping farming systems. Similar optimism surrounds the potential in Greater Central Asia (including NW China and parts of Mongolia). This does not preclude the expansion of intensive irrigated forage production from former salinized and abandoned lands in USA, Australia and elsewhere where there is demand for fodder for large-scale feedlot operations. Or indeed in the Nile Delta and elsewhere where new industries based on salt tolerant and halophytic plants can be developed on abandoned croplands thus relieving pressure on imports of fodder and generating incomes for local land users.

The need for salt-tolerant crops increases each year as a growing world population seeks to feed itself with ever-decreasing soil resources and dwindling fresh water supplies. Heightened expectations with respect to the quality of life are increasing this demand, while salinization of soils in many parts of the world is decreasing the area available for conventional agriculture. These problems are growing in the Nile Delta, Central Asia, western China, southern Australia and elsewhere. Soils that are otherwise fertile, but that have become salinized because of inadequate irrigation practices, are low in economic value but would have much of that value restored if they were used for halophyte crops. However, apart from humanitarian reasons there are economic reasons for growing halophyte crops. Countries of North Africa, the Gulf region and the Arabian peninsula, for example, import costly feedstuffs to support the burgeoning livestock inventories as human populations rise. Furthermore, there are naturally saline areas such as Egypt's Western Desert, that have never been used agriculturally, but which could be brought into production as irrigation projects are developed, if suitable salt tolerant plants become available. Other sites in other regions have potential too (Squires and El Shaer, this volume).

The problems of feeding halophytic feedstuffs (as explained in this book) seem to be largely overcome by feeding them as a supplement or as part of mixed rations and limiting the proportion of saline feedstuff to no more than 30%. Most feedstuffs require processing of some kind to make them easier to handle, more palatable to the livestock and more nutritious or to minimize the anti nutritional components.

References

Abdalla, E.B., A.S. El Hawy, M.F. El-Bassiony and H.M. El Shaer. (in press). Impact of halophytes and salt tolerant forages on animal reproduction. pp. 303–315, this volume.

Aronson, J.A. 1989. HALOPH: A Data Base of Salt Tolerant Plants of the World. Arid Land Studies.

Ashour, G., M.T. Badawy, M.F. El-Bassiony, A.S. El-Hawy and H.M. El Shaer (in press). Impact of halophytes and salt tolerant plants on physiological performance of livestock. pp. 261–286, this volume.

Attia-Ismail, S.A. (in press). Plant secondary metabolites of halophytes and salt tolerant plants. pp. 127–142, this volume.

Attia-Ismail, S.A. (in press). Nutritional and feed value of halophytes and salt tolerant plants. pp. 106–126, this volume.

Attia-Ismail, S.A. (in press). Rumen physiology under high salt stress. pp. 348–360, this volume.

Attia-Ismail, S.A. (in press). Mineral balance in animals as affected by halophyte and salt tolerance plant feeding. pp. 143–160, this volume.

Blache, D. and D.K. Revell. (in press). Short- and long-term consequences of high salt loads in breeding ruminants. pp. 316-335, this volume.

Darwish, Kh., M. Safaa, A. Momou and S.A. Saleh. 2013. Egypt: Land Degradation issues with special reference to the impact of climate change. pp. 113–136. *In*: G.A. Heshmati and Victor Squires (eds.). Combating Desertification in Asia, Africa and the Middle East: Proven Practices Springer, Dordrecht.

Degen, A.A. and V.R. Squires (in press). The rumen and its adaptation to salt. pp. 336–347, this volume.

Dudal, R. and M.F. Purnell. 1986. Land resources: salt affected soils. Reclamation and Revegetation Research 5: 1–9.

El Shaer, H.M. and S.A. Attia-Ismail (in press). Halophytic and salt tolerant feedstuffs in the Mediterranean basin and Arab region: An overview. pp. 21–36, this volume.

El Shaer, H.M. (in press). Halophytes and salt tolerant forages processing as animal feeds at farm level: Basic Guidelines. pp. 388–405, this volume.

Flowers, T.J. and T.D. Colmer. 2008. Salinity tolerance in halophytes, New Phytologist 179: 945–963.

Glenn, E.P., W. Coates, J.J. Riley, R. Kuehl and R.S. Swingle. 1992. *Salicornia bigelovii* Torr.: a seawater-irrigated forage for goats. Anim. Feed Sci. Technol. 40: 21–30.

Glenn, E., N. Hicks, J. Riley and R. Swingle. 1996. Seawater irrigation of halophytes for animal feed. pp. 221–236. *In*: R. Choukr-Allah, C.V. Malcolm and A. Hamdy (eds.). Halophytes and Biosaline Agriculture. Marcel Dekker, New York.

Grice, A. and K.H. Hodgkinson. 2002. Global Rangelands: Problems and Prospects. CABI, Wallingford, pp. 229.

Khan, M.A., B. Böer, M. Öztürk, T.Z. Al Abdessalaam, M. Clüsener-Godt and B. Gul (eds.). 2014. Sabkha Ecosystems: Volume IV: Cash Crop Halophyte and Biodiversity Conservation Series: Tasks for Vegetation Science, Vol. 47. Kluwer Academic, pp. 339.

Khidr, R.K., K. Abd El-Galil and K.R.S. Emam (in press). Utilization of halophytes and salt tolerant feedstuffs for poultry and rabbits. pp. 361–372, this volume.

Norman, H.C., E. Hulm and M.G. Wilmot. (in press). Improving the feeding value of old man saltbush for saline production systems in Australia. pp. 79–87, this volume.

O'Leary, J.W. and E.P. Glenn. 1994. Global distribution and potential of halophytes. pp. 7–17. *In*: V.R. Squires and A.T. Ayoub (eds.). Halophytes as a Resource for Livestock and for rehabilitation of degraded land. T:VS 32, Kluwer Academic, Dordrecht.

Pasternak, D. and A. San Pietro. 1985. Biosalinity in Action: Bioproduction with Saline Water Springer, Dordrecht, pp. 369.

Riley, J.J. (in press). Review of halophyte feeding trials with ruminants. pp. 177–217, this volume.

Squires, V.R. (in press). Intake and nutritive value of some salt-tolerant fodder grasses and shrubs for livestock: Selected examples from across the globe. pp. 218–246, this volume.

Squires, V.R. and A.T. Ayoub. 1995. Halophytes as resource for livestock and for rehabilitation of degraded lands., T:VS 32, Kluwer Academic, Dordrecht, pp. 316.

Squires, V.R. and H.M. El Shaer. (in press). Global perspective on distribution and abundance of sources of halophytic and salt tolerant feedstuffs. pp. 3–20, this volume.

Squires, V.R. (in press). Water requirements of livestock fed on halophytes and salt tolerant forage and fodders. pp. 287–302, this volume.

Stenhouse, J. and Jacob W. Kijne. 2011. Prospects for Productive Use of Saline Water in West Asia and North Africa. Comprehensive Assessment Research Report 11. CGIAR Initiative on Water Management. http://www.iwmi.cgiar.org/assessment/files_new/publications/CA%20Research%20 Reports/CARR%2011.pdf.

Toderich, K.N., E.V. Shuyskaya, T.F. Rajabov, I. Shoaib, Shaumrov, K. Yoshiko and E.V. Li. 2013. Uzbekistan: Rehabilitation of desert rangelands affected by salinity, to improve food security, combat desertification and maintain the natural resource base. pp. 249–278. *In*: G.A. Heshmati and Victor Squires (eds.). Combating Desertification in Asia, Africa and the Middle East: Proven Practices Springer, Dordrecht.

Yamaguchi, T. and E. Blumwald. 2005. Developing salt-tolerant crop plants: challenges and opportunities. Trends in Plant Science 10(12): 615–620.

Zerai, D.B. 2007. Halophytes for Bioremediation of Salt Affected Lands. PhD Thesis, University of Arizona, Tucson, pp. 97.

Author and Subject Index

152–154, 167, 225, 234, 236, 251, 265,
336–338, 341, 344, 348–350, 352, 367,
391, 393, 409, 410
Australia 1, 7, 9–11, 14, 28, 61, 79–82, 84, 85,
89, 111, 177, 182–186, 189, 191, 194–196,
218–222, 224, 226, 227, 230, 231, 236,
240, 291, 292, 295–267, 316, 328, 329,
340, 407, 408, 410, 413
Dryland salinity in 182, 328, 407

B

Badawy, M.T. 116, 119, 152, 261, 268, 270, 272,
273, 276, 277
Bioavailability 143, 151
Biochemical parameters 261, 266, 368
Blache, Dominique 279, 295, 305, 306, 317, 318,
327, 410
Blood (see also hemo) 114, 136, 137, 154, 261,
262, 266–276, 279, 289, 320, 321, 325,
330, 339, 350, 351, 368, 369
Chemistry, changes in 266
Bofedales 60, 73, 74
Bolivia 60, 61, 69, 70
Brackish water 11, 12, 70, 151, 190, 228, 229,
264
As drinking water 296
For irrigation 12, 188, 197, 214, 228
Brazil 60, 61, 70–72
Brevedan, Roberto 60, 61
Buffaloes 389, 404

C

Cactus 38, 40, 52–56, 69, 71, 72, 219, 239
Camels 4, 8, 14, 26, 28, 30, 32, 38, 40, 51,
52, 107, 110, 134, 151, 152, 154, 159,
162–166, 168–170, 177–179, 208–211,
222, 224, 225, 237, 247–256, 259, 263,
266–270, 272, 276–278, 280, 287, 288,
291–294, 297, 336, 339, 344, 351, 374,
389, 408
Cannulated sheep 219, 235, 236
Capybara 73, 374
Carcass traits 161–163, 365
Dressing out 163
Meat quality 161, 162
Cassia 193, 204, 219, 224, 235
Cassia in mixed diets 204, 233
Cassia sturtii 193, 204, 224, 235
Cattle 14, 27, 45, 60–64, 66, 72, 74, 109, 115,
117, 129, 134, 136, 159, 177, 178, 184–
187, 189, 190, 211–213, 220, 226–230,
237, 240, 253, 262–265, 267, 278, 287–
289, 291–294, 297, 313, 324, 336–339,
341, 342, 344, 389, 397, 402–404, 408
Cecum fermenter 73, 374

Chamaecytisus 177, 183, 184, 186, 191, 196
Chemical composition of 183
Chemistry of 123
Chemometrics 89, 97
Chenopods (see also saltbush) 29, 44, 45, 89–91,
114, 118, 121, 140, 143, 144, 373, 375,
376, 378, 381, 383, 384
Chile 60, 61, 67, 69, 70, 73, 222, 233
China 10, 11, 14, 177, 185, 188, 191, 192, 196,
197, 199, 224, 234, 287, 336, 406, 407,
409, 413
NW China 407, 409, 413
Chloris 4, 15, 62, 63, 118, 191, 193, 198, 200,
202, 204, 219, 267, 270
Coastal deserts 8, 11, 13, 74, 228, 263, 407, 410,
411
As site for seawater irrigation 11, 409
Distribution and geographical extent of 1
Conception rate 303–307, 320
Factors affecting 106
Contamination 61, 69
Coumarin 45, 127, 128, 131, 135
Crude fiber (see also dietary fiber) 27, 42, 106,
109, 115, 116, 190, 198, 205, 213, 238,
250, 255, 262
Crushed date seed meal 232, 238
Cut and carry systems 409
Cyanides 127, 128
Cynodon 5, 117, 119, 199, 200, 219, 234, 235,
298
Cynodon hay 199, 200, 234, 235, 298

D

Dairy cattle 263, 402, 403
Dairy cows 8, 225, 349, 353
Date seed 120–122, 136, 159, 163, 219, 225, 232,
238, 249
Degen, A.A. 29, 31, 54, 73, 191, 193, 203, 204,
222–224, 235, 277, 289, 293–295, 320,
336, 344, 349, 354, 362, 373, 375–381,
384, 410
Desert 3, 12, 21, 25, 26, 32, 60, 61, 72, 73, 106,
107, 127, 143, 144, 151, 153, 161, 162,
164, 166, 178, 181, 186, 187, 190, 193,
203, 205, 207, 209, 210, 212, 213, 237,
247, 248, 252, 261, 267, 291–295, 297,
303, 304, 309, 320, 336, 348, 359, 361,
362, 368, 373, 379–383, 388, 406, 409,
413
Extent and geographic distribution 1
Detoxification 128, 129, 137
Dichrostachys 219
Diet selection 27, 80, 81, 109, 181, 189, 248,
254, 323, 381
dietary requirements 152, 180, 359

Grass chopper 399
Ground water 269, 306, 307, 310, 313, 399
  Quality of 14, 29, 44, 45, 61, 64, 71, 72, 80,
    91, 97, 100, 107, 108, 111, 112, 115, 120,
    161, 162, 164, 165, 169, 171, 182, 219,
    227, 263, 294, 308, 375, 395, 413
  Salt content 14, 54, 62, 69, 96, 121, 182, 187,
    189, 190, 194, 205, 213, 236, 250, 270,
    275–277, 288, 296, 297, 317, 325, 329,
    375
Gulf Countries 21, 26
Gulf region 1, 4, 13, 14, 21, 226, 240, 407, 408,
  412, 413
  Halophytes in 15, 23, 24, 38–40, 45, 54, 56,
    61, 111, 117, 135, 151, 180, 186, 197, 240,
    274, 339, 397

H

Haloxylon 4, 13, 30, 114, 130, 146, 180, 219, 408
Hare 63, 189
Hay 4, 6, 29, 32, 51, 53, 55, 69, 71, 81, 84, 117,
  118, 121, 122, 153, 163–166, 168–173,
  183, 188, 189, 191–193, 196–207, 209–
  211, 219, 224–226, 230, 233–235, 238,
  239, 250–256, 267–270, 273–278, 297,
  298, 305–310, 312, 313, 340, 341, 389,
  391–395, 398, 399, 400, 403–405
  Berseem clover 232
  Cynodon 5, 117, 119, 199, 200, 219, 234,
    235, 298
  Rhodes grass 118, 193, 198, 200, 202, 204,
    205, 211, 219, 267, 270, 276
Heat dissipation (see also environmental stress)
  287, 290, 292, 293
Hemo-biochemical parameters 261, 266
Herbivore 16, 27, 51, 63, 65, 72, 73, 100, 213,
  248, 251, 338, 339, 374, 375, 379, 381–
  384, 408
Hordeum 10, 26, 42, 47, 49, 177, 190, 194, 196,
  207, 227, 230, 231, 271
  As grain supplement 121
  Grass seed problem 231
  Hordeum marinum 227, 230
Hormones 261, 278–280, 303, 304, 311, 312,
  317, 319, 320, 326, 367–370
  In poultry 361, 370, 371
  In sheep 54, 71, 100, 134, 135, 151, 170,
    172, 194, 225, 233, 235, 238, 266–269,
    271–273, 275–278, 280, 311, 316, 318,
    324, 325, 327, 329, 330, 339, 341–343,
    351, 354
Horse 14, 63, 66, 189, 213, 220, 226, 237, 263,
  264, 288, 294, 336, 344, 374
Hulm, Elizabeth 79, 391

I

in vitro analysis 73, 89, 90
in vivo methods 89, 93, 95, 106

India 11, 107, 219, 240, 407, 410
Indian Sea 21, 24
Inland deserts 39, 74
  Extent and geographical location 1, 21
  Potential for halophytic feedstuffs 409
Intergenerational changes 317
  Affect of high salt loads on 259
  In sheep 54, 71, 100, 134, 135, 151, 170,
    172, 194, 225, 233, 235, 238, 266–269,
    271–273, 275–278, 280, 311, 316, 318,
    324, 325, 327, 329, 330, 339, 341–343,
    351, 354
  In utero effects 323, 324
Iran 10, 11, 13, 14, 21, 24, 26, 177, 184, 186,
  212, 213
Iraq 11, 21, 24, 26
Irrigation (see also saline irrigation) 3, 10–13,
  15, 22, 25, 26, 33, 61, 71, 107, 111, 116,
  144, 148, 177, 178, 185–188, 190–193,
  197, 198, 203, 209, 212, 214, 218, 220,
  226–229, 236, 240, 352, 407–413
  Re-use of drainage water 13
  Seawater 11, 12, 33, 65, 81, 107, 143, 148,
    177, 186, 187, 189, 192, 198, 199, 203–
    205, 212, 214, 218, 226, 227, 236, 237,
    263, 264, 268, 269, 271, 410
Israel 177, 191, 193, 203, 204, 222–224, 235,
  295, 336, 373, 375

J

Japan 177, 185, 189, 203, 205
Jordan 21, 24, 26, 149, 177, 191, 193, 194, 222,
  310

K

Kallar grass 107, 184, 187, 305
Kam, Michael B. 73, 320, 344, 362, 373, 375–
  377, 379, 381
Khidr, R.K. 361–366, 410
Kidney function 261, 262, 295, 318, 325
  Impact of high salt loads 259, 316, 321, 330
Kochia 13, 25, 28, 64, 107, 116, 130, 146, 148,
  154, 166, 168, 169, 171, 179–182, 184–
  187, 212, 213, 219, 224, 231, 237, 275,
  304, 329, 339, 343, 348, 351, 390
Kuwait 24, 177, 192, 200, 201, 224, 412
  Seawater irrigation in 11, 12, 33, 107, 117,
    198, 218, 410

L

Lambing 303, 304, 306, 307
  Birth weights at 222, 232, 322
  Neonatal survival 317, 322
Land degradation 60, 61, 410
Leptochloa 6, 63, 107, 187, 219, 220, 226, 305
Leymus 188, 196, 197, 219
Libya 13, 21, 26